Edited by
Shannon S. Stahl and Paul L. Alsters

Liquid Phase Aerobic Oxidation Catalysis

Edited by
Shannon S. Stahl and Paul L. Alsters

Liquid Phase Aerobic Oxidation Catalysis

Industrial Applications and Academic Perspectives

Verlag GmbH & Co. KGaA

Editors

Prof. Dr. Shannon S. Stahl
University of Wisconsin-Madison
Department of Chemistry
1101 University Avenue
Madison WI 53706
USA

Dr. Paul L. Alsters
DSM Ahead R&D b.v.
Innovative Synthesis
P.O. Box 1066
6160 BB Geleen
The Netherlands

Cover: Erik de Graaf/iStock

All books published by **Wiley-VCH** are carefully produced. Nevertheless, authors, editors, and publisher do not warrant the information contained in these books, including this book, to be free of errors. Readers are advised to keep in mind that statements, data, illustrations, procedural details or other items may inadvertently be inaccurate.

Library of Congress Card No.: applied for

British Library Cataloguing-in-Publication Data
A catalogue record for this book is available from the British Library.

Bibliographic information published by the Deutsche Nationalbibliothek
The Deutsche Nationalbibliothek lists this publication in the Deutsche Nationalbibliografie; detailed bibliographic data are available on the Internet at <http://dnb.d-nb.de>.

© 2016 Wiley-VCH Verlag GmbH & Co. KGaA, Boschstr. 12, 69469 Weinheim, Germany

All rights reserved (including those of translation into other languages). No part of this book may be reproduced in any form – by photoprinting, microfilm, or any other means – nor transmitted or translated into a machine language without written permission from the publishers. Registered names, trademarks, etc. used in this book, even when not specifically marked as such, are not to be considered unprotected by law.

Print ISBN: 978-3-527-33781-1
ePDF ISBN: 978-3-527-69015-2
ePub ISBN: 978-3-527-69014-5
Mobi ISBN: 978-3-527-69013-8
oBook ISBN: 978-3-527-69012-1

Typesetting SPi Global, Chennai, India
Printing and Binding Markono Print Media Pte Ltd, Singapore

Printed on acid-free paper

Contents

Preface *XV*
List of Contributors *XVII*

Part I **Radical Chain Aerobic Oxidation** *1*

1 **Overview of Radical Chain Oxidation Chemistry** *3*
Ive Hermans
1.1 Introduction *3*
1.2 Chain Initiation *6*
1.3 Chain Propagation *7*
1.4 Formation of Ring-Opened By-Products in the Case of Cyclohexane Oxidation *11*
1.5 Complications in the Case of Olefin Autoxidation *12*
1.6 Summary and Conclusions *13*
References *14*

2 **Noncatalyzed Radical Chain Oxidation: Cumene Hydroperoxide** *15*
Manfred Weber, Jan-Bernd Grosse Daldrup, and Markus Weber
2.1 Introduction *15*
2.2 Chemistry and Catalysis *15*
2.2.1 Cumene Route to Phenol and Acetone: Chemistry Overview *16*
2.2.2 Thermal Decomposition of Cumene Hydroperoxide *17*
2.2.3 Oxidation of Cumene *19*
2.3 Process Technology *21*
2.3.1 Process Overview *21*
2.3.2 Reactors for the Cumene Oxidation *22*
2.3.3 Reactor Modeling *23*
2.3.4 Process Safety Aspects *26*
2.4 New Developments *27*
2.4.1 Process Intensification by Modification of the Oxidation Reaction *27*
2.4.2 Improvements of Reactor and Process Design *29*
References *30*

3	**Cyclohexane Oxidation: History of Transition from Catalyzed to Noncatalyzed** *33*
	Johan Thomas Tinge
3.1	Introduction *33*
3.2	Chemistry and Catalysis *34*
3.3	Process Technology *35*
3.3.1	The Traditional Catalyzed Cyclohexane Oxidation Process *35*
3.3.2	The Noncatalyzed DSM Oxanone® Cyclohexane Oxidation Process *37*
3.4	New Developments *38*
	Epilogue *39*
	References *39*
4	**Chemistry and Mechanism of Oxidation of *para*-Xylene to Terephthalic Acid Using Co–Mn–Br Catalyst** *41*
	Victor A. Adamian and William H. Gong
4.1	Introduction *41*
4.2	Chemistry and Catalysis *42*
4.2.1	Co–Br Catalysis *43*
4.2.2	Cobalt–Manganese–Bromide Catalysis (MC Oxidation): The Nature of Synergy between Co and Mn *48*
4.2.3	The Role and Nature of Bromine Species in MC Oxidation *50*
4.2.4	Nature of Cobalt(III) and Mn(III) Species *52*
4.2.5	Reactions of Cobalt(II) with Peroxy Radicals and the Effect of Solvent on Oxidation Rate *52*
4.2.6	Phenomenon of Manganese Precipitation *54*
4.2.7	Consolidated View of MC Oxidation Mechanism *54*
4.2.8	Oxidation By-products *56*
4.3	Process Technology *58*
4.3.1	Oxidation *58*
4.3.2	Purification *58*
4.4	New Developments *61*
4.4.1	Homogeneous Bromineless Catalysis *61*
4.4.2	Heterogeneous Bromineless Oxidation Catalysis *62*
4.4.3	Alternative Solvents *62*
4.5	Conclusions *62*
	References *63*
Part II	**Cu-Catalyzed Aerobic Oxidation** *67*
5	**Cu-Catalyzed Aerobic Oxidation: Overview and New Developments** *69*
	Damian Hruszkewycz, Scott McCann, and Shannon Stahl
5.1	Introduction *69*
5.2	Chemistry and Catalysis *70*

5.2.1	Cu-Catalyzed Oxydecarboxylative Phenol Synthesis *70*	
5.2.2	Cu-Catalyzed Oxidative Carbonylation of Methanol for the Synthesis of Dimethyl Carbonate *72*	
5.3	Process Technology *74*	
5.3.1	Cu-Catalyzed Oxydecarboxylative Phenol Synthesis *74*	
5.3.2	Cu-Catalyzed Oxidative Carbonylation of Methanol for the Synthesis of Dimethyl Carbonate *75*	
5.4	New Developments: Pharmaceutical Applications of Cu-Catalyzed Aerobic Oxidation Reactions *76*	
	References *82*	
6	**Copper-Catalyzed Aerobic Alcohol Oxidation** *85*	
	Janelle E. Steves and Shannon S. Stahl	
6.1	Introduction *85*	
6.2	Chemistry and Catalysis *86*	
6.3	Prospects for Scale-Up *91*	
6.4	Conclusions *93*	
	References *94*	
7	**Phenol Oxidations** *97*	
7.1	Polyphenylene Oxides by Oxidative Polymerization of Phenols *97*	
	Patrick Gamez	
7.1.1	Introduction *97*	
7.1.2	Chemistry and Catalysis *99*	
7.1.3	Process Technology *102*	
7.1.4	New Developments *104*	
7.2	2,3,5-Trimethylhydroquinone as a Vitamin E Intermediate via Oxidation of Methyl-Substituted Phenols *106*	
	Jan Schütz and Thomas Netscher	
	References *109*	
	Part III Pd-Catalyzed Aerobic Oxidation *113*	
8	**Pd-Catalyzed Aerobic Oxidation Reactions: Industrial Applications and New Developments** *115*	
	Dian Wang, Jonathan N. Jaworski, and Shannon S. Stahl	
8.1	Introduction *115*	
8.2	Chemistry and Catalysis: Industrial Applications *117*	
8.2.1	Acetoxylation of Alkenes to Vinyl or Allyl Acetates *117*	
8.2.2	Oxidative Carbonylation of Alcohols to Carbonates, Oxalates, and Carbamates *118*	
8.2.3	Oxidative Coupling of Arenes to Biaryl Compounds *121*	
8.3	Chemistry and Catalysis: Applications of Potential Industrial Interest *122*	
8.3.1	Oxidation of Alcohols to Aldehydes *122*	

8.3.2	Oxidation of Arenes to Phenols and Phenyl Esters	*123*
8.3.3	Benzylic Acetoxylation	*125*
8.3.4	Arene Olefination (Oxidative Heck Reaction)	*126*
8.4	Chemistry and Catalysis: New Developments and Opportunities	*128*
8.4.1	Ligand-Modulated Aerobic Oxidation Catalysis	*128*
8.4.2	Use of NO_x as Cocatalyst	*130*
8.4.3	Methane Oxidation	*132*
8.5	Conclusion	*133*
	References	*133*

9 Acetaldehyde from Ethylene and Related Wacker-Type Reactions *139*
Reinhard Jira

9.1	Introduction	*139*
9.2	Chemistry and Catalysis	*140*
9.2.1	Oxidation of Olefinic Compounds to Carbonyl Compounds	*140*
9.2.2	Kinetics and Mechanism	*140*
9.2.3	Catalytic Oxidation of Ethylene	*145*
9.2.3.1	Oxidation of Ethylene to Acetaldehyde in the Presence of $CuCl_2$	*145*
9.2.3.2	Oxidation of Ethylene to 2-Chloroethanol	*147*
9.3	Process Technology (Wacker Process)	*148*
9.3.1	Single-Stage Acetaldehyde Process from Ethylene	*148*
9.3.2	Two-Stage Acetaldehyde Process from Ethylene	*149*
9.4	Other Developments	*151*
	References	*155*
	Further Reading	*158*

10 1,4-Butanediol from 1,3-Butadiene *159*
Yusuke Izawa and Toshiharu Yokoyama

10.1	Introduction	*159*
10.2	Chemistry and Catalysis	*160*
10.2.1	Short Overview of Non-butadiene-Based Routes to 1,4-Butanediol	*160*
10.2.1.1	Acetylene-Based Reppe Process	*160*
10.2.1.2	Butane-Based Process; Selective Oxidation of Butane to Maleic Anhydride	*161*
10.2.1.3	Propylene-Based Process: Hydroformylation of Allyl Alcohol	*161*
10.2.2	Butadiene-Based Routes to 1,4-Butanediol	*162*
10.2.2.1	Oxyhalogenation of 1,3-Butadiene	*162*
10.2.2.2	Oxidative Acetoxylation of 1,3-Butadiene	*162*
10.3	Process Technology	*164*
10.3.1	Mitsubishi Chemical's 1,4-Butanediol Manufacturing Process: First-Generation Process	*165*
10.3.1.1	Oxidative Acetoxylation Step	*165*

10.3.1.2	Hydrogenation Step	*165*
10.3.1.3	Hydrolysis Step	*166*
10.3.2	Mitsubishi Chemical's 1,4-Butanediol Manufacturing Process: Second-Generation Process	*167*
10.4	New Developments	*168*
10.4.1	Improvement of the Current Process	*168*
10.4.2	Development of Alternative Processes	*169*
10.5	Summary and Conclusions	*169*
	References	*170*

11 Mitsubishi Chemicals Liquid Phase Palladium-Catalyzed Oxidation Technology: Oxidation of Cyclohexene, Acrolein, and Methyl Acrylate to Useful Industrial Chemicals *173*

Yoshiyuki Tanaka, Jun P. Takahara, Tohru Setoyama, and Hans E. B. Lempers

11.1	Introduction	*173*
11.2	Chemistry and Catalysis	*174*
11.2.1	Aerobic Palladium-Catalyzed Oxidation of Cyclohexene to 1,4-Dioxospiro-[4,5]-decane	*174*
11.2.1.1	Optimization of the Reaction Conditions	*174*
11.2.2	Aerobic Palladium-Catalyzed Oxidation of Other Types of Olefins	*176*
11.2.3	Aerobic Palladium-Catalyzed Oxidation of Acrolein to Malonaldehyde Bis-(1,3-dioxan-2-yl)-acetal Followed by Hydrolysis/Hydrogenation to 1,3-Propanediol	*178*
11.3	Prospects for Scale-Up	*180*
11.3.1	Aerobic Palladium-Catalyzed Oxidation of Methyl Acrylate (MA) to 3,3-Dimethoxy Methyl Propionate: Process Optimization and Scale-Up	*180*
11.3.2	Small-Scale Reaction Optimization	*181*
11.3.3	Large-Scale Methyl Acrylate Oxidation Reaction and Work-Up	*184*
11.3.4	Reaction Simulation Studies as Aid for Further Scale-Up	*184*
11.4	Conclusion	*187*
	References	*187*

12 Oxidative Carbonylation: Diphenyl Carbonate *189*

Grigorii L. Soloveichik

12.1	Introduction	*189*
12.1.1	Diphenyl Carbonate in the Manufacturing of Polycarbonates	*189*
12.1.2	History of Direct Diphenyl Carbonate Process at GE	*190*
12.2	Chemistry and Catalysis	*192*
12.2.1	Mechanism of Oxidative Carbonylation of Phenol	*192*
12.2.2	Catalysts for Oxidative Carbonylation of Phenol	*193*
12.2.3	Cocatalysts for Oxidative Carbonylation of Phenol	*196*
12.2.3.1	Organic Cocatalysts	*196*
12.2.3.2	Inorganic Cocatalysts	*196*

12.2.4	Multicomponent Catalytic Packages	*199*
12.2.5	Role of Bromide in Direct Synthesis of Diphenyl Carbonate	*199*
12.3	Prospects for Scale-Up *201*	
12.3.1	Catalyst Optimization *201*	
12.3.2	Water Removal in Direct Diphenyl Carbonate Process	*202*
12.3.3	Downstream Processing and Catalyst Recovery	*203*
12.4	Conclusions and Outlook *203*	
	Acknowledgments *204*	
	References *205*	

13 Aerobic Oxidative Esterification of Aldehydes with Alcohols: The Evolution from Pd–Pb Intermetallic Catalysts to Au–NiO$_x$ Nanoparticle Catalysts for the Production of Methyl Methacrylate *209*
Ken Suzuki and Setsuo Yamamatsu

13.1	Introduction *209*	
13.2	Chemistry and Catalysis *210*	
13.2.1	Discovery of the Pd–Pb Catalyst *210*	
13.2.2	Pd–Pb Intermetallic Compounds *210*	
13.2.3	Mechanism *212*	
13.2.4	The Role of Pb in the Pd–Pb Catalyst *213*	
13.2.5	Industrial Catalyst *213*	
13.3	Process Technology *214*	
13.4	New Developments *215*	
13.5	Conclusion and Outlook *217*	
	References *218*	

Part IV   Organocatalytic Aerobic Oxidation *219*

14 Quinones in Hydrogen Peroxide Synthesis and Catalytic Aerobic Oxidation Reactions *221*
Alison E. Wendlandt and Shannon S. Stahl

14.1	Introduction *221*	
14.2	Chemistry and Catalysis: Anthraquinone Oxidation (AO) Process *223*	
14.2.1	Autoxidation Process (Hydroquinone to Quinone) *223*	
14.2.2	Hydrogenation Process (Quinone to Hydroquinone) *225*	
14.3	Process Technology *227*	
14.4	Future Developments: Selective Aerobic Oxidation Reactions Catalyzed by Quinones *229*	
14.4.1	Aerobic DDQ-Catalyzed Reactions Using NO$_x$ Cocatalysts	*229*
14.4.2	Aerobic Quinone-Catalyzed Reactions Using Other Cocatalysts *230*	
14.4.3	CAO Mimics and Selective Oxidation of Amines	*231*
	References *234*	

15	**NO$_x$ Cocatalysts for Aerobic Oxidation Reactions: Application to Alcohol Oxidation** *239*	

Susan L. Zultanski and Shannon S. Stahl

- 15.1 Introduction *239*
- 15.2 Chemistry and Catalysis *241*
- 15.2.1 Aerobic Alcohol Oxidation with NO$_x$ in the Absence of Other Redox Cocatalysts *241*
- 15.2.2 Aerobic Alcohol Oxidation with NO$_x$ and Organic Nitroxyl Cocatalysts *242*
- 15.3 Prospects for Scale-Up *247*
- 15.4 Conclusions *249*
  References *249*

16 *N*-Hydroxyphthalimide (NHPI)-Organocatalyzed Aerobic Oxidations: Advantages, Limits, and Industrial Perspectives *253*

Lucio Melone and Carlo Punta

- 16.1 Introduction *253*
- 16.2 Chemistry and Catalysis *254*
- 16.2.1 Enthalpic Effect *256*
- 16.2.2 Polar Effect *256*
- 16.2.3 Entropic Effect *257*
- 16.3 Process Technology *257*
- 16.3.1 Oxidation of Adamantane to Adamantanols *257*
- 16.3.2 Oxidation of Cyclohexane to Adipic Acid *258*
- 16.3.3 Epoxidation of Olefins *259*
- 16.3.4 Oxidation of Alkylaromatics to Corresponding Hydroperoxides *260*
- 16.4 New Developments *262*
  Acknowledgments *264*
  References *264*

17 Carbon Materials as Nonmetal Catalysts for Aerobic Oxidations: The Industrial Glyphosate Process and New Developments *267*

- 17.1 Introduction *267*
  Mark Kuil and Annemarie E. W. Beers
- 17.2 Chemistry and Catalysis *268*
  Mark Kuil and Annemarie E. W. Beers
- 17.3 Process Technology *270*
  Mark Kuil and Annemarie E. W. Beers
- 17.3.1 Oxygen Pressure *271*
- 17.3.2 Oxygen Flow *271*
- 17.3.3 Activated Carbon Pore Size Distribution *271*
- 17.3.4 Activated Carbon H$_2$O$_2$ Time *271*
- 17.3.5 Activated Carbon Nitrogen Content *272*

17.4	New Developments *274*	
	Paul L. Alsters	
17.4.1	Aerobic Carbon Material Catalysis *275*	
17.4.1.1	Oxygenations and Oxidative Cleavage Reactions *275*	
17.4.1.2	Dehydrogenations and Dehydrogenative Coupling Reactions *279*	
17.4.2	Aerobic Graphitic Carbon Nitride Catalysis *280*	
17.4.2.1	Oxygenations and Oxidative Cleavage Reactions *280*	
17.4.2.2	Dehydrogenations and Dehydrogenative Coupling Reactions *281*	
17.5	Concluding Remarks *283*	
	References *283*	

Part V   Biocatalytic Aerobic Oxidation *289*

18   Enzyme Catalysis: Exploiting Biocatalysis and Aerobic Oxidations for High-Volume and High-Value Pharmaceutical Syntheses *291*
Robert L. Osborne and Erika M. Milczek

18.1	Introduction *291*
18.2	Chemistry and Catalysis *293*
18.2.1	Directed Evolution of BVMOs for the Manufacturing of Esomeprazole *295*
18.2.2	Directed Evolution and Incorporation of a Monoamine Oxidase for the Manufacturing of Boceprevir *298*
18.3	Process Technology *302*
18.4	New Developments *304*
	References *306*

Part VI   Oxidative Conversion of Renewable Feedstocks *311*

19   From Terephthalic Acid to 2,5-Furandicarboxylic Acid: An Industrial Perspective *313*
Jan C. van der Waal, Etienne Mazoyer, Hendrikus J. Baars, and Gert-Jan M. Gruter

19.1	Introduction *313*
19.1.1	The Avantium YXY Technology to Produce PEF, a Novel Renewable Polymer *314*
19.2	Chemistry and Catalysis *314*
19.2.1	Production of 2,5-Furandicarboxylic Acid Using Heterogeneous Catalysts *316*
19.2.2	Production of 2,5-Furandicarboxylic Acid Using Homogeneous Catalysts *318*
19.3	Process Technology *320*
19.3.1	Process Economics and Engineering Challenges *320*
19.3.1.1	Gas Composition Control *322*
19.3.1.2	Temperature Control *323*
19.3.1.3	Oxygen Mass Transfer Limitations *324*

19.3.1.4	Overall Safety Operation	*324*
19.4	New Developments	*325*
19.4.1	Outlook for Co/Mn/Br in the Air Oxidation of Biomass-Derived Molecules	*325*
19.5	Conclusion	*327*
	List of Abbreviations	*327*
	References	*327*

20 Azelaic Acid from Vegetable Feedstock via Oxidative Cleavage with Ozone or Oxygen *331*
Angela Köckritz

20.1	Introduction	*331*
20.1.1	Current Technical Process: Ozonolysis	*332*
20.1.1.1	Analytical Investigations of the Mechanism of Ozonolysis	*336*
20.2	Chemistry and Catalysis	*336*
20.2.1	Direct Aerobic Cleavage of the Double Bond of Oleic Acid or Methyl Oleate	*336*
20.2.2	Aerobic Oxidation Step within a Two-Stage Conversion of Oleic Acid or Methyl Oleate	*337*
20.2.3	Aerobic Oxidation Step within a Three-Stage Conversion of Oleic Acid or Methyl Oleate	*339*
20.2.4	Biocatalysis	*339*
20.3	Prospects for Scale-Up	*341*
20.4	Concluding Remarks and Perspectives	*342*
20.4.1	New Promising Developments	*342*
20.4.2	Summary	*343*
	References	*344*

21 Oxidative Conversion of Renewable Feedstock: Carbohydrate Oxidation *349*
Cristina Della Pina, Ermelinda Falletta, and Michele Rossi

21.1	Introduction	*349*
21.2	Chemistry and Catalysis	*351*
21.2.1	Oxidation of Monosaccharides	*354*
21.2.2	Oxidation of Disaccharides	*358*
21.2.3	Polysaccharide Oxidation	*361*
21.3	Prospects for Scale-Up	*362*
21.3.1	Enzymatic Process *versus* Chemical Process: Glucose Oxidation as a Model Reaction	*362*
21.3.2	Enzymatic Oxidation: Industrial Process and Prospects	*363*
21.3.3	Chemical Oxidation: Industrial Process and Prospects	*364*
21.3.3.1	Metal Catalysts: Concepts Guiding Choice and Design	*364*
21.4	Concluding Remarks and Perspectives	*366*
	References	*367*

Part VII Aerobic Oxidation with Singlet Oxygen 369

22 Industrial Prospects for the Chemical and Photochemical Singlet Oxygenation of Organic Compounds 371
Véronique Nardello-Rataj, Paul L. Alsters, and Jean-Marie Aubry

22.1 Introduction 371
22.2 Chemistry and Catalysis 373
22.2.1 Comparison of Singlet and Triplet Oxygen 373
22.2.2 Photochemical Generation of $^1O_2$ 375
22.2.3 Chemicals Sources of $^1O_2$ Based on the Catalytic Disproportionation of H_2O_2 376
22.2.4 Optimal Generation of $^1O_2$ Through the Catalytic System H_2O_2/MoO_4^{2-} 379
22.2.5 Potential Molecular Targets for Singlet Oxygenation 380
22.3 Prospects for Scale-Up 383
22.3.1 Respective Advantages and Disadvantages of "Dark" and "Luminous" Singlet Oxygenation 383
22.3.2 Choice of the Medium for Dark Singlet Oxygenation 384
22.3.2.1 Homogeneous Aqueous and Alcoholic Media 384
22.3.2.2 Single-Phase Microemulsions 385
22.3.2.3 Multiphase Microemulsions with Balanced Catalytic Surfactants 387
22.3.3 Examples of Industrialized Singlet Oxygenation 388
22.3.3.1 Synthesis of Artemisinin 389
22.3.3.2 Synthesis of Rose Oxide 391
22.4 Conclusion 392
Acknowledgments 392
References 393

Part VIII Reactor Concepts for Liquid Phase Aerobic Oxidation 397

23 Reactor Concepts for Aerobic Liquid Phase Oxidation: Microreactors and Tube Reactors 399
Hannes P. L. Gemoets, Volker Hessel, and Timothy Noël

23.1 Introduction 399
23.2 Chemistry and Catalysis 400
23.2.1 Transition Metal-Catalyzed Aerobic Oxidations in Continuous Flow 400
23.2.2 Photosensitized Singlet Oxygen Oxidations in Continuous Flow 404
23.2.3 Metal-Free Aerobic Oxidations in Continuous Flow 408
23.2.4 Aerobic Coupling Chemistry in Continuous Flow 410
23.3 Prospects for Scale-Up 413
23.4 Conclusions 417
References 417

Index 421

Preface

Oxidation reactions play a crucial role in the chemical industry, where >90% of the feedstocks derive from hydrocarbons – the most reduced organic chemicals on the planet. Sustainability concerns are demanding a greater shift toward biomass-derived feedstocks; however, oxidation methods will continue to play a major role. For example, even as this book goes to press (March 2016), BASF and Avantium have just announced plans to pursue a joint venture for the production of 2,5-furandicarboxylic acid (FDCA), an important polymer-building block derived from biomass. The proposed 50 000 metric tons per year plant will undoubtedly incorporate liquid phase aerobic oxidation chemistry similar to that described in Chapter 19 of this volume.

Molecular oxygen is the most abundant, most environmentally benign, and least expensive oxidant available. Yet, selective partial oxidation reactions with O_2 have long been dominated by the commodity chemical industry, where the scale of the processes typically dictates that O_2 or, preferably, air is the only viable oxidant. Over the past 10–15 years, an increasing number of academic research groups have embraced the challenge of developing selective oxidation reactions capable of using O_2 as the stoichiometric oxidant. These efforts have led to a growing number of catalytic oxidation methods that are compatible with complex molecules of the type encountered in the fine chemical, agrochemical, and pharmaceutical industries. Invariably, these methods are liquid phase processes, and they employ both homogeneous and heterogeneous catalysts.

We (SSS and PLA) first met at one of the triennial ADHOC conferences (Activation of Dioxygen and Homogeneous Oxidation Catalysis). The unique blend of industrial and academic attendees, who have a shared interest in oxidation chemistry and catalysis, is a special feature of this conference series. Industrial contributors to this conference regularly provide valuable context for the academic research, and often share insights that are not readily available from the open literature.

This volume was conceived with the goal of complementing the growing number of widely available academic review articles that describe research advances in the field of aerobic oxidation catalysis. Our goal was to bring together as many experts as possible, who could provide a perspective on either (i) existing liquid phase aerobic oxidation processes that are practiced in industry or (ii) emerging aerobic oxidation applications of industrial interest, including their prospects for scale-up, where appropriate. In certain cases, we encountered challenges identifying industrial contributors who were both willing to share their (company's) story and had the clearance to do so. Fortunately, we were successful in a number of cases, and the present volume documents these industrial stories and perspectives, together with selected coverage of other relevant topics by academic authors. Twenty-two chapters on aerobic oxidation reactions and processes, beginning with non-catalyzed autoxidations, are complemented by a final chapter on new reactor concepts for liquid phase aerobic oxidations. It was not possible to provide comprehensive coverage, but it is our hope that the content of this volume will provide valuable historical and industrial context for ongoing research efforts within the global research community, and also inspire further advances in the field of aerobic oxidation catalysis.

USA, March 2016 *Shannon S. Stahl and Paul L. Alsters*

List of Contributors

Victor A. Adamian
BP Petrochemicals
150 West Warrenville Rd.
MC E-1F
Naperville, IL 60563
USA

Paul L. Alsters
DSM Ahead R&D b.v.
Innovative Synthesis
P.O. Box, 1066
6160 BB Geleen
The Netherlands

Jean-Marie Aubry
Université de Lille and ENSCL
Unité de Catalyse et de Chimie
du Solide
UCCS CNRS
UMR 8181
59655 Villeneuve d'Ascq
France

Hendrikus J. Baars
Avantium Chemicals
Renewable Chemistries
Zekeringstraat 29
1014 BV Amsterdam
The Netherlands

Annemarie E. W. Beers
Cabot Norit Activated Carbon
Euro Support Catalysts
Kortegracht 26
3811 KH Amersfoort
The Netherlands

Jan-Bernd Grosse Daldrup
INEOS Phenol GmbH
Technology Department
Dechenstrasse 3
45966 Gladbeck
Germany

Ermelinda Falletta
Università degli Studi di Milano
Dipartimento di Chimica
Via G. Golgi 19
20133 Milano
Italy

Patrick Gamez
Institució Catalana de Recerca i
Estudis Avançats (ICREA)
Passeig Lluís Companys, 23
08010 Barcelona
Spain

Hannes P. L. Gemoets
Eindhoven University of Technology
Department of Chemical Engineering and Chemistry
Micro Flow Chemistry and Process Technology
Den Dolech 2
5612 AZ Eindhoven
The Netherlands

William H. Gong
1s280 Wisconsin Avenue
Lombard, IL 60147
USA

Gert-Jan M. Gruter
Avantium Chemicals
Renewable Chemistries
Zekeringstraat 29
1014 BV Amsterdam
The Netherlands

Ive Hermans
The University of Wisconsin–Madison
Department of Chemistry
1101 University Avenue
Madison, WI 53706
USA

and

The University of Wisconsin–Madison
Department of Chemical and Biological Engineering
1101 University Avenue
Madison, WI 53706
USA

Volker Hessel
Eindhoven University of Technology
Department of Chemical Engineering and Chemistry
Micro Flow Chemistry and Process Technology
Den Dolech 2
5612 AZ Eindhoven
The Netherlands

Damian Hruszkewycz
University of Wisconsin–Madison
Department of Chemistry
1101 University Avenue
Madison, WI 53706
USA

Yusuke Izawa
Mitsubishi Chemical Corporation
Petrochemicals Technology Laboratory
Toho-cho 1
Yokkaichi-shi Mie 510-8530
Japan

Jonathan N. Jaworski
University of Wisconsin–Madison
Department of Chemistry
1101 University Avenue
Madison, WI 53706
USA

Reinhard Jira
Wacker-Chemie AG
Burghausen
Germany

Angela Köckritz
Department of Heterogeneous
Catalytic Processes
Leibniz Institute for Catalysis
Albert-Einstein-Str. 29a
18059 Rostock
Germany

Mark Kuil
Cabot Norit Activated Carbon
P.O. Box 105
3800 AC Amersfoort
The Netherlands

Hans E. B. Lempers
University of Applied Sciences
Department of Chemistry
Institute of Life Sciences and
Chemistry
Heidelberglaan 7
3584CS Utrecht
The Netherlands

Etienne Mazoyer
Avantium Chemicals
Renewable Chemistries
Zekeringstraat 29
1014 BV Amsterdam
The Netherlands

Scott McCann
University of
Wisconsin–Madison
Department of Chemistry
1101 University Avenue
Madison, WI 53706
USA

Lucio Melone
Politecnico di Milano
Department of Chemistry,
Materials, and Chemical
Engineering "Giulio Natta"
Via Leonardo da Vinci 32
20131 Milano
Italy

and

Università degli Studi e-Campus
Via Isimbardi 10
22060 Novedrate
Como
Italy

Erika M. Milczek
Agem Solutions Inc.
50 Murray St. #2011
New York, NY 10007
USA

Véronique Nardello-Rataj
Université de Lille and ENSCL
Unité de Catalyse et de Chimie
du Solide
UCCS CNRS
UMR 8181
59655 Villeneuve d'Ascq
France

Thomas Netscher
DSM Nutritional Products
Research and Development
P.O. Box 2676
CH-4002 Basel
Switzerland

Timothy Noël
Eindhoven University of Technology
Department of Chemical Engineering and Chemistry
Micro Flow Chemistry and Process Technology
Den Dolech 2
5612 AZ Eindhoven
The Netherlands

Robert L. Osborne
Novozymes North America Inc.
77 Perrys Chapel Church Road
Franklinton, NC 27525
USA

Cristina Della Pina
Università degli Studi di Milano
Dipartimento di Chimica
Via G. Golgi 19
20133 Milano
Italy

Carlo Punta
Politecnico di Milano
Department of Chemistry, Materials, and Chemical Engineering "Giulio Natta"
Via Leonardo da Vinci 32
20131 Milano
Italy

Michele Rossi
Università degli Studi di Milano
Dipartimento di Chimica
Via G. Golgi 19
20133 Milano
Italy

Jan Schütz
DSM Nutritional Products
Research and Development
P.O. Box 2676
CH-4002 Basel
Switzerland

Tohru Setoyama
Mitsubishi Chemical Corporation
Department of Research & Development
1000 Kamoshida-cho
Aoba-ku
Yokohama 227-8502
Japan

Grigorii L. Soloveichik
General Electric Company
GE Global Research
One Research Circle
Niskayuna, NY 12309
USA

and

Advanced Research Project Agency—Energy (ARPA-E)
1000 Independence Ave
Washington, DC 20585
USA

Shannon S. Stahl
University of Wisconsin–Madison
Department of Chemistry
1101 University Avenue
Madison, WI 53706
USA

Janelle E. Steves
University of
Wisconsin–Madison
Department of Chemistry
1101 University Avenue
Madison, WI 53706
USA

Ken Suzuki
Asahi Kasei Chemicals
Corporation
Catalyst Laboratory
2767-11 Shionasu Kojima
Kurashiki
Okayama 711-8510
Japan

Jun P. Takahara
Mitsubishi Chemical
Corporation
Department of Research &
Development
1000 Kamoshida-cho
Aoba-ku
Yokohama 227-8502
Japan

Yoshiyuki Tanaka
Mitsubishi Chemical
Corporation
Department of Research &
Development
3-10 Ushiodori
Kurashiki
Okayama 712-8054
Japan

Johan Thomas Tinge
Research & Technology
Fibrant BV
Urmonderbaan 22
6167 RD Geleen
The Netherlands

Jan C. van der Waal
Avantium Chemicals
Renewable Chemistries
Zekeringstraat 29
1014 BV Amsterdam
The Netherlands

Dian Wang
University of
Wisconsin–Madison
Department of Chemistry
1101 University Avenue
Madison, WI 53706
USA

Manfred Weber
INEOS Phenol GmbH
Technology Department
Dechenstrasse 3
45966 Gladbeck
Germany

Markus Weber
INEOS Phenol GmbH
Technology Department
Dechenstrasse 3
45966 Gladbeck
Germany

Alison E. Wendlandt
University of
Wisconsin–Madison
Department of Chemistry
1101 University Avenue
Madison, WI 53706
USA

Setsuo Yamamatsu
Tokyo University and Technology
Department of Chemical
Engineering
2-24-16 Nakacho
Koganei
Tokyo 184-8588
Japan

Toshiharu Yokoyama
Mitsubishi Chemical Corporation
C4 Chemical Derivatives Department
1-1 Marunouchi 1-chome
Chiyoda-ku
Tokyo 100-8251
Japan

Susan L. Zultanski
University of Wisconsin–Madison
Department of Chemistry
1101 University Avenue
Madison, WI 53706
USA

Part I
Radical Chain Aerobic Oxidation

1
Overview of Radical Chain Oxidation Chemistry

Ive Hermans

1.1
Introduction

The direct reaction of triplet O_2 with hydrocarbons is usually very slow (e.g., H-abstraction), or quantum-chemically forbidden (e.g., the insertion of O_2 in a C–H bond to form a hydroperoxide). Nevertheless, those high-energy paths can be bypassed by a more efficient radical mechanism as demonstrated by the pioneering work of Bäckström [1]. Although initial studies were aimed at finding ways of preventing undesired autoxidations, it was soon recognized that the controlled oxidation of hydrocarbons can be a synthetically useful route to a wide range of oxygenated products. In subsequent chapters, some of those industrial processes are discussed in more detail. The goal of this introductory chapter is to provide a mechanistic framework for this rather complex chemistry.

In 1939, Criegee made an important contribution by showing that the primary product of cyclohexene autoxidation is the allylic hydroperoxide [2]. The classical textbook on radical pathway for the thermal autoxidation of a general hydrocarbon RH is summarized in reactions (1.1–1.5) [3–5]:

Initiation

$$ROOH \rightarrow RO^\bullet + {}^\bullet OH \quad (1.1)$$

Propagation

$$RO^\bullet/{}^\bullet OH + RH \rightarrow ROH/H_2O + R^\bullet \quad (1.2)$$

$$R^\bullet + O_2 \rightarrow ROO^\bullet \quad (1.3)$$

$$ROO^\bullet + RH \rightarrow ROOH + R^\bullet \quad (1.4)$$

Termination

$$ROO^\bullet + ROO^\bullet \rightarrow ROH + R_{-\alpha H}{=}O + O_2 \quad (1.5)$$

In general, an initiation reaction produces radicals from closed-shell molecules. Different types of initiator molecules and their initiation barriers are summarized

Liquid Phase Aerobic Oxidation Catalysis: Industrial Applications and Academic Perspectives,
First Edition. Edited by Shannon S. Stahl and Paul L. Alsters.
© 2016 Wiley-VCH Verlag GmbH & Co. KGaA. Published 2016 by Wiley-VCH Verlag GmbH & Co. KGaA.

Table 1.1 Common initiators for autoxidations and their initiation barriers.

Structure	Barrier
HO–OH	48 kcal/mol
tBuO–OH	40 kcal/mol
tBuO–OtBu	37 kcal/mol
AcO–OAc (diacetyl peroxide)	31 kcal/mol
PhC(O)O–OC(O)Ph	30 kcal/mol
tBuO–N=N–... (AIBN-type tBuO derivative)	28 kcal/mol

in Table 1.1. Under normal autoxidation conditions, initiation takes place through O–O cleavage in hydroperoxides (viz., reaction (1.2)). This rate-controlling initiation step is rapidly followed by the fast H-abstraction from the substrate by both alkoxyl and hydroxyl radicals (reaction (1.2)), producing alcohol and water, respectively.

For alkoxyl radicals, a competitive side-reaction can occur in kinetic competition with H-abstraction from the substrate, namely, β C–C cleavage to produce a carbonyl compound and an alkyl radical (viz., RCH_x-$CR'HO^{\bullet} \rightarrow RCH_x^{\bullet} + CR'H=O$). We discuss this reaction in Section 1.4.

When no initiators such as peroxides are initially added to a hydrocarbon mixture, an *induction* period is observed before oxygen consumption takes place. There are several hypotheses on what might happen during this induction–initiation stage: (i) the direct reaction of triplet oxygen with singlet closed-shell molecules (a thermodynamically disfavored reaction), (ii) the R–H homolysis (a reaction that is very slow due to typical C–H bond dissociation energies of 80–110 kcal/mol), and (iii) the termolecular reaction

$RH + O_2 + RH \rightarrow R^\bullet + H_2O_2 + R^\bullet$ (unlikely to be of any kinetic significance). A more reasonable explanation for the induction–initiation is the presence of (sub)parts per billion levels of compounds that are more easily cleaved. After the induction period, the rate increases very rapidly. Because of this accelerating effect, one refers to this type of reactions as being autocatalytic. However, until recently, the precise reason for this autocatalytic upswing was not very well understood and is discussed in more detail later.

The RO–OH cleavage can also be catalyzed by transition metal ions that are able to undergo one-electron redox switches such as cobalt and manganese, among others. The catalysis of autoxidations is discussed in more detail in chapters dealing with specific industrial processes.

Chain-propagation reactions neither increase nor decrease the radical population. Two elementary propagation reactions can be distinguished: H-abstractions (reactions (1.2) and (1.4)) and O_2 addition to a carbon-centered radical (reaction (1.3)). The latter reaction is normally diffusion controlled and hence rarely rate determining. As alkoxyl radicals are significantly more reactive than peroxyl radicals, $ROO^\bullet$ are the predominant radical species in the reaction medium and are thus the main participants in the bimolecular propagation reactions. Although the normal rule is that the weakest C–H bond is preferably attacked, the nature of the H-abstracting species also determines the selectivity. Indeed, the more reactive the H-abstracting radical, the lower the selectivity for the weakest C–H bond, as illustrated by the rate data in Table 1.2. Table 1.2 also shows the significant reactivity differences between hydroxyl, alkoxyl, and peroxyl radicals, explaining why only $ROO^\bullet$ radicals can be really termed as chain-propagating radicals.

Termination reactions such as (1.5) decrease the number of radicals and produce nonradical products such as alcohols and ketones. Mutual termination reactions of primary and secondary peroxyl radicals have near-zero activation energies but unusually low Arrhenius prefactors, suggesting a strained transition state. The exothermicity of this termination is sufficient to produce electronically excited states of either the carbonyl compound or oxygen. Besides the termination reaction (1.5), peroxyl radicals can also undergo a self-reaction without termination ($2ROO^\bullet \rightarrow 2RO^\bullet + O_2$) that is often ignored but is equally important as the termination channel itself [9, 10].

Tertiary alkylperoxyl radicals are generally assumed to combine with tetroxides (ROOOOR) that subsequently decompose to di-*tert*-alkyl peroxides or *tert*-alkoxyl radicals. The decomposition of the tertiary tetroxide has a fairly high

Table 1.2 Rate constants ($M^{-1} s^{-1}$) for hydrogen abstraction from primary, secondary, and tertiary C–H bonds by various radical species at 400 K (per H-atom) [6–8].

Attacking radical	Primary C–H	Secondary C–H	Tertiary C–H
$^\bullet OH$	8.1×10^7 (1.0)	3.7×10^8 (4.6)	1.1×10^9 (13.3)
$CH_3O^\bullet$	2.1×10^4 (1.0)	2.3×10^5 (10.8)	6.0×10^5 (28.4)
$CH_3OO^\bullet$	2.6×10^{-2} (1.0)	4.8×10^{-1} (18.5)	2.7×10^1 (1053)

activation energy, making the termination of tertiary peroxyl radicals slower than that of primary and secondary analogs [5].

The chain termination compensates the chain initiation and leads to a quasi-steady state in peroxyl radicals. Indeed, the characteristic lifetime of ROO$^\bullet$ radicals is given by $\tau = 1/\{[\text{ROO}^\bullet] \times 2k_5\}$, with k_5 being the rate constant of the termination reaction (1.5). As an example, at 0.1% cyclohexane conversion, the [CyOO$^\bullet$] concentration can be estimated at 5×10^{-8} M (derived from the product formation rate $= k_4$ [CyH] [CyOO$^\bullet$]), leading to a τ as low as ≈ 2.5 s, given $2k_{\text{term}} = 8.4 \times 10^6$ M^{-1} s^{-1}. We emphasize that τ is much smaller than the timescale of ≈ 2000 s over which the CyOO$^\bullet$ concentration changes significantly, such that [CyOO$^\bullet$] quasi-steady state will indeed be established and maintained throughout the oxidation process. This implies that the rate of chain initiation equals the rate of chain termination, or k_1 [CyOOH] $= k_5$[CyOO$^\bullet$]^2.

The radical chain mechanism outlined here avoids the ineffective direct reaction of molecular oxygen with the substrate hydrocarbon. The fast propagation reactions produce ROOH that in turn can initiate new radical chains. As the primary product of the reaction initiates new reactions, one ends up with an autocatalytic acceleration. The propagating peroxyl radicals can also mutually terminate and yield one molecule of alcohol and ketone in a one-to-one stoichiometry. The ratio between the rate of propagation and the rate of termination is referred to as the chain length and is of the order of 50–1000. As the desired chain products are more susceptible to oxidation, autoxidations are normally carried out at low conversions in order to keep the selectivity to an economically acceptable level.

In the following sections, we discuss the various reaction sequences in more detail, starting with the chain initiation.

1.2
Chain Initiation

According to reaction (1.1), the ROOH molecule yields radicals upon the unimolecular scission of the 40 kcal/mol O–O bond. However, this reaction is not only slow due to the high barrier but also very inefficient to generate free radicals. Indeed, once the O–O bond has been elongated, the nascent radicals need to diffuse away from each other (i.e., out of their solvent cage in case of liquid phase reactions) before they are really free radicals. This diffusion process faces a significantly higher barrier than the in-cage radical recombination to reform the ROOH molecule. Therefore, only a small fraction of the ROOH molecules that manage to dissociate will effectively lead to free radicals. Moreover, during cyclohexane oxidation, it was observed that the addition of a small amount of cyclohexanone not only eliminates the induction period but also enhances the oxidation rate at a given conversion (i.e., at a certain O$_2$ consumption), as shown in Figure 1.1. This observation clearly indicates that cyclohexanone plays an important role in the chain initiation, hitherto missing in the simple unimolecular mechanisms.

Based on quantum chemical calculations, a bimolecular chain initiation mechanism was proposed between the ROOH primary product and cyclohexanone

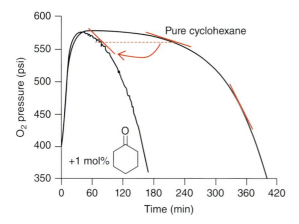

Figure 1.1 Oxygen consumption during the autoxidation of cyclohexane at 418 K with and without the addition of cyclohexanone. The red slopes signify the oxidation rate at various stages of the reaction. The initial pressure increase is due to heating, initiated at time 0.

[11–13]. In this reaction, the nascent hydroxyl radical, breaking away from the hydroperoxide molecule, abstracts the weakly bonded αH-atom from cyclohexanone, forming water and a resonance-stabilized ketonyl radical (Figure 1.2). This reaction features a significantly lower activation barrier (i.e., 28 kcal/mol) and prevents facile in-cage radical recombination. The theoretically predicted rate constant of this reaction (i.e., $0.6 \times 10^{-4}\,M^{-1}\,s^{-1}$ at 418 K) quantitatively agrees with the experimentally determined rate constant ($1 \times 10^{-4}\,M^{-1}\,s^{-1}$), corroborating this thermal initiation mechanism and explaining the autocatalytic effect of cyclohexanone [11].

The hypothesis of a bimolecular initiation reaction for liquid phase autoxidations was extended beyond cyclohexanone as a reaction partner. Also other substances featuring abstractable H-atoms are able to assist in this radical formation process. The initiation barrier was found to be linearly dependent on the C–H bond strength, ranging from 30 kcal/mol for cyclohexane to 5 kcal/mol for methyl linoleate [14, 15]. Substrates that yield autoxidation products that lack weaker C–H bonds than the substrate (e.g., ethylbenzene) do not show an exponential rate increase as the chain initiation rate is not product enhanced [16].

1.3 Chain Propagation

According to traditional textbooks, the hydroperoxide is produced in the fast propagation reaction, whereas alcohol and ketone are formed in the slow termination step. As such, one would expect the hydroperoxide selectivity to be an order of magnitude higher than the alcohol and ketone selectivity. This is, however, not in agreement with the experimental observations, indicating that (an) important source(s) of alcohol and ketone is (are) missing in this model. Moreover, the

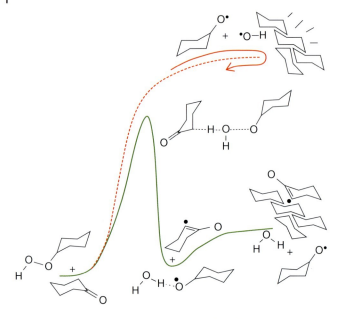

Figure 1.2 Comparison of the unimolecular RO–OH dissociation process and the bimolecular initiation between ROOH and cyclohexanone [11].

second order upswing of the alcohol and ketone as a function of the conversion indicates that these products are secondary and do not directly originate from elementary steps of the chain-carrying peroxyl radicals. In addition, one observes a maximum peroxide yield as a function of the substrate conversion, indicating that there must exist a reaction that rapidly consumes the primary hydroperoxide product. Yet, such a reaction is absent from the mechanism thus far discussed.

A careful computational analysis of the reactivity of the peroxyl radical with various substrates and their autoxidation products indicates that the abstraction of the αH-atom of the hydroperoxide is kinetically favored over other H-atoms. In the case of cyclohexane autoxidation, the ratio between the rate constants for H-abstraction from the cyclohexyl hydroperoxide versus cyclohexane is around ≈50, indicating that sequential peroxide propagation is kinetically important, even at low conversions (i.e., high CyH/CyOOH concentration ratios) [17]. The αH-abstraction immediately yields ketone plus a hydroxyl radical as the $Cy_{-\alpha H}{}^{\bullet}OOH$ species promptly dissociates without a barrier [18]. The $^{\bullet}OH$ radical rapidly abstracts an H-atom from the ubiquitous cyclohexane substrate, rendering the overall peroxide copropagation exothermic by nearly 50 kcal/mol. As shown in Figure 1.3, the nascent products {CyOOH + Cy$^{\bullet}$ + Q=O + H$_2$O} either can diffuse away from each other or can react together in a thermally activated solvent-cage-assisted reaction to form alcohol and alkoxyl radicals. Indeed, the high exothermicity of the peroxide propagation leads to a local hotspot, facilitating the reaction between the nascent Cy$^{\bullet}$ radical and CyOOH. As a consequence,

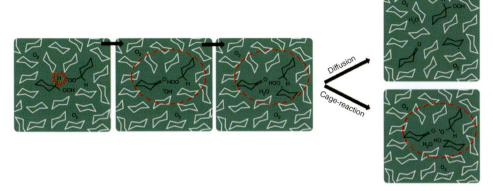

Figure 1.3 αH-abstraction from the primary cyclohexyl hydroperoxide product and subsequent chemistry.

this activated reaction can compete with the nearly temperature-independent diffusive separation, leading to a rather efficient alcohol source. As such, the consecutive cooxidation of the primary hydroperoxide product leads to the formation of ketone and alcohol in a ratio that is determined by the solvent-cage efficiency.

The cage efficiency predominantly depends on the stability of the R• radical that determines the barrier of the cage reaction. Comparing the cage efficiency observed for various substrates such as cyclohexane [17], toluene [19], and ethylbenzene [16] confirms a systematic trend and readily explains the difference in alcohol-to-ketone ratio for those substrates (see Figure 1.4).

The high reactivity of the primary hydroperoxide product toward chain-carrying peroxyl radicals readily explains the relatively high yields of alcohol

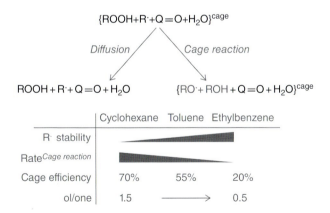

Figure 1.4 Effect of the substrate on the solvent-cage efficiency and alcohol-to-ketone ratio.

and ketone, the ratio of which is determined by the efficiency of a selectivity-controlling solvent-cage reaction. There is hence no need to ascribe the formation of alcohol and ketone to slow peroxyl radical cross-reactions.

Analogous to the hydroperoxide copropagation, one has to consider the αH-abstraction from the alcohol product (reaction (1.6)). Although slower than for the hydroperoxide, this tertiary cooxidation reaction is important as the α-hydroxylperoxyl radical, formed after O_2 addition (reaction (1.7)), readily eliminates $HO_2^\bullet$ (reaction (1.8)) [20], a known chain-terminating radical. Indeed, $HO_2^\bullet$ reacts in a head-to-tail manner with the chain-carrying peroxyl radicals to form hydroperoxide plus oxygen [21]. This termination reaction is about 500 times faster than the mutual termination reaction between secondary peroxyl radicals. As a consequence, the cooxidation of alcohol slows down the overall autoxidation reaction and explains in a very elegant manner the inhibiting effect of alcohols on the autoxidation of various substrates.

$$R - CH(R')OH + ROO^\bullet \rightarrow R - C^\bullet(R')OH + ROOH \quad (1.6)$$

$$R - C^\bullet(R')OH + O_2 \rightarrow R - C(R')(OO^\bullet)OH \quad (1.7)$$

$$R - C(R')(OO^\bullet)OH \rightarrow R - C(R') = O + HO_2^\bullet \quad (1.8)$$

$$HO_2^\bullet + ROO^\bullet \rightarrow O_2 + ROOH \quad (1.9)$$

Cooxidation of the ketone product via abstraction of a weakly bound αH-atom was, in the case of cyclohexane oxidation, assumed to be the predominant source of ring-opened by-products such as adipic acid, glutaric acid, and 6-hydroxy caproic acid [4, 5]. This hypothesis was based on two wrong assumptions. (i) It was assumed, based on Evans–Polanyi correlations between activation energy and reaction enthalpy, that the αH-abstraction from cyclohexanone would be fast due to the weak αC–H bond strength. However, the resonance stabilization, the main reason for this low bond strength, is not yet operative in the αH-abstraction transition state and hence does not contribute to a significantly lower reaction barrier compared with the parent substrate. (ii) Experiments where ^{13}C-labeled cyclohexanone was initially added to the cyclohexane substrate suggested the cyclohexanone cooxidation to be the most important source of by-products. However, in those early experiments, substantial amounts of ketone were added, that is, even more than what is formed during a normal autoxidation experiment, perturbing the kinetic network and leading to the wrong conclusion that this would be the most important source of by-products. This was later confirmed by the observation that although more by-products are formed upon initial addition of cyclohexanone, the relative increase in waste diminishes as a function of the conversion. This clearly illustrates that there is a significantly more important source of by-products than cyclohexanone [17].

1.4 Formation of Ring-Opened By-Products in the Case of Cyclohexane Oxidation

Having ruled out the overoxidation of cyclohexanone as a major source of ring-opened by-products, we turned our attention to the cyclohexoxyl radicals, coproduced in the activated cage reaction following the peroxide cooxidation (see Figures 1.3 and 1.4). As mentioned in the introduction, alkoxyl radicals not only can abstract H-atoms but also can undergo C–C bond scission [5]. In the case of cyclohexoxyl radicals, this competing unimolecular reaction leads to the formation of ω-formyl radicals that can lead either to decarbonylation under oxygen starvation conditions or to a primary peroxyl radical that can form 6-hydroperoxy- and 6-hydroxycaproic acids, that is, the primary ring-opened by-products (see Figure 1.5) [22]. No reaction mechanism can be found to explain the formation of these primary by-products from cyclohexanone. Consecutive cooxidation of these primary by-products leads to the experimentally observed secondary by-products such as adipic and glutaric acids, among several others [22]. This straightforward mechanism to explain the formation of a complex mixture of various by-products based on the co-propagation of the primary hydroperoxide product corroborates the importance of that step.

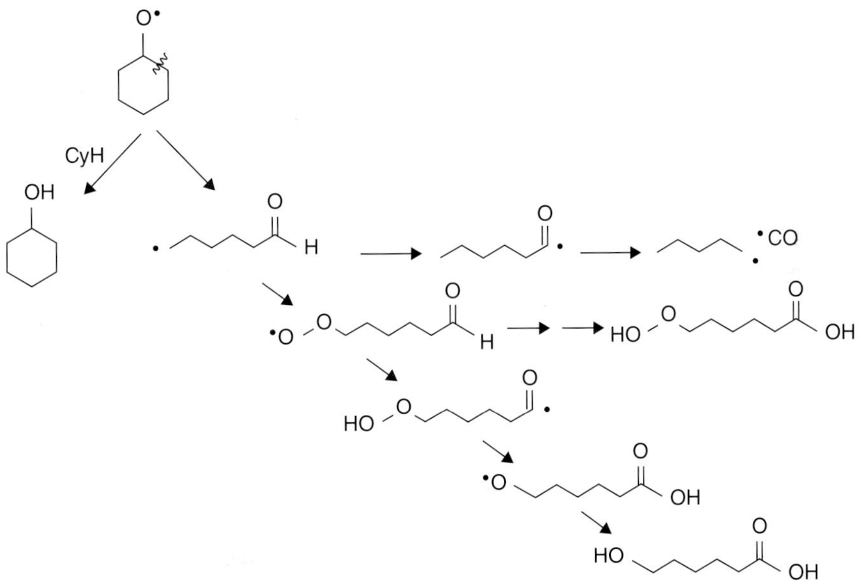

Figure 1.5 Competing reactions for the cyclohexoxyl radical: formation of the primary ring-opened by-products [22].

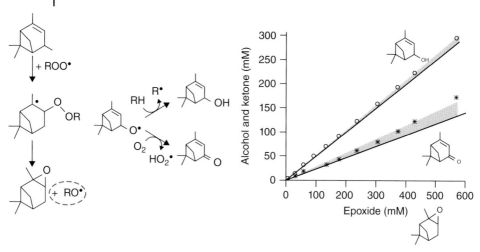

Figure 1.6 Addition of peroxyl radical to C=C bond of α-pinene and consecutive alkoxyl radical chemistry [23].

1.5
Complications in the Case of Olefin Autoxidation

The reactions discussed so far form a consistent and generic mechanism for the thermal autoxidation of saturated hydrocarbons. In the case of unsaturated compounds, an additional channel opens for the peroxyl radicals, namely, the addition to the C=C bond, followed by the formation of an epoxide upon elimination of an alkoxyl radical (see Figure 1.6). These unsaturated alkoxyl radicals are, however, special in the sense that they not only can abstract H-atoms from the substrate (forming alcohol) but also can react with O_2 to form the corresponding ketone. Experimental evidence for this is found in the nearly linear correlation between the alcohol and ketone increase as a function of the epoxide yield (see Figure 1.6) [23]. Yet, at increasing conversion, additional alcohol and ketone is formed as indicated by the green areas in Figure 1.6. This additional alcohol and ketone stems from the consecutive cooxidation of the hydroperoxide product analogs as shown in Figure 1.3 for unsaturated substrates. The 10% efficiency of the activated solvent-cage reaction matches very well with the values in Figure 1.4.

At increasing O_2 pressure, one observes a decrease in epoxide selectivity and a steady increase in reaction rate [24]. Both effects are coupled and related to the competitive O_2 addition to the ROO•-to-C=C adduct, as shown in Figure 1.7. Indeed, this leads to the formation of a new type of peroxyl radical that will ultimately yield a dialkyl peroxide that promptly dissociates to new radicals due to the very weak O–O bond (see Table 1.1).

We emphasize that the efficiency of this radical epoxidation mechanism depends on the substrate (i.e., the ease at which the chain-carrying peroxyl radicals abstract αH-atoms or add to the C=C bond) [25], and sometimes even on the precise geometric conformer [26, 27]. However, the maximum theoretical

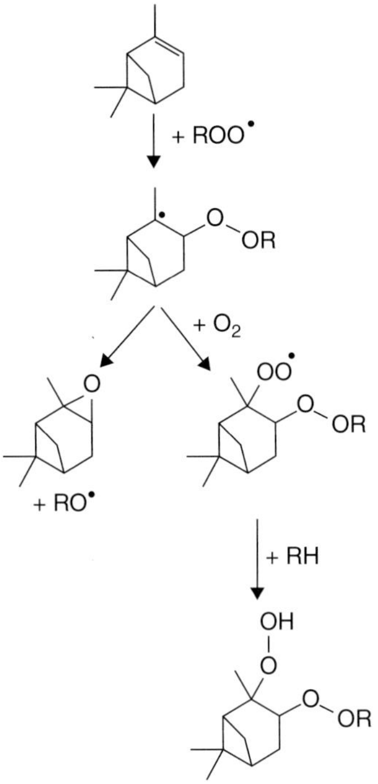

Figure 1.7 Competing O_2 addition at higher oxygen partial pressures, leading to a drop in epoxide selectivity and an increase in initiation rate via the formation of labile dialkyl peroxide.

epoxide yield in this mechanism is restricted to 50%. Yet, in a recent case study on the autoxidation of β-caryophyllene, epoxide selectivities as high as 70% were observed [28]. A detailed study revealed that the spontaneous epoxidation of β-caryophyllene constitutes a rare case of unsensitized electron transfer from an olefin to triplet oxygen under mild conditions. The formation of a caryophyllene-derived radical cation via electron transfer is proposed as the initiation mechanism. Subsequent reaction of the radical cation with O_2 yields a dioxetane via a chain mechanism with typical chain lengths that can exceed 100. The dioxetane acts as an excellent *in situ* epoxidation agent under the reaction conditions.

1.6
Summary and Conclusions

Hydrocarbon autoxidation takes place via a complex set of radical reactions, some of which were only recently identified. One of the mechanistic difficulties is that the reactions can only be indirectly investigated by monitoring the evolution of stable products. The input of quantum-chemical calculations, in combination with theoretical kinetics, turned out to be a crucial tool to construct a generic mechanism. One of the new insights is the importance of the copropagation of the primary hydroperoxide product. A solvent-cage reaction, activated by the exothermicity of this secondary step, leads to the formation of the desired alcohol and ketone products, as well as undesired ring-opened byproducts in the case of cyclohexane oxidation. In the case of unsaturated hydrocarbons, a peroxyl addition mechanism to the unsaturated bond leads to the formation of epoxides in addition to allylic products.

References

1. Bäckström, H.L.J. (1927) *J. Am. Chem. Soc.*, **49**, 1460.
2. Criegee, R., Pilz, H., and Flygare, H. (1939) *Chem. Ber.*, **72**, 1799.
3. Arpentinier, P., Cavani, F., and Trifirò, F. (2001) *The Technology of Catalytic Oxidations*, Editions Technip, Paris.
4. Bhaduri, S. and Mukesh, D. (2000) *Homogeneous Catalysis, Mechanisms and Industrial Applications*, John Wiley & Sons, Inc.
5. Tolman, C.A., Druliner, J.D., Nappa, M.J., and Herron, N. (1989) in *Activation and Functionalization of Alkanes* (ed C.L. Hill), John Wiley & Sons, Inc.
6. Atkinson, R. (1986) *Int. J. Chem. Kinet.*, 18.
7. Tsang, W. (1988) *J. Phys. Chem. Ref. Data*, **17**, 887.
8. Tsang, W. (1990) *J. Phys. Chem. Ref. Data*, 19.
9. Lightfoot, P.D., Lesclaux, R., and Veyret, B. (1990) *J. Phys. Chem.*, **94**, 700.
10. Horie, O., Crowley, J.N., and Moortgat, G.K. (1990) *J. Phys. Chem.*, **94**, 8198.
11. Hermans, I., Jacobs, P.A., and Peeters, J. (2006) *Chem. Eur. J.*, **12**, 4229.
12. Hermans, I., Peeters, J., and Jacobs, P.A. (2008) *Top. Catal.*, **48**, 41.
13. Hermans, I., Peeters, J., and Jacobs, P.A. (2008) *Top. Catal.*, **50**, 124.
14. Neuenschwander, U. and Hermans, I. (2012) *J. Catal.*, **287**, 1.
15. Turra, N., Neuenschwander, U., and Hermans, I. (2013) *ChemPhysChem*, **14**, 1666.
16. Hermans, I., Peeters, J., and Jacobs, P.A. (2007) *J. Org. Chem.*, **72**, 3057.
17. Hermans, I., Nguyen, T.L., Jacobs, P.A., and Peeters, J. (2005) *ChemPhysChem*, **6**, 637.
18. Vereecken, L., Nguyen, T.L., Hermans, I., and Peeters, J. (2004) *Chem. Phys. Lett.*, **393**, 432.
19. Hermans, I., Peeters, J., Vereecken, L., and Jacobs, P.A. (2007) *ChemPhysChem*, **8**, 2678.
20. Hermans, I., Muller, J.F., Nguyen, T.L., Jacobs, P.A., and Peeters, J. (2005) *J. Phys. Chem. A*, **109**, 4303.
21. Boyd, A.A., Flaud, P.M., Daugey, N., and Lesclaux, R. (2003) *J. Phys. Chem. A*, **107**, 818.
22. Hermans, I., Jacobs, P., and Peeters, J. (2007) *Chem. Eur. J.*, **13**, 754.
23. Neuenschwander, U., Guignard, F., and Hermans, I. (2010) *ChemSusChem*, **3**, 75.
24. Neuenschwander, U. and Hermans, I. (2010) *Phys. Chem. Chem. Phys.*, **12**, 10542.
25. Neuenschwander, U., Meier, E., and Hermans, I. (2011) *ChemSusChem*, **4**, 1613.
26. Neuenschwander, U. and Hermans, I. (2011) *J. Org. Chem.*, **76**, 10236.
27. Neuenschwander, U., Czarniecki, B., and Hermans, I. (2012) *J. Org. Chem.*, **77**, 2865.
28. Steenackers, B., Campagno, N., Fransaer, J., Hermans, I., and De Vos, D. (2014) *Chem. Eur. J.*, **20**, 1.

2
Noncatalyzed Radical Chain Oxidation: Cumene Hydroperoxide

Manfred Weber, Jan-Bernd Grosse Daldrup, and Markus Weber

2.1
Introduction

Phenol and acetone are produced from benzene and propene by the *cumene route*, which can be divided into *cumene process* and *cumene oxidation process* [1]. Benzene and propene react to cumene in the cumene process. In the cumene oxidation process, cumene is oxidized to cumene hydroperoxide (CHP), which is converted into phenol and acetone in a successive reaction, the so-called cleavage. The oxidation of cumene to CHP is described in detail in the following sections.

The cleavage of CHP into phenol and acetone was first reported by Hock and Lang [2]. Hence, the cumene oxidation process for the production of phenol and acetone is known as the *Hock process*. As claimed by Zakoshansky [3], the reaction route was supposedly discovered in parallel in the former USSR.

Minor amounts of CHP are used as an initiator in polymerization processes. However, the dominant use of CHP is as an intermediate product in the cumene oxidation process. Based on the worldwide phenol production capacity (in 2013) of approximately 11.4×10^6 t, the production capacity of CHP can be estimated to be 18.8×10^6 t/a.

2.2
Chemistry and Catalysis

An overview of the cumene route will be given first, followed by a detailed description of the thermal decomposition behavior of CHP, and finally the mechanism of the liquid phase oxidation of cumene. The oxidation of cumene to CHP is one of three major reaction steps within the cumene route.

Liquid Phase Aerobic Oxidation Catalysis: Industrial Applications and Academic Perspectives,
First Edition. Edited by Shannon S. Stahl and Paul L. Alsters.
© 2016 Wiley-VCH Verlag GmbH & Co. KGaA. Published 2016 by Wiley-VCH Verlag GmbH & Co. KGaA.

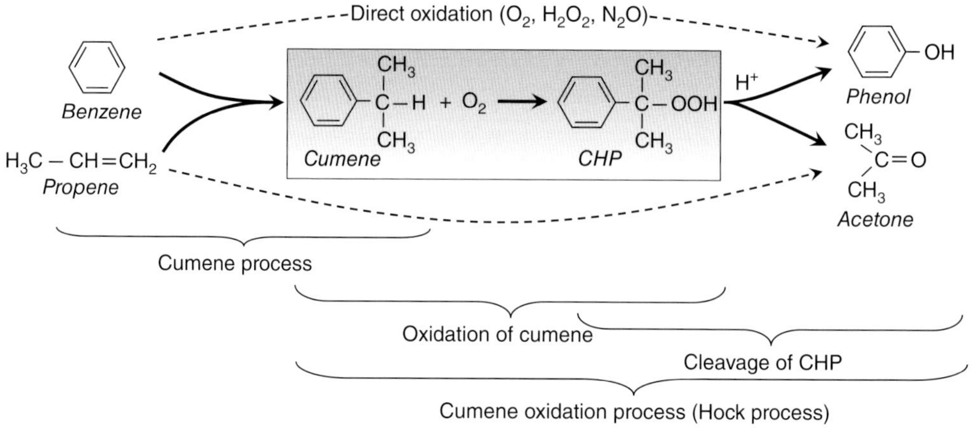

Figure 2.1 Overview of the cumene route for the production of phenol and acetone.

2.2.1
Cumene Route to Phenol and Acetone: Chemistry Overview

An overview of the production of phenol and acetone by the *cumene route* is shown in Figure 2.1 (see also [1, 4]).

In the *cumene process*, cumene is produced from benzene and propene by Friedel–Crafts alkylation. In modern cumene plants, zeolite catalysts are used with high yields of more than 99.7% at temperatures and pressures of approximately 150 °C and 30 bar, respectively. The reaction heat is 98 kJ/mol.

In the *cumene oxidation process*, in a first reaction cumene is oxidized to CHP. In a second reaction, CHP is cleaved to phenol and acetone by using a strong mineral acid as catalyst. The reaction heat is 117 kJ/mol for the oxidation and 252 kJ/mol for the cleavage. The oxidation is carried out at pressures ranging from atmospheric to approximately 7 bar and temperatures between 80 and 120 °C. The cleavage is performed around atmospheric pressure and temperatures in the range between 40 and 80 °C.

The overall process can be considered a dual oxidation [5]. Benzene is oxidized to phenol and propene is oxidized to acetone (see the dotted lines in Figure 2.1). The direct oxidation of propene as an alternative for the commercial production of acetone is used in the Wacker–Hoechst process [6]. However, the direct oxidation of benzene with oxygen, hydrogen peroxide (H_2O_2), or nitrous oxide (N_2O) is limited to a benzene conversion of only a few percent [7]. Compared to benzene, which can be also easily oxidized, the phenol formed is further oxidized more easily. It reacts very fast with oxygen or any other oxidizing agents in successive reactions to yield finally carbon dioxide and water. Therefore, any kind of direct oxidation of benzene has not yet been used for the commercial production of phenol. The cumene oxidation process takes advantage of the selective oxidation of the tertiary C atom at the propyl group of the cumene molecule. In combination with the highly selective cumene process upstream and the successive cleavage of CHP, the cumene route is the dominant route for the production of phenol and acetone.

2.2.2 Thermal Decomposition of Cumene Hydroperoxide

When storing and handling CHP at medium or even elevated temperatures, the heat release from the thermal decomposition must be efficiently removed in order to avoid any hazards from thermal explosion (runaway). Especially, in large reactors for cumene oxidation, the exothermic decomposition of CHP has to be taken into account during a shutdown process when there is no more mixing by aeration, so only limited heat removal to ambient takes place. The heat evolved from thermal decomposition is 270 kJ/mol [8, 9]. From process safety point of view, and also to understand the *auto-catalyzed* mechanism in the cumene oxidation, it is necessary to describe and quantify the thermal decomposition characteristics of CHP.

Twigg [10] was the first to present a kinetic study for the thermal decomposition of CHP in cumene. Experiments were performed at temperatures between 110 and 160 °C. Pure CHP (98–99.5%) was diluted in purified cumene down to concentrations below 10 wt%.

When simplifying the cumyl radical by $R^\bullet$ and so cumene by RH and CHP by ROOH, the proposed radical mechanism is as follows [10]:

$$\text{initiation} \quad ROOH \rightarrow RO^\bullet + OH^\bullet \quad (2.1)$$

$$\text{chain} \quad RO^\bullet + RH \rightarrow ROH + R^\bullet \quad (2.2)$$

$$R^\bullet + ROOH \rightarrow ROH + RO^\bullet \quad (2.3)$$

$$OH^\bullet + RH \rightarrow H_2O + R^\bullet \quad (2.4)$$

$$\text{termination} \quad 2R^\bullet \rightarrow R-R \quad \text{(as one example)} \quad (2.5)$$

From the chain reactions (2.2) and (2.3), ROH (dimethylbenzylalcohol or DMBA) is the major by-product:

DMBA   ACP

As a competing reaction, the $RO^\bullet$ radical decomposes into acetophenone (ACP) and a methyl radical, which mainly ends up as methane:

$$RO^\bullet \rightarrow ACP + CH_3^\bullet \tag{2.6}$$

$$CH_3^\bullet + RH \rightarrow CH_4 + R^\bullet \tag{2.7}$$

In [10], the net degradation of CHP is described by a *pseudo-first-order* reaction related to CHP:

$$d\,CHP/dt = -k \cdot CHP = -(k_P + k_C \cdot CHP^{0.5}) \cdot CHP \tag{2.8}$$

Hattori et al. [11] investigated the thermal decomposition in the following way: Cumene was oxidized in isothermal batch experiments to a certain CHP concentration between approximately 25 and 55 wt%. The temperature was varied between 110 and 130 °C. Then, the oxygen supply was stopped and the CHP decomposition was monitored over time. For the above-mentioned temperature and concentration range, a *first-order* kinetics was determined:

$$d\,CHP/dt = -k \cdot CHP \tag{2.9}$$

Hattori pointed out that at higher CHP concentrations above 60 wt%, the thermal decomposition was markedly accelerated.

Duh et al. [12, 13] used an 80 wt% CHP, which was diluted with cumene to lower concentrations down to minimum 15 wt% for various tests on the thermal decomposition behavior. In Taiwan, CHP is widely used as an initiator in polymerization. As described in [14], cumene is oxidized to 80 wt% CHP for this purpose. It is therefore assumed that an 80 wt% CHP was sampled from such a production process. The reaction order was determined to be 0.5.

$$d\,CHP/dt = -k \cdot CHP^{0.5} \tag{2.10}$$

Duh et al. pointed out that the reaction order was independent of the CHP concentration, thus implying the same mechanism of decomposition regardless of the CHP concentration.

Besides the different reaction orders, all three authors found different activation energies in the range of 104–122 kJ/mol. When calculating the thermal decomposition rates of CHP, there were significant differences between the three approaches. However, it is not surprising because the adapted reaction orders as well as the activation energies were different. There are mainly two reasons for their different findings: First, the experiments covered different CHP concentrations. Second, various CHP concentrations were prepared in different ways, by diluting pure CHP [10], by diluting an 80 wt% CHP from cumene oxidation [12, 13], or by oxidizing cumene to the specific CHP concentration [11]. These different preparation methods will lead to different concentrations of by-products and impurities such as organic acids, phenol, and other peroxides. All these components can influence the thermal decomposition of CHP. From that point of view, the preparation method used by Hattori et al. [11] is the most representative way to investigate the CHP decomposition characteristics, especially for the oxidation process.

As described above, DMBA is the major by-product from a moderate thermal decomposition of CHP in the oxidation process, due to the presence of (excess) cumene. It is obvious that the radical mechanism must change towards higher CHP concentrations as the cumene concentration largely decreases. Thus, it seems implausible to describe the thermal decomposition of CHP with only one kinetic model for the whole concentration range between 0 and 100 wt%.

As a rule of thumb, the molar ratio between DMBA and ACP from moderate thermal decomposition in an oxidation product is approximately 10:1. However, if the decomposition is accelerated because of adiabatic conditions and a complete decomposition of CHP finally occurs at high temperatures (runaway), this ratio decreases as more ACP is formed [10].

2.2.3 Oxidation of Cumene

In commercial cumene oxidation processes, the radicals from the thermal decomposition of CHP are the initiators for the free-radical chain reaction. Therefore, the reaction towards CHP in the cumene oxidation is called *autooxidation*.

The radical chain reactions are described in detail in, for example, [15–17]. The scheme of oxidation in liquid phase by molecular oxygen comprises the steps *initiation, propagation,* and *termination*.

initiation

$$ROOH \rightarrow RO^\bullet + OH^\bullet \quad (2.1)$$

$$RO^\bullet + RH \rightarrow ROH + R^\bullet \quad (2.2)$$

$$OH^\bullet + RH \rightarrow H_2O + R^\bullet \quad (2.4)$$

propagation

$$R^\bullet + O_2 \rightarrow ROO^\bullet \quad (2.11)$$

$$ROO^\bullet + RH \rightarrow ROOH + R^\bullet \quad (2.12)$$

termination (examples)

$$ROO^\bullet + ROO^\bullet \rightarrow 2RO^\bullet + O_2 \quad (2.13)$$

$$ROO^\bullet + ROO^\bullet \rightarrow ROOR + O_2 \quad (2.14)$$

Equations (2.11) and (2.12) show the chain reaction. A cumyl radical $R^\bullet$ reacts *quickly* with dissolved oxygen to form a peroxide radical $ROO^\bullet$. In the second step, the peroxide radical reacts with cumene (RH) to CHP (ROOH). This reaction is *slower* and so determines the chain reaction. A new cumyl radical $R^\bullet$ is formed, which again initiates reaction (2.11).

The more often this cycle runs through, the more selective is the process. The *chain length* is the mean number of cycles from one initiation. In industrial oxidation processes, the chain length is usually 10 [17]. Because of the termination reactions, Eqs. (2.13) and (2.14), the chain is interrupted and must be reinitiated by the thermal decomposition of CHP. Thus, DMBA (ROH) and ACP from the thermal decomposition of CHP are still the major by-products besides small amounts of dicumylperoxide (ROOR or DCP).

In industrial processes, oxygen from air is used in large bubble columns (see the next section for details). Oxygen dissolves in the cumene/CHP liquid and reacts according to Eq. (2.11). Hattori et al. [11] investigated the impact of the oxygen concentration on the CHP formation rate. At temperatures below 120 °C, the concentration of dissolved oxygen should be greater than 7×10^{-4} mol/l to ensure that the oxygen does not impact the chain reaction rate. This concentration corresponds to an equilibrium partial pressure p_{O_2} of 0.1 bar in the gas phase.

There are some approaches in the literature to set up a complex model based on these elementary radical reactions [18–20]. However, it is difficult to obtain kinetic data for these single reactions, which even do not represent the whole picture.

For a technical process, it is more practical to evaluate and use simple kinetic models to describe conversion rates and selectivities. For example, in [9], the following equations were fitted to experiments to describe overall formation rates in mol/(m³·s) for CHP, DMBA, and ACP in oxidation at $p_{O_2} > 0.1$ bar:

$$R_{CHP} = k_{CHP} \cdot CHP^{0.5} \cdot RH \tag{2.15}$$

$$R_{DMBA} = k_{DMBA} \cdot CHP \tag{2.16}$$

$$R_{ACP} = k_{ACP} \cdot CHP^{0.5} \tag{2.17}$$

where RH is the concentration of cumene. The activation energies for the formation of DMBA and ACP are higher than that for CHP formation. Therefore, the selectivity of the oxidation increases with decreasing temperatures. In order to have high space–time yields for CHP, one single back-mixed reactor would be favorable because this main reaction is autocatalytic. But to prevent DMBA formation, the CHP concentration profile should be similar to that of a plug-flow reactor because the reaction order related to CHP is higher in Eq. (2.16) than in Eq. (2.15). Thus, a cascade of back-mixed reactors increases selectivity but leads to an increase in investment costs.

If the oxygen concentration cannot be neglected, the Eq. (2.18) from Hattori et al. [11] can be used to describe the net CHP formation:

$$R_{CHP} = k_{CHP} \cdot CHP^{0.5} \cdot RH - k'_{CHP} \cdot CHP \tag{2.18}$$

The first part of Eq. (2.18) is analogous to Eq. (2.15). The reaction constant k_{CHP} depends not only on the temperature but also on the concentration of oxygen. The second part summarizes the CHP losses from thermal decomposition, but not distinguishing between DMBA and ACP.

Besides the formation of the major by-products DMBA and ACP and, in minor concentrations DCP, a lot of micro-impurities are formed in cumene oxidation. A key role is played by the methyl radical $CH_3^{\bullet}$ from the thermal decomposition of CHP. In the presence of oxygen, methyl hydroperoxide (MHP) and formaldehyde are formed (Figure 2.2). Formaldehyde is further oxidized to formic acid, which can catalyze the acidic decomposition of CHP into phenol and acetone. As

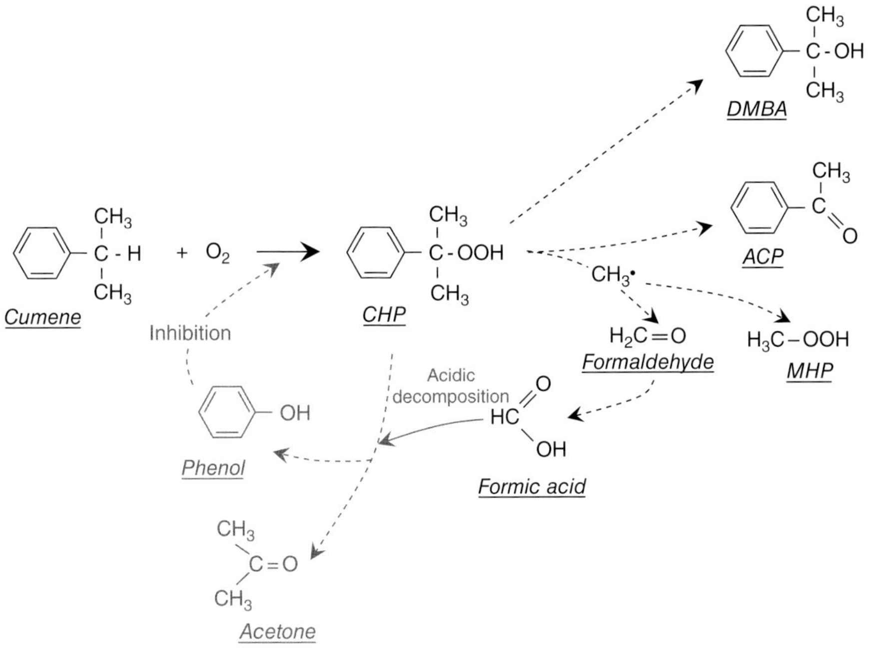

Figure 2.2 Important side reactions in the cumene oxidation.

phenol is a strong inhibitor of the radical chain reaction, these side reactions must be limited to keep the phenol concentration to below several weight parts per million. For this reason, it is most important to reduce the phenol and organic acid concentrations in the cumene feed down to "zero."

2.3 Process Technology

2.3.1 Process Overview

Figure 2.3 gives an overview of the cumene oxidation process, which includes the oxidation of cumene to CHP. A more detailed description is given in [1].

The oxidation of cumene with oxygen from air is nowadays performed in a series of bubble columns. The off-gas is treated by a combination of cooling and adsorption. The recovered cumene from the off-gas treatment is recycled to the oxidation unit. In the concentration unit, the product from oxidation with around 20–40 wt% CHP is distilled under vacuum to increase the CHP concentration to 65–90 wt%. The separated cumene is again recycled to the oxidation unit. In the cleavage, the CHP is converted into phenol and acetone using sulfuric acid as the catalyst. In parallel, the DMBA from oxidation is nearly quantitatively converted

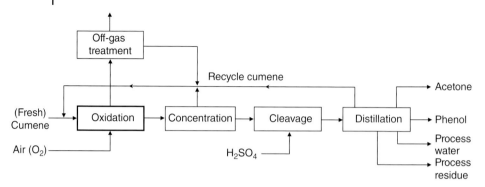

Figure 2.3 Block diagram of the cumene oxidation process.

to α-methylstyrene (AMS) and water. Smaller amounts react with phenol to high boilers such as cumylphenols. Together with the ACP from oxidation, these are the main causers for the process residue. AMS can be either separated as a pure product or hydrogenated to cumene and recovered. The product from the cleavage is finally rectified in the distillation unit. Again, a separated cumene stream is recycled to oxidation.

As already described in the previous section, great care must be taken to properly treat all cumene recycle streams. This is typically done by caustic washing to neutralize and separate organic acids and phenol; see especially [21].

2.3.2
Reactors for the Cumene Oxidation

The oxidation of cumene to CHP is a slow reaction. As a rule of thumb, the mean residence time in a continuously operated process is around 10 h at temperatures in the range of 80–120 °C. Thus, large bubble columns are preferred for this purpose with, for example, three reactors arranged in series [9]. A cascade of only two reactors [22] or even four reactors [23] is also used in commercial plants. The bubble columns are operated at high pressures ranging from atmospheric to approximately 700 kPa.

Figure 2.4 shows a typical setup of a bubble column for the oxidation of cumene [9, 24].

Cumene or cumene/CHP from a reactor upstream is combined with an external circulation flow and enters the bubble column at the bottom. Air is fed through a gas distributor. The off-gas and the liquid product can leave the reactor at the top. The liquid phase in the reactor is well mixed. Therefore, the heat from reaction can be easily removed by an external heat exchanger, as described in [20].

The superficial gas velocity v_{G0}, which is related to the temperature and pressure at the bottom, is typically about 1–2 cm/s. With $v_{G0} < 5$ cm/s, the bubble column operates in the homogeneous regime with no or little formation of large bubbles.

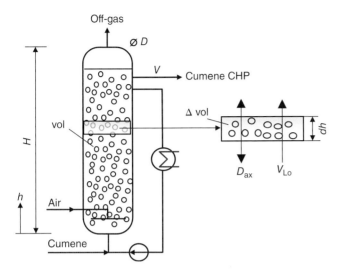

Figure 2.4 Bubble column for the oxidation of cumene.

In the beginning of the cumene oxidation process, oxidation was carried out as a three-phase reaction [25–27], the so-called *wet oxidation*. In addition to cumene and air, an aqueous sodium carbonate solution was continuously added to the reactors to extract and neutralize organic acids that are formed during oxidation. Phenolchemie, now INEOS Phenol, was the first to operate reactors without adding any caustic soda or sodium carbonate [28], which is called *dry oxidation*. Such a process is easier to handle and even leads to higher yields.

Bubble columns for the cumene oxidation may be equipped with internals such as trays [29] or an inner shaft for *airlift loop* operation [18, 20, 30]. However, the impact of any internals on the so-called *self-accelerating decomposition temperature (SADT)* and on an effective emergency cooling must be considered carefully (see Section 2.3.4).

For more details on general bubble column design, refer to [31–34].

2.3.3
Reactor Modeling

A model is used for reactor design and optimization regarding, for example, the number of reactors in series and the temperature profile. Based on the requirement and informational value, a model can be simple or relatively complex. The purpose of this section is to present some of our own experience with models of different level in accuracy for cumene oxidation.

If, for example, the *liquid phase in the bubble column is considered ideally mixed (Model I)* and the impact from mass transfer ($p_{O_2} > 0.1$ bar) is neglected, simple kinetic approaches such as Eqs. (2.15)–(2.17) can be used. The conversion rates are calculated for the mean residence time $\tau(s)$ of the liquid phase:

$$\tau = \text{vol} \cdot (1 - \varepsilon)/V \tag{2.19}$$

Here, vol is the volume of the aerated liquid (m³), ε is the gas holdup (no unit), and V is the volumetric flow rate of the liquid (m³/s) (Figure 2.4). In the homogeneous flow regime, the gas holdup is proportional to the superficial gas velocity v_{G0}. For cumene oxidation in reactors, as shown in Figure 2.4, a gas holdup of 0.108 (10.8 vol%) was measured at $v_{G0} = 2$ cm/s [9].

To take into account a limited mixing behavior in the liquid phase, the *axial dispersion model with a dispersion coefficient* D_{ax} can be used (see Figure 2.4). The back-mixing of the liquid superposes the net liquid flow (superficial liquid velocity v_{L0}). The mass balance for the individual component i in the volume element Δvol gives

$$dc/dt = v_{L0}/(1-\varepsilon) \cdot dc/dh + D_{ax}/(1-\varepsilon) \cdot d^2c/dh^2 + R_i \text{ in mol}/(m^3 \cdot s) \tag{2.20}$$

with R_i the reaction rates according to Eqs. (2.15)–(2.17).

For large bubble columns up to several meters in diameter, dispersion coefficient in the homogeneous regime can be calculated from correlations in the literature (see, e.g., [33, 35, 36]) to be approximately 2 m²/s. However, our own measurements on a production reactor with a diameter of 4.6 m by *step function responses* from tracer experiments [24] gave a dispersion coefficient of only approximately 0.3 m²/s. But even when applying the dispersion model with this lower mixing coefficient, the resulting concentrations of the components in the liquid phase (CHP, DMBA, ACP) are more or less constant over the reactor height. Thus, it is not necessary to use a dispersion model when considering the concentrations of liquid components only. Using simple Model I is sufficient for this purpose.

However, *if the impact of the oxygen concentration on the reactions is to be checked, the axial dispersion model is a good approach* (Model II). For this, the mass balance in the liquid phase must be extended for the dissolved oxygen. In addition, the mass balance for oxygen in the rising gas bubbles must be established. Both mass balances are coupled by the *mass transfer* of oxygen from the gas bubbles into the liquid. Finally, the net formation rate of CHP must be described by Eq. (2.18) to include the impact of the oxygen concentration.

Figure 2.5 shows the well-known *film model* for the mass transfer of oxygen. If the mass balance is, for example, in kg/s, the concentration of dissolved oxygen is in kg/m³.

$c^*_{O_2}$ is in equilibrium with the local partial pressure of oxygen in the gas bubbles. The function $c^*_{O_2} = f(p_{O_2})$ equates the oxygen solubility in cumene/CHP. From [37], it is 0.233 kg/(m³·bar).

The mass transfer coefficient k_L for an estimated bubble diameter d_B of 4 mm is approximately 0.000 35 m/s [38]. The volume specific surface of the bubbles (a, in m²/m³) is calculated from

$$a = 6 \cdot \varepsilon/d_B \tag{2.21}$$

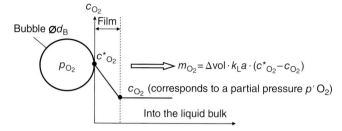

Figure 2.5 Mass transfer of oxygen (film model).

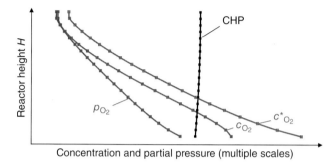

Figure 2.6 Concentration profiles in a cumene oxidation reactor.

With, for example, $\varepsilon = 0.1$, it finally leads to a typical *volume-specific mass transfer coefficient* of $k_L a = 0.05 \text{ s}^{-1}$.

For all vertical positions in the bubble column, the criterion for solving all mass balance equations at stationary conditions is *oxygen by mass transfer = oxygen consumption by reaction*. Note that, when applying Hattori's kinetic according to Eq. (2.18), the concentration of dissolved oxygen in the liquid bulk or the corresponding partial pressure (p'_{O_2}) should be used (Figure 2.5).

Figure 2.6 shows the qualitative concentration profiles in an industrial pressurized reactor for cumene oxidation with the parameters estimated above. For the gas phase, plug flow is assumed, while for the liquid phase a dispersion coefficient D_{ax} of 0.3 m²/s is used. As described, such a back-mixing already causes more or less constant concentrations over the reactor height for liquid components such as CHP. However, this is not the case for oxygen. If oxygen leaves the reactor at the top with a partial pressure of less than 0.1 bar, there is a linear increase in the oxygen concentration when going downward. At the bottom, this equals the concentration in air. Hence, from such a result it would make sense to lower the residual concentration of oxygen to <0.1 bar. That would only affect the overall reaction rate in a small region at the top of the reactor but would significantly reduce the power consumption for air compression.

For improving the reactor simulation further, the so-called *compartment modeling (Model III)* can be used [24, 39]. The total reactor volume is *split* or

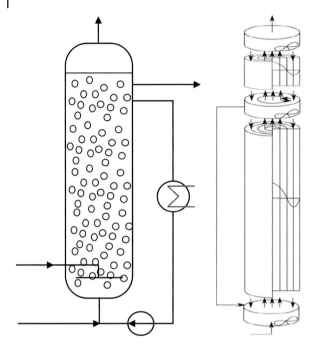

Figure 2.7 Compartment model.

compartmentalized into various zones such as ring cylinders and ideal mixed regions (Figure 2.7).

The *time-averaged* velocities and gas holdups in the compartments, as well as the fluid interactions between the zones, are first calculated by *computational fluid dynamics* (CFD). Then, balance equations for heat and mass transfer and for chemical reactions are evaluated and solved using appropriate software. First results from a simulation of a cumene oxidation reactor on an industrial scale were impressive, as they matched real temperature and concentration fields.

A complete and meaningful modeling of the two-phase system in a *large* cumene oxidation reactor exclusively by CFD is, in the authors' opinion, still not possible but may be possible in near future with further increase in computing power.

2.3.4
Process Safety Aspects

In case of power failure, there is no more heat removal from the normal cooling installation as, for example, shown in Figure 2.4. Although the air flow, and so the oxidation reaction, is stopped, there is still some heat generation from thermal decomposition of CHP. Without any countermeasure, the temperature would increase and, within a time frame of several hours, runaway would finally occur. Therefore, the reactor must be cooled down by an emergency cooling device [22], or the pressure must be relieved via adequate relief devices to a safe location [23], or the content of the reactor must be drained out to a *blowdown* tank where sufficient coolant is available for cooling [40].

It is important to know to which temperature a reactor has to be cooled down to avoid a subsequent heating-up from thermal decomposition. At a certain temperature, the heat that is generated by thermal decomposition of CHP is equal to the heat that is lost by heat losses from the reactor. The SADT depends, of course, on weather conditions, size of the reactor, CHP concentration, but especially on the intensity of free thermal convection in the nonmixed reactor. The determination of the SADT must be included in every safety concept for cumene oxidizers [8]. It should be pointed out that the efficiency of the free convection in the whole reactor will be lowered by inserts such as trays or an inner shaft in an airlift loop reactor. Thus, these kind of reactors need to be cooled down to lower temperature compared to those of equal size but without any inserts.

Special attention needs to be given to the design of the air distributor (sparger) (see [41] for details). The major safety aspect is to avoid any backflow of liquid into the sparger line, which could cause uncontrolled reactions. As already proposed by Alexander [42], the air/oxygen outlets should be at the bottom of the sparger for "self-draining" to avoid any holdup of liquid within the sparger. The outlets may be holes of several millimeters in diameter. A minimum flow through the holes must be maintained to guarantee a uniform flow through all the holes and thus to prevent any backflow of the oxidate into the sparger system. The typical gas flow rate at operation conditions should be slightly higher than the minimum flow.

Finally, oxygen conversion in the reactor should be in such a way that the residual oxygen concentration in the off-gas is sufficiently below the *limiting oxygen concentration* (LOC). The LOC for cumene is, for example, for a temperature of 100 °C and pressure of 5 bar 9.8 vol% (see [41]).

2.4 New Developments

In the recent past, the focus of new developments was on alternative phenol processes that overcome the disadvantage of the coupled product acetone in the cumene oxidation process. These processes are based on the oxidation of benzene with nitrous oxide or hydrogen peroxide [7]. The main research on the cumene oxidation process is process intensification by improving the oxidation reaction and improved process and reactor design.

2.4.1 Process Intensification by Modification of the Oxidation Reaction

The oxidation reaction is a radical chain reaction, which can be influenced by radical scavengers, such as phenol. As phenol is produced from CHP, the risk of phenol recycling within the cumene oxidation process is always present. To reduce the phenol concentration within the cumene feed to the oxidation reactor, the feed is treated with caustic lye [7]. Another source for phenol is the cleavage of CHP in the oxidation reactor by organic acids that are formed by oxygenation of formaldehyde (Figure 2.2). To overcome the internal generation of phenol, acids are neutralized with an aqueous solution of sodium carbonate in the so-called wet oxidation process [43]. Another way to reduce acid concentration is the addition of ammonia. In wet oxidation processes, it can neutralize existing acids and stabilize hydroperoxide [44].

Ammonia can also be added in the dry oxidation process. Here, it can reduce acid formation by chemical capturing of formaldehyde [3]. Thus ammonia addition increases selectivity in dry and wet oxidation.

The previous described work was focused on the optimization of the radical reaction by the removal of inhibitors (phenol) or sources for inhibitors (acids). Further research is directed towards the use of reaction promoters that are added to the reaction mixture.

These additives interact with the radical chain reaction. Matsui and Fujita [45] proposed two strategies for an improved reaction:

A) Increasing CHP formation without decreasing selectivity (Strategy "A")
B) Increasing selectivity without decreasing CHP formation (Strategy "B").

Strategy "A" is about boosting the radical propagation reaction by enhancing a specific reaction step. The mentioned reaction step is believed to be the addition of oxygen to the cumyl radical (see Eq. (2.11)). This step can be enhanced by the addition of oxygen activators, for example, metal complexes. Kinetic studies presented in [45] show an increased CHP formation rate, which, according to the author, is attributed to the oxygen-activating effect of the used metal complexes.

On the other hand, strategy "B" tries to reduce termination reactions, for example, dicumyl formation from two cumyl radicals (see Eq. (2.5)). This has been achieved by the addition of TEMPO (2,2,6,6,-tetramethyl-1-piperidinyloxy), a radical component that reacts reversibly with the cumyl radical, thereby stabilizing it. TEMPO allows the cumyl radical to proceed to the oxidation reaction step instead of reaction with another cumyl radical to dicumyl.

Other additives that are investigated in more detail are carbon nanotubes (CNTs). One advantage of pure carbon CNTs compared to metal salts is disposal, as they can be incinerated without waste formation. Additional functional groups, for example, nitrogen-containing, pose the risk of unwanted emissions, for example, nitrous oxides.

CNTs improve radical initiation from pure cumene, increase decomposition rate of CHP, and allow oxidation at lower temperatures, with the advantage of higher space–time yield. However, any improved selectivity could not be proved [46, 47]. Because of the increased CHP decomposition rate, new safety issues arise, as an uncontrolled decomposition is more probable. This risk becomes even more dangerous as the decomposition rate also depends on the CNT concentration, that is, an accumulation of CNTs is a serious risk, which can lead to runaway explosion. Similar observations were made with supported gold and silver nanoparticles, which increase the decomposition rate of CHP and thus, again, would allow operation at lower temperatures [48].

To summarize the current research on additives in cumene oxidation, despite the progress in understanding and manipulating radical reactions, several issues are not yet solved. It has been shown that many additives promote the decomposition of CHP; nevertheless, any conclusion on improved selectivity compared to an additive-free oxidation is doubtful, as most additives are not tested under industrial conditions (temperature: 80–120 °C and CHP concentrations ranging from 10 to 40 wt%).

One drawback with most additives is safety, as the reaction rates for the decomposition of CHP increase with higher additive concentrations, so the risk of uncontrolled reactions by the accumulated additive should be managed.

A point for further research could be to broaden the view on the whole industrial production process; for example, the effect of additives on the following process steps is barely known and their disposal has to be dealt with.

2.4.2
Improvements of Reactor and Process Design

Over the last decades, the reactor design for cumene oxidation made tremendous progress. In the beginning, aerated stirred-tank reactors were used for cumene oxidation [49], at present state-of-the-art bubble column reactors are used.

Because of the complex mixing behavior within the bubble column and the complex radical reaction, there is still room for improvement.

General improvements of oxidation reactors (e.g., for terephthalic acid) include optimized feed points for gas and liquid, internals such as spargers, mixers, draft tubes, or trays, or an optimized setup with a secondary reactor [50, 51].

Another improvement in reactors for cumene oxidation is the use of a draft tube, which allows internal gas recycling. Therefore, air bubbles from the upflow, whose oxygen content is consumed, are dragged along the liquid current in the downflow and are recycled to the gas inlet. This recycling is claimed to increase selectivity [30].

Use of trays is favorable, as they promote plug flow pattern, reduce back-mixing, and redisperse the gas phase. This leads to increased selectivity by decreased CHP decomposition [29].

Besides bubble columns, other reactor types are investigated. Using a high-shear device creates a high air/oxygen–cumene interface, which increases CHP formation by increased oxygen mass transfer [52].

Another method for cumene oxidation is the nonbarbotage method, where the liquid is in contact with a continuous gas phase, similar to a counterflow trickle-bed reactor. (Barbotage method involves dispersion and percolation of a gas phase in a continuous liquid phase, e.g., bubble column.) It is claimed that this nonbarbotage method provides a significantly higher selectivity than conventional bubble columns do [53].

Besides optimization of a single reactor, there is also a possibility to optimize the whole reactor train. A single bubble column has a back-mixing behavior similar to a continuous stirred-tank reactor (CSTR). The arrangement of several bubble columns in a reactor cascade leads to a plug flow pattern and reduces back-mixing, which results in an increased selectivity by a reduced CHP decomposition [9]. Further improvement can be made by optimizing the selectivity of each bubble column by adjusting the pressure, temperature, airflow, and outlet concentration of CHP [54].

Limiting factors that counteract the choice for highest selectivity oxidation process are most often economic constraints, such as investment cost for process equipment, off-gas treatment unit, and working capital for the first fill. Furthermore, operational costs are important, for example, power consumption for mechanical agitation or utility consumption in the subsequent units in the phenol process.

Besides costs, process safety is another constraint. Trays can cause an accumulation of a gas phase beneath the tray and risk a gas-phase explosion within the reactor [43]. Furthermore, internals influence the SADT (see also Section 2.3.4).

References

1. Weber, M., Pompetzki, W., Bonmann, R., and Weber, M. (2013) Acetone, in *Ullmann's Encyclopedia of Industrial Chemistry*, 7th edn, Wiley-VCH Verlag GmbH & Co. KGaA, Weinheim.
2. Hock, H. and Lang, S. (1944) *Ber. Dtsch. Chem. Ges.*, **B77**, 257.
3. Zakoshansky, V.M. (2009) *Phenol and Acetone*, Chemizdat, St. Petersburg.
4. Schmidt, R.J. (2005) *Appl. Catal., A*, **280**, 89.
5. *The Industrial Chemist* (1960), May, p. 215.
6. Arpe, H.-J. (2010) *Industrial Organic Chemistry*, 5th edn, Wiley-VCH Verlag GmbH & Co. KGaA.
7. Weber, M., Weber, M., and Kleine-Boymann, M. (2011) Phenol, in *Ullmann's Encyclopedia of Industrial Chemistry*, 7th edn, Wiley-VCH Verlag GmbH & Co. KGaA, Weinheim.
8. Weber, M. (1999) Proceedings of the DGMK-Conference "The Future Role of Aromatics in Refining and Petrochemistry", October 13-15, 1999, Erlangen, Germany.
9. Weber, M. (2002) *Chem. Eng. Technol.*, **B25**, 553.
10. Twigg, G.H. (1962) *Erdöl Kohle Erdgas Petrochem.*, **2**, 74.
11. Hattori, K., Tanaka, Y., Suzuki, H., Ikawa, T., and Kubota, H. (1970) *J. Chem. Eng. Jpn.*, **B3**, 72.
12. Duh, Y.-S., Kao, C.-S., Lee, C., and Yu, S.W. (1997) *Trans. IChemE Part B*, **75**, 73.
13. Duh, Y.-S., Kao, C.-S., Hwang, H.-H., and Lee, W.-L. (1998) *Trans. IChemE Part B*, **76**, 271.
14. Wu, S.-H. (2012) *J. Therm. Anal. Calorim.*, **109**, 921.
15. Twigg, G.H. (1962) *Chem. Ind.*, **6**, 4.
16. Emanuel, N.M., Denisov, E.T., and Maizus, Z.K. (1967) *Liquid-Phase Oxidation of Hydrocarbons*, Plenum Press, New York.
17. Franz, G. and Sheldon, R.A. (2000) Oxidation, in *Ullmann's Encyclopedia of*

18. Andrigo, P., Caimi, A., Cavalieri d'Oro, P., Fait, A., Roberti, L., Tampieri, M., and Tartari, V. (1992) *Chem. Eng. Sci.*, **47**, 2511.
19. Bhattacharya, A. (2008) *Chem. Eng. J.*, **137**, 308.
20. Camarasa, E. (2000) Etude Hydrodynamique et Modelisation des Reacteurs a Gazosiphon D'Oxydation du Cumene. PhD thesis. Institut National Polytechnique de Lorraine, France.
21. Weber, M. and Weber, M. (2010) Phenols, in *Phenolic Resins: A Century of Progress* (ed L. Pilato), Springer-Verlag, Berlin and Heidelberg.
22. Schmidt, R.J. (2005) Sunoco/UOP phenol process, in *Handbook of Petrochemicals Production Processes* (ed R.A. Meyers), McGraw-Hill.
23. Moore, A. and Birkhoff, R. (2005) KBR phenol process, in *Handbook of Petrochemicals Production Processes* (ed R.A. Meyers), McGraw-Hill.
24. Grünewald, M., Große Daldrup, J.-B., Weber, M., Weber, M., Schlusemann, L., and Abel, N. (2013) *Chem. Ing. Tech.*, **85**, 1424.
25. Distillers Co. (1948) GB 610293.
26. Hercules Powder (1951) US 2547938.
27. Hercules Powder (1951) US 2548435.
28. Phenolchemie (1960) DE 1131674.
29. ILLA International (2014) US 8674145 B2.
30. Versalis, S.P.A. (2012) US Patent Application 2012/0283486 A1.
31. Gerstenberg, H. (1979) *Chem. Ing. Tech.*, **51**, 208.
32. Deckwer, W.-D. (1985) *Reaktionstechnik in Blasensäulen*, Otto Salle Verlag.
33. Deckwer, W.-D. and Schumpe, A. (1993) *Chem. Eng. Sci.*, **48**, 889.
34. Deen, N.G., Mudde, R.F., Kuipers, J.A.M., Zehner, P., and Kraume, M. (2010) Bubble columns, in *Ullmann's Encyclopedia of Industrial Chemistry*, 7th edn, Wiley-VCH Verlag GmbH & Co. KGaA, Weinheim.
35. Zehner, P. (1982) *Verfahrenstechnik*, **16**, 514.
36. Wendt, R., Steiff, A., and Weinspach, P.-M. (1983) *Chem. Ing. Tech.*, **55**, 796.
37. Low, D.I.R. (1967) *Can. J. Chem. Eng.*, **45**, 166.
38. Motarjemi, M. and Jameson, G.J. (1978) *Chem. Eng. Sci.*, **33**, 1415.
39. Abel, N.H., Schlusemann, L., and Grünewald, M. (2013) *Chem. Ing. Tech.*, **85**, 1112.
40. Fishwick, T. (1997) *Loss Prev. Bull.*, **137**, 10.
41. Weber, M. (2006) *Process Saf. Prog.*, **25**, 326.
42. Alexander, J.M. (1975) *IChemE Symp. Ser.*, **39**, 157.
43. Ghirardini, M. and Tampieri, M. (2005) Polimeri Europa Cumene Phenol processes, in *Handbook of Petrochemicals Production Processes* (ed R.A. Meyers), McGraw-Hill.
44. Hercules Powder (1950) US 2632026.
45. Matsui, S. and Fujita, T. (2001) *Catal. Today*, **71**, 145.
46. Zeynalov, E.B., Friedrich, J.F., Hidde, G., Ibrahimov, H.J., and Nasibova, G.G. (2012) *Oil Gas Eur. Mag.*, **38**, 45.
47. Liao, S., Peng, F., Yu, H., and Wang, H. (2014) *Appl. Catal., A*, **478**, 1.
48. Crites, C.O.L., Hallett-Tapley, G.L., Frenette, M., González-Béjar, M., Netto-Ferreira, J.C., and Scaiano, J.C. (2013) *ACS Catal.*, **3**, 2062.
49. Kropf, H. (1964) *Chem. Ing. Tech.*, **36**, 759.
50. Eastman Chemical Company (2007) US Patent Application 2006/0047142 A1.
51. Eastman Chemical Company (2007) US Patent Application 2007/0046067 A1.
52. H R D Corporation (2009) US Patent Application US2009005606.
53. ILLA International (2010) US Patent Application US2010054788 A1.
54. Borealis AG (2011) US Patent Application US2011263905 A1.

3

Cyclohexane Oxidation: History of Transition from Catalyzed to Noncatalyzed

Johan Thomas Tinge

3.1
Introduction

Cyclohexanone, a six-membered carbon ring with a ketone as functional group, is almost exclusively applied as a precursor for the production of aliphatic polyamides. Pure cyclohexanone is mainly converted, via cyclohexanone oxime and caprolactam, to nylon-6 (also called *polycaprolactam*) [1]. Mixtures of cyclohexanone and cyclohexanol, often called *KA oil*, are converted via oxidation into adipic acid that reacts with hexamethylene diamine (HMDA) to nylon-6,6 (poly-hexamethylene adipamide). Other applications of these products can be found in the field of polyurethane and polyester production.

Cyclohexanone has already been produced on a commercial scale for more than 75 years [2]. The current annual production of cyclohexanone and cyclohexanol is over 8000 kt. Nowadays, almost all cyclohexanone is commercially produced by three different routes: (i) oxidation of cyclohexane with oxygen, (ii) reduction of phenol with hydrogen gas, and (iii) hydration of cyclohexene with water to cyclohexanol, followed by dehydrogenation. Of these production technologies, the oxidation of cyclohexane with (optionally diluted or enriched) air, either under catalyzed or noncatalyzed conditions, is most common. The required cyclohexane is obtained by either liquid- or vapor-phase hydrogenation of benzene with hydrogen gas.

The following listing provides some general information for cyclohexanone:

Molecular structure:	(cyclohexanone ring with =O)
Chemical name:	cyclohexanone
Chemical formula:	$C_6H_{10}O$
CAS Number:	108-94-1
EC Number:	203-631-1
Beilstein Registry Number:	385735
Molar mass:	$98.15\,g\cdot mol^{-1}$
State (at 293 K):	liquid

Liquid Phase Aerobic Oxidation Catalysis: Industrial Applications and Academic Perspectives,
First Edition. Edited by Shannon S. Stahl and Paul L. Alsters.
© 2016 Wiley-VCH Verlag GmbH & Co. KGaA. Published 2016 by Wiley-VCH Verlag GmbH & Co. KGaA.

Melting point: 241 K (−32 °C)
Boiling point: 429 K (156 °C)
Flash point: 320 K (47 °C)
Density (at 293 K):   948 kg·m^{-3}

3.2 Chemistry and Catalysis

A generally accepted, simplified reaction scheme for the liquid phase oxidation of cyclohexane with oxygen is presented in the following figure:

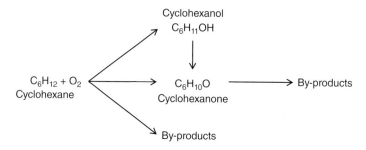

Initially, cyclohexane is oxidized to the intermediate cyclohexyl hydroperoxide, CHHP. Then, the obtained CHHP is decomposed into the desired components cyclohexanone and cyclohexanol; however, it is also partly decomposed into undesired by-products. A part of the formed cyclohexanol is further oxidized to cyclohexanone and a part of the formed cyclohexanone is converted to by-products. Part of the cyclohexane oxidation by-products are further destroyed (not shown in this figure). The by-products finally obtained include, in various amounts, acids such as adipic acid, ε-hydroxycaproic acid, glutaric acid, succinic acid, valeric acid, caproic acid, propionic acid, acetic acid, formic acid, and noncondensable gases such as CO and CO_2. In addition, several esters are formed between mainly cyclohexanol and the various carboxylic acids. The destinations of these by-products are quite diverse and depend on the producer: for example, some of these by-products are fed to combustion units for heat recovery purposes, while others are used as feedstock for chemicals such as 1,5-pentanediol, 1,6-hexanediol (HDO), and caprolactone. In general cyclohexanol is recovered from esters in a biphasic saponification step.

The selectivity of the oxidation of cyclohexane toward the desired products cyclohexanone and cyclohexanol very much depends on the degree of conversion of cyclohexane per pass. Both cyclohexanone and cyclohexanol are much more reactive than the starting material cyclohexane. And as a result, at a high degree of conversion of cyclohexane per pass, a significant fraction of the valuable products cyclohexanone and cyclohexanol are overoxidized to by-products. This overoxidation is limited at a low degree of conversion per pass. Selectivity drops up to 5% per percent conversion of cyclohexane are reported. As a consequence, for economic reasons, cyclohexane oxidation processes have to be operated at a low degree of cyclohexane conversion per pass, resulting in large recycles of unreacted cyclohexane.

Initially, all commercial processes for selective oxidation of cyclohexane were performed in the liquid phase with an oxygen-containing gas, in general air, and were catalyzed by transition metals. Both the formation of the intermediate CHHP and the decomposition of this intermediate are catalyzed by transition metals. The most popular cationic catalysts are soluble Co^{2+} and Cr^{3+} salts and mixtures thereof. Examples of such salts that are added to catalyze the oxidation are cobalt stearate and cobalt naphthenates. Applied concentrations of these transition metals range from more than 10 ppm down to sub-ppm levels. These catalysts also catalyze the decomposition of CHHP, reducing the residual concentrations of CHHP. Nevertheless, both for economic and safety reasons, the residual CHHP is decomposed to mainly cyclohexanone and cyclohexanol. This reaction is carried out in an after-reactor either in a monophasic system in cyclohexane or in a biphasic system with an aqueous caustic solution as second phase.

The mechanisms for the formation of cyclohexanone, cyclohexanol, and by-products are discussed in Chapter 1 by I. Hermans.

3.3 Process Technology

3.3.1 The Traditional Catalyzed Cyclohexane Oxidation Process

A traditional catalyzed cyclohexane oxidation process consists of an oxidation and heat recovery section, a neutralization and decomposition section, a cyclohexane recovery section, a cyclohexanone separation and purification section, and finally a cyclohexanol dehydrogenation section. A simplified diagram of such a catalyzed cyclohexane oxidation process that is operated in a continuous mode is shown in the following:

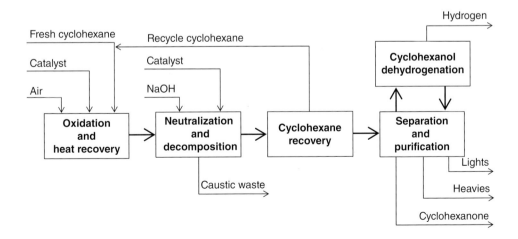

In the oxidation section, cyclohexane is selectively oxidized with oxygen in the liquid phase in a pressurized system at elevated temperatures. The pressure and temperature in this section typically range from 5 to 10 bar and 140 to 160 °C, respectively. A catalyzed cyclohexane oxidation section consists of a series of three to eight reactors or compartments that are operated in plug flow in order to minimize by-product formation. These reactors are generally equipped with heavy-duty stirrer systems to enhance the dissolution rate of oxygen. Compressed air, which is fed via sparger systems to each of the reactors, acts as oxygen source. A mixture of fresh and recycled preheated cyclohexane and dissolved catalyst is charged to the first reactor and the produced reaction mixture flows to the second reactor. The various reactors are operated in an adiabatic manner; that is, the heat of reaction is mainly removed by partial evaporation of cyclohexane. Nitrogen gas and cyclohexane are the main components of the produced hot off-gases of this section. In the heat recovery section, both heat and cyclohexane are recovered from these off-gases and recycled back into the process.

The oxidation mixture with about 5% cyclohexanone, cyclohexanol, and CHHP that is discharged from the last oxidation reactor is fed as such to the neutralization and decomposition section. In addition, an aqueous caustic solution and optionally a catalyst are fed to this section. In the resulting biphasic system, the CHHP is almost completely decomposed and the acids are neutralized. Subsequently, the biphasic system is separated into an aqueous slightly caustic waste stream and a cyclohexanol and cyclohexanone containing cyclohexane stream. The molar ratio between cyclohexanone and cyclohexanol, also called *KA ratio*, ranges from 0.3 to 0.8.

In the cyclohexane recovery section, cyclohexane is separated by overhead distillation from the resulting decomposed organic stream, whereby a highly concentrated mixture of cyclohexanone and cyclohexanol (also called *KA oil*) is obtained. The recovered cyclohexane is recycled back to the cyclohexane oxidation section.

The KA oil is partitioned in the "separation and purification" section by a sequence of distillation columns into a purified cyclohexanone flow, a cyclohexanol flow that is fed to the cyclohexanol dehydrogenation section, and by-product flows for light and heavy boiling components. The final product cyclohexanone is produced in a distillation column that is operated under high vacuum conditions and contains a large number of separation stages. The high vacuum conditions are an economic consequence of the tendency of cyclohexanone to form oligomers at elevated temperatures, and these columns are, as a result, necessarily tall in order to achieve high-purity end-product cyclohexanone regarding impurities (purity >99.9%).

Cyclohexanol is catalytically dehydrogenated at elevated temperature and almost ambient pressure in the cyclohexanol dehydrogenation section. After cooling of the gaseous reaction mixture, a liquefied KA oil and a hydrogen-containing gas phase are obtained. The KA oil is fed to the separation and purification section. Disadvantages of this section are the high energy consumption and the formation of undesired decomposition products.

Some of the major disadvantages of the traditional catalyzed cyclohexane oxidation process are its low cyclohexane selectivity (76–80%), its high caustic consumption, and high by-product formation rates (especially the caustic waste stream). All these factors result in high variable costs for the production of cyclohexanone. Another disadvantage of this process is that a part of the homogeneous catalyst that is added to the oxidation section precipitates and causes severe fouling. And as a result of this fouling problem, the cyclohexanone production has to be interrupted on a regular basis for cleaning of the oxidation reactors. The consequence of this frequent cleaning is a reduced number of effective production hours per annum, which impacts the fixed production costs.

3.3.2
The Noncatalyzed DSM Oxanone® Cyclohexane Oxidation Process

DSM developed the noncatalyzed DSM Oxanone® cyclohexane oxidation process in order to overcome the disadvantages of the traditional catalyzed cyclohexane oxidation process, namely, high cyclohexane and NaOH consumption figures and a large number of downtime hours. In addition, the DSM Oxanone® process produces, after decomposition of CHHP, a KA oil with a KA ratio of more than 1.5 that requires a cyclohexanol dehydrogenation section with just a limited capacity.

In the noncatalyzed DSM Oxanone® cyclohexane oxidation process, the step for the formation of the intermediate CHHP and the step for the decomposition of CHHP are separated to a large extent (a form of process extensification). In the oxidation section, cyclohexane is mainly converted into CHHP with limited decomposition and/or overoxidation due to the absence of catalyzing components. The temperature in a noncatalyzed oxidation section has to be increased by about 10 °C compared with the catalyzed versions in order to obtain sufficiently high productivity in the oxidation section. The oxidation mixture leaving the last oxidation reactor is fed to a modified neutralization and decomposition section, where the CHHP is decomposed in a biphasic system in the presence of a decomposition catalyst under optimal process conditions, namely, reduced temperature and an aqueous phase with a very high pH value (>13). Cooling down of the oxidate (i.e., the mixture obtained in the oxidation reactors) leaving the oxidation section can be done with indirect heat exchange and/or by flashing. The decomposition of CHHP in the decomposition section gives predominantly the desired component cyclohexanone. The resulting selectivity both in the oxidation section and in the decomposition section is above 95%. The NaOH consumption required to neutralize acids is significantly reduced due to less by-product (mainly acids) formation.

In order to reduce the caustic consumption even more, without ruining the overall selectivity, a low-temperature preneutralization step is introduced, in which acids are neutralized with spent caustic lye originating from the decomposition section. Effectively, the aqueous caustic stream and the organic oxidate stream are flowing in countercurrent mode. Fresh NaOH and preneutralized oxidate are supplied to the decomposition section that requires a high free caustic content. Spent caustic lye that is separated from the decomposed oxidate is used in the preneutralization section to neutralize organic acids, whereby an aqueous caustic waste stream is obtained with a low free caustic content. The modified neutralization and decomposition section as part of the noncatalyzed DSM Oxanone® cyclohexane oxidation process is schematically visualized in the following diagram:

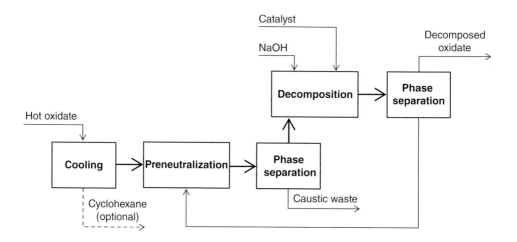

3.4 New Developments

Although the cyclohexane oxidation process has already a long commercial history, still a lot of (patent) publications are issued each year. The main topic of these (patent) publications is related to the selectivity or carbon efficiency of the oxidation and decomposition step. Other topics of these (patent) publications are related to, for example, improving the energy efficiency of the process, treating/upgrading the by-product containing streams, and the application of dedicated apparatus in the process.

It is expected that process improvements of the cyclohexane oxidation process aiming at improved variable costs such as reduced consumption of the raw materials cyclohexane and sodium hydroxide, reduced energy consumption, and upgrading the value of the various by-product streams will continue. On the other hand, it is observed that small-scale units (e.g., 45 kt/annum) are phased out and replaced by larger-scale units in order to reduce the fixed costs.

In the meantime, it is observed that two other technologies for the production of cyclohexanone are becoming more important. The first one is the hydration of cyclohexene with water to cyclohexanol that is subsequently dehydrogenated to obtain cyclohexanone. An advantage of this process is the rather high carbon efficiency. However, harsh process conditions and high investments are major disadvantages of this technology.

The second upcoming technology is based on the hydrogenation of phenol, for example, the Hydranone® process, the vapor-phase-based phenol hydrogenation technology of DSM. With this technology, it is possible to convert in a single highly exothermic reaction step over 95% of the phenol that is fed to the reaction section into mixtures with cyclohexanone to cyclohexanol ratios that are over 20. As a consequence, the process flows are much smaller than those in the competing technologies and separation and purification of cyclohexanone is less complex. A prerequisite is the availability of sufficient quantities of reasonably priced phenol compared with the price of cyclohexane. Additional advantages of this technology are its very high carbon yield (>99%), the almost absence of by-products, its (almost) energy neutrality, its ease of up-scaling to annual capacities of several hundred kilotons, and finally the very low investment costs compared with its rival technologies.

Epilogue

Effective January 18, 2016, Fibrant became the new name for DSM Caprolactam.

References

1. Ritz, J., Fuchs, H., Kieczka, H., and Moran, W.C. (2000) Caprolactam, in *Ullmann's Encyclopedia of Industrial Chemistry*, Wiley-VCH Verlag GmbH, Published Online: 15 JUN 2000, doi: 10.1002/14356007.a05_031.

2. Musser, M.T. (2011) Cyclohexanol and cyclohexanone, in *Ullmann's Encyclopedia of Industrial Chemistry*, Wiley-VCH Verlag GmbH, Published Online: 15 OCT 2011, doi: 10.1002/14356007.a08_217.pub2.

4
Chemistry and Mechanism of Oxidation of *para*-Xylene to Terephthalic Acid Using Co–Mn–Br Catalyst

Victor A. Adamian and William H. Gong

4.1
Introduction

The largest chemical manufacturing process that utilizes a homogeneous catalytic liquid phase oxidation is the production of *purified terephthalic acid (PTA)* (1,4-benzenedicarboxylic acid) from *p*-xylene (pX) (1,4-dimethylbenzene) (Eq. (4.1)) [1]. PTA is a commodity chemical with a demand of 51 million tons per year in 2014 [2] and is mainly used in the production of polyethylene terephthalate (PET), which is made by polycondensation of *PTA* with ethylene glycol (Eq. (4.2)). PET is used in numerous applications, ranging from fibers to water bottles.

$$p\text{-Xylene (pX)} \xrightarrow[\text{HOAc, H}_2\text{O}]{\text{Oxidation} \atop \text{Co(II), Mn(II), HBr}} \text{Crude terephthalic acid (TA)} + 2\text{H}_2\text{O} \xrightarrow{\text{Purification}} \text{Purified terephthalic acid (PTA)} \quad (4.1)$$

$$n\text{PTA} + n\text{HO-CH}_2\text{CH}_2\text{-OH} \longrightarrow \text{Polyethylene terephthalate (PET)} + 2n\text{H}_2\text{O} \quad (4.2)$$

The history of the *BP PTA* process began in 1955 with the discovery by *Scientific Design/Mid-Century Corporation* of a homogeneous oxidation catalyst of combinations of cobalt(II) and manganese(II) salts and a source of bromide, and thus, the oxidation process is known by the name MC oxidation [3–6]. Patents disclosing the oxidation process were later purchased by the *Standard*

Liquid Phase Aerobic Oxidation Catalysis: Industrial Applications and Academic Perspectives,
First Edition. Edited by Shannon S. Stahl and Paul L. Alsters.
© 2016 Wiley-VCH Verlag GmbH & Co. KGaA. Published 2016 by Wiley-VCH Verlag GmbH & Co. KGaA.

Oil Company, which later became *Amoco Corporation* (before the latter merged with *BP* in 1999). Terephthalic acid produced through MC oxidation could not be used directly for PET manufacturing due to the presence of impurities. By 1963, researchers at *Standard Oil* had developed new purification technology [7]. By 1965, the first *Amoco* PTA process was commercialized. Today, all commercial production of PTA is largely based on the oxidation and purification chemistry of this original *Amoco* PTA process. In addition, fundamentally the same oxidation catalyst is employed for the production of other aromatic acids (*i.e.*, isophthalic, trimellitic, and 2,6-naphthalenedicarboxylic acids). Continuous incremental improvements in PTA technology proceeded for the next 30 years, and as a result, the variable costs were gradually reduced and increased plant capacities have been achieved. *BP* is one of the largest producers of PTA in facilities located in the United States, Europe, and Asia. *BP*'s position as a leading PTA manufacturer is, to a significant degree, determined by its leadership in technology development, which is illustrated by the successful commercialization of two new generations of the *BP* PTA process within the last 10 years.

Chemistry is at the heart of *BP*'s PTA process in both oxidation and purification stages. Since the discovery and commercialization of the MC technology, research efforts have continuously occurred among diverse groups of industrial and academic researchers seeking to understand the oxidation chemistry and/or to improve the MC process, or perhaps to discover alternative catalysts. The PTA process and MC oxidation chemistry have been thoroughly reviewed over the years. For a general overview of PTA process and historical overview of process options, the reader is referred to a contribution by Sheehan in *Ullmann's Encyclopedia of Industrial Chemistry* [1]. Significant review of the chemistry of MC oxidation was conducted by Partenheimer in 1995 [8]. The chemistry and kinetics with an emphasis on oxidation process modeling were reviewed in 2013 by Tomás *et al.* [9].

This chapter is not intended to be a comprehensive literature review, but instead presents the authors' views that explain the key features of the chemistry and mechanism of cobalt–manganese–bromide-catalyzed oxidation of pX to TA, by-product formation, and purification chemistry. An overview of the engineering aspects of the process and of the purification chemistry is also provided.

4.2
Chemistry and Catalysis

The oxidation of pX to TA is conducted in one step in a continuously stirred reactor in aqueous acetic acid (HOAc). pX is oxidized by O_2 from air in the presence of $Co(OAc)_2$, $Mn(OAc)_2$, and HBr. TA is rapidly formed in very high yields (>98 mol%) from this highly exothermic reaction (3300 kcal/kg O_2 or 2×10^8 J/kg pX), typically at 180–220 °C and under 15–30 bar air. TA is nearly insoluble in aqueous acetic acid at reaction temperatures and precipitates as it is made. Following separation of the solid product from the reaction mixture, crude TA is isolated.

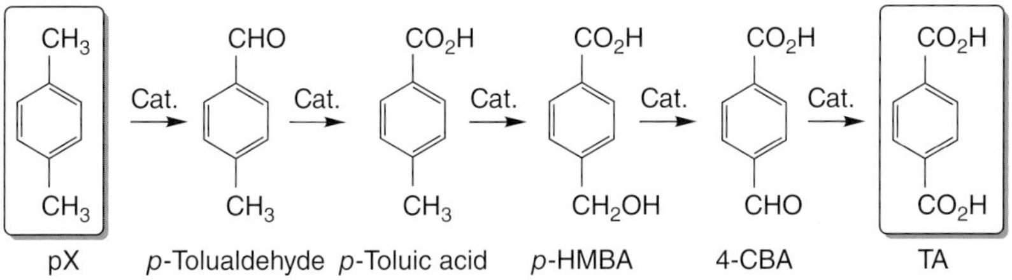

CH₃–C₆H₄–CH₃ (pX) →Cat.→ CHO–C₆H₄–CH₃ (p-Tolualdehyde) →Cat.→ CO₂H–C₆H₄–CH₃ (p-Toluic acid) →Cat.→ CO₂H–C₆H₄–CH₂OH (p-HMBA) →Cat.→ CO₂H–C₆H₄–CHO (4-CBA) →Cat.→ CO₂H–C₆H₄–CO₂H (TA)

Figure 4.1 Major oxidation intermediates from pX to TA. p-HMBA is p-hydroxymethylbenzoic acid, and 4-CBA is 4-carboxybenzaldehyde. Cat.: Co–Mn–Br.

The oxidation reaction in Eq. (4.1) is actually very complex. There are four major intermediates that are made during the sequential oxidations of methyl groups of pX (Figure 4.1). The conversion of each intermediate requires intervention of the MC catalyst.

As will be discussed later, the catalyst itself is involved intimately in intertwined redox cycles between Co, Mn, and Br and the overall oxidation occurs via at least 150 reactions in a three-phase system (gas, liquid, and solid) with complex physical behavior. In addition to the main oxidation pathway in Figure 4.1, the oxidation of pX results in the formation of a variety of by-products in low but technologically significant concentrations. Solvent HOAc also participates in the oxidation process, resulting in carbon oxides and other volatile by-products.

Understanding the nuances of the oxidation chemistry is very important for effective technology development, and this has been achieved from many years of research. The chemistry of Co–Br and Co–Mn–Br catalysis has been studied by many over the last 60 years. Our understanding of the mechanism as we herein describe is based on published research and internal research conducted over the years at the BP Amoco Naperville Center and in collaborations with partners.

From a mechanistic point of view, the MC catalyst, that is, Co–Mn–Br, is essentially a cobalt–bromide catalyst (Co–Br) promoted by Mn ions. The reason for this definition is that the Co–Br catalyst exhibits all properties of the Co–Mn–Br catalyst, while the Mn–Br catalyst is much less active and has significant mechanistic differences from the Co–Br catalyst. For this reason, we first discuss the nature of the Co–Br catalysis mechanism and then the mechanism of the Co–Mn–Br system.

4.2.1
Co–Br Catalysis

In very general terms, the Co–Br-catalyzed oxidation is a particular case of the free radical chain oxidation, common for all liquid phase oxidations of hydrocarbons [8–10]. The free radical chain oxidation occurs with four types of free radicals: alkyl, alkoxy, alkylperoxy, and acylperoxy radicals [11, 12]. Other key active intermediates are hydroperoxides and peracids [11, 12]. The nomenclature and structures are displayed in Figure 4.2.

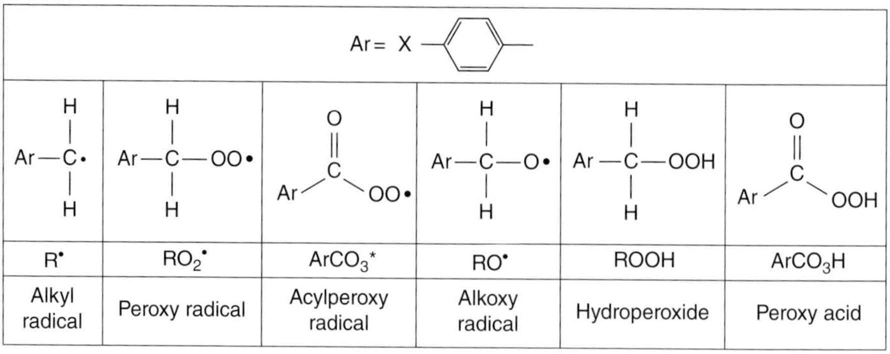

			Ar = X —⟨ ⟩—			
Ar—C• H / H	Ar—C—OO• H / H	Ar—C(=O)—OO•	Ar—C—O• H / H	Ar—C—OOH H / H	Ar—C(=O)—OOH	
R•	RO₂•	ArCO₃*	RO•	ROOH	ArCO₃H	
Alkyl radical	Peroxy radical	Acylperoxy radical	Alkoxy radical	Hydroperoxide	Peroxy acid	

Figure 4.2 Reactive intermediates in liquid phase oxidation of hydrocarbons.

In most cases, free radical oxidations of hydrocarbons can be described by the reactions of the peroxy radical ($RO_2^•$) with a hydrocarbon (RH), leading to the formation of an alkyl hydroperoxide (ROOH) (Eqs. (4.4) and (4.5)) [11–14]. The chain oxidation is initiated by usually very slow initiation reaction (Eq. (4.3)) which may occur via different mechanisms and is only important in the absence of ROOH at the very beginning of oxidation. Once trace amounts of ROOH are formed in reaction (4.5), its decomposition becomes the main source of radicals and reaction (4.3) becomes irrelevant.

In the absence of a transition metal catalyst, the thermal homolytic cleavage of ROOH in Eq. (4.6) provides the chain branching that leads to an increase in the concentration of $RO_2^•$ and, consequently, to an increase in the oxidation rate until a maximum (steady-state) rate of oxidation is reached [11–14]. The maximum rate of oxidation is limited by the termination reaction (4.9) between two peroxy radicals [15]. The theoretical maximum rate of oxidation is described by Eq. (4-I) [13]. Rate constant values in Eqs. (4.3–4.9) are those determined for pX at 70 °C [10b, 11–13].

$$R - H \longrightarrow R^• \tag{4.3}$$

$$R^• + O_2 \xrightarrow{k_4 \sim 10^9 \, l \cdot mol^{-1} \cdot s^{-1}} RO_2^• \tag{4.4}$$

$$RO_2^• + R - H \xrightarrow{k_5 \sim 5 \, l \cdot mol^{-1} \cdot s^{-1}} ROOH + R^• \tag{4.5}$$

$$ROOH \xrightarrow{k_6 \sim 10^{-8} - 10^{-6} \, s^{-1}} RO^• + {}^•OH \tag{4.6}$$

$$RO^• + RH \longrightarrow ROH + R^• \tag{4.7}$$

$${}^•OH + RH \longrightarrow H_2O + R^• \tag{4.8}$$

$$RO_2^\bullet + RO_2^\bullet \xrightarrow{k_9 \sim 10^8 \text{ l·mol}^{-1}\cdot\text{s}^{-1}} ROH + \text{Ar-CHO} + O_2 + H_2O \quad (4.9)$$

$$\text{rate} = -\frac{d[O_2]}{dt} = \frac{2k_5^2[RH]^2}{k_9} \quad (4\text{-I})$$

Since the thermal decomposition of ROOH in reaction (4.6) is a slow process, the initial rates of oxidation could be very low, leading to long induction periods.

Transition metal ions (Co, in particular) catalyze decomposition of ROOH with formation of alkoxy and peroxy radicals, as shown in reactions (4.10) and (4.11) [11, 12]. We would like to note here that the decomposition of hydroperoxides and peracids on cobalt(II) and (III) is more complex than what are shown in Eqs. (4.10) and (4.11) and involves several steps and two equivalents of Co(II) [16, 17, 18a, 19].

Since the catalytic decomposition of ROOH with Co(II) is much faster than the thermal decomposition of ROOH, the latter reaction becomes almost irrelevant in the overall kinetics of oxidation. The steady-state concentrations of $RO_2^\bullet$ and ROOH are reached much faster, significantly reducing the observed induction period compared with uncatalyzed oxidation. The theoretical maximum rate in this case is described by Eq. (4-II) [11, 13]. A comparison between Eqs. (4-I) and (4-II) shows that the theoretical maximum rate of uncatalyzed oxidation is four times higher than that of cobalt-catalyzed oxidation. Practically, the theoretical maximum rate of oxidation is rarely reached due to very long induction periods.

$$ROOH + Co(II) \longrightarrow RO^\bullet + {}^\ominus OH + Co(III) \quad (4.10)$$

$$ROOH + Co(III) \rightleftharpoons ROO^\bullet + Co(II) \quad (4.11)$$

$$\text{rate} = -\frac{d[O_2]}{dt} = \frac{k_5^2[RH]^2}{2k_9} \quad (4\text{-II})$$

The reduction of Co(III) by ROOH is a relatively slow reaction and, therefore, Co(III) is a predominant state of cobalt in a typical cobalt-catalyzed oxidation.

The addition of Br^- to a Co-only catalyst increases the oxidation rate by two to three orders of magnitude [3–10]. Br^- also enables the catalyst to oxidize a hydrocarbon substrate in very high selectivity (up to 98 mol%, depending on substrate) at high temperature while avoiding excessive levels of solvent combustion [8, 9]. In explaining the Co–Br (and Co–Mn–Br) catalysis, one needs to explain two key features: very high rates of oxidation and very high selectivity to the dicarboxylic acid product. The core mechanism that the authors of this chapter have adopted is based on the research by Zakharov [10, 16, 20–23], Partenheimer [8, 24–28], Espenson [29–34], and many early works that laid the foundation for our current knowledge [17, 35–40].

The addition of bromide to cobalt-catalyzed oxidation (reactions (4.3–4.11)) results in all Co(III) formed during oxidation being now reduced by inorganic bromide in a fast reaction (4.12), forming bromine radical species and Co(II). We shall elaborate on the nature of the Br radical species in Section 4.2.3. The bromine radical species abstracts hydrogen from a methyl group of pX to yield an alkyl radical that is quickly converted into a peroxy radical in the presence of O_2 (reaction (4.13))

$$Co(III) + Br^- \longrightarrow Co(II) + \text{``Br}^{\bullet}\text{''} \tag{4.12}$$

$$\text{``Br}^{\bullet}\text{''} + \text{R-H} \xrightarrow{\quad} \text{R}^{\bullet} \xrightarrow{O_2} RO_2^{\bullet} \tag{4.13}$$
$$\downarrow \text{HBr}$$

Thus, the net reaction (Eqs. (4.12) and (4.13)) is a fast bromide-catalyzed reduction of Co(III) by a hydrocarbon to generate Co(II) and $RO_2^{\bullet}$ (Eq. (4.14)).

$$Co(III) + \text{R-H} \xrightarrow{Br^- \quad Br^{\bullet}} Co(II) + R^{\bullet} \xrightarrow{O_2} RO_2^{\bullet} \tag{4.14}$$

As a consequence of fast reduction of Co(III) by bromide, the direct reaction (4.11) now does not contribute to the overall kinetics, but its reverse reaction (4.15) becomes very important [10, 21, 22].

$$RO_2^{\bullet} + Co(II) \xrightarrow[H^+]{k_{15}} ROOH + Co(III) \tag{4.15}$$

The rapid reaction (4.15) between peroxy radicals and Co(II) in the presence of a protic acid leads to the formation of ROOH and Co(III). The protic acid is necessary for the formation of a ROOH and the uncovering of the critical role of a carboxylic acid solvent such as acetic acid.

The overall effect of addition of reactions (4.12–4.15) to the free-radical oxidation of hydrocarbons (reactions (4.3–4.9)) is that the fast reaction (4.15) between a $RO_2^{\bullet}$ and Co(II) leading to ROOH triggers a chain of fast reactions, leading to the formation of three new $RO_2^{\bullet}$ in a *branching sequence*, as illustrated in Figure 4.3 for the general case of Co–Mn–Br catalysis. Indeed, the ROOH formed in Eq. (4.15) reacts with another equivalent of Co(II) in reaction (4.10) to release an alkoxy radical, which reacts with the hydrocarbon to form an alcohol and alkyl radical. Two Co(III) species generated in reactions (4.10) and (4.15) are then reduced by a hydrocarbon in bromide-catalyzed reaction (4.14) to generate two alkyl radicals. All three alkyl radicals react with O_2 to yield three peroxy radicals. Thus, in the presence of Co(II) and Br^-, a new branching sequence is realized, leading to higher oxidation rates that are only limited by the bimolecular recombination of peroxy radicals. The branching sequence was first proposed by Zakharov in the 1980s based on his mechanistic studies using, among other techniques, a chemiluminescent method that permits direct monitoring of $RO_2^{\bullet}$ concentration [10, 21, 22].

Maximum rate of oxidation with Co–Br catalyst is achieved when all intermediate species reach quasi-steady-state concentrations, as described

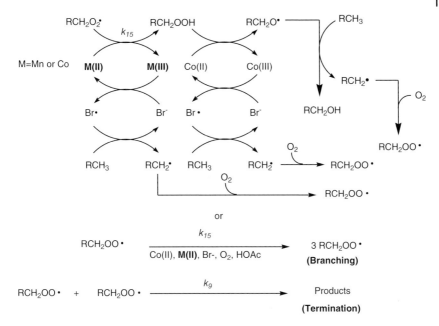

Figure 4.3 Branching sequence in Co–Br and Co–Mn–Br catalysis.

by Eq. (4-III) [10, 21, 22].

$$\text{rate} = -\frac{d[O_2]}{dt} = \frac{2k_{15}^2[Co^{(II)}]^2}{k_9} \qquad (4\text{-III})$$

A comparison of the rate laws in Eq. (4-III) with Eqs. (4-I) and (4-II) for noncatalyzed and Co-only-catalyzed oxidation illustrates the key advantages of the Co–Br system. In the absence of Br, a rather slow ($k_5 \sim 5$ l/mol/s at 70 °C) reaction of $RO_2^\bullet$ with RH is the rate-limiting step, as reflected in Eq. (4-II) [11–13]. In the case of the Co–Br catalyst, the rate-limiting step is a much faster ($k \sim 10^2 – 10^3$ l/mol/s) reaction between $RO_2^\bullet$ and Co(II) and the rate is zeroth order on RH [10, 21, 22].[1] These features allow for higher rates and very high conversions of hydrocarbon at its low concentration, thus allowing commercial operations using continuous oxidation to achieve very high conversions in a single step at relatively low residence times.

The high selectivity of Co–Br (and Co–Mn–Br) catalysis is explained by the high activity of the "Br radical" that easily abstracts hydrogen from a deactivated methyl group of *p*-toluic acid [31]. An alkylperoxy radical, in contrast, cannot

---

1) This is true for oxidations at low temperature and low concentration of hydrocarbon. At high concentration of hydrocarbon or high temperatures, there is a contribution to the chain branching from the hydrocarbon as discussed in Ref. [10].

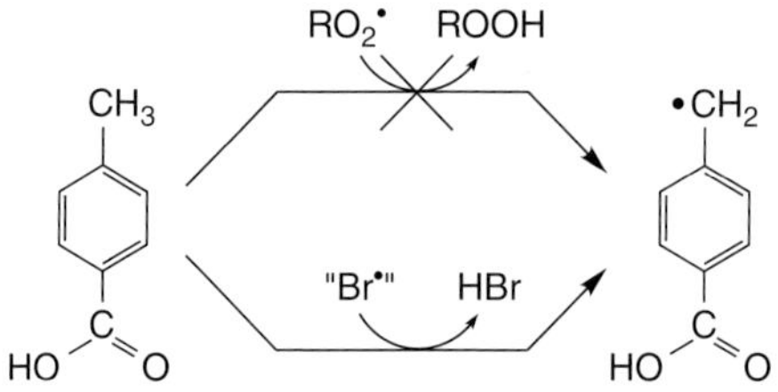

Figure 4.4 Hydrogen abstraction reactions in Co-only and Co–Br-catalyzed oxidation of p-toluic acid.

abstract hydrogen from the methyl group of *p*-toluic acid (Figure 4.4). This limitation explains why Co-only-catalyzed oxidations do not produce TA in high yield and the oxidation stops mainly at *p*-toluic acid even at high temperature and pressure conditions in addition to a lengthy reaction time [41].

4.2.2
Cobalt–Manganese–Bromide Catalysis (MC Oxidation): The Nature of Synergy between Co and Mn

The synergy of Mn in the Co–Br system is one of the most puzzling and complicated aspects of MC-catalyzed oxidation, and the reactions of Mn, particularly under commercial conditions and concentrations, are still not fully understood. The synergistic effect is seen from a significant increase in oxidation rate (up to three to five times) upon addition of small amounts of Mn (Mn/Co ranging from 0.01 to 0.1) to the Co–Br catalyst. Under industrial conditions, the synergy is also observed in significant improvements of oxidation selectivity to TA and a decrease in acetic acid solvent combustion.

In our view, there are two main reaction sets that explain the observed synergy. The first one is a fast reaction between peroxy radicals and Mn(II), leading to the formation of a hydroperoxide and Mn(III) species [10, 42–44]. The second set of reactions are Mn(II) and Mn(II)Br$_2$-catalyzed reductions of Co(III) and Mn(III) [25, 40, 45].

The increase in the rate of oxidation upon addition of Mn was explained by Zakharov by a rapid reaction of Mn(II) with RO$_2^\bullet$, leading to ROOH when the reaction is conducted in an acidic solvent such as acetic acid (Eq. (4.16)) [10, 42–44].

$$\text{RO}_2^\bullet + \text{Mn(II)} \xrightarrow[\text{H}^+]{k_{16}} \text{ROOH} + \text{Mn(III)} \qquad (4.16)$$

At Mn(II) concentrations of less than 10^{-4} M, the rate constant k_{16} is approximately 10^5 l·mol^{-1}·s^{-1} [42, 43]. This is much larger than the value of the rate constant k_{15} for the same reaction involving Co(II) (Eq. (4.15)), which is approximately $5 \times 10^2 - 5 \times 10^3$ l·mol^{-1}·s^{-1} at a Co concentration between 10^{-3} and 10^{-2} M (all rate constants were obtained at 70 °C) [21–23, 42, 43]. Therefore, the introduction of Mn(II) results in an increase in the rate of formation of ROOH and, ultimately, in faster branching sequence (and oxidation rate). At the same time, Mn(II) is much less active in a radical decomposition of a hydroperoxide – a second key reaction in a branching sequence [46]. The much slower reaction with hydroperoxides explains why Mn–Br catalysts are less active than Co–Br as branching does not occur. Thus, Mn(II) replaces Co(II) in the reaction with $RO_2^•$, while Co(II) is still vital for the radical decomposition of a ROOH (Eq. (4.10)), leading to the branching sequence for the Co–M–Br catalyst, as shown in Figure 4.3, where M = Mn. The observed rate constant in reaction (4.16) is a function of Mn concentration and decreases with the increase in Mn concentration, indicating that the reaction is complex and most likely involves another Mn(II) species [42, 43]. This phenomenon explains why the synergistic effect of Mn on the rate of oxidation is observed mainly at relatively low Mn concentrations. The effect of Mn(II) is further complicated by the facile interaction between Mn(III) and Mn(II) or Co(II) species and by the inhibiting properties that Mn(II) exhibits at higher concentration [42]. Interactions between Mn(II), Mn(III), and Co(II) have been studied by Espenson's group who demonstrated the formation of Mn(III)–Mn(II) and Mn(II)–Mn(II) complexes that determine the kinetics of Mn-catalyzed reduction of Co(III) [29]. Overall, metal–metal interactions, especially involving Mn, could be very important in MC oxidations. Such interactions may explain very complex kinetics and difficulties in modeling the MC oxidation process.

The second set of reactions responsible for the synergistic effect is a Mn(II) and $Mn(II)Br_2$-catalyzed reduction of Co(III) and Mn(III), studied independently by Jones and Partenheimer and, later, by Espenson [25, 29, 39, 40]. It was found that Co(III) oxidizes Mn(II) much more quickly than Co(III) oxidizes Br^-. In turn, Mn(III) oxidizes bromide faster than Co(III) does. This creates a cascade of reactions where the reaction between Co(III) and Br^- is catalyzed by Mn(II), as shown in Figure 4.5.

The Mn(II) reduction of Co(III) results in a decrease of steady-state concentration of Co(III), which, at commercial temperatures, is the main oxidant affecting the decomposition of acetic acid solvent. Participation of Mn(II) in the Co(III) reduction in Figure 4.5 decreases the steady-state concentration of Co(III). However, the overall rate of oxidation is determined by the rate of the reaction between peroxy radicals and Mn(II) and Co(II) since the subsequent cascade of reactions in Figure 4.5 is fast.

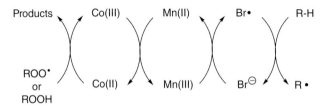

Figure 4.5 Mn-catalyzed reduction of Co(III) by bromide.

The discussion thus far has sought to explain the key phenomena of MC oxidation on a conceptual level. While the equations and schemes correctly represented the key reactions and formal oxidation states in a branching sequence, they do not fully represent the actual reacting species. The following sections focus on the natures of the Co, Mn, and Br species in MC catalysis.

4.2.3
The Role and Nature of Bromine Species in MC Oxidation

Fundamentally, Br's role in Co–Br and Co–Mn–Br catalysis is that it catalyzes the oxidation of alkylaromatics by Co(III) and Mn(III). Since the discovery of Co–Br catalysis, it has been postulated that $Br^•$ is formed from Br^- added initially as a salt ($CoBr_2$, NaBr) or HBr [35, 36, 38]. Organic bromides could also serve as bromine sources, but in order to be active, the organic bromide must be converted into inorganic bromide [8, 38]. Thus, there are two questions related to the nature of the bromine species in Co–Mn–Br catalysis: what is the major catalytically active form of bromide and what is the active bromine radical species?

The addition of a bromide to $Co(OAc)_2$ in HOAc system produces an equilibrium mixture of mixed acetates–bromides, as shown in Figure 4.6 [45, 47–49].

The equilibria in Figure 4.6 are affected by the concentration of water and other solvent components, by anions other than Br^- and AcO^- that may be present in solution, and by the temperature. The values of $K_a - K_d$ have been determined under varying conditions [45, 47–49]. The major conclusion is that in glacial or nearly glacial HOAc and at the [Co]:[Br] > 3, all Br^- are bound to Co(II), mainly as the Co(OAc)Br. The latter was assumed to be the major catalytically active form of bromide (and of Co(II)) in the MC oxidation. The reduction of Co(III) and Mn(III) species would then occur via Eq. (4.17), and the active bromine species was believed to be $Br^•$ [8, 10, 36–38]. However, it was also suggested by Jones that $Br_2^{•-}$ is an active species for hydrogen abstraction, but he proposed that $Br^•$ was

$$Co(OAc)_2 + YBr \underset{}{\overset{K_a}{\rightleftharpoons}} Co(OAc)Br + YOAc$$

$$Co(OAc)Br + YBr \overset{K_b}{\rightleftharpoons} CoBr_2 + YOAc$$

$$CoBr_2 + Br^- \overset{K_c}{\rightleftharpoons} CoBr_3^-$$

$$CoBr_3^- + Br^- \overset{K_d}{\rightleftharpoons} CoBr_4^{2-}$$

Y = H or metal ion

Figure 4.6 Equilibria in $Co(OAc)_2$–HOAc–HBr system. Solvent ligands surrounding octahedral Co(II) species are not shown.

initially formed before it combined with an equivalent of Br^- to produce $Br_2^{\bullet-}$ (Eq. (4.18)) [40].

$$Co(III) + Co(II)Br \longrightarrow 2Co(II) + Br^{\bullet} \qquad (4.17)$$

$$Br^{\bullet} + Br^- \longrightarrow Br_2^{\bullet-} \qquad (4.18)$$

The nature of the bromine radical species was recently reinvestigated by the Espenson group [29, 31]. Based on detailed kinetic study, it was proposed that the actual reducing agents for metal(III) species are $CoBr_2$ and $MnBr_2$ and that $Br_2^{\bullet-}$ is formed directly, as shown in Figure 4.7. This proposal was based on the investigation of Mn(III) oxidation of HBr, where the dependence on $[HBr]^2$ or $[MBr]^2$ was consistent with the formation of a dibromo radical anion ($Br_2^{\bullet-}$) or its conjugate acid, $HBr_2^{\bullet}$ [29, 31]. Direct formation of $HBr_2^{\bullet}$ avoids the generation of the significantly higher energy "free bromine atom $Br^{\bullet}$." The same reduction mechanism is applicable to Co(III) species, as shown in Figure 4.7 [29].

If the active bromine species is indeed the $Br_2^{\bullet-}$, a critical question must be asked: can $HBr_2^{\bullet}$ abstract benzylic hydrogens? This question was addressed by Metelski and Espenson in a study where second-order rate constants between $Br_2^{\bullet-}$ and RH were measured for a variety of alkylaromatics, ranging from monosubstituted monoaromatics to polysubstituted naphthalenes [31].

In summary, our current view is that the main active bromide species are metal(II) dibromides, which, after oxidation by metal(III) species, undergo spontaneous reductive elimination to a dibromo radical anion. The latter is the hydrogen-abstracting species that selectively reacts with a hydrocarbon to generate an alkyl radical.

Another important form of bromine in MC oxidation is benzylic bromides. In the process of oxidation, inorganic bromides are rapidly converted to benzylic bromides in proportions ranging from 10% to 90% until the oxidation has been completed [8, 26, 50]. Partenheimer has reported that benzylic bromides are catalytically inactive and that their accumulation is responsible for decelerating the oxidation rate. Solvolysis of benzylic bromides liberates catalytically active inorganic bromides, thus establishing the observed equilibrium between the two forms of bromine [26]. However, it was later uncovered that benzylic bromides can be easily oxidized in the presence of oxygen and cobalt(II), and this oxidation appears to be the main route in which inorganic bromides are restored [30, 31].

Our view is that benzylic bromides are continuously formed and destroyed in the MC oxidation of hydrocarbons, which leads to a steady-state concentration

Figure 4.7 Reduction of Co(III) in its reaction with MBr_2, (M – Co, Mn).

of benzylic bromides. The main reaction that leads to the formation of benzylic bromides is, most likely, the reaction between alkyl radicals and Co(III) and Mn(III) in the presence of bromide ions [33].

4.2.4
Nature of Cobalt(III) and Mn(III) Species

Cobalt species cycle through $2^+/3^+$ oxidation states, resulting in a steady-state concentration of Co(III) species. There are at least three different Co(III) species: dimeric Co(III)a and Co(III)s, shown in Figure 4.8, and trimeric Co(III)c with the formula $[Co_2^{(III)} Co^{(II)} (\mu_3\text{-}O)(\mu\text{-}OAc)_6(HOAc)_3]$. These species have varying reactivities toward the oxidation of inorganic bromides [8, 17, 39, 46, 51–54]. The most active form, Co(III)a, was assigned to an oxo-bridged dimer, while the less active Co(III)s was assigned to a hydroxo-bridge dimer [17, 39, 53]. It has been proposed that Co(III)a is the active form in MC oxidation, while the existence of monomeric Co(III) species has not been demonstrated [29].

In most of the published studies, Co(III) species were generated in model systems. Therefore, it is difficult to ascertain the precise composition of Co(III) in the commercial process. Regardless of the forms of Co(III), the steady-state concentration of this species during the oxidation process should be minimized in order to limit undesired side reactions.

Less attention was paid to the nature of Mn(III) in MC oxidation, however. The kinetic properties of *in situ* generated Mn(III) were found to be different from those of aged Mn(III) [45]. Aged Mn(III) is believed to be a dimeric, acetate-bridged species [55]. Metelski and Espenson have concluded that *in situ* generated Mn(III) is monomeric in structure and this is the form that is made in an MC oxidation. One also cannot exclude the potential formation of oxidatively active Mn(IV) species [56].

4.2.5
Reactions of Cobalt(II) with Peroxy Radicals and the Effect of Solvent on Oxidation Rate

Composition of the reaction solvent and, more specifically, the H_2O concentration has a major effect on the rate and selectivity of MC oxidations. H_2O is formed as a

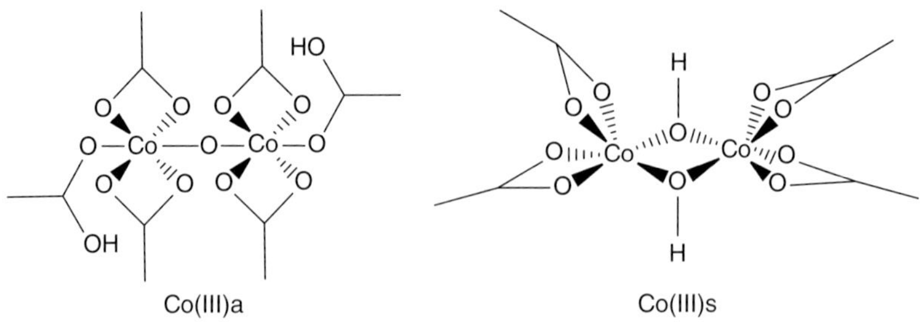

Figure 4.8 Structures of Co(III)a and Co(III)s (based on Ref. [29]).

coproduct and inhibits oxidation at its high levels. Therefore, continuous removal of H_2O from the oxidation reactor is necessary in order to maintain its concentration between 5 and 15 wt% in a typical commercial process.

H_2O affects the MC oxidation by influencing the coordination environment of Co(II) and Mn(II) species, which in turn impacts their participation in two critical steps of the branching reaction: rate-determining reactions with $RO_2^\bullet$ (Eqs. (4.15) and (4.16)) [23, 57] and the reactions between metal(III) species and metal(II) dibromides (Figure 4.7) [27].

To account for the different ligand environment, one must rewrite Eq. (4.15) to the following form (Eq. (4.19)):

$$RO_2^\bullet + Co(II)(OAc)_2(H_2O)_n(HOAc)_{4-n} \longrightarrow ROOH + Co(III)(OAc)_3(H_2O)_n(HOAc)_{3-n} \quad (4.19)$$

The rate constants have been measured by Zakharov et al. for different alkylaromatic peroxy radicals at temperatures ranging from 60 to 95 °C. The dependence of the rate constant on the solution composition H_2O–HOAc–RH has been examined, as well as the effect of a strong ligand such as pyridine [23, 57]. It was concluded that the main components of the solvent, HOAc and H_2O, have dual roles. They are proton donors necessary for the formation of hydroperoxide, and they are ligands that can block access of $RO_2^\bullet$ to the Co(II) coordination sphere. Therefore, the observed rate constant in reaction (4.19) shows a bell-shaped dependence on the concentration of H_2O in HOAc solution. At lower concentrations, H_2O increases the rate constant, but at higher H_2O concentration, the rate constant decreases. The same bell-shaped dependence is also observed for the overall oxidation rate, in accordance with the theory that reaction (4.19) is the rate-limiting stage [23]. The effect of H_2O has a strong temperature dependence: at low temperatures (50–90 °C), very low H_2O concentration can completely inhibit the oxidation, while at 180–200 °C, the H_2O concentration could be as high as 20 wt%.

A very interesting aspect of H_2O dependence is that it increases the oxidation rate at its low levels [23, 57]. This may be explained by decreases in the reduction potential of Co(III)/Co(II), which would facilitate the reaction with $RO_2^\bullet$. A similar effect (i.e., promotion at low concentrations and inhibition at higher concentrations) was observed for pyridine addition to the MC oxidation [57]. It is likely the same effect that explains the action of guanidine and other strong ligands [58]. Addition of pyridine or guanidine results in oxidation systems that can tolerate relatively higher H_2O concentrations in comparison to the amine-free aqueous HOAc.

The second reaction that H_2O affects is the equilibrium between metal(II) and Br^- ions [27]. At H_2O concentrations greater than 5 wt%, approximately 12% of total bromide is coordinated to cobalt and only approximately 6% is coordinated to Mn at room temperature. Thus, the metal–bromide equilibria in Figure 4.6 are now all shifted toward $Co(OAc)_2 \cdot 4H_2O$. It has been suggested that the bromide is still ion-paired to the metal ion, which makes the reaction between metal(III) and bromide more difficult in comparison to the case of lower H_2O concentration in

HOAc. This effect would result in higher steady-state concentrations of Co(III) and Mn(III) species, resulting in higher rates of HOAc decomposition and by-product formation.

4.2.6
Phenomenon of Manganese Precipitation

Use of Br-containing catalysts poses challenges such as high rates of corrosion. As such, commercial oxidation reactors and equipment that come in contact with the MC catalyst solution must be cladded with titanium in order to minimize corrosion. This desire to discover an alternative halogen-free homogeneous oxidation catalyst to the MC catalyst has been a significant motivation for many.

One approach to mitigating the corrosion by Br$^-$ in an MC oxidation is to use less of it while keeping the Co(II) and Mn(II) concentrations constant. However, this approach has created a secondary challenge such as lower oxidation activity. An unexpected crude TA product quality consequence ensued from the use of lower Br/Co + Mn atom ratios. Reports began surfacing from as early as the 1970s that "gray TA" was made. The gray color was eventually traced to the presence of black finely divided Mn dioxide (MnO_2), identified by electron microscopy [59, 60]. Investigators from the *Mitsubishi Gas Chemical Co.* have disclosed that addition of metals from the lanthanide series was beneficial in preventing manganese dioxide production [61].

Amoco researchers began to research various catalytic additives that could prevent MnO_2 precipitation while permitting one to operate at a lower Br/Co + Mn ratio. Rare earth metal salts, particularly $Ce(OAc)_3$ at Co/Ce = 10 : 1, were discovered to inhibit MnO_2 precipitation even when very low Br/(Co + Mn) ratio was employed (<0.5) [62, 63]. A fundamental study of the unwanted production of MnO_2 was later pursued by the group of Bakac [56]. Mn indeed can access the 4+ oxidation state by either direct oxidation of Mn(III) or reduction from Mn(VII). Mn(IV) can also oxidize bromides and Mn(II).

4.2.7
Consolidated View of MC Oxidation Mechanism

As a conclusion to Sections 4.2.1–4.2.4, Figure 4.9 displays our most current consolidated view of the MC catalytic mechanism that expands a more conceptual mechanism in Figure 4.3.

The MC oxidation mechanism shown in Figure 4.9 was developed from studies of the initial stages of hydrocarbon oxidations, that is, conversion of the first methyl groups of pX to alcohols (*p*-methylbenzyl alcohol) and aldehydes (*p*-tolualdehyde). During the complete oxidation of pX to TA, the initially formed *p*-methylbenzyl alcohol and *p*-tolualdehyde are further oxidized to *p*-toluic acid. Once this conversion is complete (i.e., *p*-toluic acid is formed), the oxidation of

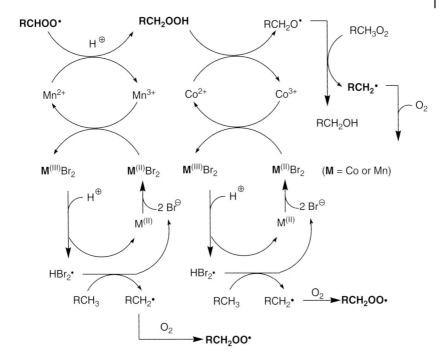

Figure 4.9 Current view of MC catalytic mechanism for hydrocarbon oxidation.

the methyl group of *p*-toluic acid to TA begins following principally the same mechanism.

Oxidation of *p*-methylbenzyl alcohol to *p*-tolualdehyde is a relatively fast reaction in the presence of an MC catalyst, and *p*-methylbenzyl alcohol is typically not detected or observed at low concentrations, especially under commercial conditions. Very few details are known about the mechanism of MC-catalyzed oxidation of benzylic alcohols. The final stage of the oxidation of the first methyl group is the oxidation of *p*-tolualdehyde to *p*-toluic acid. We believe that fundamentally the same catalytic cycle as shown in Figure 4.9 is operational in each step, although the nature and reactivity of peroxy radical are very different in the row of alkyl, alkylhydroxyperoxy, and acylperoxy radicals (see Figure 4.2), which are the main chain propagating species in oxidation of alkyl, alcohol, and aldehyde group, respectively [12, 19, 64, 65].

The mechanism of oxidation of *p*-toluic acid with MC catalyst has not been studied in great detail since the early paper by Ravens (which was one of the first papers on MC cobalt–bromide oxidation) [66]. We assume that the mechanism of oxidation of the second methyl group is the same as the first one, but the reaction rate constants may be very different for *p*-toluic acid-derived radicals and hydroperoxides [33]. This is an area that deserves further investigation.

4.2.8
Oxidation By-products

High temperatures of oxidation and low partial pressure of O_2 result in formation of by-products from decomposition of HOAc and from side reactions of alkylaromatic radicals with catalyst components. Oxidation by-products can be subdivided into *volatile* and *nonvolatile* varieties. Both classes of impurities are managed differently in order to meet crude TA and PTA product specifications.

Volatile by-products include methyl acetate, methyl formate, methanol, formic acid, methane, carbon oxides, and methyl bromide. The sources of carbons for the formation of volatiles result from overoxidation (i.e., combustion) of both HOAc and pX [18b, 67]. Partenheimer has written that the steady-state Co(III) concentration in the reaction is correlated with the level of overoxidations, resulting in these volatile by-products [8]. Very little research has been published on Co(III)-catalyzed overoxidation side reactions. We speculate that the reduction of Co(III) species can occur either via oxidation of $M^{(II)}Br_2$ (k_a) or via oxidation of its own acetate ligand (k_b) to produce a methyl radical (Figure 4.10). Various factors control the relative rates of desired and unwanted reactions, one of the factors being availability of $M^{(II)}Br_2$. If its concentration is too low, the oxidation favors acetate oxidation.

Once methyl radicals are generated, two main pathways can occur, depending on whether the environment is O_2 deficient or O_2 rich. In an O_2-deficient environment, methyl radicals are likely to abstract benzylic hydrogens to produce methane and the corresponding benzylic radicals. Methyl bromide, methyl acetate, and methanol can also be made from reactions involving methyl radicals (Figure 4.11).

An alternative route for methyl bromide formation may be via an *in situ* formation of bromoacetic acid by a separate reaction of Br_2 and HOAc, and

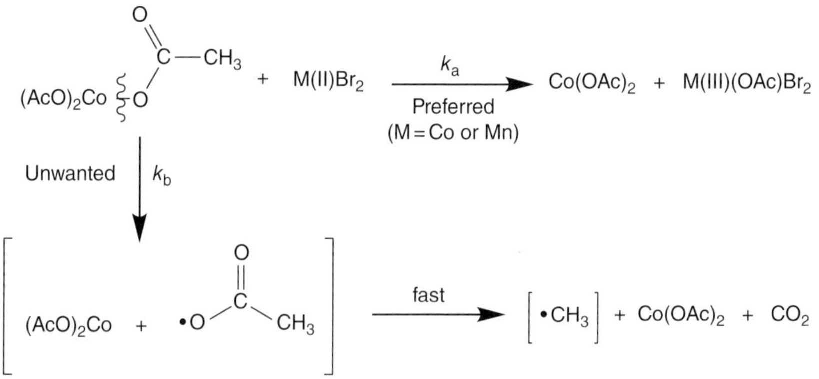

Figure 4.10 Mechanism for Co(III)-catalyzed acetic acid combustion.

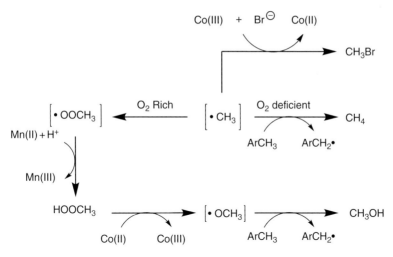

Figure 4.11 Subsequent reactions of methyl radicals produced in O_2-deficient or O_2-rich environment.

Co(III)-catalyzed decarboxylation of bromoacetic acid followed by hydrogen abstraction can result in methyl bromide.

Volatile organics are destroyed via combustion that converts oxygenated organics into carbon oxides and water over a heterogeneous catalyst. Methyl bromide is an ozone-depleting chemical, and it is subjected to additional treatment, where methyl bromide is converted into carbon dioxide, H_2O, and NaBr.

Nonvolatile by-products range from mono- to polycarboxylated monoaromatics, and diaromatic compounds are found in the reaction mixture. Among the monoaromatic varieties are benzoic acid and trimellitic acid (TMLA or 1,2,4-benzenetricarboxylic acid). Orthophthalic and isophthalic acids can also be formed, particularly from *o*-xylene and *m*-xylene impurities in pX feed. Polycarboxylated diaromatic by-products are further classified as biphenyl- or diphenylmethane-based compounds. Biphenyl-based by-products are 2,4′,5-tricarboxybiphenyl (TCBi) and 2,6-dicarboxyfluorenone (DCF). Diphenylmethane-based by-products include 2,4′,5-tricarboxybenzophenone (TCBen) and 2,6-dicarboxyanthraquinone (DCAq). Detailed discussion of by-products formed in pX to TA oxidation process can be found in the review by Tomás *et al.* and references therein [9].

Because the pX oxidation process is a continuous one, coupled to very high rates of HOAc recycle, even trace level production of these by-products can result in rapid accumulation of these impurities and product quality issue. Benzoic acid, TMLA, TCBi, and TCBen are white compounds and are relatively soluble materials. These by-products do not impact the crude TA product color, and their concentrations can be controlled by purging a small amount of the recycled mother liquor. DCF and DCAq are comparatively less soluble and are removed in the purification process described in the next section.

4.3 Process Technology

The key in successful technology development is the use of engineering solutions in combination with chemistry in order to deliver the most efficient processes, as we show in this section using the PTA process as an example. PTA is produced by a two-step process: (i) homogeneous catalytic oxidation of pX to crude TA and (ii) heterogeneous catalytic purification of the crude TA to PTA.

4.3.1 Oxidation

The oxidation is conducted at 180–220 °C under approximately 15 bar pressure. Oxidation of pX to TA results in two other major products: heat and water. The reaction occurs in a boiling solvent so that the heat of reaction is removed by the vaporization of the solvent. The vapor leaving the reactor passes through a staged condensing system, where virtually all of the solvent is condensed and the heat is recovered as low-pressure steam (about 5 barg) that can be used either directly to provide heat for HOAc dehydration or to generate electricity, typically via a condensing steam turbine. Utilization of the reaction heat and H_2O management can greatly impact the economics of the *PTA* process. Processes that provide the dehydration of H_2O from HOAc solvent with less low-pressure steam use have more low-pressure steam available to generate electricity. This has led to extensive use of azeotropic distillation of acetic acid and H_2O in the industry, which requires less heat to effect the H_2O separation from HOAc.

TA is almost completely insoluble in aqueous acetic acid even at reaction temperature and rapidly precipitates as it is made in the oxidation reactor. In conventional PTA technology, crude TA is isolated, dried, and then stored in product silos before it is conveyed to the purification section (Figure 4.12). This technology has been implemented on a very large scale with the highest oxidation reactor capacity at approximately 1 million metric tons of *PTA* per year.

4.3.2 Purification

Crude TA contains several thousand ppm of impurities by weight. Particularly, 4-CBA (4-carboxybenzaldehyde) cocrystallizes with TA in a solid solution and these two solids cannot be readily separated by crystallization. Furthermore, high concentrations of 4-CBA can render crude TA unsuitable for PET manufacturing. A purification stage is necessary to chemically transform these by-products and a physical process is required to remove the transformed by-products [7].

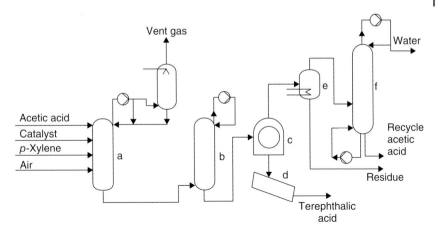

Figure 4.12 Process diagram for the oxidation section of conventional BP PTA process. (From Ref. [1].) (a) Oxidation reactor; (b) surge vessel; (c) filter; (d) dryer; (e) residue still; and (f) dehydration column.

In the purification process, crude TA is dissolved in H_2O to at least 15 wt% at approximately 290 °C and treated with high-pressure H_2 in excess of the vapor pressure of H_2O, over a fixed catalyst of Pd/C (Figure 4.13). Under these reductive conditions, these impurities undergo a hydrogenolysis reaction, resulting in the reduction of carbonyl groups to methyl/methylene groups. 4-CBA is reduced to relatively more soluble *p*-toluic acid, and DCF and DCAq are converted to secondary, non-yellow by-products, which are also relatively more soluble (Eqs. (4.20–4.22)). These impurities are then partitioned into the aqueous solvent during a multistage recrystallization process and are removed from PTA by filtration. The resulting white PTA, which contains less than 25 ppmw of 4-CBA, is now suitable for PET manufacturing. This hydrogenolysis reaction is highly selective

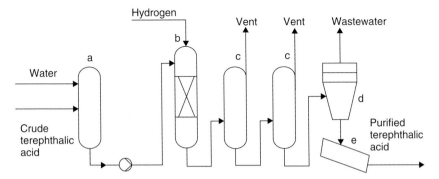

Figure 4.13 Process diagram for the purification section of conventional BP PTA process. (From Ref. [1].) (a) Slurry drum; (b) hydrogenation reactor; (c) crystallizers; (d) centrifuge; and (e) dryer.

for these impurities as loss of TA from carboxylic acid group hydrogenolysis and aromatic ring saturations is limited to less than 1 wt%.

$$HO_2C-C_6H_4-CHO \xrightarrow[H_2O]{2H_2 \ Pd/C \ cat.} HO_2C-C_6H_4-CH_3 + H_2O \quad (4.20)$$

4-CBA → p-Toluic acid

2,6-Dicarboxyfluorenone (DCF) $\xrightarrow[H_2O]{2H_2 \ Pd/C \ Cat}$ 2,6-Dicarboxyfluorene + H_2O (4.21)

2,6-Dicarboxyanthraquinone (DCAq) $\xrightarrow[H_2O]{4H_2 \ Pd/C \ cat.}$ 2,6-Dicarboxy-9,10-dihydroanthracene + $2H_2O$ (4.22)

As with the oxidation section, management of heat in purification is necessary for good process economics. The heat is released as flashing vapor from the multiple-step pressure letdown after the hydrogenation reactor is used as the heat source in some of the heat exchangers used to heat the crude TA slurry from near ambient conditions to the temperature of the purification reactor inlet. By doing so, approximately 70% of the total heat, required to dissolve crude TA into solution and to bring the solution to PTA reactor temperature, can be recovered.

In the most recent generations of *BP* technology, a highly integrated process allows for even more efficient utilization of the reaction heat while greatly improving H_2O management. In a conventional system described earlier (Figures 4.12 and 4.13), the oxidation and purification sections are completely separate. Crude TA from the oxidation section is dried and stored in a silo before its conveyance to the purification section. The new generation of *BP* technology removes many of the steps of the conventional production method and now couples the oxidation and purification stages. New proprietary solid/liquid separation technology has also permitted direct transfer of the crude TA from the oxidation section to the purification reactor. Most of the water required to dissolve crude TA is now sourced from the condensed vapor that was employed to produce electricity in the oxidation section. The liquid separated from the product cake in the purification is used as reflux for the dehydration tower in the oxidation section. Thus, a complete internal recycle of H_2O has been achieved, resulting in a very small H_2O purge, which significantly reduces the wastewater treatment load.

This process integration has resulted in significant capital savings. The new generation of *BP* PTA process is also much greener and is estimated to achieve greater than 60% reduction in greenhouse gas emissions in comparison to the conventional technology.

4.4 New Developments

Since the discovery of the MC catalyst for the oxidation of alkylaromatic hydrocarbons, significant research activities have been dedicated toward the development of alternatives to this catalyst [9, 68]. One motivation for the research is to eliminate Br^- from the process. This would allow one to reduce unit corrosion and downgrade metallurgical requirements for the process equipment. A bromide-free catalyst could also eliminate the formation of CH_3Br (discussed in Section 4.2.8). Another motivation for the alternative oxidation catalyst is to reduce or eliminate usage of HOAc and, therefore, reduce variable cost. Despite the extensive research efforts, however, no known alternative catalysts could generate the oxidation rates and product selectivity observed from the MC system.

A relatively new area of research (which is outside of the scope of this chapter but important to mention) is the development of methods to produce so-called bio-based PTA from renewable carbon sources, which was recently reviewed [69].

4.4.1 Homogeneous Bromineless Catalysis

The earliest disclosure (1941) of a bromineless catalyst used high air pressure (700 psig) for oxidation of mixed xylenes (68% conversion) to aromatic dicarboxylic acids (2 mol%). This oxidation relied on a combination of Co(II) and Mn(II) plus an organic promoter (methyl ethyl ketone MEK) [41]. In addition, the oxidation also coproduced tarry by-products. Other modifications to this catalyst package were soon disclosed, including the use of acetaldehyde and α-methylbutyric acid as a promoter [70–72]. A downside to the use of promoters is their conversion to coproducts including carbon oxides. Unless one can sell these coproducts, they represent a significant cost to the process. Promoted oxidations are much slower than a Co–Br-catalyzed oxidation and require a combination of lower temperature and higher pressure, requiring an alternate cooling scheme.

Cobalt-based catalyst packages that do not use sacrificial promoters but augment the Co salt with other transition metals have received considerable attention since the 1950s. Zr(IV) was discovered to be an activator to Co(II) for the oxidation of alkylaromatics [73, 74]. The use of a combination of various metal ions with cobalt and of variously ligated forms of Co(II) has been also reported [75–78]. Bromineless catalysts have also been combined with CO_2-expanded solvents such as acetic acid or water [79–81].

Another approach to develop a bromineless catalyst is in the replacement of bromine with a nonhalogen substitute. In 1996, Y. Ishii first reported that an organic compound, N-hydroxyphthalimide (NHPI), could replace bromide in the MC catalyst [82, 83]. Numerous investigators have since tested the scope and limitation of these imides [84–89]. Other groups have probed the kinetics and mechanisms of these catalysts [88, 90, 91]. The reader is referred to Chapter 16 for a detailed discussion of this chemistry.

4.4.2
Heterogeneous Bromineless Oxidation Catalysis

Numerous reports of heterogeneous catalysis active for alkylaromatic oxidations have appeared. These include an encapsulation of metal ions by zeolites or polymers [92–95]. Non-Co, Pd-based heterogeneous catalysts have been discovered by *BP* researchers [96–98]. Very recently, nanocrystalline ceria (CeO_2) has been discovered to be a highly active heterogeneous catalyst for oxidation of pX in water to TA [99, 100].

4.4.3
Alternative Solvents

Oxidation processes that could potentially eliminate reliance on a valuable solvent have been explored by a number of researchers. In these cases, the homogeneous catalyst must be modified so that it is soluble in a hydrocarbon, which serves simultaneously as a feedstock [101]. As discussed earlier, expanded solvents such as CO_2 and HOAc have also been explored with conventional MC-type catalysts [79, 102, 103]. Supercritical H_2O as a solvent was explored for Mn–Br catalyst for pX oxidation [104–109]. Combinations of ionic liquids and HOAc were recently disclosed by *UOP* investigators for pX oxidation to TA [110]. Lastly, the use of HOAc in a spray reactor in combination with the MC catalyst for pX oxidation was explored by Li. *et al.* [111].

4.5
Conclusions

Since its discovery over 50 years ago, much has been learned about the MC catalyst, and this overview presented the current view of the authors on key aspects of MC oxidation chemistry. The unmasking of the inner workings of this catalyst has revealed that each of the catalytic species plays unique roles in complex catalytic cycles that lead to fast and highly selective conversion of pX to TA. The extremely high efficiency of the oxidation process makes the MC catalysis the best known system for TA manufacturing.

References

1. Sheehan, R.J. (2011) Terephthalic acid, dimethyl terephthalate, and isophthalic acid, in *Ullmann's Encyclopedia of Industrial Chemistry*, John Wiley & Sons.
2. IHS Chemical World Analysis (2014) World Terephthalates and Polyester Analysis.
3. Landau, R. and Saffer, A. (1968) *Chem. Eng. Prog.*, **64**, 20.
4. Saffer, A. and Barker, R.S. (1958) Preparation of aromatic carboxylic acids. US Patent 2,833,816, May 6, 1958.
5. Saffer, A. and Barker, R.S. (1959) Process for the production of aromatic carboxylic acids. GB 807,091, Jan. 7, 1959.
6. Saffer, A. and Barker, R.S. (1963) Oxidation chemical process. US Patent 3,089,906, May 14, 1963.
7. Meyer, D.H. (1971) Fiber-grade terephthalic acid by catalytic hydrogen treatment of dissolved impure terephthalic acid. US patent 3,584,039, Jun. 8, 1971.
8. Partenheimer, W. (1995) *Catal. Today*, **23**, 69.
9. Tomás, R.A.F., Bordado, J.C.M., and Gomes, J.F.P. (2013) *Chem. Rev.*, **113**, 7421.
10. (a) Zakharov, I.V. and Geletti, Y.V. (1986) *Petrol. Chem. USSR*, **26**, 234; (b) Geletii, Y.V. and Zakharov, I.V. (1984) *Oxid. Commun.*, **6**, 23.
11. Emanuel, N.M., Denisov, E.T., and Maizus, Z.K. (1967) *Liquid-Phase Oxidation of Hydrocarbons*, Plenum Press, New York.
12. Denisov, E.T., Mitskevich, N.I., and Agabekov, V.E. (1977) *Liquid-Phase Oxidation of Oxygen-Containing Compounds*, Consultants Bureau, New York.
13. Walling, C. (1969) *J. Am. Chem. Soc.*, **91**, 7590.
14. Woodward, A.E. and Mesrobian, R.B. (1953) *J. Am. Chem. Soc.*, **75**, 6189.
15. Russell, G.A. (1957) *J. Am. Chem. Soc.*, **79**, 3871.
16. Zakharov, I.V. and Kumpan, Y.V. (1993) *Kinet. Katal.*, **34**, 1026.
17. Jones, G.H. (1979) *p*-Xylene autoxidation studies. *J. Chem. Soc., Chem. Commun.*, 536.
18. (a) Ariko, N.G., Kornilova, N.N., and Mitskevich, N.I. (1985) *Kinet. Katal.*, **26**, 859; (b) Kenigsberg, T.P., Ariko, N.G., Mitskevich, N.I., and Nazimok, V.F. (1985) *Kinet. Katal.*, **26**, 1485.
19. Hendriks, C.F., van Beek, H.C.A., and Heertjes, P.M. (1979) *Ind. Eng. Chem., Prod. Res. Dev.*, **18**, 38.
20. Geletii, Y.V., Zakharov, I.V., and Strizhak, P.E. (1985) *Kinet. Katal.*, **26**, 852.
21. Zakharov, I.V., Geletii, Y.V., and Adamyan, V.A. (1986) *Kinet. Katal.*, **27**, 1128.
22. Zakharov, I.V., Geletii, Y.V., and Adamyan, V.A. (1988) *Kinet. Katal.*, **29**, 1072.
23. Zakharov, I.V., Geletii, Y.V., and Adamyan, V.A. (1991) *Kinet. Katal.*, **32**, 39.
24. Partenheimer, W. and Gipe, R.K. (1993) *ACS Symp. Ser.*, **523**, 81.
25. Partenheimer, W. (1993) in *Catalytic Selective Oxidation*, Chapter 7 (eds S.T. Oyama and J.W. Hightower), ACS, pp. 81–88.
26. Partenheimer, W. (2004) *Adv. Synth. Catal.*, **346**, 297.
27. Partenheimer, W. (2001) *J. Mol. Catal. A: Chem.*, **174**, 29.
28. (a) Partenheimer, W. (2014) *Appl. Catal. A: Gen.*, **481**, 183; (b) Partenheimer, W. (2014) *Appl. Catal. A: Gen.*, **481**, 190.
29. Jiao, X.-D. and Espenson, J.H. (2000) *Inorg. Chem.*, **39**, 1549.
30. Metelski, P.D., Adamian, V.A., and Espenson, J.H. (2000) *Inorg. Chem.*, **39**, 2434.
31. Metelski, P.D. and Espenson, J.H. (2001) *J. Phys. Chem. A*, **105**, 5881.

32. Saha, B. and Espenson, J.H. (2005) *J. Mol. Catal. A: Chem.*, **241**, 33.
33. Espenson, J.H. and Yiu, D.T.-Y. (2005) *Int. J. Chem. Kinet.*, **37**, 599.
34. Saha, B. and Espenson, J.H. (2007) *J. Mol. Catal. A: Chem.*, **271**, 1.
35. Kamiya, Y. (1966) *Tetrahedron*, **22**, 2029.
36. Kamiya, Y., Nakajima, T., and Sakota, K. (1966) *Bull. Chem. Soc. Jpn.*, **39**, 2211.
37. Sakota, K., Kamiya, Y., and Ohta, N. (1968) *Bull. Chem. Soc. Jpn.*, **41**, 641.
38. Hay, A.S. and Blanchard, H.S. (1965) *Can. J. Chem.*, **43**, 1306.
39. Jones, G.H. (1981) *J. Chem. Res. (S)*, **1981**, 228.
40. (a) Jones, G.H. (1982) *J. Chem. Res. (S)*, **1982** 2137–2163; (b) Jones, G.H. (1982) *J. Chem. Res. M*, 2137.
41. Loder, D.J. (1941) Catalytic oxidation of alkyl substituted aromatic compounds. US Patent 2,245,528, Jun. 10, 1941.
42. Zakharov, I.V. (1998) *Kinet. Katal.*, **39**, 485.
43. Adamyan, V.A., Geletii, Y.V., Popova, M., and Zakharov, I.V. (1991) *Kinet. Katal.*, **31**, 1490.
44. De Klein, W.J. and Kooyman, E.C. (1965) *J. Catal.*, **4**, 626.
45. Jiao, X.-D., Metelski, P.D., and Espenson, J.H. (2001) *Inorg. Chem.*, **40**, 3228.
46. Szymanska-Buzar, T. and Ziolkowski, J.J. (1979) *J. Mol. Catal.*, **5**, 341.
47. Proll, P.J. and Sutcliffe, L.H. (1961) *J. Phys. Chem.*, **65**, 1993.
48. Sowada, K. and Tanaka, M. (1977) *J. Inorg. Nucl. Chem.*, **39**, 339.
49. Akai, T., Okuda, M., and Nomura, M. (1999) *Bull. Chem. Soc. Jpn.*, **72**, 1239.
50. Dugmore, G.M., Powels, G.J., and Zeelie, B. (1995) *J. Mol. Catal. A: Chemical*, **99**, 1.
51. Sumner, C.E. Jr., and Steinmetz, G.R. (1985) *J. Am. Chem. Soc.*, **107**, 6124.
52. Ziolkowski, J.J., Pruchnik, F., and Szymanska-Buzar, T. (1973) *Inorg. Chim. Acta*, **7**, 473.
53. Chipperfield, J.R., Lau, S., and Webster, D.E. (1992) *J. Mol. Catal.*, **75**, 123.
54. Blake, A.B., Chipperfield, J.R., Lau, S., and Webster, D.E. (1992) *J. Chem. Soc., Dalton Trans.*, 3719.
55. Anderson, J.M. and Kochi, J.K. (1970) *J. Am. Chem. Soc.*, **92**, 2450.
56. Jee, J.-E., Pestovsky, O., and Bakac, A. (2010) *Dalton Trans.*, **39**, 11636.
57. Adamyan, V.A., Geletii, Y.V., Hronec, M., and Zakharov, I.V. (1993) *Kinet. Katal.*, **34**, 646.
58. Cheng, Y., Li, X., Wang, Q., and Wang, L. (2005) *Ind. Eng. Chem. Res.*, **44**, 7756.
59. Jhung, S.H. and Park, Y.-S. (2002) *Bull. Korean Chem. Soc.*, **23**, 369.
60. Partenheimer, W. (2003) *J. Mol. Catal. A: Chem.*, **206**, 131.
61. Komatsu, M., Ohta, T., Tanaka, T., and Akagi, K. (1980) Process for producing terephthalic acid. US Patent 4,211,882, Jul. 8, 1980.
62. Broeker, J.L., Partenheimer, W., and Rosen, B.I. (1995) Process for the manufacturing of aromatic dicarboxylic acids utilizing cerium to facilitate a low bromine to metals catalyst ratio. US Patent 5,453,538, Sept. 26, 1995.
63. Broeker, J.L. and Gong, W.H. (1996) Low bromine *p*-xylene oxidations. 6th International Symposium on the Activation of Dioxygen and Homogeneous Catalytic Oxidation, Noordwijkherhout, The Netherlands, April 14–19, 1996.
64. (a) Hendriks, C.F., van Beek, H.C.A., and Heertjes, P.M. (1977) *Ind. Eng. Chem., Prod. Res. Dev.*, **16**, 270–275; (b) Hendriks, C.F., van Beek, H.C.A., and Heertjes, P.M. (1978) *Ind. Eng. Chem., Prod. Res. Dev.*, **17**, 260–264.
65. Sajus, L. and Sérée De Roch, I. (1980) in *Liquid-Phase Oxidation*, Comprehensive Chemical Kinetics, vol. **16** (eds C.H. Bamford and C.F.H. Tipper), Elsevier, pp. 89–124.
66. Ravens, D.A.S. (1959) *Trans. Faraday Soc.*, **55**, 1768.
67. Roffia, P., Perangelo, C., and Tonti, S. (1988) *Ind. Eng. Chem. Res.*, **27**, 765.
68. Fadzil, N.A.M., Ab Rahim, M.H., and Maniam, G.P. (2014) *Chin. J. Catal.*, **35**, 1641–1652.
69. Collias, D.I., Harris, A.M., Nagpal, V., Cottrell, I.W., and Schutheis, M.W. (2014) *Ind. Biotechnol.*, **10**, 91.
70. Brill, W. (1960) *Ind. Eng. Chem.*, **52**, 837.

71. Bryant, H.S., Duval, C.A., McKakin, L.E., and Savoca, J.I. (1971) *Chem. Eng. Prog.*, **67**, 69.
72. Ichikawa, Y., Yamashita, G., Tokashiki, M., and Yamaji, T. (1970) *Ind. Eng. Chem.*, **62**, 38.
73. Chester, A.W., Landis, P.S., and Scott, E.J.Y. (1978) Oxidize aromatics over doped cobalt. *CHEMTECH*, **366**, p. 367.
74. Scott, E.J.Y. and Landis, P.S. (1988) *J. Catal.*, **46**, 308.
75. Codignola, F. and Moro, A. (2000) Process for the production of aromatic acids. WO 2000058257, Oct. 5, 2000.
76. Saxena, M.P., Sharma, S.K., Gupta, A.K., Kumar, K., and Bangwal, D.P. (2005) A new strategy for preparation of terephthalic acid by oxidation of *p*-xylene in aqueous medium. GDGMK/SCI-Conference – Oxidation and Functionalization: Classical and Alternative Routes and Sources, p. 201.
77. Jiang, Q., Xiao, Y., Tan, Z., Li, Q.H., and Guo, C.C. (2008) *J. Mol. Catal. A: Chem.*, **285**, 162.
78. Sheldon, R.A. and Kochi, J.K. (1981) *Metal-Catalyzed Oxidation of Organic Compounds*, Academic Press, New York, pp. 319–320.
79. Zuo, X., Niu, F., Snavely, K., Subramaniam, B., and Busch, D.H. (2010) *Green Chem.*, **12**, 260–267.
80. Ma, C., Givens, R.S., Busch, D.H., and Subramaniam, B. (2007) Xylene uses a cobalt/initiator catalyst in CO_2 expanded solvents (CXLs). Abstract of Papers, 233rd ACS National Meeting, Chicago, IL, March 25–29, 2007.
81. Kwak, J.W., Lee, J.S., and Lee, K.H. (2009) *Appl. Catal. A: Gen.*, **358**, 54.
82. Ishii, Y. (1996) A novel catalysis of *N*-hydroxyphthalimide (NHPI) in the oxidation of organic substrates with molecular oxygen. 6th International Symposium on the Activation of Dioxygen and Homogeneous Catalysis, April 14–19, 1996.
83. Ishii, Y. and Sakaguchi, S. (2006) *Catal. Today*, **117**, 105.
84. Tashiro, Y., Iwahama, T., Sakaguchi, S., and Ishii, Y. (2001) *Adv. Synth. Catal.*, **343**, 220.
85. Hirai, N., Tatsukawa, Y., Kameda, M., Sakaguchi, S., and Ishii, Y. (2006) *Tetrahedron*, **62**, 6695.
86. Hirai, N., Sawatari, N., Nakamura, N., Sakaguchi, S., and Ishii, Y. (2003) *J. Org. Chem.*, **68**, 6587.
87. Xu, J. (2005) *Selective Oxidation of Hydrocarbons*, Dalian Institute of Chemical Physics, Chinese Academy of Sciences, May 2005.
88. Saha, B., Koshino, N., and Espenson, J.H. (2004) *J. Phys. Chem. A*, **108**, 425.
89. Yoshino, Y., Hayashi, Y., Iwahawa, T., Sakaguchi, S., and Ishii, Y. (1997) *J. Org. Chem.*, **62**, 6810.
90. Cai, Y., Koshino, N., Saha, B., and Espenson, J.H. (2005) *J. Org. Chem.*, **70**, 238.
91. Falcon, H., Compas-Martin, J.M., Al-Zahrani, S.M. and Fierro, J.L.G. (2010) *Catal. Commun.*, **12**, 5.
92. Chavan, S.A., Srinivas, D., and Ratnasamy, P. (2001) *J. Catal.*, **204**, 409.
93. Ratnasamy, P. and Srinivas, D. (2009) *Catal. Today*, **141**, 3.
94. Lei, Z., Han, X., Hu, Y., Wang, R., and Wang, Y. (2000) *J. Appl. Polym. Sci.*, **75**, 1068.
95. Ghiaci, M., Mostajeran, M., and Gil, A. (2012) *Ind. Eng. Chem. Res.*, **51**, 15821.
96. Schammel, W.P., Adamian, V.A., Brugge, S.P., Gong, W.H., Metalski, P.D., Nubel, P.O., and Zhou, C. (2012) Process and catalyst for oxidizing aromatic compounds. US Patent 8163954, Apr. 24, 2012.
97. Schammel, W.P., Huggins, B.J., Kulzick, M.A., Nubel, P.O., Rabatic B.M., Zhou, C., Adamian, V.A., Gong, W.H., Metelski, P.D., and Miller, J.T. (2014) Process and catalyst for oxidizing aromatic compounds. US Patent 8624055, Jan. 7, 2014.
98. Gong, W.H., Schammel, W.P., and Adamian, V.A. (2008) Process for the production of non-aromatic carboxylic acids. WO 2008/136857, Nov. 13, 2008.
99. Deori, K., Gupta, D., Saha, B., Awasthi, S.K., and Deka, S. (2013) *J. Mater. Chem. A.*, **1**, 7091.

100. Deori, K., Gupta, D., Saha, B., and Deka, S. (2014) *ACS Catal.*, **4**, 3169.
101. Xiao, Y., Zhang, X.Y., Wang, Q.B., Tan, Z., and Guo, C.C. (2011) *Chin. Chem. Lett.*, **22**, 135.
102. Rajagopalan, B., Subramaniam, B., and Busch, D.H. (2007) Oxidation of *p*-xylene to terephthalic acid in CO_2 based reaction medium. 2007 AiChE Annual Meeting.
103. Kim, D.S., Shin, Y.H., and Lee, Y.-W. (2015) *Chem. Eng. Commun.*, **202**, 78.
104. Pérez, E., Fraga-Dubreuil, J., García-Verdugo, E., Hamley, P.A., Thomas, M.L., Yan, C., Thomas, W.B., Houseley, D., Partenheimer, W., and Poliakoff, M. (2011) *Green Chem.*, **13**, 2397.
105. Dunn, J.B. and Savage, P.E. (2002) *Ind. Eng. Chem. Res.*, **41**, 4460.
106. Dunn, J.B. and Savage, P.E. (2003) *Green Chem.*, **5** (5), 649.
107. Dunn, J.B. and Savage, P.E. (2005) *Environ. Sci. Technol.*, **39**, 5427.
108. Osada, M. and Savage, P.E. (2009) *AIChE J.*, **55**, 710.
109. Osada, M. and Savage, P.E. (2009) *AIChE J.*, **55**, 1530.
110. Bhattacharyya, A. and Walenga, J.T. (2012) Process for oxidizing alkyl aromatic compounds. US Patent 20120004448, Jan. 5, 2012.
111. Li, M., Niu, F., Zuo, X., Metelski, P.D., Busch, D.H., and Subramanian, B. (2013) *Chem. Eng. Sci.*, **104**, 93–102.

Part II
Cu-Catalyzed Aerobic Oxidation

Liquid Phase Aerobic Oxidation Catalysis: Industrial Applications and Academic Perspectives,
First Edition. Edited by Shannon S. Stahl and Paul L. Alsters.
© 2016 Wiley-VCH Verlag GmbH & Co. KGaA. Published 2016 by Wiley-VCH Verlag GmbH & Co. KGaA.

5
Cu-Catalyzed Aerobic Oxidation: Overview and New Developments

Damian Hruszkewycz, Scott McCann, and Shannon Stahl

5.1
Introduction

Copper(II) salts are versatile one-electron oxidants in organic chemistry, and, in many cases, the Cu^I by-product of these reactions may be reoxidized to Cu^{II} with O_2. This reactivity provides the basis for selective aerobic oxidation of organic molecules with Cu catalysts [1]. Cu-containing active sites are present in a wide range of oxygenase and oxidase enzymes in nature, and these systems have motivated extensive studies of the fundamental reactivity of Cu^I complexes with O_2 [2]. In addition to the many academic studies of these topics, Cu-catalyzed aerobic oxidation reactions have been applied in a number of important industrial processes, several of which are the focus of dedicated chapters in this volume. Copper chloride salts are commonly used as cocatalysts in the Wacker process and related Pd-catalyzed methods for aerobic oxidation of alkenes (see Chapters 9 and 11). These applications exploit the facile oxidation of Cu^I to Cu^{II} by oxygen gas, and the Cu^{II} salts oxidize Pd^0 to Pd^{II} to promote catalytic turnover. In many other cases, however, Cu catalysts are used directly to achieve oxidation of the organic molecules (Scheme 5.1). Chapter 6 presents Cu-catalyzed methods for aerobic alcohol oxidation, which are finding increasing utility in pharmaceutical and related complex molecule syntheses (Scheme 5.1a). Chapter 7 describes the industrial use of Cu-catalyzed oxidations of phenols for the preparation of the high-performance thermoplastic poly(*p*-phenylene oxide) and 2,3,5-trimethyl-1,4-benzoquinone, a precursor to vitamin E (Scheme 5.1b,c). As a prelude to these discussions, this chapter highlights two large-scale liquid phase applications of Cu aerobic oxidation catalysis, the conversion of benzoic acid to phenol (Scheme 5.1d), and the production of dimethyl carbonate (DMC) via oxidative carbonation of methanol (Scheme 5.1e). Then, it concludes by highlighting a number of Cu-catalyzed aerobic oxidations that have begun finding utility in the pharmaceutical industry. Overall, the reactions illustrate the diverse reactivity of Cu catalysts and their potential to impact chemical synthesis, from commodity and fine chemicals to agrochemicals and pharmaceuticals.

Liquid Phase Aerobic Oxidation Catalysis: Industrial Applications and Academic Perspectives,
First Edition. Edited by Shannon S. Stahl and Paul L. Alsters.
© 2016 Wiley-VCH Verlag GmbH & Co. KGaA. Published 2016 by Wiley-VCH Verlag GmbH & Co. KGaA.

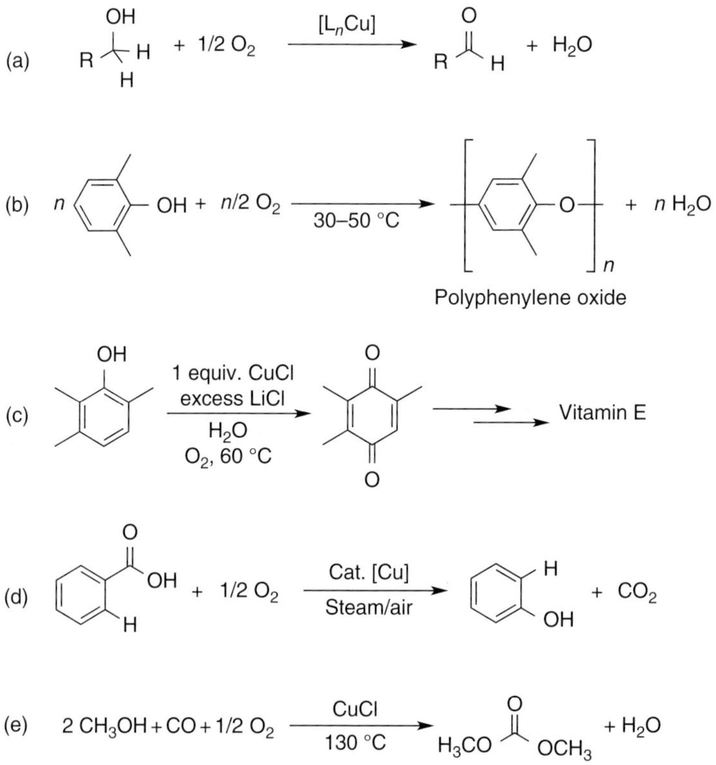

(a) $R_2CHOH + 1/2\ O_2 \xrightarrow{[L_nCu]} RC(O)H + H_2O$

(b) $n\ \text{(2,6-dimethylphenol)}\text{-OH} + n/2\ O_2 \xrightarrow{30-50\ °C}$ Polyphenylene oxide $+ n\ H_2O$

(c) 2,3,6-trimethylphenol $\xrightarrow[H_2O,\ O_2,\ 60\ °C]{\text{1 equiv. CuCl, excess LiCl}}$ 2,3,5-trimethylbenzoquinone $\longrightarrow$ Vitamin E

(d) benzaldehyde-2-carboxylic acid $+ 1/2\ O_2 \xrightarrow[\text{Steam/air}]{\text{Cat. [Cu]}}$ phenol $+ CO_2$

(e) $2\ CH_3OH + CO + 1/2\ O_2 \xrightarrow[130\ °C]{CuCl} H_3CO-C(O)-OCH_3 + H_2O$

Scheme 5.1 Industrially important liquid phase Cu-catalyzed aerobic oxidation reactions.

5.2
Chemistry and Catalysis

5.2.1
Cu-Catalyzed Oxydecarboxylative Phenol Synthesis

Phenol has been produced industrially by a number of different routes, the most significant of which now is the cumene (or Hock) process, involving autoxidation of cumene to cumene hydroperoxide, followed by acid-catalyzed rearrangement to acetone and phenol [3]. Another route, developed by the California Research Corporation [4] and Dow Chemical in the 1950s [5], and later by DSM [6], involves Cu-catalyzed oxydecarboxylation of benzoic acid. The economics of this process are unfavorable relative to the cumene process, and, to our knowledge, it is no longer practiced industrially. However, the first commercial application of this process began in the 1960s, and, as recently as 2002, it accounted for greater than 250 000 t phenol annually. The benzoic acid starting material is accessible from liquid phase aerobic oxidation of toluene via a catalyzed radical chain process similar to the Mid-Century process used for production of terephthalic acid (cf. Chapter 4). Cu-catalyzed conversion of benzoic acid to phenol results in the production of CO_2 and installation of a hydroxyl group adjacent (*ortho*) to the position originally occupied by the carboxylic acid (Eq. (5.1)):

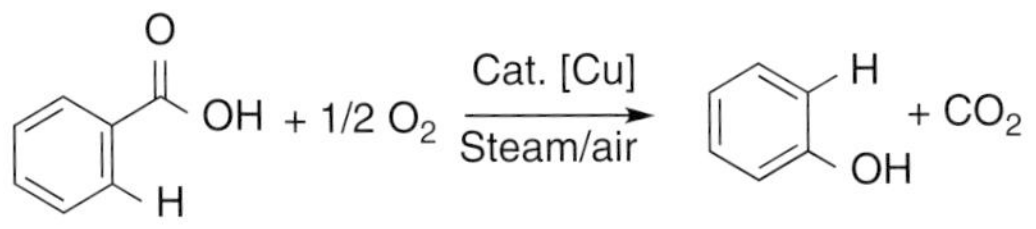

O
OH + 1/2 O2
H
Cat. [Cu]
Steam/air
H
OH
+ CO2

(5.1)

The mechanism of the oxidative decarboxylation process is not fully understood, but a proposed mechanism is depicted in Scheme 5.2. A $Cu^{II}(OBz)_2$ species is believed to form benzoyl salicylic acid through a process that has been interpreted according to single-electron transfer/radical [7] or electrophilic aromatic substitution [8] pathways. Recent studies of Cu-catalyzed C–H oxidation pathways raise the possibility that the mechanism could involve an "organometallic" chelate-directed C–H activation pathway, analogous to Pd-catalyzed C–H activation [1b, 9]. Benzoyl salicyclic acid is the product of Cu-mediated oxidative coupling of 2 equiv. of benzoic acid (i.e., *ortho* C–H benzoyloxylation of benzoic acid), and salicylic acid will arise from hydrolysis of this product. Control experiments show that salicylic acid undergoes rapid decarboxylation under the reaction conditions [5e]. Aerobic oxidation of the Cu^I species in the presence of benzoic acid regenerates the active $Cu^{II}(OBz)_2$ species to complete the catalytic cycle:

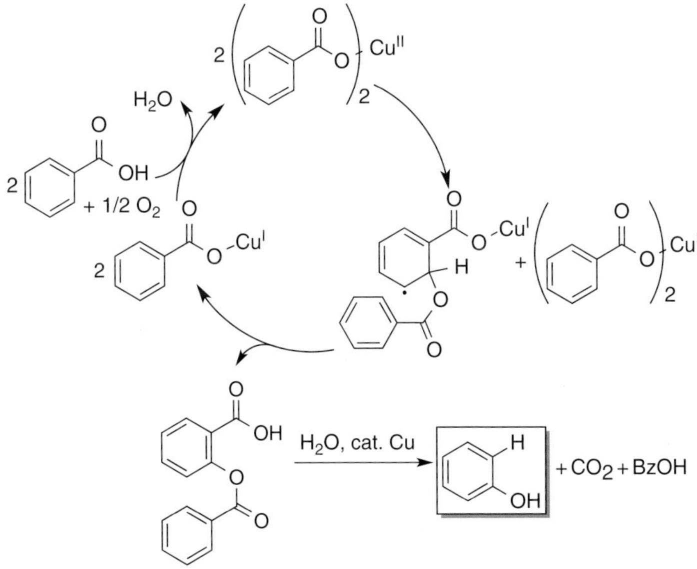

2
CuII
H2O
2
OH
+ 1/2 O2
2
CuI
CuI
H
+
CuII
OH
H2O, cat. Cu
H
OH
+ CO2 + BzOH

Scheme 5.2 Proposed mechanism for Cu-catalyzed aerobic decarboxylation of benzoic acid.

Addition of cobalt benzoate accelerates the rate of oxydecarboxylation; however, it also leads to increased tar formation. Use of sodium and potassium benzoate promotes nonoxidative decarboxylation of benzoic acid, producing significant amounts of benzene, while magnesium benzoate enhances the desired oxydecarboxylation process and reduces tar formation [5e]. The effect of Mg promoters has been interrogated by EPR spectroscopy. Mg salts contribute to the formation of mononuclear copper species, whereas a mixture of mononuclear, binuclear (i.e., "paddlewheel" complexes), and polymeric copper species is present in the absence of Mg promoters [8].

5.2.2
Cu-Catalyzed Oxidative Carbonylation of Methanol for the Synthesis of Dimethyl Carbonate

DMC has attracted significant industrial interest because of its potential use as a fuel additive, monomer for the production of polycarbonates, commercial solvent, battery electrolyte, and alkylating agent. Its low toxicity and environmental impact offer significant advantages over other chemicals commonly used in these applications, such as phosgene and methyl halides [10–12]. One of the most effective routes for the production of DMC involves CuCl-catalyzed liquid phase, aerobic oxidative carbonylation of methanol, which produces water as the sole by-product (Eq. (5.2)). This route, developed by Enichem (now Eni-Versalis), is complemented by a vapor-phase Pd/NO oxidative carbonylation process of Ube and Bayer and nonoxidative routes, involving transesterification of ethylene or propylene carbonate with methanol [11, 12]:

$$2CH_3OH + CO + \tfrac{1}{2}O_2 \rightarrow (CH_3O)_2CO + H_2O \qquad (5.2)$$

The simplified sequence of half-reactions for oxidative carbonylation of methanol, shown in Eqs. (5.3) and (5.4), consists of aerobic oxidation of CuCl in the presence of methanol to afford $Cu(OCH_3)Cl$ and reaction of CO with 2 equiv. of this Cu^{II} species to afford DMC and two CuCl. A more detailed depiction of the catalytic mechanism that accounts for various experimental observations is illustrated in Scheme 5.3 [13]. $Cu(OCH_3)Cl$ will polymerize in methanol and has low solubility. Water enhances the rate of DMC formation up to a concentration of 3 wt%, probably via partial hydrolysis of polymeric Cu species, but too much water can contribute to the formation of insoluble Cu hydroxy chlorides $Cu(Cl)x(OH)\cdot nH_2O$ [12]. It also promotes a competitive side reaction, the combustion of CO to CO_2 (Eq. (5.5)) [13]. The reaction in anhydrous methanol is also accelerated by the addition of a small amount of soluble $CuCl_2$ [14]. Principal steps in the proposed mechanism in Scheme 5.3 include (a) aerobic oxidation of CuCl in methanol solvent to form (oligomeric) Cu^{II} methoxychloride $[Cu(OCH_3)Cl]_n$, (b) HCl-induced formation of soluble $CuCl_2$, (c) binding of CO to $CuCl_2$, (d) carbonylation of methanol via nucleophilic addition to a Cu^{II}-coordinated carbonyl ligand, and (e) reductive coupling of the Cu^{II}-acyl species with methoxide derived from a $Cu(OCH_3)Cl$ species.

$$2CH_3OH + 2CuCl + \tfrac{1}{2}O_2 \rightarrow 2Cu(OCH_3)Cl + H_2O \qquad (5.3)$$

$$2Cu(OCH_3)Cl + CO \rightarrow (CH_3O)_2CO + 2CuCl \qquad (5.4)$$

$$2CuCl + CO + H_2O \rightarrow 2Cu + CO_2 + 2HCl \qquad (5.5)$$

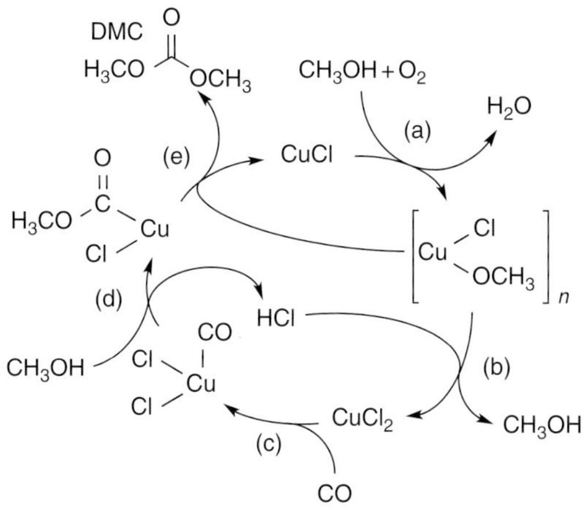

Scheme 5.3 Proposed mechanism of DMC formation in the Enichem/Versalis process.

The reaction is sensitive to the concentration of chloride ions, and the rate and selectivity are maximized at a 1:1 Cu:Cl ratio. Increased formation of chloromethane (CH_3Cl) and dimethyl ether (CH_3OCH_3) is observed at both higher and lower chloride concentrations [15]. The presence of chloride in the reaction mixture, even at these low levels, contributes to the formation of HCl, which requires the use of corrosion-resistant reactors [12]. Considerable research has been performed to identify catalytic systems that are less sensitive to water and/or less corrosive [16]. Use of mono- and bidentate nitrogen-donor ligands to CuCl-based catalysts enhances the rate and selectivity of DMC formation and decreases the corrosiveness of the system [17]. A chloride-free 2,2′-bipyrimidine-Cu(OAc)$_2$ catalyst was identified that exhibits similar activity to CuCl (Figure 5.1a) [18]. Other metal complexes, such as Co^{II}-Schiff base complexes (Figure 5.1b), have also been shown to be effective catalysts for the oxidative carbonylation of methanol [19]. Polymer-immobilized ligands have been investigated as a means to heterogenize the Cu catalyst [20]. These efforts complement considerable investigation of more traditional heterogeneous Cu-based catalysts using activated carbon or zeolite frameworks to support active Cu species in both liquid- and gas-phase processes [16].

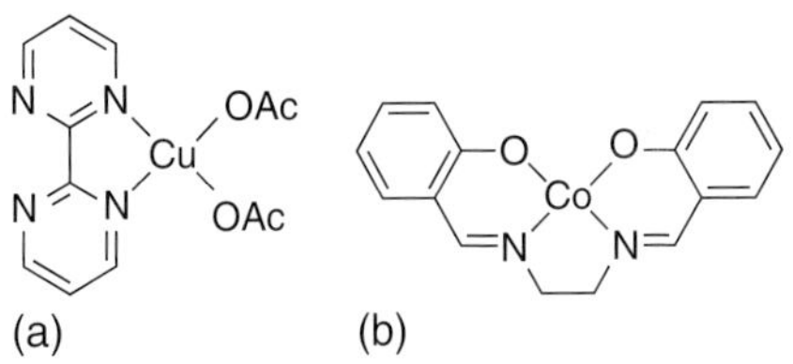

Figure 5.1 Illustration of chloride-free catalysts for the oxidative carbonylation of methanol: (a) 2,2′-bipyrimidine-Cu(OAc)$_2$ and (b) a Co^{II}-Schiff base complex.

5.3
Process Technology

5.3.1
Cu-Catalyzed Oxydecarboxylative Phenol Synthesis

Commercial application of Cu-catalyzed oxydecarboxylation of benzoic acid was initiated in the 1960s by Dow Canada. At least three different companies operated similar processes before they were discontinued in recent years [3c]. The process uses toluene as a feedstock and consists of two main reactors (Figure 5.2) [3a]. Toluene is oxidized in the first reactor to approximately 40% conversion and 90% selectivity for benzoic acid. Toluene is recycled and intermediate benzaldehyde is isolated or recycled. Benzyl benzoate, methylbiphenyls, formic acid, and acetic

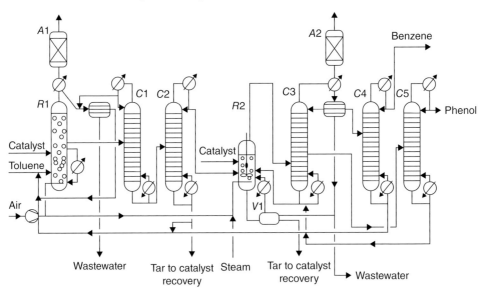

Figure 5.2 Reactor scheme for toluene to phenol oxidation. R1: toluene oxidation reactor, A1: adsorption unit, C1: toluene column, C2: benzoic acid, R2 benzoic acid oxidation reactor, A2 adsorption unit, C3: crude phenol column, C4: benzene column, C5: pure phenol column, and V1: tar extraction unit. (Reactor diagram reproduced from [3a].)

acid are formed as by-products, as well as smaller amounts of carbon dioxide and carbon monoxide. The purified (technical grade) benzoic acid is passed into a second bubble column reactor, where the copper catalyst is introduced. The reaction is typically performed at ambient pressure and 250 °C in molten benzoic acid [3c]. Steam may be added to the air stream to facilitate hydrolysis of intermediate benzoyl salicyclic acid, and catalytic amounts of magnesium benzoate salts may be used as promoters to enhance the selectivity for phenol. The conversion of benzoic acid is 70–80%, and the selectivity to phenol is approximately 90% [3]. One unattractive feature of the liquid phase oxidation of benzoic acid is the formation of tars, which can be problematic in some cases [3a]. The Lummus Company disclosed a process for benzoic acid in the vapor phase over a heterogeneous copper-containing catalyst (Cu/Zr/alkali metal); the reported selectivity for phenol is approximately 90% [21]. They report no tar formation, possibly reflecting more rapid removal of product from the reaction zone relative to the liquid phase process.

5.3.2
Cu-Catalyzed Oxidative Carbonylation of Methanol for the Synthesis of Dimethyl Carbonate

The liquid phase Enichem process for DMC production was first commercialized in 1983 [22], and the worldwide production capacity of this process was estimated to be over 70 000 t per year in 2001 [10b]. The reaction is performed in neat methanol solvent at 24 bar and 130 °C using an excess of CO and a continuous feed of a low pressure of O_2, which is the limiting reagent and is maintained below the limiting O_2 concentration [11]. High-purity CO is not necessary for the success of the reaction. Enichem has even demonstrated the use of syngas as a feedstock; the H_2 does not interfere with the reaction [23]. The process is performed in a slurry reactor (Figure 5.3) consisting of two columns that circulate the contents because of a density difference that results from the upstream of gas in one of the columns [24]. As the contents circulate, a DMC/methanol/water mixture is vaporized and removed from the reactor. The significant exothermocity of the oxidative carbonylation reaction ($\Delta H_{r(500°C)} = -318$ kJ/mol) provides the heat for this vaporization [12]. Some HCl is also released in these vapors, and HCl is continually fed into the reactor to maintain the optimal 1 : 1 Cu:Cl ratio [15]. After product purification, a methanol/DMC azeotrope is circulated back into the reactor, together with the recovered CO gas.

Several complications in the Enichem process constrain the scale of the reactor volume [12]. The reaction is so exothermic that the reactor needs to be cooled to maintain the optimal 130 °C temperature. Vaporization of the reaction mixture removes only a fraction of the water by-product, and concentration of the water in the reactor hampers the selectivity and rate of the reaction. Therefore, the methanol conversion per run is limited to 20%. The presence of HCl necessitates the use of corrosion-resistant reactors, and either glass linings or high-nickel alloys are used.

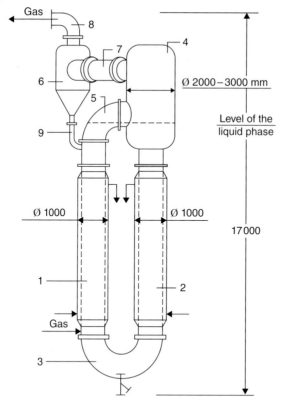

Figure 5.3 Schematic view of Enichem's slurry reactor. (Reactor diagram reproduced from [24].)

5.4
New Developments: Pharmaceutical Applications of Cu-Catalyzed Aerobic Oxidation Reactions

The processes described here, as well as the reactions described in Chapter 7, highlight the large-scale viability of liquid phase copper-catalyzed aerobic oxidation reactions. Alongside these developments, there has been growing academic interest in the development of new oxidation reactions for organic chemistry that may be performed under mild conditions. Copper-catalyzed aerobic oxidation reactions of this type have significant potential to impact the production of pharmaceuticals, agrochemicals, and related products due to the low cost and toxicity of copper and broad functional group tolerance often exhibited by copper catalysts [1]. This section highlights such applications, emphasizing those reported by pharmaceutical companies and performed on scales larger than typical laboratory applications.

5.4 New Developments: Pharmaceutical Applications of Cu-Catalyzed Aerobic Oxidation Reactions

Chan–Lam coupling reactions use Cu-based catalysts in the oxidative coupling of boronic acids (and related derivatives) with heteroatom nucleophiles (Scheme 5.4) [25]. These reactions are believed to proceed by an "organometallic" mechanism involving four key steps [26]: (a) transmetallation of an aryl group from the arylboronic acid to Cu^{II}; (b) oxidation of the aryl-Cu^{II} species by a second equivalent of CuX_2, resulting in the formation of an aryl-Cu^{III} species and Cu^{I}; (c) reductive elimination of Nu-Ar; and (d) aerobic oxidation to regenerate CuX_2. An application of this reaction was reported in 2014 by researchers at Bristol-Myers Squibb for the preparation of dicyclopropylamine (DCPA; Scheme 5.5) [27]. This secondary amine is particularly sensitive, and the success of the reaction highlights the mildness of the Chan–Lam coupling conditions. The reaction was performed on 60–80 kg scale and afforded high product yields (85–90%). The reactions were performed by bubbling a 5% O_2/N_2 gas mixture (to stay well below the limiting oxygen concentration [28]) into the acetonitrile solution while vigorously stirring the reaction mixture at 45 °C with a 50 mol% loading of $Cu(OAc)_2$ and 2,2′-bipyridyl (bpy) ligand. Portionwise addition of 5 × 0.3 equiv. of cyclopropylboronic acid gave a better yield than the single addition of 1.5 equiv. of cyclopropylboronic acid:

$$R\text{-B(OH)}_2 + \text{Nu-H} + 1/2\, O_2 \xrightarrow{Cu^{II}X_2} R\text{-Nu} + B(OH)_3$$

R = aryl, vinyl, Nu-H = alcohol or amine-derived nucleophile

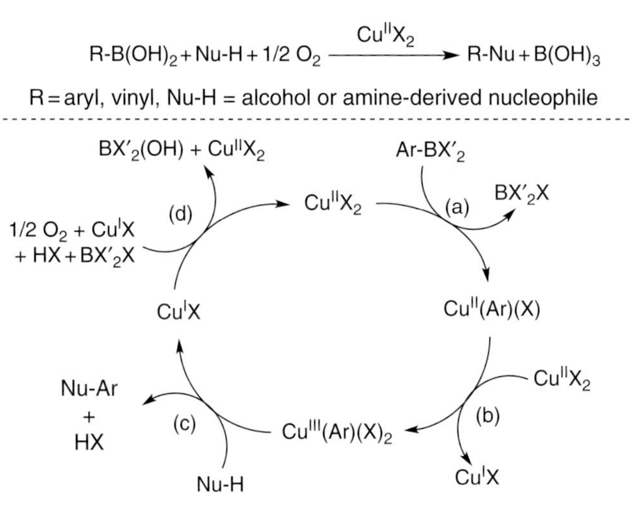

Scheme 5.4 Chan–Lam coupling reactions and a simplified catalytic cycle.

In 2013, AstraZeneca researchers reported an application of copper-catalyzed aerobic dehydrogenative aromatization in the 120 g scale synthesis of AZD8926, a drug candidate for the treatment of central nervous system disorders (Scheme 5.6) [29]. This aerobic dehydrogenation process was determined to be superior for large-scale implementation over other oxidation methods, such as a benzoquinone-mediated oxidation and various protocols using peroxide-based

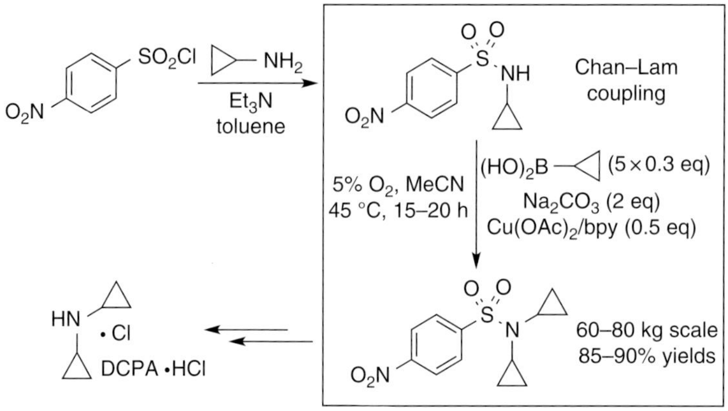

Scheme 5.5 Efficient preparation of DCPA•HCl outlined by researchers at Bristol-Myers Squibb.

oxidants. The reaction was performed by bubbling 5% O₂ through a stirring acetonitrile slurry containing the *in situ*-generated dihydropyrimidine intermediate, 5 mol% Cu(OAc)$_2$, and 1 equiv. of triethylamine at 10 °C. High-purity pyrimidine product was obtained in 60% yield on a 120 g scale over the three steps shown in Scheme 5.6:

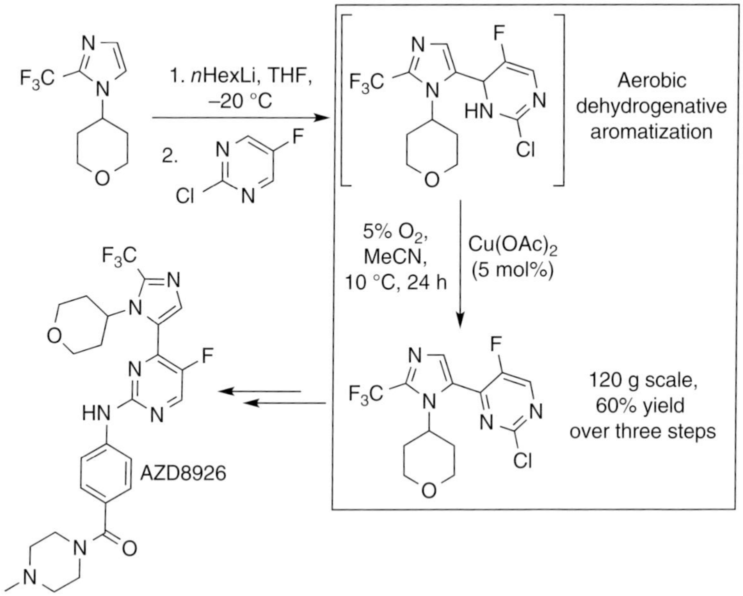

Scheme 5.6 Scalable process for the synthesis of AZD8926 outlined by researchers at AstraZeneca.

5.4 New Developments: Pharmaceutical Applications of Cu-Catalyzed Aerobic Oxidation Reactions

Inspired by a method developed by Nagasawa and coworkers (Eq. (5.6)) [30], researchers at Hoffmann-La Roche developed a scalable copper-catalyzed aerobic oxidative heterocyclization method to prepare 1,2,4-triazoles from benzonitrile and 2-aminopyridine derivatives (Schemes 5.7 and 5.8) [31]. Synthesis of the 7-bromo product (Scheme 5.7b) was performed on 120 g scale by stirring the reaction in neat benzonitrile at 130 °C for 23 h under a stream of 8% O_2. The protocol developed at Hoffmann-La Roche addressed two drawbacks of the originally reported reaction conditions: (a) the use of 1,2-dichlorobenzene as a solvent and (b) the use of a ZnI_2 cocatalyst, which led to difficult chromatographic purification of the product. The modified reaction conditions use neat benzonitrile as both a reagent and solvent, and the ZnI_2-free conditions enabled the product to be obtained in high yield and purity without chromatography:

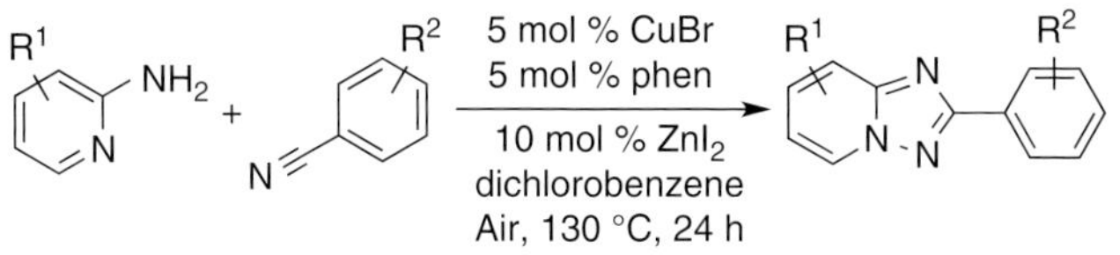

$$\text{(5.6)}$$

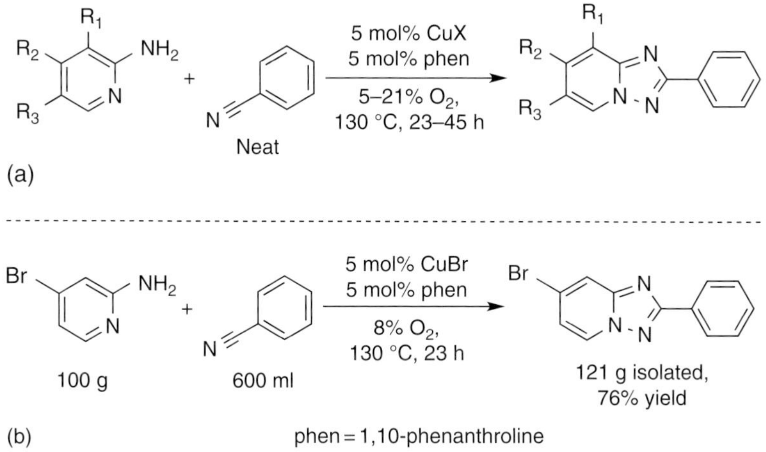

Scheme 5.7 Scalable synthesis of 1,2,4-triazoles developed at Hoffman-La Roche: (a) general reaction conditions and (b) scale-up example.

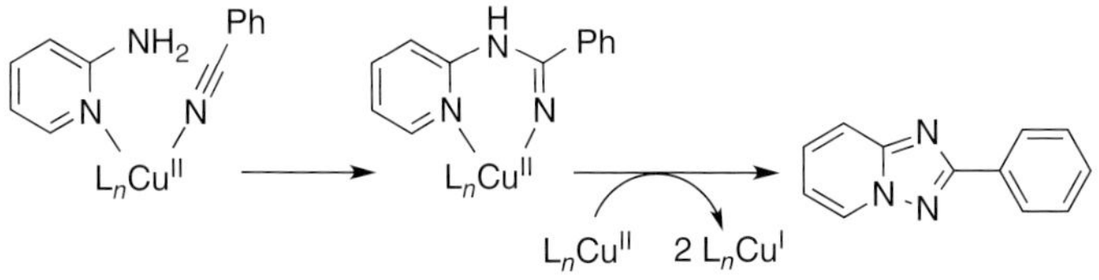

phen = 1,10-phenanthroline

Scheme 5.8 Proposed mechanism of 1,2,4-triazole formation.

New copper-catalyzed aerobic transformations continue to be developed in academia, and the developments described here highlight prospects for scale-up. In an application closely resembling Nagasawa's 1,2,4-triazole synthesis, Hajra and coworkers reported the synthesis of imidazo[1,2-a]pyridine (Schemes 5.9 and 5.10) [32]. They applied this method to a gram-scale synthesis of the acid reflux drug zolimidine (Scheme 5.9b). In a different application, MacMillan and coworkers reported a copper-catalyzed aerobic oxidative method for net α-amination of aldehydes, ketones, and esters (Schemes 5.11 and 5.12) [33]. The method features initial bromination of the α position, followed by *in situ* displacement of the bromide by various amine nucleophiles. The utility of this reaction was highlighted in one-pot routes to racemic analogs of two pharmaceutical compounds, Plavix and amfepramone (Scheme 5.11c,d):

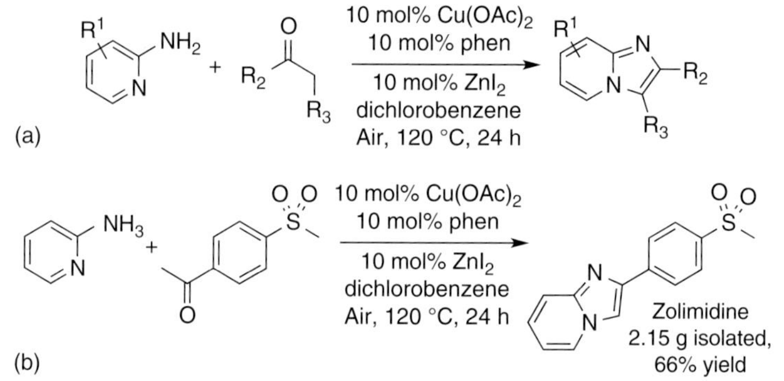

Scheme 5.9 Efficient synthesis of imidazo[1,2-a]pyridines: (a) general reaction conditions and (b) gram-scale synthesis of zolimidine.

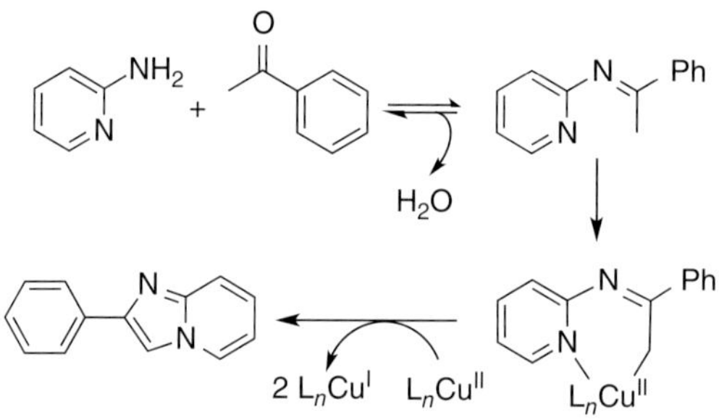

Scheme 5.10 Presumed mechanism of imidazo[1,2-a]pyridine formation.

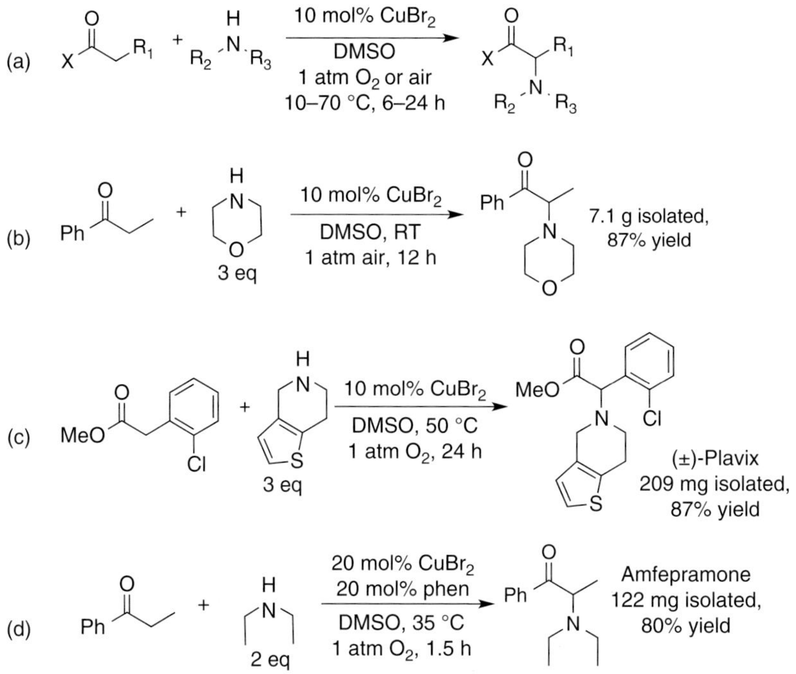

5.4 New Developments: Pharmaceutical Applications of Cu-Catalyzed Aerobic Oxidation Reactions | 81

Scheme 5.11 Aerobic Cu-catalyzed α-amination of aldehydes, ketones, and esters: (a) general reaction conditions, (b) multigram-scale coupling, (c) synthesis of Plavix, and (d) synthesis of amfepramone.

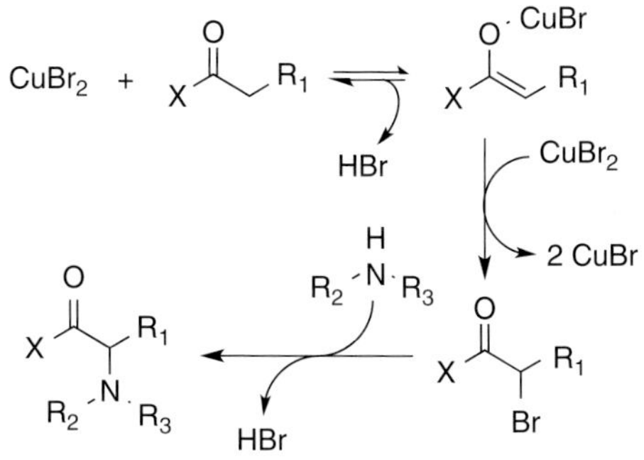

Scheme 5.12 Proposed mechanism for the aerobic α-amination of aldehydes, ketones, and esters.

References

1. (a) Punniyamurthy, T. and Rout, L. (2008) *Coord. Chem. Rev.*, **252**, 134–154; (b) Wendlandt, A.E., Suess, A.M., and Stahl, S.S. (2011) *Angew. Chem. Int. Ed.*, **50**, 11062–11087; (c) Allen, S.E., Walvoord, R.R., Padilla-Salinas, R., and Kozlowski, M.C. (2013) *Chem. Rev.*, **113**, 6234–6458.
2. (a) Mirica, L.M., Ottenwaelder, X., and Stack, T.D.P. (2004) *Chem. Rev.*, **104**, 1013–1045; (b) Lewis, E.A. and Tolman, W.B. (2004) *Chem. Rev.*, **104**, 1047–1076; (c) Hatcher, L.Q. and Karlin, K.D. (2006) *Adv. Inorg. Chem.*, **58**, 131–184.
3. (a) Weber, M., Weber, M., and Kleine-Boymann, M. (2012) in *Ullmann's Encyclopedia of Industrial Chemistry*, 5th edn (eds W. Gerhartz, Y.S. Yamamoto, F.T. Campbell, R. Pfefferkorn, and J.F. Rounasaville), Wiley-VCH Verlag GmbH, pp. 503–519; (b) Wallace, J. (2004) in *Kirk-Othmer Encyclopedia of Chemical Technology*, 5th edn (ed A. Seidel), John Wiley & Sons, Inc., pp. 747–756; (c) Weissermel, K. and Arpe, H.-J. (2008) *Industrial Organic Chemistry*, 4th edn, John Wiley & Sons, Inc., pp. 337–385.
4. (a) Toland, W.G. (1956) Manufacture of Monohydroxy Aromatic Compounds from Aromatic Carboxylic Acids. US Patent 2,762,838; (b) Toland, W.G. (1961) *J. Am. Chem. Soc.*, **83**, 2507–2512.
5. (a) Pearlman, M.B. (1955) Phenols from Aromatic Carboxylic Acids. US Patent 2,727,924; (b) Kaeding, W.W. and Lindblom, R.O. (1955) Catalytic Oxidation of Aromatic Carboxylic Acids to Phenols. US Patent 2,727,926; (c) Barnard, R.D. and Meyer, R.H. (1958) Production of Phenols from Aromatic Carboxylic Acids. US Patent 2,852,567; (d) Lam, C.T. and Shannon, D.M. (1986) Production of Phenols and Catalysts Therefor. US Patent 4,567,157; (e) Kaeding, W.W., Lindblom, R.O., and Temple, R.G. (1961) *Ind. Eng. Chem.*, **53**, 805–808.
6. Buijs, W., Frijns, L.H.B., and Offermanns, M.R.J. (1993) Method for the Preparation of a Phenol. US Patent 5,210,331.
7. (a) Kaeding, W.W. (1961) *J. Org. Chem.*, **26**, 3144–3148; (b) Kaeding, W.W. and Collins, G.R. (1965) *J. Org. Chem.*, **30**, 3750–3754; (c) Kaeding, W.W., Kerlinger, H.O., and Collins, G.R. (1965) *J. Org. Chem.*, **30**, 3754–3759.
8. Buijs, W. (1999) *J. Mol. Catal. A: Chem.*, **146**, 237–246.
9. Suess, A.M., Ertem, M.Z., Cramer, C.J., and Stahl, S.S. (2013) *J. Am. Chem. Soc.*, **135**, 9797–9804.
10. (a) Tundo, P. and Selva, M. (2002) *Acc. Chem. Res.*, **35**, 706–716; (b) Delledonne, D., Rivetti, F., and Romano, U. (2001) *Appl. Catal., A*, **221**, 241–251; (c) Keller, N., Rebmann, G., and Keller, V. (2010) *J. Mol. Catal. A: Chem.*, **317**, 1–18; (d) Chankeshwara, S.V. (2008) *Synlett*, **4**, 624–625.
11. Pacheco, M.A. and Marshall, C.L. (1997) *Energy Fuels*, **11**, 2–29.
12. (a) Buysch, H.J. (2012) in *Ullmann's Encyclopedia of Industrial Chemistry*, 5th edn (eds W. Gerhartz, Y.S. Yamamoto, F.T. Campbell, R. Pfefferkorn, and J.F. Rounasaville), Wiley-VCH Verlag GmbH, pp. 45–71; (b) Kreutzberger, C.B. (2004) in *Kirk-Othmer Encyclopedia of Chemical Technology*, 5th edn (ed A. Seidel), John Wiley & Sons, Inc., pp. 290–323.
13. Romano, U., Tesei, R., Mauri, M.M., and Rebora, P. (1980) *Ind. Eng. Chem. Prod. Res. Dev.*, **19**, 396–403.
14. Romano, U. and Rivetti, F. (1990) Improved process for preparing di-alkyl carbonates. Patent EP0366177B1.
15. Rivetti, F. and Romano, U. (1997) Procedure for the production of alkyl carbonates. US Patent 5,686,644.
16. Huang, S., Yan, B., Wang, S., and Ma, X. (2015) *Chem. Soc. Rev.*, **44**, 3079–3116.
17. (a) Bhattacharya, A.K. (1988) Pyridine ligands for preparation of organic carbonates. US Patent 4,761,467; (b) Raab, V., Merz, M., and Sundermeyer, J. (2001) *J. Mol. Catal. A: Chem.*, **175**, 51–63; (c) Mo, W., Xiong, H., Li, T., Guo, X., and Li, G. (2006) *J. Mol. Catal. A: Chem.*, **247**, 227–232; (d) Mo, W., Xiong, H., Hu, J., Ni, Y., and Li, G. (2010) *Appl. Organomet. Chem.*, **24**, 576–580.

18. Csihony, S., Mika, L.T., Vlad, G., Barta, K., Mehnert, C.P., and Horvath, I.T. (2007) *Collect. Czech. Chem. Commun.*, **72**, 1094–1106.
19. (a) Delledonne, D., Rivetti, F., and Romano, U. (1995) *J. Organomet. Chem.*, **488**, C15–C19; (b) Delledonne, D., Rivetti, F., and Romano, U. (1995) Catalytic procedure for the preparation of organic carbonates. US Patent 5,457,213.
20. (a) Curnutt, G.L. (1986) Process of preparing dihydrocarbyl carbonates using a nitrogen-containing coordination compound supoprted on activated carbon. US Patent 4,625,044; (b) Hu, J.-C., Cao, Y., Yang, P., Deng, J.-F., and Fan, K.-N. (2002) *J. Mol. Catal. A: Chem.*, **185**, 1–9; (c) Mo, W., Liu, H., Xiong, H., Li, M., and Li, G. (2007) *Appl. Catal., A*, **333**, 172–176; (d) Cao, Y., Hu, J.-C., Yang, P., Dai, W.-L., and Fan, K.-N. (2003) *Chem. Commun.*, 908–909.
21. Gelbein, A.P. and Khonsari, A.M. (1981) Catalyst and Process for Producing Aromatic Hydroxy Compounds. US Patent 4,277,630.
22. Romano, U., Tesei, R., Cipriani, G., and Micucci, L. (1980) Method for the preparation of esters of carbonic acid. US Patent 4,218,391.
23. Romano, U., Rivetti, F., and Muzio, N. (1982) Process for producing dimethylcarbonate. US Patent 4,318,862.
24. Paret, G., Donati, G., and Ghirardini, M. (1996) Process for producing dimethyl carbonate and apparatus suitable for such purpose. US Patent 5,536,864.
25. Qiao, J.X. and Lam, P.Y.S. (2011) *Synthesis*, **6**, 829–856.
26. King, A.E., Brunold, T.C., and Stahl, S.S. (2009) *J. Am. Chem. Soc.*, **131**, 5044–5045.
27. Mudryk, B., Zheng, B., Chen, K., and Eastgate, M.D. (2014) *Org. Process Res. Dev.*, **18**, 520–527.
28. Osterberg, P. M.; Niemeier, J. K.; Welch, C. J.; Hawkins, J. M.; Martinelli, J. R.; Johnson, T. E.; Root, T. W; Stahl, S. S. (2015) *Org. Process Res. Dev.* **19**, doi: 10.1021/op500328f
29. Witt, A., Teodorovic, P., Linderberg, M., Johansson, P., and Minidis, A. (2013) *Org. Process Res. Dev.*, **17**, 672–678.
30. Ueda, S. and Nagasawa, H. (2009) *J. Am. Chem. Soc.*, **131**, 15080–15081.
31. (a) Bartels, B., Fantasia, S.M., Flohr, A., Puentener, K., and Want, S. (2013) A process for the preparation of phenyl-triazolopyridine derivatives. Patent WO2013117610A1; (b) Bartels, B., Bolas, C.G., Cueni, P., Fantasia, S., Gaeng, N., and Trita, A.S. (2015) *J. Org. Chem.*, **80**, 1249–1257.
32. Bagdi, A.K., Rahman, M., Santra, S., Majee, A., and Hajra, A. (2013) *Adv. Synth. Catal.*, **355**, 1741–1747.
33. Evans, R.W., Zbieg, J.R., Zhu, S., Li, W., and MacMillan, D.W.C. (2013) *J. Am. Chem. Soc.*, **135**, 16074–16077.

6
Copper-Catalyzed Aerobic Alcohol Oxidation
Janelle E. Steves and Shannon S. Stahl

6.1
Introduction

The oxidation of alcohols to carbonyl compounds is one of the most common classes of oxidation reactions in organic chemistry. Various traditional reagents, such as Cr and Mn oxides, offer broad utility, practicality, and selectivity, but they also have undesirable features such as high cost, toxicity, and/or poor atom economy that represent significant drawbacks for large-scale applications [1]. Alternative reagents, such as TEMPO/NaOCl (TEMPO = 2,2,6,6-tetramethylpiperidinyl-*N*-oxyl) or pyridine•SO_3 [2], tend to be used on a large scale, but these methods also have drawbacks. For example, NaOCl is often supplied as a dilute aqueous solution that contributes to the reaction volume and process waste and/or interferes with downstream chemistry. Use of NaOCl can also lead to unwanted chlorination of organic compounds, such as alkenes or carbonyl compounds, introducing toxic impurities into pharmaceuticals or influencing the olfactory properties of fragrance and flavoring compounds. Aerobic alcohol oxidation is an attractive alternative to these methods because the use of O_2 as an oxidant typically affords only H_2O as a stoichiometric by-product [3]. Efforts to address safety hazards associated with mixtures of O_2 and organic solvents [4] make such methods increasingly viable for large-scale chemistry. In this chapter, we highlight copper-catalyzed methods for aerobic alcohol oxidation that rival traditional reagents and common process methods with respect to synthetic scope, chemoselectivity, and reaction efficiency. Complementary methods that use organic nitroxyl/NO_x-based cocatalyst systems and heterogeneous Pd, Pt, and Au catalysts for aerobic alcohol oxidation are presented in Chapters 15 [5] and 21 [6].

Many reports of aerobic alcohol oxidation involve homogeneous Pd- [7] and Ru-based [8] catalysts that are effective with benzylic, allylic, and aliphatic primary and secondary alcohols. These catalysts are often inhibited by coordinating functional groups such as heterocycles, amines, and oxygen- or sulfur-containing moieties and are capable of oxidizing alkenes. Efforts to develop scalable applications of Pd-based catalysts raised concerns about large-scale prospects for these

Liquid Phase Aerobic Oxidation Catalysis: Industrial Applications and Academic Perspectives,
First Edition. Edited by Shannon S. Stahl and Paul L. Alsters.
© 2016 Wiley-VCH Verlag GmbH & Co. KGaA. Published 2016 by Wiley-VCH Verlag GmbH & Co. KGaA.

methods [9]. On the other hand, significant advances have been made in copper-based catalyst systems that appear to overcome many of the limitations evident with Pd [10]. Broad synthetic utility, practicality, and low cost are key features of many of the Cu-based catalysts, suggesting they have the potential to find use in large-scale applications within the pharmaceutical, fragrance and flavoring, and fine chemicals industries.

6.2
Chemistry and Catalysis

Early reports of Cu-catalyzed aerobic alcohol oxidation focused on simple homogeneous complexes [11] with nitrogenous ligands, such as 2,2′-bipyridine and 1,10-phenanthroline. However, copper-catalyzed methods that include nitroxyl radical or azodicarboxylate cocatalysts exhibit significant synthetic advantages and have been widely investigated in the literature (Figure 6.1).

The synthetic scope of the homogeneous copper-only catalysts (i.e., that do not use a redox-active organic cocatalyst) is often limited to simple alkanols. Only recently have homogeneous [11e] and heterogeneous [12] catalyst systems been reported that show good utility with heteroatom-functionalized substrates. A noteworthy example was reported by Lumb, Arndtsen, and coworkers and involves a Cu^I/DBED/DMAP catalyst system (DBED = N,N'-di-*tert*-butyl-ethylenediamine; DMAP = 4-dimethylaminopyridine) (Figure 6.2) [11e]. This protocol offers mild and selective oxidation of primary and secondary alcohols to aldehydes and ketones, without overoxidation of aldehydes to carboxylic acids. Recent mechanistic studies suggest that this may not be a true Cu-only catalyst system. *In situ* oxidation of the DBED ligand appears to generate an open-shell nitroxyl species that contributes to the efficacy of these reactions (S.D. McCann et al., unpublished results).

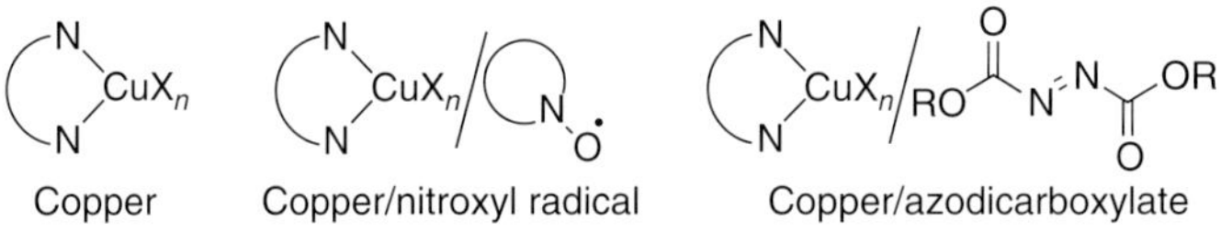

Figure 6.1 Catalyst types developed for copper-catalyzed aerobic alcohol oxidation.

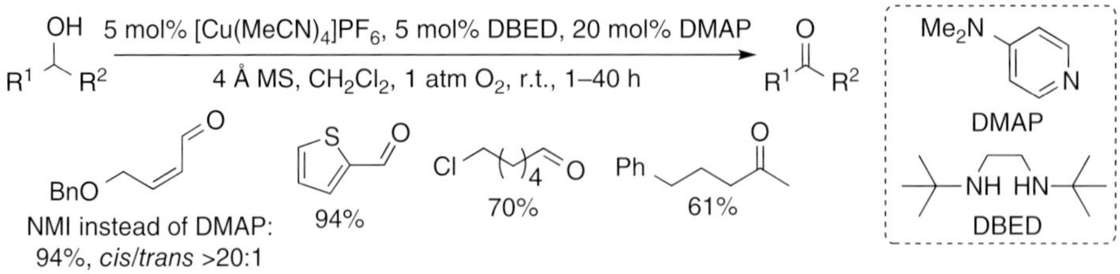

Figure 6.2 Scope of copper(I)/DBED-catalyzed aerobic alcohol oxidation.

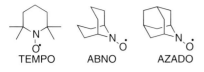

Figure 6.3 Common nitroxyl radicals used in copper-catalyzed aerobic alcohol oxidation.

Copper salts used in combination with redox-active organic molecules have proven to be highly effective catalyst systems for aerobic alcohol oxidation [3a, i]. TEMPO has been the most widely used cocatalyst for this purpose, while unique opportunities have been identified with less sterically encumbered organic nitroxyls, such as ABNO and AZADO (Figure 6.3; discussed later in detail). These catalysts are practically appealing because they may be generated *in situ* by combining the appropriate Cu source, ligand, TEMPO, and any other additives.

In the 1960s, Brackmann and Gaasbeek demonstrated that a (phen)Cu^{II}/di-*tert*-butyl nitroxide catalyst could promote the oxidation of methanol to formaldehyde in alkaline solution [13]. It was nearly two decades later that Semmelhack reported the first synthetic methodological study of Cu/TEMPO-catalyzed aerobic alcohol oxidation using CuCl as the copper source in DMF as the solvent [14]. This Cu/TEMPO catalyst system is effective with activated (i.e., benzylic, allylic) primary alcohols, but aliphatic alcohols require stoichiometric Cu/TEMPO. Knochel [15], Sheldon [16], and Punniyamurthy [17] then demonstrated the utility of nitrogen-based ligands, such as 2,2′-bipyridine, and other solvents to stabilize the copper catalyst and enable moderate catalytic activity with aliphatic alcohols. Following these and other literature reports [3i], Kumpulainen and Koskinen [18] and Hoover and Stahl [19] reported Cu/TEMPO catalyst systems that exhibit broad scope and functional group compatibility and show good activity with both activated and unactivated primary alcohols. Hoover and Stahl showed that the best activity is obtained by using a Cu^{I} source instead of Cu^{II}, and they traced the origin of this observation to the *in situ* generation of a reactive Cu^{II}–OH species upon oxidation of Cu^{I} by O$_2$. Collectively, the copper/TEMPO catalysts exhibit very high selectivity for primary alcohol oxidation. Steves and Stahl then showed that the replacement of TEMPO with a less sterically hindered nitroxyl, such as ABNO (cf. Figure 6.3), enables very efficient oxidation of both primary and secondary alcohols [20]. Iwabuchi and coworkers expanded upon these observations by using AZADO as the nitroxyl cocatalyst [21].

A collection of results obtained with the most effective catalyst systems is summarized in Figure 6.4. Noteworthy examples include the oxidation of α-aminoalcohols with no loss in enantiopurity and the oxidation of *cis*-allylic alcohols without *Z:E* isomerization. Stahl also demonstrated the chemoselective oxidation of primary diols to form lactones [22]. While ABNO provides efficient oxidation of symmetric diols, TEMPO discriminates between subtle steric differences in nonsymmetrical substituted diols. Cu/nitroxyl catalysts have also been applied to a variety of tandem reactions [23], perhaps the most noteworthy of which is the conversion of primary alcohols to nitriles via *in situ* condensation of ammonia with the aldehyde and subsequent dehydrogenation of the primary imine to the nitrile (Figure 6.5) [24].

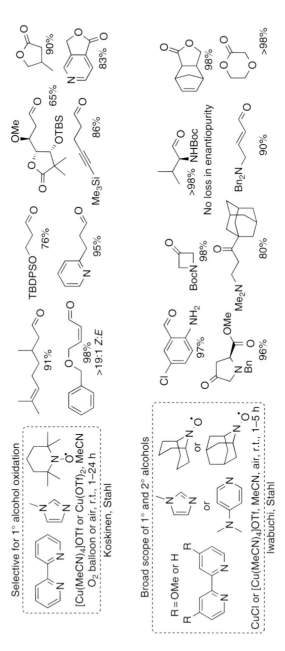

Figure 6.4 Broad-scope Cu/nitroxyl catalyst systems for aerobic alcohol oxidation.

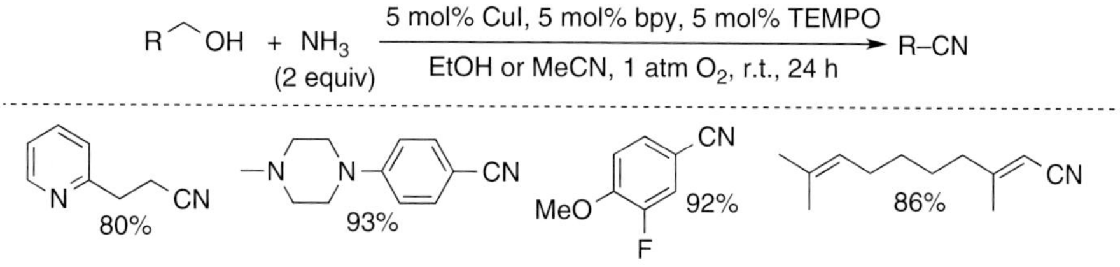

Figure 6.5 Aerobic oxidation of primary alcohols to nitriles by Huang and coworkers [24b].

Markó and coworkers have shown that di-*tert*-butyl azodicarboxylate (DBAD) is a useful redox cocatalyst in combination with Cu for aerobic alcohol oxidation [25]. Overall, the conditions and reaction scope are not quite as favorable as those observed with the more recent Cu/nitroxyl systems, but primary and secondary activated and unactivated alcohols with diverse functional groups undergo effective oxidation using a catalyst consisting of (phen)Cu^I/DBAD (Figure 6.6).

The Cu/TEMPO catalyst system has been the subject of considerable mechanistic investigation. Initial reports of the Cu/TEMPO catalyst drew mechanistic analogies to the enzyme galactose oxidase, in which a coordinated tyrosyl radical ligand mediates H-atom abstraction from a Cu-bound alkoxide (Figure 6.7) [26]. Early mechanistic studies [27] showed that kinetic isotope effect (KIE) values with Cu/TEMPO catalysts and galactose oxidase are similar and led to the proposal that Cu/TEMPO-mediated alcohol oxidation proceeds via intramolecular abstraction of an H atom by and of a η^2-coordinated TEMPO.

Subsequent kinetic, spectroscopic, and computational studies, however, support a different mechanistic pathway [28, 29]. A simplified version of the mechanism features oxidation of Cu^I and TEMPOH by O$_2$ to afford a Cu^{II}–OH species and TEMPO (Scheme 6.1). The copper hydroxide and nitroxyl radical

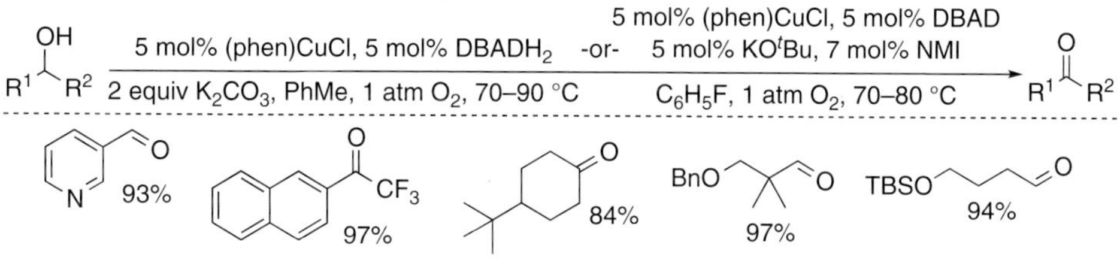

Figure 6.6 Cu/DBAD-catalyzed aerobic oxidation reported by Markó.

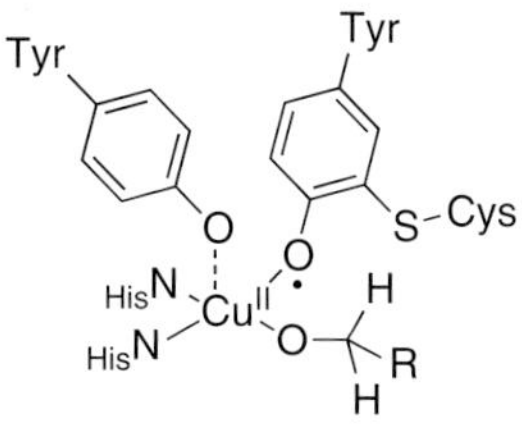

Figure 6.7 Active site structure of galactose oxidase.

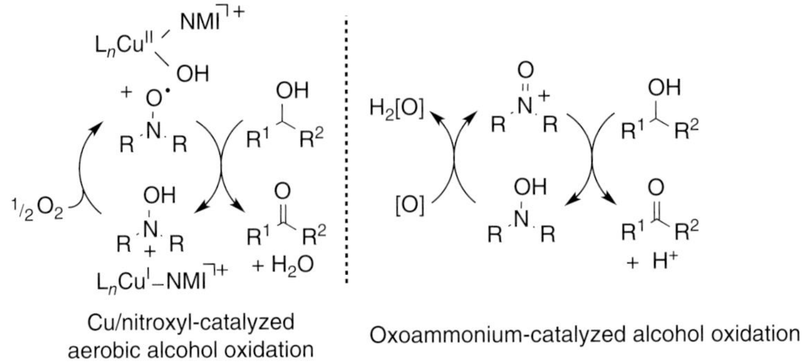

Scheme 6.1 Simplified mechanism of Cu/TEMPO- and oxoammonium-catalyzed aerobic alcohol oxidation.

then serve as cooperative one-electron oxidants in the alcohol oxidation step, forming H_2O as the by-product and regenerating Cu^I and TEMPOH. This mechanism is conceptually related to classical oxoammonium-catalyzed alcohol oxidations [30]; however, the nitroxyl does not need to be oxidized to the oxoammonium redox state to promote alcohol oxidation. The latter feature is noteworthy because it enables O_2 to serve as an efficient stoichiometric oxidant.

Experimental and computational studies suggest that the alcohol oxidation step involves intramolecular hydrogen transfer from an alkoxide ligand to an η^1 O-bound TEMPO ligand [28c, 29]. The lowest-energy six-membered transition state resembles an Oppenauer oxidation pathway, and the calculated barriers are consistent with selective 1° alcohol oxidation with Cu/TEMPO (Figure 6.8) [28c]. For example, the barrier for oxidation of isopropanol is 6.3 kcal/mol higher than for *n*-propanol. Use of ABNO instead of TEMPO results in much lower transition state energies and nearly identical barriers for both 1° and 2° alcohol oxidation (Figure 6.8). These observations are consistent with experimental data for the Cu/ABNO catalyst system, which is highly effective for the oxidation of both 1° and 2° alcohols.

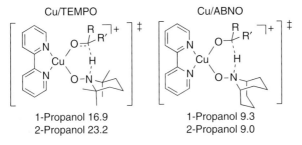

Figure 6.8 Transition state energies providing insight into the selectivity differences between TEMPO and ABNO.

6.3 Prospects for Scale-Up

Cu-only-, Cu/nitroxyl-, and Cu/azodicarboxylate-catalyzed aerobic alcohol oxidation methods have been the subject of interest within the industry. To our knowledge, however, the only commercial-scale copper-catalyzed alcohol oxidation has been applied to the production of vanillin. Vanillin may be isolated directly from vanilla beans and is broadly used in food applications, the fragrance industry, and the pharmaceutical and agrochemical applications [31]. Worldwide demand has contributed to the development of synthetic routes to vanillin. Most synthetic vanillin is currently produced by Rhodia (formerly Rhône-Poulenc), starting with guaiacol. Regioselective $S_E Ar$ reaction of guaiacol and glyoxylic acid produces a mandelic acid intermediate, which then undergoes Cu-catalyzed aerobic alcohol oxidation and decarboxylation to produce vanillin (Scheme 6.2) [32]. The guaiacol process yields vanillin with consistent high quality and sufficient volume to meet demand. At present, this process is advantageous relative to alternate routes from waste lignosulfonates obtained from the pulp and paper industry and an *ortho*-chloronitrobenzene (OCNB) process that has environmental concerns [33].

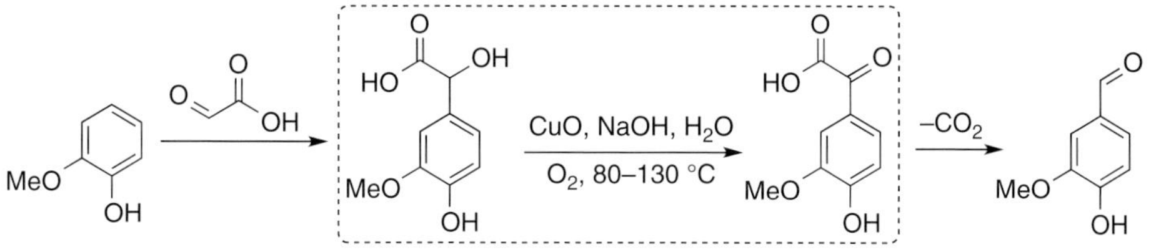

Scheme 6.2 Production of vanillin via the guaiacol process.

The selectivity, activity, and broad scope of recently reported Cu/nitroxyl-catalyzed aerobic alcohol oxidation protocols make these methods well suited for application on both laboratory and process scale in the pharmaceutical and fine chemicals industry. The early catalyst systems of Semmelhack et al. [14] and Sheldon [16] have attracted industrial interest, albeit not necessarily for commercial production.

BASF is a major producer of vitamin A, a carotenoid which is found naturally in various foods and is important for vision and for tissue growth and differentiation. Carotenoid derivatives have also been used as pharmaceuticals and as colorants for a wide variety of foods. The synthesis of vitamin A by BASF involves the Wittig coupling of an ylide and aldehyde to form the desired vitamin A product (Scheme 6.3) [34], and BASF explored variations of the Semmelhack catalyst system for the oxidation of carotenoid precursors [35]. The high copper and TEMPO loading originally reported by Semmelhack (10 mol%) were lowered to make the process more suitable for industrial scale, while mild temperatures, bubbling O_2, and the use of DMF as a solvent were maintained in the modified procedure

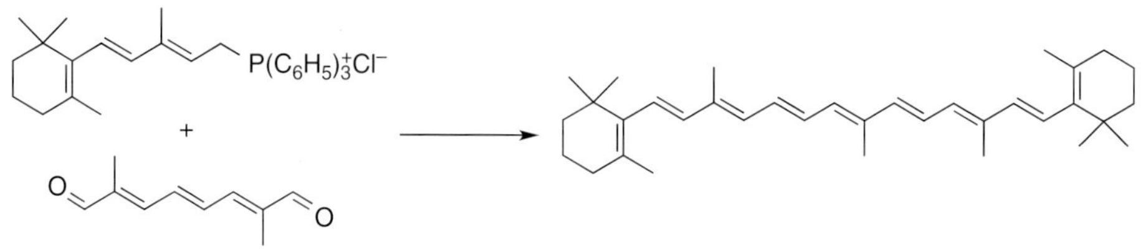

Scheme 6.3 BASF synthesis of vitamin A.

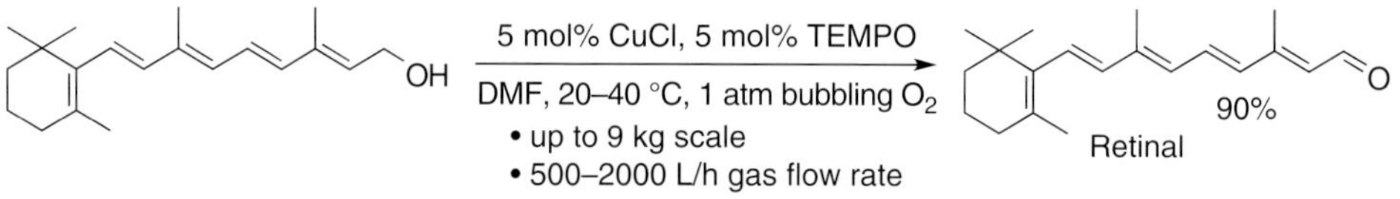

Scheme 6.4 BASF adaptation of Cu/TEMPO-catalyzed aerobic alcohol oxidation.

(Scheme 6.4). Various acetal alcohols, including a retinol precursor en route to vitamin A, were oxidized in very good to excellent yield on a scale up to 9 kg. Prior to product isolation, nitrogen was introduced to the reaction mixture to remove oxygen and avoid flammability of the organic layer during the aqueous workup.

DSM explored optimization of the Sheldon method in 2006, prior to reports of the improved Cu/nitroxyl catalyst systems described earlier [36]. The Cu loading was lowered relative to the original report, and $CuBr_2$ was replaced with metallic Cu powder. In addition, bpy loading was also lowered to 2.5 mol%, and a less expensive base, NaOH, was employed instead of KO^{t}Bu. It was necessary to increase the TEMPO loading, however, to maintain good catalyst activity. Various benzylic, allylic, and aliphatic alcohols were oxidized on 50 mmol scale in fair to excellent yield under ambient conditions (Figure 6.9). While reactions with electron-rich benzylic alcohols proceeded to the corresponding aldehydes in high conversion and selectivity, reactions with electron-poor benzylic alcohols afforded lower selectivity and overoxidized products.

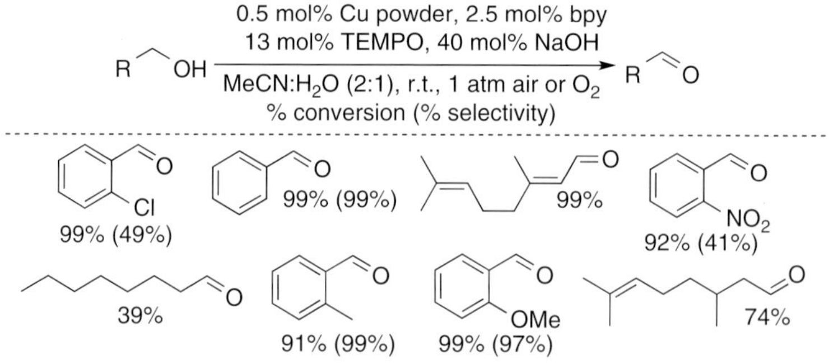

Figure 6.9 Cu/TEMPO-catalyzed aerobic alcohol oxidation by DSM.

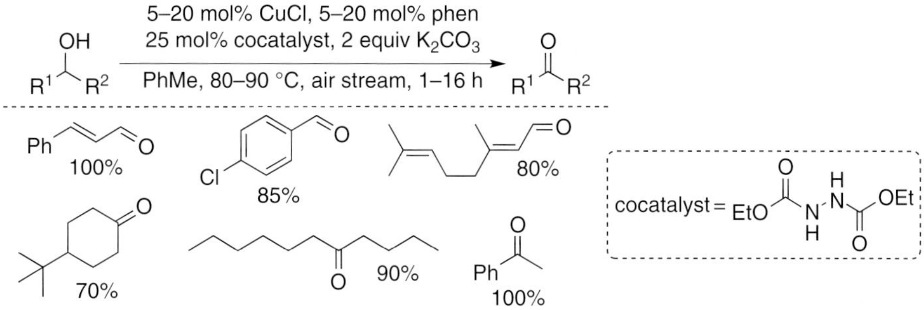

Figure 6.10 Zeneca application of Cu/DBAD-catalyzed aerobic alcohol oxidation.

Markó's Cu/DBAD catalyst [25] has also been explored for the oxidation of primary and secondary alcohols to aldehydes, acids, and ketones. By employing K_2CO_3 as an inexpensive alternative to KO^{t}Bu and increasing the catalyst loading, Zeneca Pharmaceuticals (now AstraZeneca) achieved oxidation of primary and secondary benzylic and aliphatic alcohols to aldehydes and ketones under a stream of air instead of 1 atm of pure O_2 (Figure 6.10) [37a]. Reactions were performed on up to 35–50 g scale. Zeneca also reported the oxidation of primary alcohols to acids under similar conditions without DBAD [37b, c].

Future large-scale applications will benefit from recent studies focused on addressing safety concerns associated with mixtures of organic solvents and oxygen gas. A team of academic and industrial researchers from the pharmaceutical industry has reported limiting oxygen concentrations (LOCs) for a series of nine commonly used organic solvents [4]. These values establish useful benchmarks for consideration of partial oxygen pressures that would be safe to use on a large scale. In many cases, safety hazards are readily addressed by using continuous-flow reaction methods (e.g., by facilitating the use of high pressures of dilute oxygen gas). Several recent studies have performed Cu-catalyzed aerobic alcohol oxidation reactions in continuous flow [10], and Chapter 23 highlights diverse applications of continuous liquid phase aerobic oxidation reactions. Batch reactions have been performed under nonflammable conditions either by using a dilute source of O_2, thereby remaining below the LOC [38], or by using a high-flash-point solvent (e.g., N-methylpyrrolidone) and maintaining the reaction temperature well below the flash point of the solvent [39].

6.4 Conclusions

Copper-catalyzed aerobic alcohol oxidation has proven to be a key step in the production of vanillin, but other commercial-scale applications have not yet been realized. Nevertheless, the industrial interest in some of the early copper/cocatalyst systems for aerobic oxidation bodes well for future applications. Recent academic studies have led to significantly improved catalyst systems for aerobic alcohol oxidation. The synthetic scope, selectivity, functional group compatibility, and efficiency of these methods now match or exceed traditional oxidation methods. The complementary Cu/TEMPO and Cu/ABNO catalyst systems are particularly promising, and it seems only a matter of time before these or related methods are used in large-scale applications within the pharmaceutical and fine chemicals industries.

References

1. (a) Tojo, G. and Fernández, M. (2010) *Oxidation of Alcohols to Aldehydes and Ketones*, Springer, New York; (b) Tojo, G. and Fernández, M. (2010) *Oxidation of Primary Alcohols to Carboxylic Acids*, Springer, New York.
2. (a) Brown Ripin, D.H., Caron, S., Dugger, R.W., Ruggeri, S.G., and Ragan, J.A. (2006) *Chem. Rev.*, **106**, 2943–2989; (b) Carey, J.S., Laffan, D., Thomson, C., and Williams, M.T. (2006) *Org. Biomol. Chem.*, **4**, 2337–2347; (c) Alfonsi, K., Colberg, J., Dunn, P.J., Fevig, T., Jennings, S., Johnson, T.A., Kleine, H.P., Knight, C., Nagy, M.A., Perry, D.A., and Stefaniak, M. (2008) *Green Chem.*, **10**, 31–36.
3. For reviews of aerobic alcohol oxidation, see: (a) Sheldon, R.A., Arends, I.W.C.E., ten Brink, G.-J., and Dijksman, A. (2002) *Acc. Chem. Res.*, **35**, 774–781; (b) Mallat, T. and Baiker, A. (2004) *Chem. Rev.*, **104**, 3037–3058; (c) Stahl, S.S. (2004) *Angew. Chem. Int. Ed.*, **43**, 3400–3420; (d) Zhan, B.-Z. and Thompson, A. (2004) *Tetrahedron*, **60**, 2917–2935; (e) Schultz, M.J. and Sigman, M.S. (2006) *Tetrahedron*, **62**, 8227–8241; (f) Matsumoto, T., Ueno, M., Wang, N., and Kobayashi, S. (2008) *Chem. Asian J.*, **3**, 196–214; (g) Gligorich, K.M. and Sigman, M.S. (2009) *Chem. Commun.*, 3854–3867; (h) Parmeggiani, C. and Cardona, F. (2012) *Green Chem.*, **14**, 547–564; (i) Ryland, B.L. and Stahl, S.S. (2014) *Angew. Chem. Int. Ed.*, **53**, 8824–8838.
4. See Chapter 23 by Gemoets *et al.* [Noel: currently chapter 24] and the following: Osterberg, P.M., Niemeier, J.K., Welch, C.J., Hawkins, J.M., Martinelli, J.R., Johnson, T.E., Root, T.W., and Stahl, S.S. (2015) *Org. Process Res. Dev.*, **19**, 1537–1543.
5. See Chapter 15 by Zultanski, and Stahl.
6. See Chapter 21 by Della Pina, Falletta, and Rossi.
7. (a) Nishimura, T., Onoue, T., Ohe, K., and Uemura, S. (1999) *J. Org. Chem.*, **64**, 6750–6755; (b) Schultz, M.J., Park, C.C., and Sigman, M.S. (2002) *Chem. Commun.*, 3034–3035; (c) ten Brink, G.J., Arends, I.W.C.E., Hoogenraad, M., Verspui, G., and Sheldon, R.A. (2003) *Adv. Synth. Catal.*, **345**, 1341–1352; (d) Mueller, J.A., Goller, C.P., and Sigman, M.S. (2004) *J. Am. Chem. Soc.*, **126**, 9724–9734; (e) Schultz, M.J., Hamilton, S.S., Jensen, D.R., and Sigman, M.S. (2005) *J. Org. Chem.*, **70**, 3343–3352; (f) Bailie, D.S., Clendenning, G.M.A., McNamee, L., and Muldoon, M.J. (2010) *Chem. Commun.*, **46**, 7238–7240.
8. (a) Dijksman, A., Marino-Gonzalez, A., Payeras, A.M.I., Arends, I.W.C.E., and Sheldon, R.A. (2001) *J. Am. Chem. Soc.*, **123**, 6826–6833; (b) Lenz, R. and Ley, S.V. (1997) *J. Chem. Soc., Perkin Trans. 1*, 3291–3292; (c) Hasan, M., Musawir, M., Davey, P.N., and Kozhevnikov, I.V. (2002) *J. Mol. Catal. A: Chem.*, **180**, 77–84; (d) Mizuno, N. and Yamaguchi, K. (2008) *Catal. Today*, **132**, 18–26.
9. Ye, X., Johnson, M.D., Diao, T., Yates, M.H., and Stahl, S.S. (2010) *Green Chem.*, **12**, 1180–1186.
10. For example, a comparison of process-scale issues with Pd and Cu is evident from ref. 9 and the following: (a) Greene, J.F., Hoover, J.M., Mannel, D.S., Root, T.W., and Stahl, S.S. (2013) *Org. Process Res. Dev.*, **17**, 1247–1251; (b) Greene, J.F., Preger, Y., Root, T.W., and Stahl, S.S. (2015) *Org. Process Res. Dev.*, **19**, 858–864.

11. See, for example: (a) Liu, X., Qiu, A.M., and Sawyer, D.T. (1993) *J. Am. Chem. Soc.*, **115**, 3239–3243 (b) Tsai, W.W., Liu, Y.H., Peng, S.M., and Liu, S.T. (2005) *J. Organomet. Chem.*, **690**, 415–421; (c) Cho, C.S. and Oh, S.G. (2007) *J. Mol. Catal. A: Chem.*, **276**, 205–210; (d) Liang, L., Rao, G., Sun, H.-L., and Zhang, J.-L. (2010) *Adv. Synth. Catal.*, **352**, 2371–2377; (e) Xu, B., Lumb, J.-P., and Arndtsen, B.A. (2015) *Angew. Chem. Int. Ed.*, **54**, 4208–4211.
12. See, for example:(a) Naik, R., Joshi, P., and Deshpande, R.K. (2004) *Catal. Commun.*, **5**, 195–198; (b) Haider, P. and Baiker, A. (2007) *J. Catal.*, **248**, 175–187; (c) Kantam, M.L., Arundhathi, R., Likhar, P.R., and Damodara, D. (2009) *Adv. Synth. Catal.*, **351**, 2633–2637; (d) Han, C.Y., Yu, M., Sun, W.J., and Yao, X.Q. (2011) *Synlett*, **16**, 2363–2368; (e) Kalbasi, R.J., Nourbakhsh, A.A., and Zia, M.Y. (2012) *J. Inorg. Organomet. Polym.*, **22**, 536–542.
13. Brackmann, W. and Gaasbeek, C.J. (1966) *Recl. Trav. Chim. Pays-Bas*, **85**, 221–256.
14. Semmelhack, M.F., Schmid, C.R., Cortes, D.A., and Chou, C.S. (1984) *J. Am. Chem. Soc.*, **106**, 3374–3376.
15. Betzemeier, B., Cavazzini, M., Quici, S., and Knochel, P. (2000) *Tetrahedron Lett.*, **41**, 4343–4346.
16. Gamez, P., Arends, I.W.C.E., Sheldon, R.A., and Reedijk, J. (2004) *Adv. Synth. Catal.*, **346**, 805–811.
17. Velusamy, S., Srinivasan, A., and Punniyamurthy, T. (2006) *Tetrahedron Lett.*, **47**, 923–926.
18. Kumpulainen, E.T.T. and Koskinen, A.M.P. (2009) *Chem. Eur. J.*, **15**, 10901–10911.
19. Hoover, J. M. and Stahl, S. S. (2011) *J. Am. Chem. Soc.* **133**, 16901-16910.
20. Steves, J.E. and Stahl, S.S. (2013) *J. Am. Chem. Soc.*, **135**, 15742–15745.
21. Sasano, Y., Nagasawa, S., Yamazaki, M., Shibuya, M., Park, J., and Iwabuchi, Y. (2014) *Angew. Chem. Int. Ed.*, **53**, 3236–3240.
22. Xie, X. and Stahl, S.S. (2015) *J. Am. Chem. Soc.*, **137**, 3767–3770.
23. For thorough presentation of such applications, see ref. 3i.
24. (a) Tao, C., Liu, F., Zhu, Y., Liu, W., and Cao, Z. (2013) *Org. Biomol. Chem.*, **11**, 3349–3354; (b) Yin, W., Wang, C., and Huang, Y. (2013) *Org. Lett.*, **15**, 1850–1853; (c) Dornan, L.M., Cao, Q., Flanagan, J.C.A., Crawford, J.J., Cook, M.J., and Muldoon, M.J. (2013) *Chem. Commun.*, **49**, 6030–6032.
25. (a) Markó, I.E., Giles, P.R., Tsukazaki, M., Brown, S.M., and Urch, C.J. (1996) *Science*, **274**, 2044–2046; (b) Markó, I., Gautier, A., Chellé-Regnaut, I., Giles, P.R., Tsukazaki, M., Urch, C.J., and Brown, S.M. (1998) *J. Org. Chem.*, **63**, 7576–7577; (c) Markó, I.E., Gautier, A., Dumeunier, R.L., Doda, K., Philippart, F., Brown, S.M., and Urch, C.J. (2004) *Angew. Chem. Int. Ed.*, **43**, 1588–1591; (d) Nishii, T., Ouchi, T., Matsuda, A., Matsubara, Y., Haraguchi, Y., Kawano, T., Kaku, H., Horikawa, M., and Tsunoda, T. (2012) *Tetrahedron Lett.*, **53**, 5880–5882.
26. Whittaker, J.W. (2003) *Chem. Rev.*, **103**, 2347–2363.
27. Dijksman, A., Arends, I.W.C.E., and Sheldon, R.A. (2003) *Org. Biomol. Chem.*, **1**, 3232–3237.
28. (a) Hoover, J.M., Ryland, B.L., and Stahl, S.S. (2013) *J. Am. Chem. Soc.*, **135**, 2357–2367; (b) Hoover, J.M., Ryland, B.L., and Stahl, S.S. (2013) *ACS Catal.*, **3**, 2599–2605; (c) Ryland, B.L., McCann, S.D., Brunold, T.C., and Stahl, S.S. (2014) *J. Am. Chem. Soc.*, **136**, 12166–12173.
29. (a) Michel, C., Belanzoni, P., Gamez, P., Reedijk, J., and Baerends, E.J. (2009) *Inorg. Chem.*, **48**, 11909–11920; (b) Belanzoni, P., Michel, C., and Baerends, E.J. (2011) *Inorg. Chem.*, **50**, 11896–11904.
30. See Chapter 15 by Zultanski and Stahl, and the following: (a) Bailey, W.F., Bobbitt, J.M., and Wiberg, K.B. (2007) *J. Org. Chem.*, **72**, 4504–4509; (b) Bobbitt, J.M., Bartelson, A.L., Bailey, W.F., Hamlin, T.A., and Kelly, C.B. (2014) *J. Org. Chem.*, **79**, 1055–1067.
31. Vidal, J.-P. (2006) in *Kirk-Othmer Encyclopedia of Chemical Technology* (eds A. Seidel and M. Bickford), John Wiley & Sons, Inc., pp. 1–10.

32. (a) Bjørsvik, H.-R., Liguori, L., and Minisci, F. (2000) *Org. Process Res. Dev.*, **4**, 534–543; (b) Buddoo, S. (2003) Process for the preparation of vanillin from a mixed *m*-cresol/*p*-cresol stream. Dissertation. Port Elizabeth Technikon.
33. Giannotta, D. (2004) Vanillin: The OCNB route—a sustainable option? Proceedings of the IFEAT International Conference, Portugal, Europe, October, 2004.
34. Van Arnum, S.D. (2000) in *Kirk-Othmer Encyclopedia of Chemical Technology* (eds A. Seidel and M. Bickford), John Wiley & Sons, Inc., pp. 1–9.
35. (a) Knaus, G. H. and Paust, J. (1987) Process for the preparation of polyene aldehydes. DE Patent 3,705,785, filed Feb. 24, 1987, and published on Sep. 1, 1988; (b) Krause, W. and Paust, J. (1995) Process for preparing retinal by oxidation of retinol with oxygen in the presence of 4-hydroxy-2,2,6,6-tetramethyl-piperidin-1-oxyl and copper(I) chloride. EP Patent 718,283, filed Dec. 18, 1995, and published on Jun. 26, 1996; (c) Ernst, H. and Heinrich, K. (2002) Process for preparing 2,7-dimethyl-2,4,6-octatrienal monoacetals. US Patent 128,519, filed Mar. 12, 2002, and published Sep. 12, 2002; (d) Ernst, H. and Heinrich, K. (2002) Process for preparing polyene aldehyde monoacetals. EP Patent 325,919, filed Dec. 13, 2002, and published on Jul. 9, 2003.
36. DSM Fine Chemicals Austria NFG GMBH & Co. KG. (2006) A process for producing aldehydes from alcohols by copper-catalyzed oxidation. AT Patent 501,929, filed on May 17, 2005, and published on Dec. 15, 2006.
37. (a) Urch, C. J., Brown, S.M., Markó, I.E., Giles, P.R., and Tuskazaki, M. (1996) Preparation of aldehydes or ketones from alcohols. WO Patent 3033, filed Jul. 4, 1996, and published on Jan. 30, 1997; (b) Markó, I.E., Giles, P.R., and Tuskazaki, M. (1996) Oxidative process and catalysts for the preparation of carboxylic acids from alcohols. GB Patent 2,312,208, filed Apr. 19, 1996, and published on Oct. 22, 1997; (c) Urch, C.J., Brown, S.M., Markó, I.E., Giles, P.R., and Tuskazaki, M. (1997) Catalytic oxidative process for preparing carboxylic acids from alcohols or aldehydes. WO Patent 51,654, filed May 14, 1997, and published on Nov. 19, 1998.
38. The following reference describes a Cu-catalyzed aerobic oxidative coupling reaction: Mudryk, B., Zheng, B., Chen, K., and Eastgate, M.D. (2014) *Org. Process Res. Dev.*, **18**, 520–527.
39. Steves, J.E., Preger, Y., Martinelli, J.R., Welch, C.J., Root, T.W., Hawkins, J.M., and Stahl, S.S. (2015) *Org. Process Res. Dev.*, **19**, 1548–1553.

7
Phenol Oxidations

7.1
Polyphenylene Oxides by Oxidative Polymerization of Phenols
Patrick Gamez

7.1.1
Introduction

In 1955, Terent'ev and Mogilyanskii reported the catalytic oxidative coupling of aniline to azobenzene with a yield of 88%, mediated by copper(I) chloride in pyridine (which acts as both a metal ligand and solvent) in the presence of molecular oxygen [1]. This system was subsequently used for the generation of various conjugated and nonconjugated main-chain aromatic azo polymers from primary aromatic diamines [2]. In contrast, the use of $CuCl-O_2$ with phenol produces tars [3] as a result of the inherent properties of this reactant. Indeed, while the metal salt produces N-centered aniline radicals that dimerize to form azobenzene [4], it gives rise to the creation of several phenol radical species (Figure 7.1a) that couple to yield insoluble black materials through C–C and C–O coupling reactions, hydroxylations, and oxidations of substituted phenols (Figure 7.1b) [5]. In 1956, Süs and coworkers [6], followed by Dewar and James 2 years later [7], used *ortho*-protected phenols in attempts to generate poly-2,6-disubstituted-1,4-phenylene ethers by thermal decomposition of the corresponding benzene-1,4-diazooxides; however, this method only functioned with limited success. Concurrently, Staffin and Price, inspired by an earlier work reported by Hunter and Morse [8], developed a procedure to produce poly-2,6-dimethyl-1,4-phenylene ether via displacement of the bromine in 4-bromo-2,6-dimethylphenol, but with low molecular weights [9].

In 1959, Hay and coworkers described the first efficient route for the production of high-molecular-weight poly-2,6-disubstituted-1,4-phenylene ethers [10]. For instance, the reaction of 2,6-dimethylphenol (DMP) in the presence

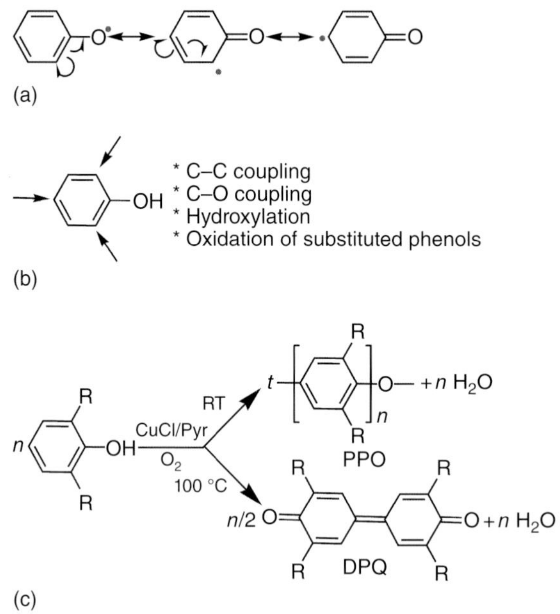

Figure 7.1 (a) Resonance structures for the phenoxyl radical; (b) phenol carbon atoms that undergo C–C or C–O coupling reactions; and (c) reaction scheme for the copper-catalyzed oxidative coupling of DMP (R = CH$_3$).

of CuCl/pyridine (5 mol% Cu) under molecular oxygen at room temperature generates poly-2,6-dimethyl-1,4-phenylene ether (PPO™ resin) [11], with a yield of 84% and a molecular weight of 28 000 Da (Figure 7.1c; R = CH$_3$). When the 2,6-substituents R are small (e.g., R = CH$_3$), the polymer is obtained, whereas for bulky R groups (like *tert*-butyl), the diphenoquinone (DPQ) is the sole product of the reaction (Figure 7.1c) [3]. For intermediate cases, for example, for 2-methyl-6-isopropylphenol, a mixture of the C–C (polymer) and C–O (quinone) compounds is achieved. The rate of monomer oxidation is first-order in catalyst, first-order in oxygen pressure, and zero-order in monomer [12].

PPO™, that is, poly-2,6-dimethyl-1,4-phenylene ether or PPE, is a high-performance thermoplastic obtained by polymerization of DMP, which exhibits particular physical properties such as high thermal stability, elevated mechanical strength, and excellent hydrolytic stability (Table 7.1) [14].

PPO™-based resins, for example, NORYL™, have many industrial uses, namely, computer and business machine housings, fluid handling and water pump housings, structural and interior components in electrical/electronic devices, automotive underhood components and instrument panels, microwaveable food packaging, telecommunications, appliances and housewares, building and construction, frames, and ventilating parts.

Table 7.1 Physical properties of PPO™ [13].

Specific gravity at 25 °C	1.06
Glass-transition temperature (T_g)	210 °C
Heat distortion temperature (HDT)	179 °C at 0.46 MPa (66 psi)
	174 °C at 1.80 MPa (264 psi)
Tensile strength	80 MPa (11 600 psi) at 25 °C
	55 MPa (8000 psi) at 93 °C
Tensile modulus (Young's modulus)	2689 MPa (390 000 psi) at 25 °C
	2482 MPa (360 000 psi) at 93 °C

7.1.2
Chemistry and Catalysis

The mechanism of this polymerization reaction has been investigated thoroughly [15, 16]. However, even after more than 50 years since its commercial development, the mechanism of PPO™ formation is not fully unraveled [17, 18]. In fact, the nature of the initial active species is still a matter of debate. Indeed, while some researchers consider that the C–O coupling reaction occurs via phenoxyl radicals (Figure 7.2b) [21], others believe that phenoxonium cations are involved in the polymerization process (Figure 7.2a) [22].

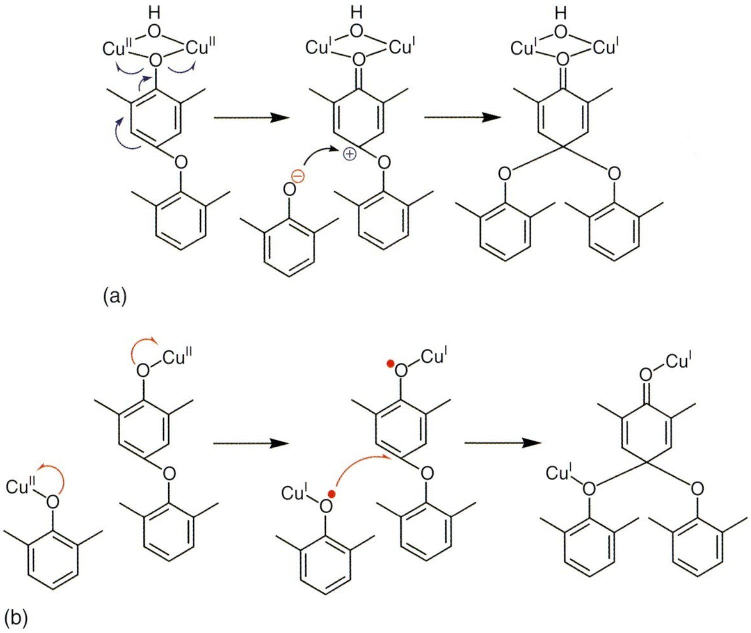

Figure 7.2 Schematic representation of the proposed possible initial steps for the oxidative coupling of DMP: (a) ionic pathway [19] and (b) radical pathway [20].

The two different pathways that both involve copper ions are equally possible and lead to the formation of a quinone-ketal intermediate (Figure 7.2), which is universally accepted [23]. Thus, the critical issue consists in the way in which these quinone-ketal species are generated. Most researchers studying this polymerization reaction believe that its coupling mechanism involves phenoxyl radicals [19], which are produced by mononuclear copper species (homolytic or radical pathway; Figure 7.2b). In contrast, other groups propose that a biomimetic dinuclear copper catalyst mediates the oxidative coupling of DMP molecules (heterolytic or ionic pathway; Figure 7.2a). Therefore, pathway (b) is achieved through a one-electron process, whereas pathway (a) involves a two-electron process (each copper ion of the dinuclear unit acting as a one-electron oxidant).

It has been shown that the reaction of DMP with a one-electron oxidant such as $K_3Fe(CN)_6$ yields mainly DPQ (Figure 7.1c), which is obtained by C–C coupling of two phenoxyl radicals [24]. These data tend to suggest that the C–O coupling reaction is a two-electron process and therefore point toward pathway (a) for the formation of PPO™. Moreover, for bulky 2,6-substituents (for instance, for R = *tert*-butyl; Figure 7.1c), the corresponding polymer cannot be formed and the DPQ is obtained in high yields [25]. Most likely the steric hindrance due to the *tert*-butyl groups precludes the formation of the dicopper species, allowing a two-electron process (Figure 7.2a), and the one-electron C–C coupling pathway is favored.

Once the quinone-ketal species are created, two events may take place, namely, rearrangement [16] (Figure 7.3) and redistribution [26] (Figure 7.4). The polymer growth occurs via a Claisen-type rearrangement (Figures 7.3 and 7.4), which is a concerted sigmatropic transposition [27]. Note that the extension of the polymer does not happen at the polymer ends by addition of a monomer (such as for chain-growth polymerization); instead, the formation of the poly-1,4-phenylene ether is a stepwise process where polymer chains of different sizes combine to ultimately generate the polymeric material (step-growth polymerization). For instance, in Figure 7.4, a tetramer and a trimer produce a heptamer. The formation of the quinone-ketal species is reversible; hence, the quinone-ketal obtained by the combination of a tetramer and a trimer may regenerate these two units, but it can also yield a pentamer and a dimer (Figure 7.4).

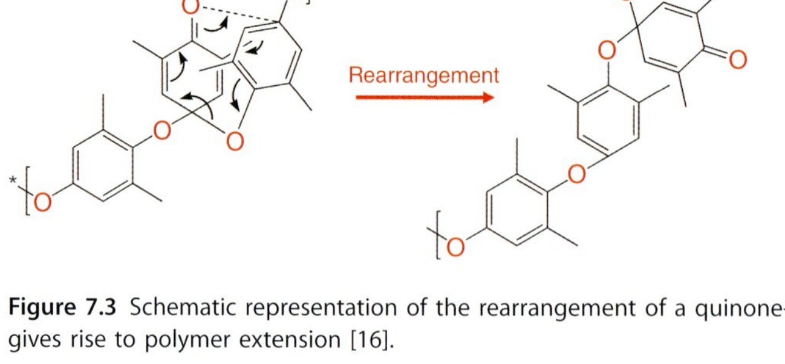

Figure 7.3 Schematic representation of the rearrangement of a quinone-ketal moiety that gives rise to polymer extension [16].

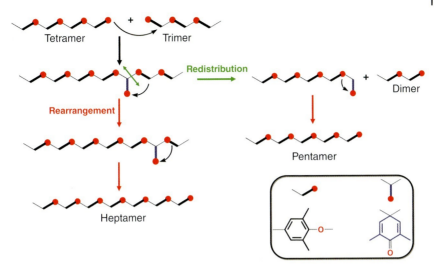

Figure 7.4 Schematic representation of the quinone-ketal rearrangement and redistribution [26].

Since the initial CuCl/pyridine catalytic system (with pyridine acting as both ligand and solvent), a number of other N-donor ligands have been used with the objective to develop more efficient polymerization catalysts [11, 28]. The catalytic activities (in moles of DMP converted per mole of Cu) of different copper complexes are listed in Table 7.2. Furthermore, it was established that for monodentate ligands, the selectivity toward C–O coupling of the resulting catalysts was largely dependent on the ligand-to-copper ratio (for instance, in the case of pyridine, the optimum selectivity is achieved with a ligand/Cu ratio of 100/1 [17]). In contrast, with bidentate diamine ligands, highly selective catalytic species are obtained with a one-to-one ligand/copper ratio. It was also shown that in nonpolar solvents such as toluene, the catalytically active system Cu/TMEDA (N,N,N',N'-tetramethyl-ethane-1,2-diamine) only mediates the formation of low-molecular-weight materials. Moreover, a drying agent, namely, anhydrous magnesium sulfate, is required to eliminate the water produced during the reaction as it deactivates the catalyst. With more polar solvents such as chlorobenzene or *ortho*-dichlorobenzene, higher-molecular-weight polymers can be obtained.

The necessity of a drying agent (i.e., $MgSO_4$ or molecular sieves) when applying the highly efficient TMEDA/Cu system makes it impractical on a large (industrial) scale. Hence, the use of a large excess of di-*n*-butylamine as monodentate ligand allowed the production of high-molecular-weight polymers without any drying agent. This system was commercialized; however, the isolation of PPO™ and the recycling of both the solvent and the amine turned out to be a too difficult and costly procedure [29]. In the mid-1970s, an hydrolytically stable catalyst was developed from N,N'-di-*tert*-butyl-ethane-1,2-diamine (DTBEDA; Table 7.2)

Table 7.2 Catalytic activity of various copper/amine complexes.

N-Donor ligand	[DMP]/[Cu]
Pyridine	40
Diethylamine	200
Di-n-butylamine	200
N,N,N′,N′-Tetramethyl-ethane-1,2-diamine (TMEDA)	750
N,N′-Di-tert-butyl-ethane-1,2-diamine (DTBEDA)	900
DTBEDA/N,N-Dimethyl-n-butylamine (DMBA)	1500

[30, 31]. Moreover, a very high activity could be achieved by adding N,N-dimethyl-n-butylamine (DMBA) to this Cu/DTBEDA system [32]; indeed, only 0.07 mol% of copper could be employed to yield high-molecular-weight polymers (Table 7.2). Therefore, the typical industrial "catalytic mixture" used for this phenol-coupling reaction is composed of 1 equiv. of copper(I) bromide, 1 equiv. of the bidentate ligand DTBEDA, 10 equiv. of the monodentate amine DMBA, and 9 equiv. of HBr (as a source of bromide anions) [32]. Actually, it has been shown that Br^- ions stabilize the copper complex, avoiding the formation of inactive copper-hydroxide species (which degrade through the generation of insoluble copper(II) compounds) [31]. Using this catalyst, a DMP/Cu ratio of 1500/1 can be used (Table 7.2), allowing the production of PPO™ with a degree of polymerization that is suitable for molding (i.e., with a number average molecular weight within the range of about 3000–40 000).

7.1.3
Process Technology

The commercial development of PPO™ started in the 1960s, and the first priority was to find an efficient and low-cost procedure to produce the monomer, namely, DMP. Thus, DMP is obtained from cyclohexanone; hence, the reaction of cyclohexanone with formaldehyde at 350 °C in the presence of tricalcium phosphate generates DMP with a reasonable yield [33]. DMP can also be produced by the reaction of phenol with methanol at 350 °C, with magnesium oxide as catalyst [34]. The latter method rendered the manufacture of the polymer economically attractive; accordingly, from 1964, its commercialization accelerated.

Typically, the polymerization reaction [35] is initiated by the addition of a solution of DMP in toluene to a toluene solution of the catalyst system (discussed earlier) in a reactor equipped with an internal cooling coil and an inlet tube for dioxygen. The reaction mixture is stirred vigorously at 30 °C for 2 h. The copper catalyst is then removed from the polymer solution containing 16 wt% of PPO™ and recycled. To this end, an aqueous solution of trisodium ethylenediaminetetraacetate (Na_3EDTA; 38 wt%) is added to the toluene solution, producing a biphasic mixture comprising about 4 vol% of an aqueous phase containing copper–EDTA (copper extracted from the organic phase) and free EDTA. Subsequently, the polymer (with a weight average M_w in the range 20 000–80 000) can be obtained by precipitation through the addition of methanol to this mixture, followed by filtration and drying under reduced pressure. The filtrate comprises about 60 wt% of methanol, 35 wt% of toluene, and 5 wt% of water containing copper(II) ions. Two equivalents of sodium bromide are added to the aqueous solution of the copper–EDTA complex, which is then treated with alum, that is, $Al_2(SO_4)_3$. Next, 2.2 equiv. of a $NaHSO_3/Na_2S_2O_5$ mixture (effective SO_2 content of 58.5%) is added. These reducing agents convert copper(II) into copper(I), which is recovered as a white crystalline material, that is, CuBr, that can be recycled to the PPO™ polymerization process [35]. The organic solvents (viz., methanol and toluene) are recuperated by distillation.

As already mentioned earlier (see Section 7.1.1), because of its high glass-transition temperature, that is, $T_g = 208\,°C$, PPO™ has to be melt-processed at elevated temperatures. As a result, degradation of the polymer may occur at such temperatures (particularly through oxidation reactions at the methyl substituents). Furthermore, upon cooling from the liquid to the rubber state, two unwanted events can take place: (i) the polymer crystallizes and (ii) the molecular motions are frozen and the rubbery polymer turns to a glass. As a consequence, the material becomes brittle and cannot be used for practical applications. Fortunately, PPO™ exhibits unusual and remarkable blending properties [36]. For instance, it has been shown that the T_g can be significantly reduced by mixing PPO™ with increasing amounts of polystyrene (Figure 7.5a) [37]. For example, the T_g is around 129 °C for 50/50 mixture of PPO™/PS (Figure 7.5a). Actually, this polymer blend, known as Noryl®SE 100 (which has a Vicat softening temperature of 120 °C), is an excellent material for use in capacitor housings, primarily because it provides high-dimensional stability and high resistance to heat deformation. As the processing temperatures of the PPO™–PS blends are much lower than for pure PPO™, oxidative degradation is avoided; in fact, Noryl® resins are commonly processed through injection molding, and a significant price/property gap between commodity plastics and engineering plastics could be identified (Figure 7.5b). By changing the PPO™-to-PS ratio, a large number of polymeric materials bearing different (tunable) properties could be produced, and a series of blends were introduced to the marketplace [38].

A limitation is, however, encountered with Noryl® resins; indeed, because PS is a very brittle material, their impact strength decreases when the percentage of PS is raised [37]. To circumvent this problem, rubbers are added to Noryl® resins to increase the toughness [39]. Thanks to the exceptional blending properties of PPO™, numerous mixed materials have been developed, for instance, with diallyl phthalates, polysulfone, acrylates, coumarone–indene, or polyvinyl chloride (PVC) [39]. It has also been shown that PPO™ could be blended with styrene–butadiene block copolymers, hence allowing to expand the temperature of use of the resulting materials [40]. Accordingly, the combination of various

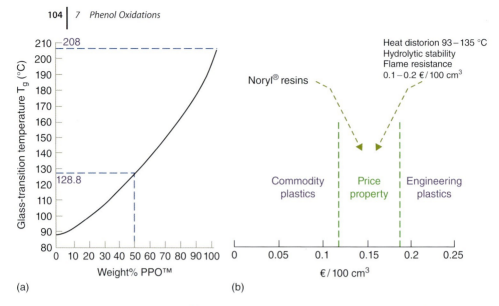

Figure 7.5 (a) T_g of PPO™–PS blends [37] and (b) marketing chart for Noryl® resins.

PPO™–PS levels with other additives provides a myriad of resins that cover a very wide range of physical and thermomechanical properties. Their general characteristics include high heat resistance, excellent electrical properties over a wide range of temperatures and frequencies, low density, high hydrolytic stability, chemical resistance to most acids, dimensional stability, low mold shrinkage, and very low creep behavior at elevated temperatures.

Since the discovery of PPO™ by Allan S. Hay of the General Electric Company in the 1960s [3, 41], over 1500 patents have been issued in the field of oxidative coupling polymerization, and the total sales worldwide of engineering plastics based on PPO™ to date represent about US$ 1 billion per year. In fact, Noryl®-modified PPO™ resins have become the world's most successful and best-known polymer blends and alloys.

7.1.4
New Developments

Because the methyl substituents of PPO™ are subjected to oxidative degradation during the processing of the material, new polymers based on the 2,6-diphenylphenol moiety (R = phenyl in Figure 7.1c) have been developed, with the idea that the aryl substituents will exhibit improved degradation-resistant properties [42].

A high-molecular-weight polymer that shows a T_g of 220 °C can be obtained at 60 °C. This polymer has a melting temperature T_m of 481 °C and crystallizes when heated above its T_g [43]. As expected, the compound displays very good oxidative stability (no changes are observed in air at 175 °C); as a matter of fact, it can be solution-processed to give thermoxidatively stable fibers and films. However, this polymer cannot be used as molding resin owing to its high T_m. The superior electrical properties of the material led to its commercial development under the name Tenax® [44] as a high-temperature resistant synthetic paper for extra-high-voltage (EHV) cables. Tenax® only reached the pilot plant scale for this application. Nowadays, Tenax® is used in personal exposure monitoring, hazardous organic emission measurements, volatile plant component analysis, analysis of organic compounds in water, and measurements of volatile compounds in human breath/serum [45].

One important aspect that should be considered in future industrial processes to produce polyphenylene ethers is the use of greener solvents. In that context, some studies have been carried out to try to generate PPO™ in water [46]. Thus, Nishide and coworkers have developed a procedure where an aqueous solution of copper(II) complex (10 mol%), that is, $CuCl_2$/DTPA (DTPA = diethylenetriamine-N,N,N',N'',N''-pentaacetic acid), is mixed with an aqueous solution of DMP/NaOH 1/1, containing sodium n-dodecyl sulfate. After vigorous stirring at 50 °C for 12 h under an atmosphere of molecular oxygen, PPO™ is precipitated by the addition of sodium chloride (yield = 70%). The PPO™ thus obtained has an M_w of 8100 and an M_w/M_n of 2.0 [47]. More recently, Horie and coworkers described a related system but using natural copper ligands, such as arginine [48]. Hence, the reaction between an aqueous solution of $CuBr_2$/arginine (10 mol%) and an aqueous solution of DMP/NaOH 1/1 containing 10 mol% of sodium dodecyl sulfate (SDS), at 50 °C for 12 h under an atmosphere of dioxygen, produces PPO™ (yield = 72%), which is isolated through the addition of methanol after reaction quenching with HCl. The polymer obtained has an M_n of 4400 and M_w/M_n of 1.7.

Another possibility is to use a supercritical fluid as green solvent. Actually, a method for the polymerization of DMP in supercritical carbon dioxide (scCO$_2$) had been developed nearly two decades ago [49]. Thus, copper(I) bromide and DMP are added to a high-pressure reactor purged with argon beforehand. Then, dioxygen is added to reach a pressure of 180 psi (1240 kPa), and carbon dioxide is introduced up to a final pressure of 700 psi (4826 kPa). The resulting mixture is stirred while pyridine is added through an injector loop (to reach a Cu:pyridine:DMP ratio of 1 : 44 : 75), resulting in a pressure of 3000 psi (20.7 MPa) inside the reactor. The high-pressure reactor is heated to 40 °C, and CO_2 is added to achieve a pressure of 4700 psi (32.4 MPa) at this temperature (pressure and temperature conditions for which CO_2 is supercritical). After stirring for 20 h at 40 °C, the CO_2 is vented and the product washed with 1% HCl in methanol (to remove the catalyst). The crude polymer is then dissolved in toluene and precipitated in methanol. PPO™ is obtained with a yield of 67%, an average molecular weight M_n of 6140, and a polydispersity M_w/M_n of 1.63.

7.2
2,3,5-Trimethylhydroquinone as a Vitamin E Intermediate via Oxidation of Methyl-Substituted Phenols

Jan Schütz and Thomas Netscher

Vitamin E is the most important lipid-soluble antioxidant in biological systems. (all-*rac*)-α-Tocopherol (synthetic vitamin E, **1**) is the economically most important product industrially prepared on a multi-10 000 t/year scale and mainly used in animal nutrition [50]. In the large-scale syntheses of **1**, 2,3,5-trimethylhydroquinone (**6**) is used as the aromatic key building block, which is condensed with isophytol (**2**) to yield **1** by all producers worldwide (Scheme 7.1). Trimethylhydroquinone (TMHQ, **6**), in turn, is obtained from trimethylquinone (TMQ, **5**) by reduction procedures, in particular catalytic hydrogenation. Besides other possibilities to access TMHQ (**6**), this route is generally preferred, and efficient oxidation processes for the production of quinone **5** from alkylated phenols are, therefore, of high interest [51].

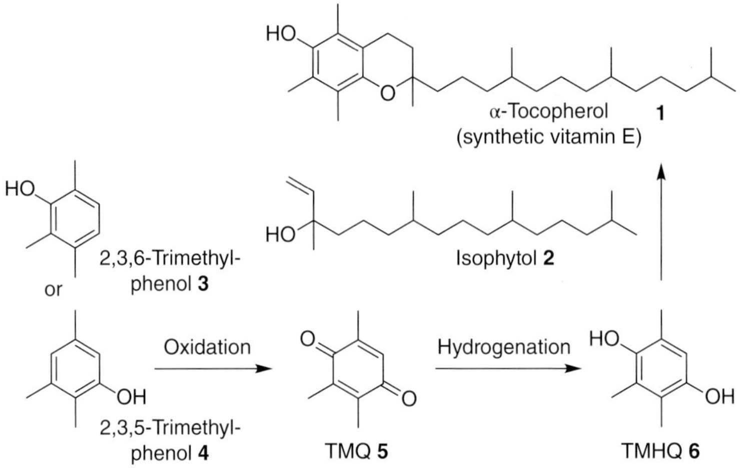

Scheme 7.1 Synthesis of (all-*rac*)-α-tocopherol (**1**) via phenol oxidation as a key step.

Most industrially operated routes start from 2,3,6-trimethylphenol (**3**), but also 2,3,5-trimethylphenol (**4**) and DMP (**8**, Scheme 7.3) are possible starting materials. Differences between the starting phenols are discussed with regard to reactivity, selectivity, and by-product formation. From an industrial point of view, careful selection of the various oxidation catalyst systems and reaction conditions is of utmost importance for the successful large-scale production of economically important compounds in life science industry [52].

For the oxidation of 2,3,6-trimethylphenol (**3**) or 2,3,5-trimethylphenol (**4**), a variety of oxidants can be used. The oxidation of phenol **3** leads to TMQ (**5**), with the formation of the dimer 2,2′,3,3′,6,6′-hexamethylbiphenyl-4,4′diol (**7**, Scheme 7.2), depending on the conditions [53, 54].

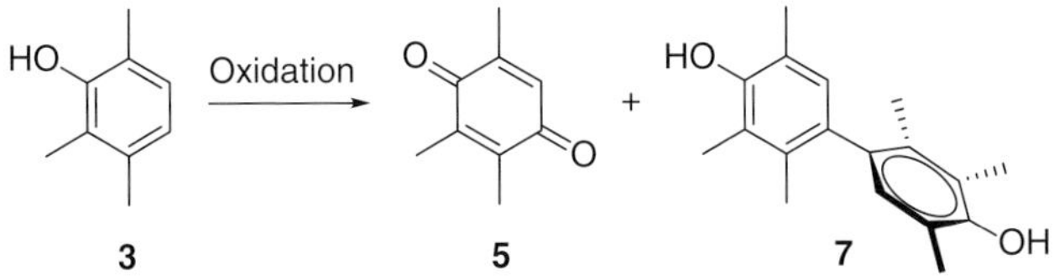

Scheme 7.2 Products formed in the oxidation of 2,3,6-TMP (**3**).

Inorganic acids or salts are used as oxidants, as well as molecular oxygen, air, or H_2O_2 combined with homogeneous or heterogeneous catalysts. From an economic and environmental point of view, however, the use of stoichiometric amounts of inorganic salts, and thus generating stoichiometric amounts of inorganic waste material, is not acceptable. Processes using H_2O_2, oxygen, or air as oxidants in the presence of catalysts are advantageous since less salt waste is generated. In this regard, aqueous hydrogen peroxide is claimed as an environmentally friendly oxidizing reagent for this transformation [55]. Due to process economics, however, the application of oxygen (or air) in combination with a suitable catalyst system is clearly favored for this transformation operated on large scale.

In the case of applying oxygen in the gas phase, its mass transfer to the liquid phase containing starting material and catalyst can be rate controlling. Gas dispersion is, therefore, critical [56]. A major drawback in using oxygen as an oxidant is the risk of explosion of highly flammable organic solvents at reaction temperatures above their flash point. Hazardous reaction conditions can be suppressed by using a biphasic reaction medium consisting of water and a longer-chain alcohol and/or an aromatic solvent [57] or in biphasic reaction mixtures consisting of a long-chain carboxylic acid and water [58].

Homogeneous catalyst systems include transition metals, for example, Schiff base complexes with cobalt [59]. Unfortunately, there is no protocol known to recover and recycle the expensive Schiff base. Many copper complexes have been employed for the oxidation of phenols to 1,4-benzoquinones [60]. Copper salts such as $CuCl_2$ and $CuBr_2$ are widely used for the oxidation of trimethylphenol with molecular oxygen at larger scale [61]. In addition, applications of the halides of Cr, Mn, Fe, Ni, and Zn as catalysts have been described [58]. Copper halides have also been combined with earth metal halides, for example, $MgCl_2$ [62] or LiCl, in the presence of a phase transfer catalyst [63].

In general, high copper salt loadings are necessary to reach an acceptable conversion due to the deactivation of the catalyst. The addition of stabilizing agents such as complexing agents, hydroxylamines, oximes, and amines is beneficial to lower the amount of copper salt while maintaining high conversion of TMP and selectivity to TMQ [64]. Also, the use of ionic liquids as stabilizing agents serves this goal. Furthermore, only catalytic amounts of catalyst are necessary in the case of $CuCl_2$, and separation of catalyst and product and recycling of the catalyst are facilitated [65].

Heteropolyacids and the corresponding polyoxometalates have been frequently applied in the oxidation of 2,3,6-TMP (**3**) with molecular oxygen or H_2O_2. Kholdeeva *et al.* studied the kinetics of 2,3,6-TMP oxidation over heteropolyacids $H_{3+n}PMo_{12-n}V_nO_{40}$ with molecular oxygen. They showed that the catalytically active species is the VO_2^+ ion [54]. Fujibayashi *et al.* found that the activity of molybdovanadophosphate was greatly enhanced by supporting on charcoal [66], when 2,3,6-TMP is oxidized with molecular oxygen over catalytic amounts of heteropolyoxometalates containing molybdenum and vanadium in a solvent mixture of water and acetic acid. The heteropolyacid can be easily recycled using a biphasic system containing a water–acetic acid phase and an organic solvent nonmiscible with water [67]. Also, the oxidation of 2,3,6-TMP (**3**) with molecular oxygen in the presence of non-Keggin-type Mo-V-phosphoric heteropoly acids has been investigated [68].

In comparison of the two regioisomers, phenol **3** is easily converted to quinone **5** than phenol **4** (cf. Scheme 7.1) when using catalytic protocols, most likely due to less steric hindrance around the *para* position to be oxidized. The photooxidation of 2,3,5-TMP (**4**) has been described using porphyrins or metallophthalocyanins [69]. The combination of $CuCo_2O_4$ with hypocrellins, which are naturally occurring photosensitizers, was found to mediate the oxidation under irradiation (<400 nm) and air [70]. The use of a complex consisting of $Co(OAc)_2$ and a multidentate *N*-heterocyclic podant ligand yielded quinone **5** in 70% yield [71].

Since improved and cheap routes to the aromatic key building block TMHQ (**6**) are still needed, an alternative access starting from easily available DMP (**8**) has been worked out. In the oxidation of phenol **8** to quinone **9**, the formation of the corresponding DPQ (**10**) as a typical by-product is often the prevailing reaction (Scheme 7.3). Retarding this undesired oxidative coupling is of interest, since 2,6-dimethylhydroquinone (**11**) obtained after catalytic hydrogenation of the

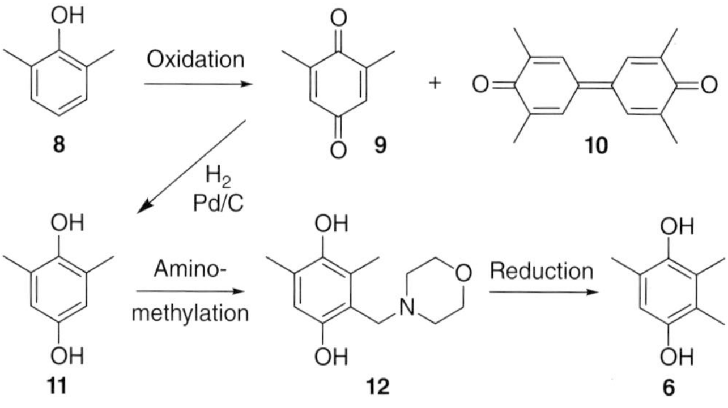

Scheme 7.3 2,6-Dimethylphenol (DMP, **8**) as a starting material for an alternative access to TMHQ (**6**).

quinone offers this pathway to TMHQ (**6**) via selective aminomethylation to **12** followed by reduction [72].

References

1. Terent' ev, A.P. and Mogilyanskii, Y.D. (1955) *Dokl. Akad. Nauk SSSR*, **103**, 91.
2. (a) Bach, H.C. (1966) *Polym. Prepr. (Am. Chem. Soc., Div. Polym. Chem.)*, **7**, 576; (b) Bach, H.C. (1967) *Polym. Prepr. (Am. Chem. Soc., Div. Polym. Chem.)*, **8**, 610.
3. Hay, A.S. (1962) *J. Polym. Sci.*, **58**, 581–591.
4. Terent' ev, A.P. and Mogilyanskii, Y.D. (1961) *J. Gen. Chem. USSR*, **31**, 298.
5. Cosgrove, S.L. and Waters, W.A. (1951) *J. Chem. Soc.*, 1726–1730.
6. Süs, O., Möller, K., and Heiss, H. (1956) *Liebigs Ann. Chem.*, **598**, 123–138.
7. Dewar, M.J.S. and James, A.N. (1958) *J. Chem. Soc.*, 917–922.
8. Hunter, W.H. and Morse, M.L. (1933) *J. Am. Chem. Soc.*, **55**, 3701–3705.
9. Staffin, G.D. and Price, C.C. (1960) *J. Am. Chem. Soc.*, **82**, 3632–3634.
10. Hay, A.S., Blanchard, H.S., Endres, G.F., and Eustance, J.W. (1959) *J. Am. Chem. Soc.*, **81**, 6335–6336.
11. Hay, A.S. (1967) US Patent 3306875.
12. Price, C.C. and Nakaoka, K. (1971) *Macromolecules*, **4**, 363–369.
13. Zoller, P. and Hoehn, H.H. (1982) *J. Polym. Sci., Part B: Polym. Phys.*, **20**, 1385–1397.
14. Mark, J.E. (2007) *Physical Properties of Polymers Handbook*, 2nd edn, Springer-Verlag, New York.
15. (a) Li, K.T. (1994) *J. Appl. Polym. Sci.*, **54**, 1339–1351; (b) Endres, G.F. and Kwiatek, J. (1962) *J. Polym. Sci.*, **58**, 593–609; (c) Viersen, F.J., Challa, G., and Reedijk, J. (1990) *Polymer*, **31**, 1368–1373.
16. Mijs, W.J., van Lohuizen, O.E., Bussink, J., and Vollbracat, L. (1967) *Tetrahedron*, **23**, 2253–2264.
17. Finkbeiner, H., Hay, A.S., Blanchard, H.S., and Endres, G.F. (1966) *J. Org. Chem.*, **31**, 549–555.
18. Baesjou, P.J., Driessen, W.L., Challa, G., and Reedijk, J. (1999) *Macromolecules*, **32**, 270–276.
19. van Aert, H.A.M., van Genderen, M.H.P., van Steenpaal, G.J.M.L., Nelissen, L., Meijer, E.W., and Liska, J. (1997) *Macromolecules*, **30**, 6056–6066.
20. Baesjou, P.J., Driessen, W.L., Challa, G., and Reedijk, J. (1996) *J. Mol. Catal. A: Chem.*, **110**, 195–210.
21. Tsuchida, E., Kaneko, M., and Nishide, H. (1972) *Makromol. Chem.*, **151**, 221–234.
22. Baesjou, P.J., Driessen, W.L., Challa, G., and Reedijk, J. (1997) *J. Am. Chem. Soc.*, **119**, 12590–12594.
23. Saito, K., Masuyama, T., Oyaizu, K., and Nishide, H. (2003) *Chem. Eur. J.*, **9**, 4240–4246.
24. Haynes, C.G., Turner, A.H., and Waters, W.A. (1956) *J. Chem. Soc.*, 2823–2831.
25. Hay, A.S. (1969) *J. Org. Chem.*, **34**, 1160–1161.
26. Gupta, S., van Dijk, J., Gamez, P., Challa, G., and Reedijk, J. (2007) *Appl. Catal., A*, **319**, 163–170.
27. Ionescu, M. and Mihis, B. (1995) *J. Macromol. Sci., Pure Appl. Chem.*, **A32**, 1661–1680.
28. Hay, A.S. (1967) US Patent 3306874.
29. Hay, A.S. (1998) *J. Polym. Sci., Part A: Polym. Chem.*, **36**, 505–517.
30. Hay, A.S. (1975) US Patent 3914266.
31. Hay, A.S. (1977) US Patent 4028341.
32. Mobley, D.P. (1984) *J. Polym. Sci. Polym. Chem. Ed.*, **22**, 3203–3215.
33. Hamilton, S.B. and Hay, A.S. (1966) US Patent 3280201.
34. Hamilton, S.B. (1969) US Patent 3446856.
35. Edema, J.J.H. (1997) US Patent 5621066.
36. Boldebuck, E.M. (1962) US Patent 3063872.
37. Cizek, E.P. (1968) US Patent 3383435.
38. Peters, E.N. (2006) in *Engineering Plastics Handbook* (ed J.M. Margolis), The McGraw-Hill Companies, Inc., Toronto, pp. 181–220.
39. Cooper, G.D. (1978) US Patent 4101503.
40. Kambour, R.F. (1972) US Patent 3639508.

41. Hay, A.S. (1967) *Adv. Polym. Sci.*, **4**, 496–527.
42. Hay, A.S. (1969) US Patent 3432466.
43. Wrasidlo, W. (1971) *Macromolecules*, **4**, 642–648.
44. Vermeer, J., Boone, W., Bussink, J., and Brakel, H. (1971) in *Conference on Electrical Insulation and Dielectric Phenomena*, CA, vol. **75**, p. 153114p.
45. Buchem BV http://www.isotope.nl/index.php?pageName=home# (accessed October 2014).
46. (a) Saito, K., Tago, T., Masuyama, T., and Nishide, H. (2004) *Angew. Chem. Int. Ed.*, **43**, 730–733; (b) Saito, K., Kuwashiro, N., and Nishide, H. (2006) *Polymer*, **47**, 6581–6584.
47. Nishide, H. and Saito, K. (2010) US Patent 7649075 B2.
48. Chen, C.W., Lin, I.H., Lin, C.C., Lin, J.L., and Horie, M. (2013) *Polymer*, **54**, 5684–5690.
49. Desimone, J.M., Kapellen, K.K., and Mistele, C.D. (1996) Patent WO1996037535 A1.
50. Baldenius, K.-U., von dem Bussche-Hünnefeld, L., Hilgemann, E., Hoppe, P., and Stürmer, R. (1996) *Ullmann's Encyclopedia of Industrial Chemistry*, vol. **A27**, VCH Verlagsgesellschaft, Weinheim.
51. (a) Eggersdorfer, M., Laudert, D., Letinois, U., McClymont, T., Medlock, J., Netscher, T., and Bonrath, W. (2012) *Angew. Chem. Int. Ed.*, **51**, 12960–12990; (b) Netscher, T. (2007) in *Vitamin E: Vitamins and Hormones Advances in Research and Applications*, vol. **76** (ed G. Litwack), Elsevier Academic Press Inc., San Diego, CA, pp. 155–202.
52. Bonrath, W., Eggersdorfer, M., and Netscher, T. (2007) *Catal. Today*, **121**, 45–57.
53. (a) Takehira, K., Shimizu, M., Watanabe, Y., Orita, H., and Hayakawa, T. (1989) *Tetrahedron Lett.*, **30**, 6691–6692; (b) Takehira, K., Shimizu, M., Watanabe, Y., Orita, H., and Hayakawa, T. (1989) *J. Chem. Soc., Chem. Commun.*, 1705–1706; (c) Kholdeeva, O.A., Golovin, A.V., and Kozhevnikov, I.V. (1992) *React. Kinet. Catal. Lett.*, **46**, 107–113; (d) Mizukami, F., Satoh, K., Niwa, S., Tsuchiya, T., Shimizu, K., and Imamura, J. (1985) *Sekiyu Gakkaishi*, **28**, 293–299 (Chem. Abstr., (1986), **105**, 97088).
54. Kholdeeva, O.A., Golovin, A.V., Maksimovskaya, R.I., and Kozhevnikov, I.V. (1992) *J. Mol. Catal.*, **75**, 235–244.
55. (a) Kholdeeva, O.A., Ivanchikova, I.D., Guidotti, M., Pirovano, C., Ravasio, N., Barmatova, M.V., and Chesalov, Y.A. (2009) *Adv. Synth. Catal.*, **351**, 1877–1889; (b) Kholdeeva, O.A., Ivanchikova, I.D., Guidotti, M., and Ravasio, N. (2007) *Green Chem.*, **9**, 731–733.
56. Hook, B.D. (2013) in *Liquid Phase Oxidation via Heterogeneous Catalysis. Organic Synthesis and Industrial Applications* (eds M.G. Clerici and O.A. Kholdeeva), John Wiley & Sons, Inc., Hoboken, NJ, pp. 496–506.
57. (a) Bockstiegel, B., Hoercher, U., and Laas, H. (1992) Patent DE 4029198; (b) Isshiki, T., Yui, T., Abe, M., Jono, M., and Uno, H. (1989) US Patent 4828762.
58. Maassen, R., Krill, S., and Huthmacher, K. (2001) Patent EP 1132367.
59. (a) Diehl, H. and Hach, C.C. (1950) *Inorg. Synth.*, **3**, 196–201; (b) Jouffret, M. (1975) Patent DE 2450908; (c) Jouffret, M. (1970) Patent FR 2015576.
60. Allen, S.E., Walvoord, R.R., Padilla-Salinas, R., and Kozlowski, M.C. (2013) *Chem. Rev.*, **113**, 6234–6458.
61. (a) Thoemel, F. and Hoffmann, W. (1983) Patent DE 3215095; (b) Bartoldus, D. and Lohri, B. (1981) Patent EP 35635.
62. (a) Brenner, W. (1972) Patent DE 2221624; (b) Hirose, N., Hamamura, K., and Inai, Y. (1988) Patent EP 0294584; (c) Hsu, C.Y. and Lyons, J.E. (1984) Patent EP 107427.
63. Liu, D.-M., Lu, X.-Y., and Mi, F.-Y. (2011) Patent CN 102108047A.
64. (a) Takehiro, K., Orita, H., and Shimizu, M. (1990) Patent EP 0369823; (b) Bodnar, Z., Mallat, T., and Baiker, A. (1996) *J. Mol. Catal. A: Chem.*, **110**, 55–63.
65. (a) Sun, H.J., Harms, K., and Sundermeyer, J. (2004) *J. Am. Chem. Soc.*, **126**, 9550–9551; (b) Guan, W.H., Wang, C.M., Yun, X., Hu, X.B., Wang, Y., and Li, H.R. (2008) *Catal. Commun.*, **9**, 1979–1981; (c) Wang, C.M., Guan,

W.H., Xie, P.H., Yun, X., Li, H.R., Hu, X.B., and Wang, Y. (2009) *Catal. Commun.*, **10**, 725–727; (d) Sun, H.J., Li, X.Y., and Sundermeyer, J. (2005) *J. Mol. Catal. A: Chem.*, **240**, 119–122; (e) Li, H.-R., Guan, W.-H., Wang, C.-M., and Hu, X.-B. (2008) Patent CN 101113131A; (f) Li, H.-R., Chen, Z.-R., Hu, X.-B., Shen, Q.-C., Wang, C.-M., and Hu, B.-Y. (2008) Patent CN 101260030A; (g) Li, H.-R., Chen, Z.-R., Guan, W.-H., Hu, X.-B., Wang, C.-M., Hu, B.-Y., and Zhang, Z.-X. (2007) Patent CN 1986513A.

66. Fujibayashi, S., Nakayama, K., Hamamoto, M., Sakaguchi, S., Nishiyama, Y., and Ishii, Y. (1996) *J. Mol. Catal. A: Chem.*, **110**, 105–117.
67. Vandewalle, M.-F. (1998) Patent WO 9818746.
68. Rodikova, Y.A. and Zhizhina, E.G. (2013) *J. Chem. Chem. Eng.*, 7, 808–820.
69. (a) Murtinho, D., Pineiro, M., Pereira, M.M., Gonsalves, A.M.D., Arnaut, L.G., Miguel, M.D., and Burrows, H.D. (2000) *J. Chem. Soc., Perkin Trans. 2*, 2441–2447; (b) Sorokin, A.B. and Tuel, A. (1999) *New J. Chem.*, **23**, 473–476; (c) Sorokin, A.B., Mangematin, S., and Pergrale, C. (2002) *J. Mol. Catal. A: Chem.*, **182**, 267–281.
70. Li, Y., Liu, W., Wu, M.Z., Yi, Z.Z., and Zhang, J.C. (2010) *Mendeleev Commun.*, **20**, 218–219.
71. Moriuchi, T., Hirao, T., Ishikawa, T., Ohshiro, Y., and Ikeda, I. (1995) *J. Mol. Catal. A: Chem.*, **95**, L1–L5.
72. (a) Bonrath, W., Netscher, T., Schütz, J., and Wüstenberg, B. (2012) Patent WO 2012025587; (b) Ogura, S. and Hiramine, T. (2006) Patent JP 2006249036A.

Part III
Pd-Catalyzed Aerobic Oxidation

8
Pd-Catalyzed Aerobic Oxidation Reactions: Industrial Applications and New Developments

Dian Wang, Jonathan N. Jaworski, and Shannon S. Stahl

8.1
Introduction

Palladium catalysts are widely used in liquid phase aerobic oxidations, and numerous examples have been employed for large-scale chemical production (Scheme 8.1). Several industrially important examples are the focus of dedicated chapters in this book: Wacker and Wacker-type oxidation of alkenes into aldehydes, ketones, and acetals (Scheme 8.1a; Chapters 9 and 11), 1,4-diacetoxylation of 1,3-butadiene (Scheme 8.1b; Chapter 10), and oxidative esterification of methacrolein to methyl methacrylate (Scheme 8.1c; Chapter 13). In this introductory chapter, we survey a number of other Pd-catalyzed oxidation reactions that have industrial significance, including acetoxylation of ethylene to vinyl acetate (Scheme 8.1d), oxidative carbonylation of alcohols to dialkyl oxalates and carbonates (Scheme 8.1e), and oxidative coupling of dimethyl phthalate to 3,3′,4,4′-tetramethyl biphenylcarboxylate (Scheme 8.1f).

In addition to the industrial applications, in Scheme 8.1, other reactions have been the focus of extensive research and development. For example, Chapter 12 surveys the research efforts directed toward Pd-catalyzed oxidative carbonylation of phenol affords the important monomer, diphenyl carbonate (Scheme 8.2a). Other reactions of potential industrial significance highlighted in this chapter include the oxidation of alcohols to aldehydes and ketones (Scheme 8.2b), oxidative coupling of arenes and carboxylic acids to afford aryl esters (Scheme 8.2c), benzylic acetoxylation (Scheme 8.2d), and oxidative Heck reactions (Scheme 8.2e). The chapter concludes by highlighting a number of newer research developments, including ligand-modulated catalytic oxidations, Pd/NO$_x$ cocatalysis, and alkane oxidation.

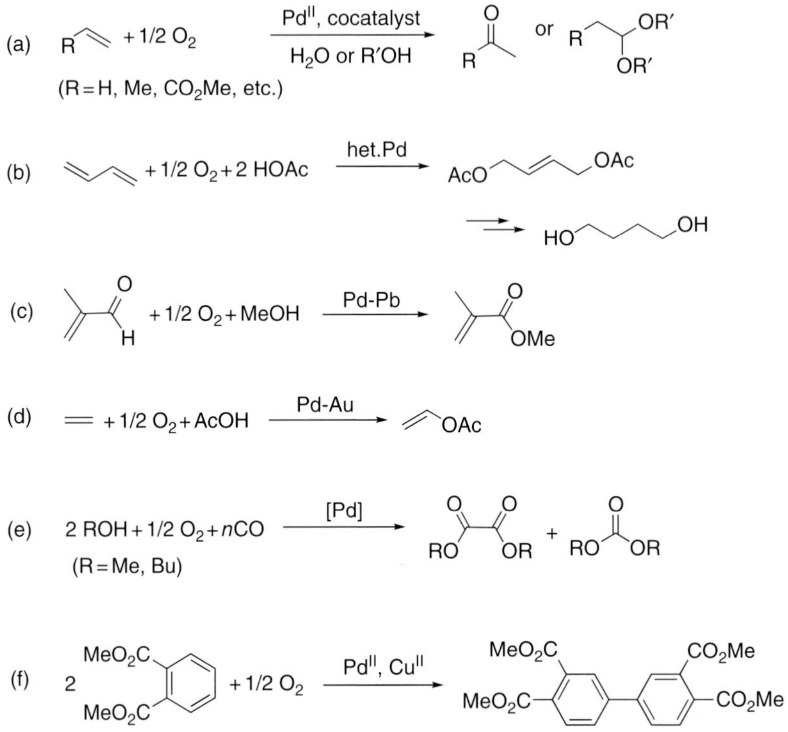

Scheme 8.1

(a) R⎯⎯+ 1/2 O$_2$ $\xrightarrow{\text{Pd}^{II}, \text{cocatalyst}}_{\text{H}_2\text{O or R'OH}}$ R−C(=O)−CH$_3$ or R−CH(OR')−CH$_2$OR'

(R = H, Me, CO$_2$Me, etc.)

(b) CH$_2$=CH−CH=CH$_2$ + 1/2 O$_2$ + 2 HOAc $\xrightarrow{\text{het.Pd}}$ AcO−CH$_2$−CH=CH−CH$_2$−OAc

⟶ HO−CH$_2$−CH$_2$−CH$_2$−CH$_2$−OH

(c) CH$_2$=C(CH$_3$)−CHO + 1/2 O$_2$ + MeOH $\xrightarrow{\text{Pd-Pb}}$ CH$_2$=C(CH$_3$)−C(=O)OMe

(d) CH$_2$=CH$_2$ + 1/2 O$_2$ + AcOH $\xrightarrow{\text{Pd-Au}}$ CH$_2$=CH−OAc

(e) 2 ROH + 1/2 O$_2$ + nCO $\xrightarrow{\text{[Pd]}}$ RO−C(=O)−C(=O)−OR + RO−C(=O)−OR

(R = Me, Bu)

(f) 2 (MeO$_2$C)$_2$C$_6$H$_4$ + 1/2 O$_2$ $\xrightarrow{\text{Pd}^{II}, \text{Cu}^{II}}$ biphenyl-tetracarboxylate (MeO$_2$C, CO$_2$Me on each ring)

Scheme 8.1 Industrial examples of Pd-catalyzed aerobic oxidation.

Scheme 8.2

(a) 2 PhOH + 1/2 O$_2$ + CO $\xrightarrow{\text{[Pd]}}$ PhO−C(=O)−OPh

(b) RCH$_2$OH + 1/2 O$_2$ $\xrightarrow{\text{[Pd]}}$ RCHO

(R = aryl, vinyl)

(c) C$_6$H$_6$ + 1/2 O$_2$ + RCO$_2$H $\xrightarrow{\text{[Pd]}}$ Ph−O$_2$CR

(d) PhCH$_3$ + 1/2 O$_2$ + HOAc $\xrightarrow{\text{[Pd]}}$ PhCH$_2$−OAc

(e) C$_6$H$_6$ + 1/2 O$_2$ + CH$_2$=CHR $\xrightarrow{\text{[Pd]}}$ Ph−CH=CH−R

(R = Ph, COOEt, etc.)

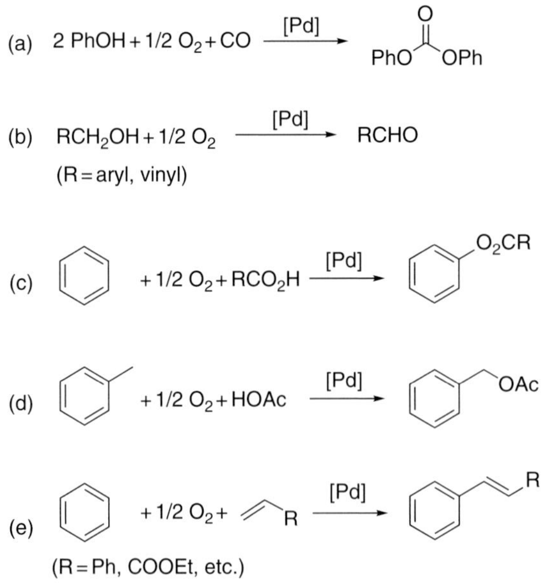

Scheme 8.2 Additional Pd-catalyzed aerobic oxidations that have been targets of extensive research.

8.2
Chemistry and Catalysis: Industrial Applications

8.2.1
Acetoxylation of Alkenes to Vinyl or Allyl Acetates

Pd-catalyzed acetoxylation of ethylene with acetic acid and O_2 accounts for approximately 80% of today's vinyl acetate production [1]. Vinyl acetate has a worldwide production capacity of about 6 million tons/year (2007) and is used to prepare a number of important polymers (e.g., PVA, EVA, PVCA).

Oxidative acetoxylation of ethylene was originally developed as a liquid phase process using homogenous $PdCl_2/CuCl_2$ catalysts similar to the Wacker process; however, corrosion problems originating from the HOAc solvent motivated the development of new catalysts and process technology. Today, the reaction is conducted in the gas phase using a process developed by researchers at Bayer [2] and Hoechst [3]. A typical process operates at 140 °C and employs a catalyst system containing metallic Pd and alkali acetate supported on silica, alumina, or activated charcoal [1]. Modern catalysts employ noble metals (mostly Au) as additional promoters (Eq. (8.1)).

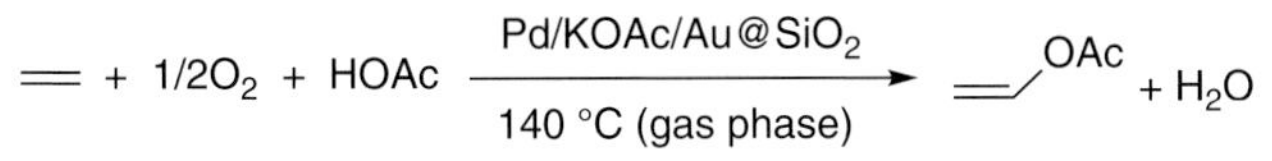

$$= + \ 1/2 \ O_2 + HOAc \xrightarrow[\text{140 °C (gas phase)}]{Pd/KOAc/Au@SiO_2} =\!\!\diagup^{OAc} + H_2O \quad (8.1)$$

Various gas–solid mechanisms have been proposed for vinyl acetate synthesis [4]; however, there is evidence that a liquid phase mechanism also occurrs in the reaction. First Samanos *et al.* [5] and later MacGowan [6] demonstrated that under typical plant conditions, condensation of HOAc on silica and alumina supports can be substantial, resulting in about three monolayers of adsorbed HOAc [6]. Lerou *et al.* noted that the presence of the HOAc layer is consistent with the observed promoting effects from KOAc [7]. The proposed roles of KOAc in the liquid phase include (i) helping to immobilize HOAc by the formation of $KH(OAc)_2$ and (ii) reacting with $Pd(OAc)_2$, formed via oxidation of metallic Pd, to the more active $Pd(OAc)_3^-$, $Pd(OAc)_4^{2-}$ species (Scheme 8.3). Recently, Lercher *et al.* demonstrated that the liquid phase plays a key role in the self-organization of supported Pd/Au catalysts during vinyl acetate synthesis. Reconstruction of bimetallic Pd/Au particles into a PdAu phase and dispersed Pd was observed, which was attributed to the extraction of Pd into the liquid layer consisting of HOAc and H_2O [8].

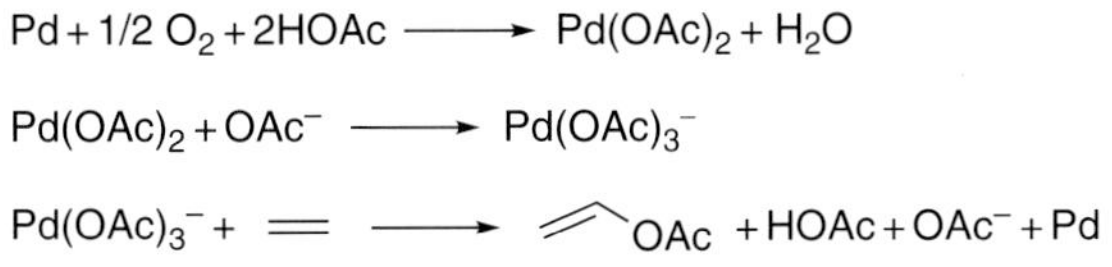

$Pd + 1/2 \ O_2 + 2HOAc \longrightarrow Pd(OAc)_2 + H_2O$

$Pd(OAc)_2 + OAc^- \longrightarrow Pd(OAc)_3^-$

$Pd(OAc)_3^- + = \longrightarrow \diagup\!\!^{OAc} + HOAc + OAc^- + Pd$

Scheme 8.3 Proposed liquid phase type mechanism in vinyl acetate synthesis [7]

It should be noted that despite the attention given to the liquid HOAc layer, more traditional surface-mediated mechanisms remain the conventional model in the literature. For example, recent kinetic studies on vinyl acetate synthesis over Pd-only [9] and Pd/Au [10] catalysts suggest that the enhanced activity of the Pd/Au catalyst originates from its ability to provide more binding sites for O_2 and thus increase the O_2 mobility on the catalyst surface.

A gas-phase process has also been developed for the conversion of propene to allyl acetate. This process involves the reaction of propylene with HOAc and O_2 over heterogeneous Pd (Pd-Cu-KOAc/SiO_2) at 100–300 °C. The allyl acetate product is then hydrolyzed to afford allyl alcohol (Eqs. (8.2) and (8.3)) [11]. Pd-catalyzed acetoxylation of propylene affords propenyl and isopropenyl acetate as the major products [12]. Showa Denko K. K. had used this route to produce allyl alcohol since 1985 [13]. This process contributes significantly to allyl alcohol production (~70 000 t/year) and has the potential to be the main route in the future [13].

$$\text{propene} + 1/2\,O_2 + \text{HOAc} \xrightarrow[\text{gas phase}]{\text{Pd/KOAc/Cu@SiO}_2} \text{allyl-OAc} + H_2O \quad (8.2)$$

$$\text{allyl-OAc} + H_2O \xrightleftharpoons{H^+} \text{allyl-OH} + \text{HOAc} \quad (8.3)$$

8.2.2
Oxidative Carbonylation of Alcohols to Carbonates, Oxalates, and Carbamates

Oxidative carbonylation of alcohols in the presence of CO provides an economically viable route to dialkyl carbonates and/or oxalates (Eqs. (8.4) and (8.5)), both of which have important industrial applications. Dialkyl carbonates (e.g., dimethyl carbonate, propylene carbonate) are excellent solvents for a variety of organic substances [14]. Dialkyl oxalates have utility as solvents, C2 building blocks in fine chemicals synthesis, and intermediates in the manufacture of oxamide (as a fertilizer) [15]. Hydrogenation of dialkyl oxalates provides an alternative route to ethylene glycol that is independent of oil-derived resources [15, 16].

$$2\,ROH + CO + 1/2\,O_2 \xrightarrow{[Pd]} RO-C(=O)-OR \quad (8.4)$$

$$2\,ROH + 2\,CO + 1/2\,O_2 \xrightarrow{[Pd]} RO-C(=O)-C(=O)-OR \quad (8.5)$$

A major challenge in Pd-catalyzed oxidative carbonylation reactions is the decomposition of catalysts into inactive metallic Pd, promoted by the strong reducing CO reagent. Successful examples of Pd-catalyzed oxidative carbonylation typically

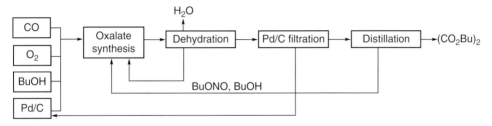

Figure 8.1 Flow diagram of Ube liquid phase process of dibutyl oxalate [18].

employ redox-active cocatalysts (e.g., Cu^{2+}, Fe^{3+}, benzoquinone (BQ)) and/or ligands to prevent Pd decomposition [17]. Extensive efforts toward the identification of a commercially viable Pd-catalyzed route to diphenyl carbonate are elaborated in Chapter 12. A Cu-based process has been developed as the main route to dimethyl carbonate production (see Chapter 5 for a discussion of this process).

Ube Industries has used a liquid phase Pd-catalyzed process for the manufacture of dibutyl oxalates since 1978 at a capacity of several thousand tons per year [15, 18]. The process employs a Pd/C catalyst and ^{n}BuONO as the redox mediator and typically operates at 90–110 °C [19]. Under these conditions, the catalyst exhibits high activity (TOF > 10 000 h^{-1}, which decreases gradually because of the formation of trace cyanide). Although many details of the mechanism are not understood, the reaction is considered to consist of two general reaction steps (Eqs. (8.6–8.8)): (i) Pd-mediated oxidative coupling of 2 equiv. of carbon monoxide and the ^{n}BuO fragments of ^{n}BuONO (Eqs. (8.6)) and (ii) aerobic oxidative coupling of ^{n}BuOH and NO to generate ^{n}BuONO (Eq. (8.7)). Water is removed azeotropically with alkyl nitrite to avoid water inhibition of the process (Figure 8.1). For this purpose, butyl nitrite is selected as the most favorable nitrite source

$$2CO_2 + 2\,^n BuONO \xrightarrow{[Pd]} (COO^n Bu)_2 + 2NO \quad (8.6)$$

$$2NO + 2\,^n BuOH + 1/2 O_2 \rightarrow 2\,^n BuONO + H_2 O \quad (8.7)$$

$$\overline{2CO + 2\,^n BuOH + 1/2 O_2 \rightarrow (COO^n Bu)_2 + H_2 O} \quad (8.8)$$

Dibutyl carbonate is a by-product of the Ube dibutyl oxalate process, and conditions that lead to a higher oxalate:carbonate ratio include lower reaction temperatures, lower concentrations of butanol, and higher CO pressure. A reaction mechanism for dibutyl oxalate and carbonate formation has been proposed based on the observed trends for selectivity and kinetic studies (Scheme 8.4) [18, 20]. Pd^0 species are believed to favor oxalate formation, while Pd^{II} species favor carbonate formation (step 1a/2a). Oxalate formation starts with facile oxidative addition of ^{n}BuONO to Pd^0 (step 1b) and is followed by the rate-limiting CO insertion (step 1c).

Ube has also developed Pd/MeONO-based gas-phase processes for the production of dimethyl oxalate and dimethyl carbonate (Eqs. (8.9) and (8.10)) using

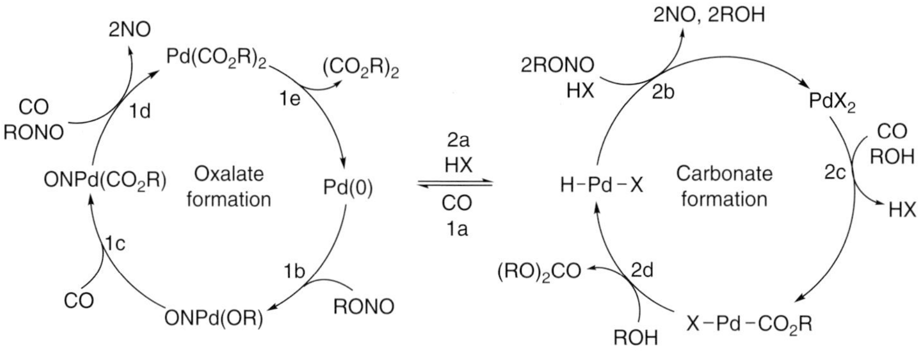

Scheme 8.4 Proposed mechanism for oxidative carbonylation of alcohols using alkyl nitrites.

heterogeneous Pd catalysts. Pd^0 on activated carbon affords dimethyl oxalate selectively. On the other hand, dimethyl carbonate is formed when the same catalyst is treated with HCl [18]. Other effective supports include activated carbon, silica, alumina, NaY zeolite, and silica-alumina. The HCl effect may originate from the oxidation of Pd^0 to Pd^{II} (Scheme 8.4, step 2a). As of 2005, Ube was producing dimethyl carbonate at a capacity of approximately 10 000 t/year and was planning to produce dimethyl oxalate at a comparable scale [21].

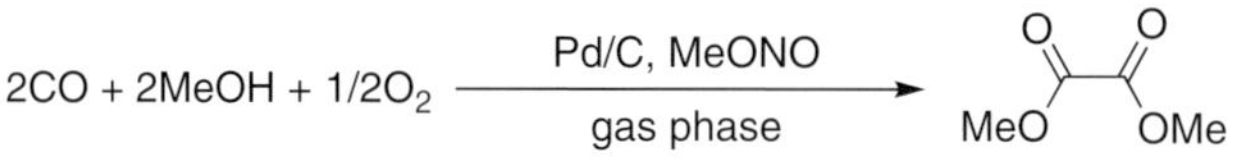

$$2CO + 2MeOH + 1/2 O_2 \xrightarrow[\text{gas phase}]{\text{Pd/C, MeONO}} \text{MeO-C(O)-C(O)-OMe} \quad (8.9)$$

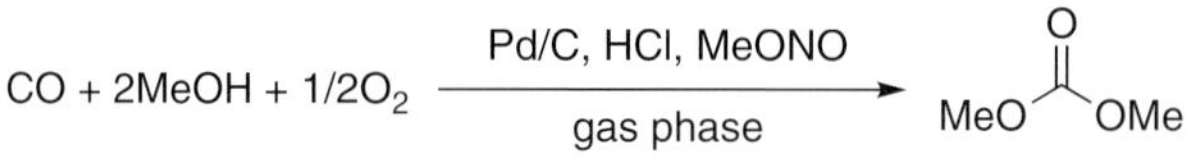

$$CO + 2MeOH + 1/2 O_2 \xrightarrow[\text{gas phase}]{\text{Pd/C, HCl, MeONO}} \text{MeO-C(O)-OMe} \quad (8.10)$$

In the presence of O_2, aniline and alcohol undergo oxidative carbonylation with CO to produce N-phenyl carbamates (Eq. (8.11)). A homogenous process was developed by researchers at Bayer with a catalyst system containing $PdCl_2$ and Fe^{III} [22, 23]. The same transformation can also proceed with heterogeneous Pd/C, as reported by Asahi Chemical (Japan) [23–25]. The reaction is significantly promoted by adding KI [26]. The resulting carbamate product can undergo further transformation to methylene diphenyl diisocyanate (Eq. (8.12)), which is one of the most important intermediates for the manufacture of polyurethanes [25].

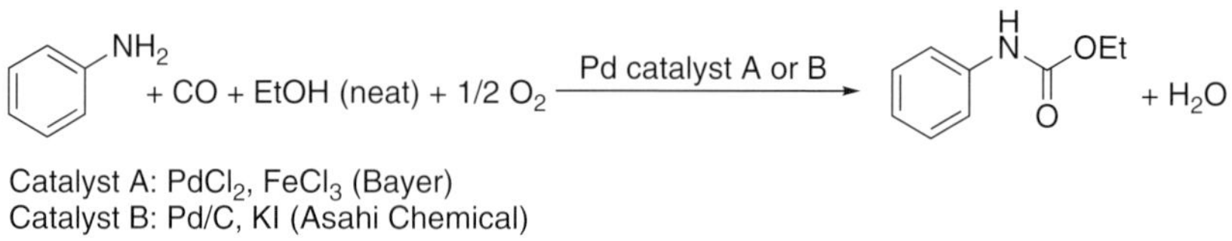

PhNH$_2$ + CO + EtOH (neat) + 1/2 O_2 $\xrightarrow{\text{Pd catalyst A or B}}$ PhNH-C(O)-OEt + H_2O

Catalyst A: $PdCl_2$, $FeCl_3$ (Bayer)
Catalyst B: Pd/C, KI (Asahi Chemical)

(8.11)

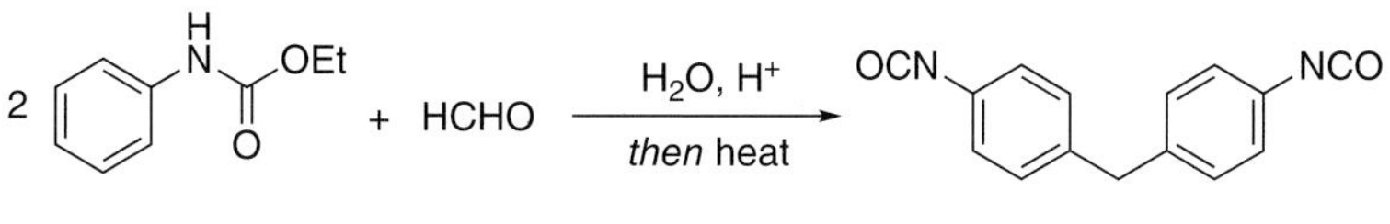

$$2 \;\; \text{PhNHC(O)OEt} + \text{HCHO} \xrightarrow[\text{then heat}]{\text{H}_2\text{O, H}^+} \text{OCN-C}_6\text{H}_4\text{-CH}_2\text{-C}_6\text{H}_4\text{-NCO} \qquad (8.12)$$

8.2.3
Oxidative Coupling of Arenes to Biaryl Compounds

Pd-catalyzed aerobic oxidative coupling of arenes provides efficient access to biaryl compounds. The reaction exhibits near-ideal atom economy with water as the sole by-product; however, it typically suffers from low regioselectivity and the biaryl products are often susceptible to further oxidation to afford oligomers.

Pd-catalyzed oxidative coupling of dimethyl phthalate has been developed and commercialized by Ube Industries (Japan) [27]. This reaction is a key step in the preparation of the monomeric precursor to Upilex, a high-performance polyimide resin (Scheme 8.5) [28]. Using $Pd(OAc)_2/Cu(OAc)_2$ as the catalyst system, the reaction affords 3,4,3′,4′-tetramethyl biphenyltetracarboxylate with high regioselectivity (93 : 7 vs. 2,3,3′,4′-isomer) and good biaryl selectivity (83%) [29]. It was noted that $Cu(OAc)_2$ does not change the initial coupling rate, but elevates the lifetime of the Pd catalyst and allows for the use of lower O_2 pressure [30]. Addition of phenanthroline contributes to higher regioselectivity. Although the reported turnover numbers (TONs) of the Pd catalyst are only modest (~300) [31], the patent literature implies that the process is viable as a result of an efficient method for Pd recycling involving hydrogenation work-up of the reaction mixture to yield metallic palladium without arene reduction [32].

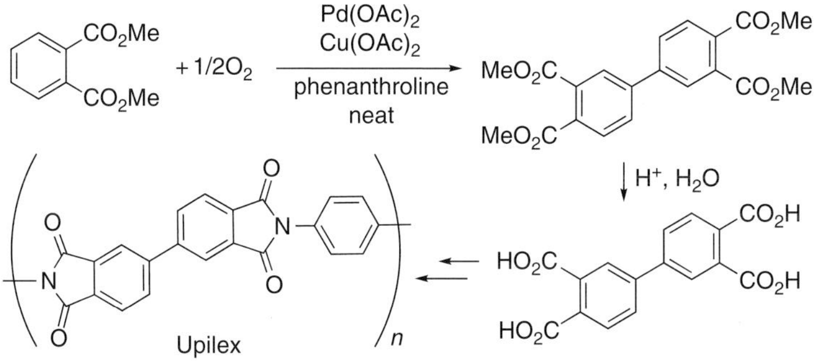

Scheme 8.5 Pd-catalyzed aerobic oxidative coupling of dimethyl phthalate.

The preparation of biphenyltetracarboxylic acid monomer in Upilex production could potentially be streamlined by carrying out the homocoupling of o-xylene, followed by oxidation of the benzylic methyl groups (Eq. (8.13)) [33]. The major drawback associated with the coupling of o-xylene is the reduced bixylyl selectivity arising from benzylic oxidation to benzyl esters and significant production of high-boiling oligoaromatic by-products [34]. It has been shown recently, however, that the use of 2-fluoropyridine as a ligand and Cu(OTf)$_2$ as a cocatalyst improves the bixylyl selectivity to greater than 90% [35].

$$2 \; \text{(o-xylene)} \xrightarrow{\text{[Pd], O}_2} \text{(bixylyl)} \xrightarrow{\text{Co/Mn/Br}^-} \text{(tetracarboxylic acid)} \tag{8.13}$$

Oxidative coupling of other arenes (e.g., toluene, benzene, methyl benzoate) has also been investigated. The activity of Pd^{II} catalysts is enhanced by various additives, including β-diketones [36], Zr^{IV} [37], MoO$_2$(acac)$_2$ [38], Zr^{IV}/Co^{II}/Mn^{II} [39], HPMoV-type heteropolyacids (HPA) [40], and triflic acid [41]. Acetic acid is a common solvent for these reactions. It was disclosed recently that supported nanoparticle Au catalysts can be applied as catalysts in reactions of this type [42].

8.3
Chemistry and Catalysis: Applications of Potential Industrial Interest

8.3.1
Oxidation of Alcohols to Aldehydes

Heterogeneous and homogeneous Pd catalysts for aerobic oxidation of alcohols to aldehydes, ketones, and carboxylic acids have been the subject of considerable academic study [43]. To our knowledge, few of these have been applied commercially. Aerobic oxidation of 2-hydroxybenzyl alcohol has received considerable industrial interest because the aldehyde product, salicylaldehyde, is used in perfumes and is a precursor to coumarin, and the industrial process employs a Pt/Pb/C catalyst [44]. Other routes employing Pd-based catalysts have also been investigated. In a process developed by Rhone-Poulenc [45], the oxidation of 2-hydroxybenzyl alcohol occurs over a catalyst system consisting of Pd/charcoal and Bi$_2$(SO$_4$)$_3$ (Eq. (8.14)). Addition of bismuth led to improved reaction rate (8 times faster) and product yields (10% increase) and allowed 10 times reduction in Pt loading. Recent studies on Pd/Bi-catalyzed oxidation of 1-phenylethanol have suggested that bismuth can suppress decarbonylation of benzaldehyde and over-oxidation to benzoic acid [46]. In another process developed by Dow Chemical [47], oxidation of 2-hydroxybenzaldehyde takes place in a multiphase system composed of Pd/charcoal and aq. NaOH in a H$_2$O/CH$_2$Cl$_2$ mixture. The multiphase protocol allows efficient isolation of the aldehyde product from the organic phase and thus eliminates the need for neutralization.

$$\text{2-hydroxybenzyl alcohol} + 1/2\,O_2 \xrightarrow[\text{Bi}_2(\text{SO}_4)_3, \text{ aq. NaOH}]{\text{Pd on charcoal}} \text{salicylaldehyde} + H_2O \tag{8.14}$$

There are also examples of gas-phase oxidation of allyl alcohol using Pd-Cu or Pd-Ag catalysts (Eq. (8.15)) [48]. The protocol was part of a multistep propylene-oxidation process, and the allyl alcohol starting material was produced from Pd-catalyzed acetoxylation of propylene followed by hydrolysis of allyl acetate (cf. Eqs. (8.2) and (8.3)) [48]. The presence of water improves the conversion of alcohol to acrylic products.

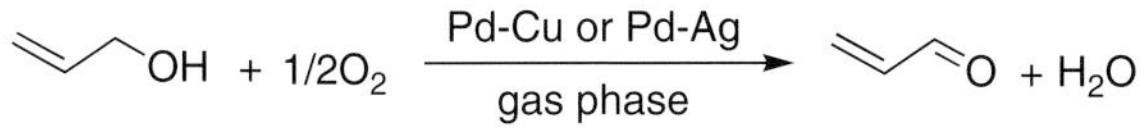

$$\text{CH}_2\text{=CHCH}_2\text{OH} + 1/2\,\text{O}_2 \xrightarrow[\text{gas phase}]{\text{Pd-Cu or Pd-Ag}} \text{CH}_2\text{=CHCHO} + \text{H}_2\text{O} \qquad (8.15)$$

8.3.2
Oxidation of Arenes to Phenols and Phenyl Esters

Oxidative esterification of arenes with carboxylic acids produces aryl esters, which can be used as precursors to valuable phenol derivatives (Scheme 8.6). Commercial production of phenol involves the aerobic oxidation of cumene to cumene hydroperoxide, followed by conversion to acetone and phenol under acidic conditions (Hock process) [49]. Aerobic acetoxylation of benzene to phenyl acetate provides a potential alternative route to phenol, and Pd-catalyzed methods for this transformation have been the focus of considerable effort. None of these methods are yet commercially viable, however.

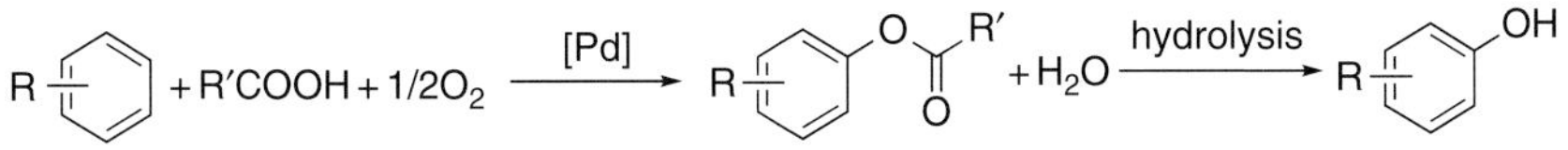

$$\text{R-C}_6\text{H}_5 + \text{R'COOH} + 1/2\,\text{O}_2 \xrightarrow{[\text{Pd}]} \text{R-C}_6\text{H}_4\text{-OC(O)R'} + \text{H}_2\text{O} \xrightarrow{\text{hydrolysis}} \text{R-C}_6\text{H}_4\text{-OH}$$

Scheme 8.6 Pd-catalyzed oxidative esterification of arenes to aryl esters and phenol derivatives.

In the first report of benzene acetoxylation by Davidson and Triggs in 1966, phenyl acetate was observed as a by-product in the formation of biphenyl with stoichiometric Pd(OAc)$_2$. Addition of LiOAc improved the yield of PhOAc, suggesting an acetate attack on a Pd-aryl species [50]. Investigation with NaNO$_2$ afforded a mixture of phenyl acetate and nitrobenzene with up to 60 TONs after 3 days with the catalyst remaining active with no Pd-black formation [51]. Selectivity for benzene acetoxylation improved with the addition of HNO$_3$ and O$_2$ in stoichiometric reactivity [52], and phenyl acetate formation can be promoted by using more nucleophilic carboxylic acids (e.g., pivalic acid vs. dichloroacetic acid) [53]. A recent variation of this chemistry has achieved 136 TONs of Pd at 1 atm of O$_2$, with PhOAc:PhNO$_2$ ratios up to 40:1 [54]. Patents also describe the use of NO$_x$ additives and either homogeneous or supported Pd for the conversion of benzene to phenyl acetate, although low TONs are reported (<10) [55]. Attention has also been given to the synthesis of supported Pd catalysts for use in either liquid or vapor phase [56]. Additives such as Bi [57], Au, Dy, Cd [58], Zn, and Sb [59] have been incorporated into these catalysts.

Ashland Oil, Inc. has developed solution-phase methods for the Pd-catalyzed oxidative esterification of benzene [60], naphthalene [61], and styrene [62]. Acetic acid and low-molecular-weight carboxylic acids were substituted with heavier carboxylic acids such as lauric acid because of corrosion problems, carboxylic acid stability, and ease of recyclability. In their studies, they noted that catalytic amount of fluorinated carboxylic acids was beneficial [60a] through the formation of a more for preventing cationic Pd center *in situ* [63]. Dean–Stark conditions were found to be beneficial by preventing the formation of phenol via hydrolysis of the aryl ester. The highest TON obtained was 90.4 for benzene [60b] and 69 for naphthalene [61b]. The synthesis of naphthol from tetralin gave 55.6 TONs, and the inventors remarked that this method allows one to bypass the difficulties associated with the separation of tetralone and tetralol from naphthol [64]. It was also found that there was very little decarboxylation or biaryl formation. Additives including $Sb(OAc)_3$, $Cr(OAc)_3$, $Mn(OAc)_3$, and $Zn(OAc)_2$ were also incorporated in the reaction mixture. The role of metal acetate additives in Pd-catalyzed acetoxylation reactions has been investigated, and the formation of $PdM(OAc)_4$ (M = Cu, Pb, Zn, K) dimers and the palladate complex $[K_2Pd(OAc)_4]$ [65] was observed. Tosoh Corporation has disclosed the use of a supported Pd catalyst that includes early transition-metal or lanthanide oxides with main group additives [66] and also observed that addition of alcohols, aldehydes, cyclic hydrocarbons, and formic acid can improve catalyst lifetime [67].

Direct synthesis of phenol is also possible by the introduction of water as a cosolvent, and the reaction is significantly improved by using molybdovanadophosphoric heteropoly acid (HPA) as a cocatalyst for the reoxidation of Pd (Eq. (8.16)). By using a solvent $HOAc/H_2O$ (1:2), HPAs with catalytic $Pd(OAc)_2$ and very high O_2 pressure (60 bar of O_2), phenol was obtained in 8% yield (TON = 800) [68]. At much lower O_2 pressure (3 bar) in the presence of CO (7 bar) as reductant in aqueous NaOAc/AcOH, catalytic Pd/Al_2O_3 (0.1 mol% Pd) with $H_5PV_2Mo_{10}O_{40}$ as cocatalyst hydroxylates toluene to cresols with only minor benzylic oxidation and over-oxidation to 2-methylquinone, despite the very high conversion (80%) [69]. The unusually high conversion to phenols without significant quinone formation may reflect the nucleophilic character of the active oxygenating species, making the electron-rich phenolic product less prone to arene oxidation than the toluene substrate. The use of CO as a co-substrate (sacrificial reductant) resembles an earlier Pd/phenanthroline catalyst system that achieved multiple turnovers in the oxidation of benzene to phenol under elevated pressures (15 bar O_2 and CO) [70].

$$\text{C}_6\text{H}_6 + 1/2\,\text{O}_2 \;(60\,\text{bar}) \xrightarrow[\text{HOAc/H}_2\text{O (1:2)},\; 130\,°\text{C},\; 4\,\text{h}]{\substack{0.01\,\text{mol\% Pd(OAc)}_2 \\ 10\,\text{mol\% LiOAc} \\ \text{H}_{3+x}\text{PMo}_{12-x}\text{V}_x\text{O}_{40}}} \text{C}_6\text{H}_5\text{OH} \quad \text{TON} = 800 \tag{8.16}$$

8.3.3
Benzylic Acetoxylation

Benzyl acetates are important precursors to fragrances, and their hydrolyzed alcohol products are valuable synthons. Benzylic acetoxylation of toluene to benzyl acetate is seen as a potential route for commercial production. The major existing route to benzyl acetate proceeds via benzyl chlorides [71]. Pd-catalyzed aerobic acetoxylation toluene and other methyl arenes would offer an appealing alternate route to these products (Scheme 8.7).

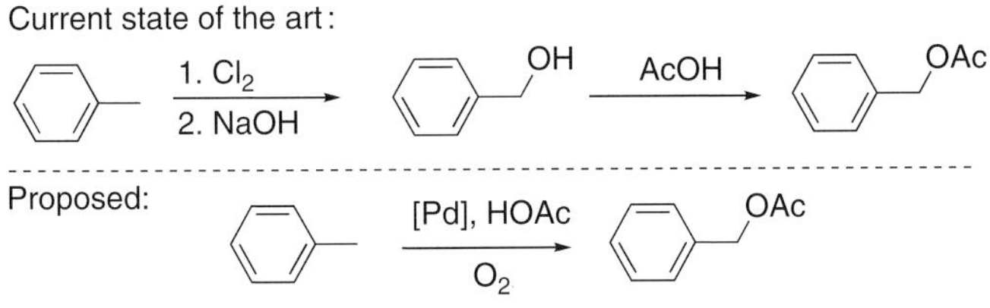

Current state of the art:

$$\text{PhCH}_3 \xrightarrow[\text{2. NaOH}]{\text{1. Cl}_2} \text{PhCH}_2\text{OH} \xrightarrow{\text{AcOH}} \text{PhCH}_2\text{OAc}$$

Proposed:

$$\text{PhCH}_3 \xrightarrow[\text{O}_2]{\text{[Pd], HOAc}} \text{PhCH}_2\text{OAc}$$

Scheme 8.7 Routes for benzyl acetate production.

Early observations of benzylic acetoxylation were made in the study of arene acetoxylation and biaryl coupling when toluene was used as a substrate. In 1966, the reaction between stoichiometric Pd(OAc)$_2$ and toluene to give benzyl acetate as the major product was disclosed [72]. Two years later, acetoxylation of toluene with catalytic Pd salts was reported by Union Carbide by using phosphines or a combination of Sn(OAc)$_2$, charcoal, and air as oxidant to give 96 TONs [73]. Additional metal acetates such as KOAc are beneficial for the reaction [74]. These acetoxylation methods were further applied to other arenes [75] (e.g., benzene, cyclohexene) and the synthesis of benzyl diacetate [76] (a precursor to benzaldehyde).

Multiple researchers have observed an induction period [77, 78], which is implicated to correspond with the *in situ* reduction of Pd(OAc)$_2$ and formation of a heterogeneous catalyst [79]. All additives (charcoal, Sn(OAc)$_2$, KOAc) contributed to the reduction of Pd^{II} *in situ*. However, catalysts prepared *ex situ* were found to be less active. This could be due to the impregnation of K atoms into the heterogeneous scaffold, since K$^+$ may help render the Pd center more electrophilic. High O$_2$ pressure inhibited the reaction due to the formation of PdO. The addition of exogenous Sn(OAc)$_2$ resuscitated catalyst activity by removing the surface oxide layer on Pd. It was also suggested that tin could facilitate oxidation of Pd hydrides, for example, by SnO$_2$ reacting with neighboring Pd hydrides to form water (with an equivalent of H$^+$) (Eq. (8.17)). The formed SnO can then reduce surface PdO via oxygen atom migration from Pd to Sn (Eq. (8.18)) [79b].

$$Pd_x - H^- + SnO_2 + H^+ \rightarrow Pd_x + SnO + H_2O \qquad (8.17)$$

$$PdO + SnO \rightarrow Pd + SnO_2 \tag{8.18}$$

Characterization of *in situ* and *ex situ* synthesized catalysts on a silica support confirms the presence of tin oxides and tin hydroxyl species [80]. In the same report, the authors determined that a 1 : 2 Pd:Sn stoichiometry is optimal and implicated PdSn$_2$ particles as being important for effective catalysis. It was proposed that *in situ* catalyst formation is a two-step process that involves the formation of a Pd^{II}Sn^{II}$_2$(OAc)$_6$ complex followed by the decomposition of this complex to give oxygenated PdSn$_2$ clusters. Other Pd-based catalysts have also been developed. The addition of Bi [71, 81], persulfate/Sn (Phillips Petroleum Co.) [82], Sn/Sb (BP) [83], and ultrafine Au [84] have been shown to be beneficial.

Two generally accepted reaction mechanisms for benzylic acetoxylation are shown in Scheme 8.8 [79b]. One proceeds with the rupture of a benzylic C–H bond to give surface-bound benzyl and hydride species. The Pd-benzyl species is then attacked by acetate. The other mechanism involves a concerted substitution to add an acetate ion and release of hydride to the Pd surface. This field continues to be an area of active research [85–88, 88], although high yields of benzyl acetate remain elusive [89].

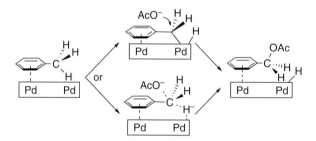

Scheme 8.8 Proposed mechanism for Pd-based acetoxylation of toluene [79b].

8.3.4
Arene Olefination (Oxidative Heck Reaction)

The formation of styrene via oxidative coupling of ethylene and benzene has been a long-standing target in the catalysis community. Currently, styrene is produced via alkylation of benzene with ethylene followed by thermal dehydrogenation. Drawbacks of this process include the involvement of two unit operations and the formation of side products such as ethylene oligomers, ethylene oxide, toluene, and products from multiple alkylation on the same benzene ring [90]. Pd-catalyzed C–C coupling of aryl halide and alkene [91] (Mizoroki-Heck reaction) has been a versatile protocol for the synthesis of styrene derivatives in the pharmaceutical and fine chemicals industries, but it generates stoichiometric amounts of inorganic waste and would not be applicable on the scale of styrene production. Pd-catalyzed oxidative couplings of arenes and alkenes (oxidative Heck reactions) operate at near-ideal atom economy to afford styrene derivatives (Eq. (8.19)), and it has been an active area of research in both academia and industry.

$$\text{C}_6\text{H}_6 + \text{CH}_2=\text{CHR} + [\text{O}] \xrightarrow{[\text{Pd}]} \text{PhCH=CHR} + [\text{O}]\text{H}_2 \quad (8.19)$$

The initial discovery of the oxidative Heck reaction was reported by Fujiwara and Moritani in 1967 when they disclosed the coupling of styrene with benzene in the presence of acetic acid and PdCl$_2$ to give *trans*-stilbene and α-methylbenzyl acetate (Eq. (8.20)) [92]. Attempts to achieve catalytic turnover were made by adding Cu or Ag salts, but only 1–2 TONs were obtained [93].

$$\text{C}_6\text{H}_6 + \text{CH}_2=\text{CHPh} + \text{PdCl}_2 \xrightarrow[\text{8 h, N}_2]{\text{HOAc, 65 °C}} \text{trans-stilbene (26\%)} + \text{PhCH(OAc)CH}_3 \text{ (13\%)} \quad (8.20)$$

In the 1970s, researchers at Phillips demonstrated the first catalytic oxidative Heck reaction with O$_2$ as the terminal oxidant and achieved up to 11 TONs [94]. Asahi investigated a process for the Pd-catalyzed coupling between ethylene and various arenes with Ag, Ni, U, Li, Ag, Hg, Co, Tl, and Cd additives on different supports [95]. Up to 7.8 TONs with chlorobenzene (Pd(OAc)$_2$/NiO/SiO$_2$) and 8.4 TONs for *m*-xylene (Pd-Ag/Mn/SiO$_2$) were obtained. Ishii *et al.* demonstrated that polyoxometalates promote the coupling of benzene and ethyl acrylate under air [96]. Ishii and co-workers also observed that by adding acetylacetone (acacH) they were able to improve their product selectivity. The beneficial effect of added acacH had been previous observed by researchers at Nippon Mitsubishi Oil Corp for Pd(OAc)$_2$ and various rhodium catalysts [97]. Jacobs *et al.* reported a method that achieved 762 TONs, 90% conversion, and 95% selectivity in the coupling of anisole and ethyl cinnamate (Eq. (8.21)) [97]. Tsuji and Nagashima obtained 35 TONs in the coupling of benzene and methyl acrylate using catalytic Pd(OAc)$_2$ and BQ rather than O$_2$ as oxidant [98]. Fujiwara *et al.* obtained 280 TONs in the coupling of benzene and ethyl cinnamate using BQ/^{t}BuOOH as the oxidant [99].

$$\text{MeO-C}_6\text{H}_5 \text{ (neat)} + \text{CH}_2=\text{CHCO}_2\text{Et} \xrightarrow[\text{24 h, 110 °C, 7.9 atm O}_2]{\text{0.12 mol\% Pd(OAc)}_2, \text{ 0.18 mmol BzOH}} \text{MeO-C}_6\text{H}_4\text{-CH=CH-CO}_2\text{Et}$$

(o/m/p 35:15:50)
TON = 762
TOF = 73 h^{-1}

(8.21)

Aerobic oxidative Heck reactions also proceed between olefins and other aryl nucleophiles such as aryl tin and aryl boron reagents (Eq. (8.22)). This field started by utilizing aryl tin reagents and electron-deficient alkenes with stoichiometric base additives such as NaOAc [100], but has been significantly improved by using boronic acid and boronic ester [101] derivatives to couple with alkenes of various electronic and steric profiles. Many of these reactions employ 2,9-dimethylphenanthroline as ligand [102]. These applications offer complementary scope to traditional Mizoroki-Heck reactions and often can be performed at room temperature; however, they also form undesirable stoichiometric by-products.

$$\text{Ph-B(OH)}_2 + \text{CH}_2=\text{CHR} + 1/2\,O_2 \xrightarrow{Pd^{II}/L} \text{Ph-CH=CHR} + B(OH)_3 \quad (8.22)$$

8.4
Chemistry and Catalysis: New Developments and Opportunities

There are several serious challenges associated with Pd-catalyzed aerobic oxidations. These include (i) controlling the chemo- and regioselectivity, (ii) promoting the oxidizing ability of Pd^{II}, particularly in challenging oxidation reactions, such as C–H oxidations, and (iii) enhancing the Pd catalyst lifetime. In this section, we describe a few new developments in Pd oxidation catalysis that point toward potential solutions to the aforementioned challenges.

8.4.1
Ligand-Modulated Aerobic Oxidation Catalysis

One strategy to promote aerobic Pd-catalyzed oxidation reactions and to improve catalyst lifetime is to use ancillary ligands to stabilize Pd^0. Phosphine ligands play an important role in traditional Pd-catalyzed cross-coupling reactions [103] by promoting key oxidative addition and reductive elimination steps, but these ligands are unstable under aerobic oxidation conditions [104]. Most ligands that have exhibited success in Pd-catalyzed aerobic oxidation reactions consist of sulfoxides [105] and mono- and bidentate nitrogen ligands, such as pyridine [106], and phenanthroline [107] derivatives. These oxidatively stable ligands help minimize Pd^0 aggregation into Pd black and modulate the reactivity and selectivity of Pd^{II} in the substrate oxidation steps (Scheme 8.9).

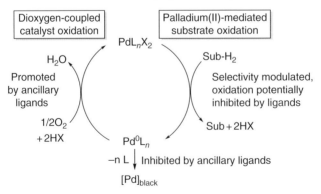

Scheme 8.9 Roles of ligands in Pd-catalyzed aerobic oxidations.

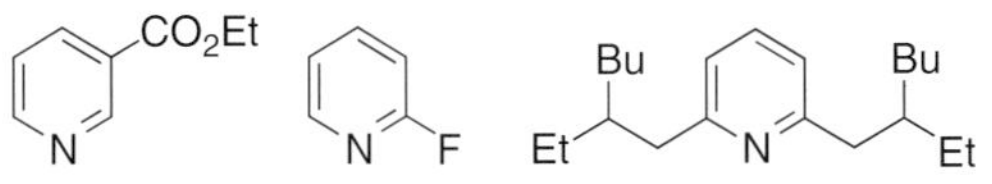

Figure 8.2 Representative effective pyridine-based ligands in Pd-catalyzed aerobic oxidations.

Weakly coordinating ligands have shown considerable success, including pyridines bearing electron-withdrawing groups (e.g., ethyl nicotinate [108] and 2-fluoropyridine [35]) and sterically bulky 2,6-dialkyl pyridine [109] (Figure 8.2). In some reactions, such as alcohol oxidation and intramolecular Wacker-type cyclization reactions, pyridine is an effective ligand. Mechanistic studies show that the optimal rate is achieved at a 1 : 1 pyridine/Pd ratio, but catalyst lifetime is improved in the presence of excess pyridine [110]. These observations can be explained by an active catalyst that has a single pyridine ligand in the turnover-limiting step(s), together with the presence of catalyst decomposition pathways (e.g., aggregation of Pd^0) that are inhibited by excess ligand equivalents. N-heterocyclic carbene ligands have also been used in aerobic palladium catalysis; however, their strong σ-donor ability can attenuate the rate of substrate oxidation by Pd^{II} [111].

Bidentate nitrogen ligands often inhibit Pd-catalyzed aerobic oxidation reactions in nonpolar solvents because the bidentate ligand is nonlabile and blocks access to coordination sites, while displacement of anionic ligands generates charged intermediates that are energetically unfavorable. A unique ligand, however, is 4,5-diazafluoren-9-one (DAF) (Scheme 8.10) [112], which promotes allylic acetoxylation of terminal alkenes. Most other examples of Pd-catalyzed allylic acetoxylation employ BQ as a cocatalyst or stoichiometric oxidant [113] or other strong oxidants (e.g., hypervalent iodine, peroxides) [114]. The mechanistic basis for the utility of DAF is not fully understood but has been attributed to the wide ligand bite angle of DAF that destabilizes Pd^{II} and promotes functionalization of the π-allyl-Pd^{II} intermediate [112, 115].

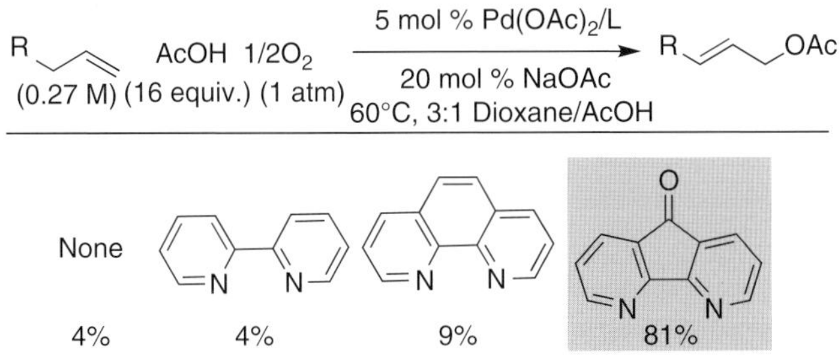

Scheme 8.10 Pd-catalyzed aerobic allylic acetoxylation with DAF.

Valuable ligand effects have been observed in oxidative Heck reactions. Arenes with weak directing groups benefit from the use of amino acid ligands such as acetylated isoleucine (Ac Ile OH), which promote C–H activation via proton transfer steps. The ligand operates in concert with the acetate ligand and shuttles between a bidentate monoanionic ligand and a protonate neutral ligand [116], and 455 TONs were obtained at low catalyst loadings under O_2 (Scheme 8.11) [117].

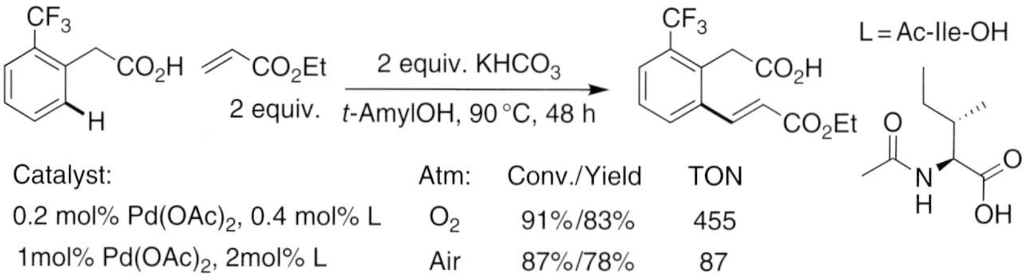

Scheme 8.11 Amino acid–promoted aerobic C–H olefination of phenylacetic acid derivatives.

Ligands also have been shown to influence the regioselectivity of oxidative Heck reactions (Scheme 8.12). Ligand-free oxidative Heck reactions between styrenyl-boronic acid and electronically unbiased alkenes favor linear dienes [118], while the use of 2,9-dimethyl-1,10-phenanthroline as ligand results in a switch in the reaction regioselectivity, significantly favoring branched diene formation [119]. The methyl substituents on the ligand are crucial to this switch in selectivity, as poor selectivity is observed with 1,10-phenanthroline and 2,2′-bipyridyl.

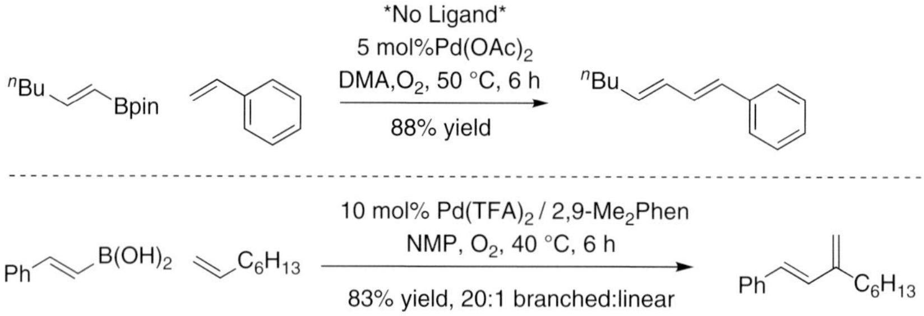

Scheme 8.12 Ligand-modulated regioselective oxidative Heck reaction for diene synthesis.

8.4.2
Use of NO_x as Cocatalyst

Nitrogen oxide (NO_x) cocatalysts [120] have received industrial interest in Pd-catalyzed aerobic oxidations such as oxidative carbonylation (see Section 8.2.2) [18], alkene oxidation [121], and arene acetoxylation [55]. Recent studies from academic literature have provided new insights into the roles of NO_x in these reactions. Pd-catalyzed aerobic alkene oxidation (Wacker reaction) typically affords methyl ketones arising from Markovnikov addition of water (or hydroxide) to an alkene, whereas anti-Markovnikov aldehyde products are favored in the presence of a NO_x source (e.g., metal nitrites) (Scheme 8.13) [122, 123]. ^{18}O labeling studies are consistent with a mechanism involving metal-mediated addition of NO_2 to the olefin [123a]. The addition is believed to favor the terminal attack because it forms a more stable secondary radical. The hypothesis was supported by a computational study [124].

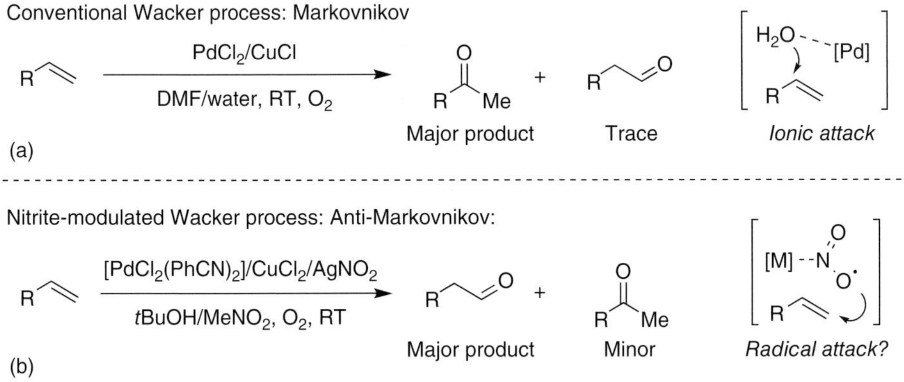

Scheme 8.13 Pd-catalyzed aldehyde-selective alkene oxidation and proposed radical model.

(a) Conventional Wacker process: Markovnikov — $PdCl_2/CuCl$, DMF/water, RT, O_2; Major product (methyl ketone) + Trace (aldehyde); Ionic attack via $[H_2O\cdots[Pd], R\text{-alkene}]$.

(b) Nitrite-modulated Wacker process: Anti-Markovnikov — $[PdCl_2(PhCN)_2]/CuCl_2/AgNO_2$, $tBuOH/MeNO_2$, O_2, RT; Major product (aldehyde) + Minor (methyl ketone); Radical attack? via $[M]\text{-}NO_2$ adding to $R\text{-alkene}$.

As described in Section 8.3.2, Pd^{II} and NO_x cocatalysts mediate the acetoxylation of benzene to phenyl acetate. Nitrobenzene is a common by-product for this reaction. The nitrobenzene to phenyl acetate ratio can be tuned via modification of the reaction condition (Scheme 8.14) [51–55]. The proposed role of NO_x in the acetoxylation reaction is to oxidize Pd^{II} intermediates to Pd^{IV} species to facilitate the otherwise difficult Ar–X reductive elimination (X = O, N). This hypothesis is supported by the fundamental studies by Cámpora et al. where they showed that a Pd^{II} center bearing alkyl and aryl groups is oxidized by NO_2 to Pd^{IV} species at room

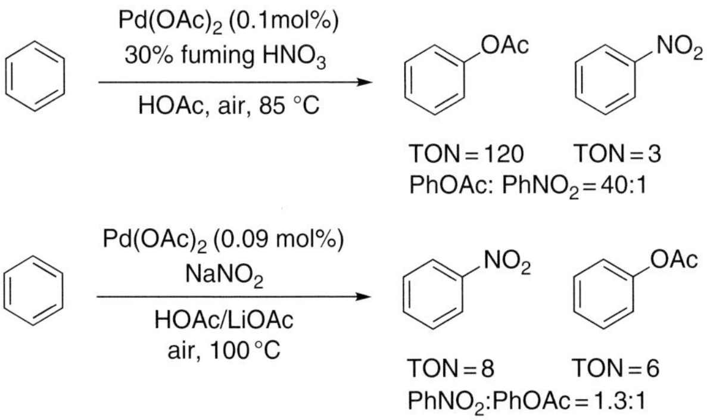

Benzene → $Pd(OAc)_2$ (0.1 mol%), 30% fuming HNO_3, HOAc, air, 85 °C → PhOAc (TON = 120) + $PhNO_2$ (TON = 3); PhOAc:$PhNO_2$ = 40:1

Benzene → $Pd(OAc)_2$ (0.09 mol%), $NaNO_2$, HOAc/LiOAc, air, 100 °C → $PhNO_2$ (TON = 8) + PhOAc (TON = 6); $PhNO_2$:PhOAc = 1.3:1

Scheme 8.14 Pd-catalyzed aerobic benzene acetoxylation and nitration in the presence of NO_x [51, 54].

temperature [125]. The application of NO_x in aerobic Pd^{II}/Pd^{IV} catalysis has been further extended to other C–O bond-forming reactions such as alkene dioxygenation [126] and directed sp^3 C–H acetoxylation [127].

8.4.3
Methane Oxidation

Selective oxidation of methane is a subject of significant interest owing to the increased availability of natural gas as a feedstock.

Building on the pioneering work of Shilov more than 40 years ago in the development of homogeneous Pt-catalyzed oxidation of methane to methanol and methyl chloride [128], a number of groups have explored Pd-catalyzed methods for the selective oxidation of methane. For example, methane oxidation using catalytic Pd/C in the presence of CF_3CO_2H and H_2 under aerobic conditions gives formic acid with a TON of 80–160 (Eq. (8.23)) [129]. It was suggested that H_2 and O_2 form H_2O_2 *in situ*, which is the oxidant used for methane oxidation.

$$CH_4 + 3/2\,O_2 \xrightarrow[\substack{H_2(100\text{ psi}),\\ CF_3CO_2H,\,85\,°C}]{Pd/C} HCO_2H + H_2O \quad (8.23)$$
(100 psi)

$PdSO_4$ in concentrated H_2SO_4 catalyzes the oxidative coupling of methane to acetic acid (Eq. (8.24)) [130]. Up to 90% HOAc selectivity and 12% yield (18 TONs) were obtained. Sulfuric acid is the oxidant under these conditions, but, in principle, the SO_2 by-product could be reoxidized aerobically to H_2SO_4. ^{13}C-labeling experiments confirmed that both carbons in the HOAc products come from CH_4. $Pd(OAc)_2$ catalyzes the aerobic oxidation of CH_4 in CF_3COOH to afford CF_3COOCH_3 with BQ and NO_x as redox cocatalysts (7 TONs; Eq. (8.25)) [131]. The same transformation was revisited using BQ and PMoV-based heteropolyacids as cocatalysts. Use of C_8F_{18} as solvent to increase O_2 solubility resulted in more than 100 TONs [132]. Finally, a Pd^{II}/N-heterocarbene (NHC)-type catalyst system has been reported for oxidation of methane and propane (Eq. (8.26)) [133]. $Pd(NHC)Br_2$-catalyzed oxidation of propane affords 2-trifluoroacetoxy propane with more than 90% selectivity [133c]. A Pd^{II}/Pd^{IV} catalytic cycle has been proposed, involving cationic Pd intermediates and *in situ* generated Br_2 to oxidize Pd to +4 oxidation state.

$$2\,CH_4 + 4\,H_2SO_4 \xrightarrow[\substack{180\,°C\\ \text{TON}=18}]{PdSO_4} CH_3COOH + 4\,SO_2 + 6\,H_2O \quad (8.24)$$
27.2 atm

$$2\,CH_4 + 1/2\,O_2 + CF_3COOH \xrightarrow[\substack{80\,°C\\ \text{TON}=7}]{Pd/BQ/NaNO_2} CF_3COOCH_3 + H_2O \quad (8.25)$$
54 atm 1 atm

$$CH_4 + K_2S_2O_8 + CF_3COOH \xrightarrow[90\,°C]{PdLBr_2} CF_3COOCH_3 + 2\,KHSO_4 \quad (8.26)$$
30 bar up to 30 TONs

$$L = \underset{\substack{| \\ Me}}{N}\!\!\diagup\!\!\underset{\substack{| \\ Me}}{N}\,\cdots\,N\!\!\diagdown\!\!N$$

8.5
Conclusion

Pd-catalyzed aerobic oxidations have been used by the chemical industry for over 50 years, originating with the Wacker process in 1959. As summarized here and discussed in subsequent chapters, both homogeneous and heterogeneous catalysts are particularly versatile for the selective aerobic oxidation of organic molecules, among which are alkenes, carbon monoxide, and arenes. The key to success in these reactions is the ability to achieve efficient aerobic catalytic turnover, minimize catalyst deactivation, and control the reaction selectivity. The extensive recent academic efforts in this field suggest that new industrially relevant applications of this chemistry will continue to emerge. In particular, the use of ancillary ligands and various cocatalysts offer potential solutions to some of the key challenges that have faced the field historically. Mechanistic understanding of the roles of the ligands and additives should provide guidance for the future directions of this field.

References

1. Roshcer, G. (2012) Vinyl esters, in *Ullmann's Encyclopedia of Industrial Chemistry*, Wiley-VCH Verlag GmbH, Weinheim, DOI: 10.1002/14356007.a27_419.
2. Holzrichter, H., Kronig, W., and Frenz, B. (1966) US Patent 3275680.
3. Fernholz, H., Schmidt, H.-J., and Wunder, F. (1967) US Patent 3625998.
4. (a) Seisiro, N. and Teruo, Y. (1970) *J. Catal.*, **17**, 366; (b) Debellefontaine, H. and Besombes-Vailhe, J. (1978) *J. Chim. Phys.*, **75**, 801; (c) Vargaftik, M.I., Zagorodnikov, V.P., and Moiseev, I.I. (1981) *Kinet. Katal.*, **22**, 951; (d) Augustine, S.M. and Blitz, J.P. (1993) *J. Catal.*, **142**, 312.
5. Samanos, B., Boutrym, P., and Montarnal, R. (1971) *J. Catal.*, **23**, 19.
6. Crathorne, E.A., MacGowan, D., Morris, S.R., and Rawlinson, A.P. (1994) *J. Catal.*, **149**, 254.
7. Reilly, C.R. and Lerou, J.J. (1998) *Catal. Today*, **41**, 433.
8. Simson, S., Jentys, A., and Lercher, J.A. (2013) *J. Phys. Chem. C*, **117**, 8161.
9. Yan, Y.F., Kumar, D., Sivadinarayana, C., and Goodman, D.W. (2004) *J. Catal.*, **224**, 60.
10. Yan, Y.F., Wang, J.H., Kumar, D., Yan, Z., and Goodman, D.W. (2005) *J. Catal.*, **232**, 467.
11. Nagato, N., Mori, H., Maki, K., and Ishioka, R. (1987) US Patent 4634784.
12. (a) Moiseev, I.I., Belov, A.P., and Syrkin, Y.K. (1963) *Bull. Acad. of Sci. USSR, Div. Chem. Sci.*, **12**, 1395. DOI: 10.1007/BF00847826; (b) Kitching, W., Rappoport, Z., Winstein, S., and Young, W.G. (1966) *J. Am. Chem. Soc.*, **88**, 2054.
13. Nagato, N. (1999–2014) Allyl alcohol and monoallyl derivatives, in *Kirk-Othmer Encyclopedia of Chemical Technology*, (eds A. Seidel and M. Bickford), John Wiley & Sons, Inc. (online version).
14. Buysch, H.-J. (2012) Carbonic esters, in *Ullmann's Encyclopedia of Industrial Chemistry*, Wiley-VCH Verlag GmbH, Weinheim.
15. Cornils, B. and Herrmann, W.A. (1996) *Applied Homogeneous Catalysis with Organometallic Compounds*, VCH Verlagsgesellschaft mbH, Weinheim, p. 179.
16. Wittcoff, H.A., Reuben, B.G., and Plotkin, J.S. (2013) Chemicals and polymers from ethylene, in *Industrial Organic Chemicals,*

17. Bertleff, W., Roeper, M., and Sava, X. (2012) Carbonylation, in *Ullmann's Encyclopedia of Industrial Chemistry*, Wiley-VCH Verlag GmbH, Weinheim 10.1002/14356007.a05_217.pub2.
18. Uchiumi, S.-I., Ataka, K., and Matsuzaki, T. (1999) *J. Organomet. Chem.*, **576**, 279.
19. Nishimura, K., Uchiumi, S., Fujii, K., Nishihira, K., and Yamashita, M., and H. Itatani (1980) US Patent 4229589.
20. Waller, F.J. (1985) *J. Mol. Catal.*, **31**, 123.
21. Ube Industries, Ltd. http://www.ube-ind.co.jp/english/news/2005/2005_11.htm (accessed 13 April 2015).
22. Baker, R., Grolig, J., and Rasp, C. (1981) US Patent 4297501.
23. Parshall, G.W. and Ittel, S.D. (1992) *Homogeneous Catalysis: The Applications and Chemistry of Catalysis by Soluble Transition Metal Complexes*, 2nd edn, John Wiley Sons, Inc., New York, p. 117.
24. Fukuoka, S., Chono, M., and Kohno, M. (1984) *J. Org. Chem.*, **49**, 1458.
25. Cornils, B. and Herrmann, W.A. (1996) *Applied Homogeneous Catalysis with Organometallic Compounds*, VCH Verlagsgesellschaft mbH, Weinheim, p. 183.
26. Such iodide effect has been observed in Rh-catalyzed carbonylation of methanol to acetic acid (Monsanto process), where iodide is invoked to facilitate several key steps. See: Maitlis, P.M., Haynes, A., James, B.R., Catellani, M., and Chiusoli, G.P. (2004) *Dalton Trans.*, 3409.
27. Tsuji, J. (2004) *Palladium Reagents and Catalysts—New Perspectives for the 21st Century*, John Wiley & Sons, Ltd, New York, p. 77.
28. Kreuz, J.A. and Edman, J.R. (1998) *Adv. Mater.*, **10**, 1229.
29. Itatani, H., Shiotani, A., and Fujimoto, M. (1986) US Patent 4581469.
30. Itatani, H., Shiotani, A., and Yokota, A. (1982) US Patent 4338456.
31. Nishio, M. and Imajima, K. (2011) Japanese Patent 2011213666.
32. (a) Imatani, K., Ishikawa, A., and Kashibe, M. (1988) UK Patent GB2205765; (b) Sherman, S.C., Iretskii, A.V., White, M.G., and Schiraldi, D.A. (2000) *Chem. Innov.*, **30**, 25.
33. Itatani, H., Kashima, M., Matsuda, M., Yoshimoto, H., and Yamamoto, H. (1976) US Patent 3940426.
34. Itatani, H., Yoshimoto, H., Shiotani, A., Yokota, A., and Yoshikiyo, M. (1981) US Patent 4294976.
35. Izawa, Y. and Stahl, S.S. (2010) *Adv. Synth. Catal.*, **352**, 3223.
36. Itatani, H. and Yoshimoto, H. (1975) US Patent 3895055
37. Ichikawa, Y. and Yamaji, T. (1976) US Patent 3963787.
38. Okamoto, M. and Yamaji, T. (2001) *Chem. Lett.*, 212.
39. Mukhopadhyay, S., Rothenberg, G., Lando, G., Agbaria, K., Kazanci, M., and Sasson, Y. (2001) *Adv. Synth. Catal.*, **343**, 455.
40. Yokota, T., Sakaguchi, S., and Ishii, Y. (2002) *Adv. Synth. Catal.*, **344**, 849.
41. (a) Schiraldi, D.A., Sherman, S.C., Sood, D.S., and White, M.G. (2000) US Patent 6103919; (b) Xu, B.-Q., Sood, D., Iretskii, A.V., and White, M.G. (1999) *J. Catal.*, **187**, 358; (c) Iretskii, A.V., Sherman, S.C., White, M.G., Kevin, J.C., and Schiraldi, D.A. (2000) *J. Catal.*, **193**, 49.
42. (a) Serna, P. and Corma, A. (2014) *ChemSusChem*, **7**, 2136; (b) Ishida, T., Aikawa, S., Mise, Y., Akebi, R., Hamasaki, A., Homma, T., Ohashi, H., Tsuji, T., Yamamoto, Y., Miyasaka, M., Yokoyama, T., and Tokunaga, M. (2015) *ChemSusChem*, **8**, 695.
43. For reviews, see: a Mallat, T. and Baiker, A. (2004) *Chem. Rev.*, **104**, 3037; (b) Schultz, M.J. and Sigman, M.S. (2006) *Tetrahedron*, **62**, 8227; (c) Parmeggiani, C. and Cardona, F. (2012) *Green Chem.*, **14**, 547–564.
44. Brunhe, F. and Wright, E. (2012) Benzaldehyde, in *Ullmann's Encyclopedia of Industrial Chemistry*, Wiley-VCH Verlag GmbH, Weinheim, DOI: 10.1002/14356007.a03_463.pub2.

45. Le Ludec, J. (1977) US Patent 4026950.
46. Mondelli, C., Ferri, D., Grunwaldt, J.-D., Krumeich, F., Mangold, S., Psaro, R., and Baiker, A. (2007) *J. Catal.*, **252**, 77.
47. Ma, K.W. (1981) US Patent 4306083.
48. Murib, J.H. (1977) US Patent 4051181.
49. Weber, M., Weber, M., and Kleine-Boymann, M. (2012) Phenol, in *Ullmann's Encyclopedia of Industrial Chemistry*, Wiley-VCH Verlag GmbH & Co. KGaA. DOI: 10.1002/14356007.a19_299.pub2
50. Davidson, J.M. and Triggs, C. (1966) *Chem. Ind.*, 457.
51. Tisue, T. and Downs, W.J. (1969) *J. Chem. Soc. D.: Chem. Commun.*, 410.
52. Ichikawa, K., Uemura, S., and Okada, T. (1969) *Nippon Kagaku Zassi*, **90**, 212.
53. Norman, R.O.C., Parr, W.J.E., and Thomas, C.B. (1974) *J. Chem. Soc., Perkin Trans. 1*, 369.
54. Zultanski, S.L. and Stahl, S.S. (2015) *J. Organomet. Chem.*, **793**, 263–268.
55. (a) Selwitz, C.M. (1970) US Patent 3542852; (b) Kominami, N., Tamura, N., and Yamamoto, E. (1975) US Patent 3887608; (c) Kominami, N. and Tamura, N. (1973) US Patent 3772383.
56. Hornig, L., Arpe, H.J., and Boldt, M. (1972) US Patent 3651127.
57. (a) Boldt, M., Arpe, H.J., and Hornig, L. (1972) US Patent 3644486 A; (b) Eberson, L. and Jonsson, L. (1974) *Acta Chem. Scand. Ser. B-Org. Chem. Biochem.*, **B 28**, 597.
58. Arpe, H.J., Boldt, M., and Hornig, L. (1972) US Patent 3651101 A.
59. (a) Onoda, T., Wada, K., and Otake, M. (1976) US Patent 3959352 A; (b) Onoda, T. and Wada, K. (1976) US Patent 3959354 A.
60. (a) Goel, A.B. and Grimm, R.A. (1984) US Patent 4456555; (b) Goel, A.B. and Throckmorton, P.E. (1984) US Patent 4465633 A; (c) Goel, A.B. and Throckmorton, P.E. (1985) US Patent 4507243 A; (d) Goel, A.B. and Grimm, R.A. (1987) US Patent 455974.
61. (a) Goel, A.B. and Pettiford, M.E. (1984) US Patent 4477385 A; (b) Goel, A.B. (1984) US Patent 4464303 A.
62. Goel, A.B. (1985) US Patent 4508653 A.
63. Sen, A., Gretz, E., Oliver, T.F., and Jiang, Z. (1989) *New J. Chem.*, **13**, 755.
64. Goel, A.B. (1985) US Patent 4512927 A.
65. (a) Goel, A.B. (1986) *Inorg. Chim. Acta*, **121**, L11; (b) Goel, A.B., Throckmorton, P.E., and Grimm, R.A. (1986) *Inorg. Chim. Acta*, **117**, L15; (c) Goel, A.B. (1984) *Inorg. Chim. Acta*, **90**, L15; (d) Goel, A.B. (1987) *Inorg. Chim. Acta*, **129**, L31.
66. (a) Mori, Y., Doi, T., Asakawa, T., and Miyake, T. (2003) US Patent 6528679; (b) Mori, Y., Doi, T., Asakawa, T., and Miyake, T. (2005) US Patent 6958407.
67. Mori, Y. (2002) US Patent 6342620 B1.
68. Passoni, L.C., Cruz, A.T., Buffon, R., and Schuchardt, U. (1997) *J. Mol. Catal., A*, **120**, 117.
69. Goldberg, H., Kaminker, I., Goldfarb, D., and Neumann, R. (2009) *Inorg. Chem.*, **48**, 7947.
70. Jintoku, T., Taniguchi, H., and Fujiwara, Y. (1987) *Chem. Lett.*, 1865–1868.
71. (a) Miyake, T. and Asakawa, T. (2005) *Appl. Catal., A*, **280**, 47; (b) Miyake, T., Hattori, A., Hanaya, M., Tokumaru, S., Hamaji, H., and Okada, T. (2000) *Top. Catal.*, **13**, 243.
72. (a) Davidson, J.M. and Triggs, C. (1968) *J. Chem. Soc. A*, 1324; (b) Davidson, J.M. and Triggs, C. (1968) *J. Chem. Soc. A*, 1331.
73. Bryant, D.R., McKeon, J.E., and Ream, B.C. (1968) *J. Org. Chem.*, **33**, 4123.
74. Bryant, D.R., McKeon, J.E., and Ream, B.C. (1968) *Tetrahedron Lett.*, **9**, 3371.
75. Ohishi, T., Tanaka, Y., Yamada, J., Tago, H., Tanaka, M., and Yamashita, M. (1997) *Appl. Organomet. Chem.*, **11**, 941.
76. Bryant, D.R., McKeon, J.E., and Ream, B.C. (1969) *J. Org. Chem.*, **34**, 1106.
77. (a) McKeon, J.E. and Bryant, D.R. (1970) US Patent 3547982A; (b) Selwitz, C.M. (1970) US Patent 3493605.
78. (a) Katsman, L.A., Vargaftik, M.N., Starchevskii, M.K., Grigor'eva, A.A., and Moiseev, I.I. (1981) *Kinet. Catal.*, **22**, 739; (b) Eberson, L. and Gomezgon, L. (1973) *Acta Chem. Scand.*, **27**, 1162; (c) Ivanov, S.K. and Tanielyan, S.K. (1984) *Oxid. Commun.*, **7**, 69.
79. (a) Benazzi, E., Cameron, C.J., and Mimoun, H. (1991) *J. Mol. Catal.*, **69**, 299; (b) Benazzi, E., Mimoun, H., and

Cameron, C.J. (1993) *J. Catal.*, **140**, 311.

80. (a) Augustine, R.L. and Tanielyan, S.K. (1993) US Patent 5189006A; (b) Augustine, R.L. and Tanielyan, S.K. (1993) US Patent 5206423A.

81. Scheben, J.A. and Hinnenkamp, J.A. (1985) US Patent 4499298 A.

82. (a) Tso, C. C. and Wynn, R. L. (1994) US Patent 5280001A; (b) Tso, C. C. and Wynn, R. L. (1992) US Patent 5183931A.

83. (a) Onoda, T., Wada, K., and Otake, M. (1977) US Patent 4016200; (b) Schammel, W.P., Adamian, V., Brugge, S.P., Gong, W.H., Metelski, P.D., Nubel, P.O., and Zhou, C. (2007) WO2007133978 A3; (c) Gong, W.H., Adamian, V., Brugge, S.P., Metelski, P.D., Nubel, P.O., Schammel, W.P., and Zhou, C. (2007) WO2007133978 A2; (d) Schammel, W.P., Adamian, V., Brugge, S.P., Gong, W.H., Metelski, P.D., Nubel, P.O., Zhou, C. (2012) US Patent 8163954; (e) Schammel, W.P., Huggins, B.J., Kulzick, M.A., Nubel, P.O., Rabatic, B.M., Zhou, C., Adamian, V. A., Gong, W.H., Metelski, P.D., Miller, J.T. (2014) US Patent 8624055.

84. Hayashi, T., Iida, T., Satoh, Y., and Tatsumi, J. (1999) EP 0965383 A1.

85. (a) Benhmid, A., Narayana, K.V., Martin, A., Lucke, B., and Pohl, M.M. (2004) *Chem. Lett.*, **33**, 1238; (b) Benhmid, A., Narayana, K.V., Martin, A., Lucke, B., and Pohl, M.M. (2004) *Chem. Commun.*, 2416; (c) Benhmid, A., Narayana, K.V., Martin, A., and Lucke, B. (2004) *Chem. Commun.*, 2118.

86. Komatsu, T., Inaba, K., Uezono, T., Onda, A., and Yashima, T. (2003) *Appl. Catal., A,*, **251**, 315.

87. (a) Madaan, N., Gatla, S., Kalevaru, V.N., Radnik, J., Pohl, M.-M., Luecke, B., Brueckner, A., and Martin, A. (2013) *ChemCatChem*, **5**, 185; (b) Gatla, S., Madaan, N., Radnik, J., Kalevaru, V.N., Pohl, M.M., Lücke, B., Martin, A., Bentrup, U., and Brückner, A. (2013) *J. Catal.*, **297**, 256.

88. Madaan, N., Gatla, S., Kalevaru, V.N., Radnik, J., Lücke, B., Brückner, A., and Martin, A. (2011) *J. Catal.*, **282**, 103.

89. Radnik, J., Benhmid, A., Kalevaru, V.N., Pohl, M.-M., Martin, A., Lücke, B., and Dingerdissen, U. (2005) *Angew. Chem. Int. Ed.*, **44**, 6771.

90. (a) Hwang, S.Y. and Chen, S.-S. (2000) Cumene, in *Kirk-Othmer Encyclopedia of Chemical Technology*, vol. **23**, (eds A. Seidel and M. Bickford), John Wiley & Sons, Inc; (b) James, D.H. and Castor, W.M. (2012) Styrene, in *Ullmann's Encyclopedia of Industrial Chemistry*, Wiley-VCH Verlag GmbH & Co. KGaA. DOI: 10.1002/14356007.a25_329.pub2

91. Heck, R.F. (1979) *Acc. Chem. Res.*, **12**, 146.

92. Moritanl, I. and Fujiwara, Y. (1967) *Tetrahedron Lett.*, **8**, 1119.

93. (a) Fujiwara, Y., Noritani, I., Danno, S., Asano, R., and Teranishi, S. (1969) *J. Am. Chem. Soc.*, **91**, 7166; (b) Fujiwara, Y., Moritani, I., Matsuda, M., and Teranishi, S. (1968) *Tetrahedron Lett.*, **9**, 3863; (c) Moritani, I. and Fujiwara, Y. (1973) *Synthesis-Stuttgart*, 524.

94. (a) Shue, R.S. (1971) *J. Chem. Soc. D*, **1971**, 1510; (b) Shue, R.S. (1972) *J. Catal.*, **26**, 112; (c) Shue, R.S. (1973) US Patent 3775511.

95. (a) Kominami, N., Tamura, N., and Yamamoto, E. (1973) US Patent 3767714; (b) Kominami,N., Tamura, N., and Yamamoto, E. (1972) US Patent 3689583.

96. Yokota, T., Tani, M., Sakaguchi, S., and Ishii, Y. (2003) *J. Am. Chem. Soc.*, **125**, 1476.

97. (a) Dams, M., De Vos, D. E., Celen, S. and Jacobs, P. A. (2003) *Angew. Chem. Int. Ed.*, **42**, 3512; (b) Matsumoto, T., Periana, R.A., Taube, D.J., and Yoshida, H. (2002) *J. Catal.*, **206**, 272.

98. Tsuji, J. and Nagashima, H. (1984) *Tetrahedron*, **40**, 2699.

99. Jia, C., Lu, W., Kitamura, T., and Fujiwara, Y. (1999) *Org. Lett.*, **1**, 2097.

100. Parrish, J.P., Jung, Y.C., Shin, S.I., and Jung, K.W. (2002) *J. Org. Chem.*, **67**, 7127.

101. (a) Yoo, K.S., Park, C.P., Yoon, C.H., Sakaguchi, S., O'Neil, J., and Jung, K.W. (2007) *Org. Lett.*, **9**, 3933; (b) Werner, E.W. and Sigman, M.S. (2010) *J. Am. Chem. Soc.*, **132**, 13981.

102. (a) Andappan, M.M.S., Nilsson, P., and Larhed, M. (2004) *Chem. Commun.*, 218; (b) Andappan, M.M.S., Nilsson, P., von Schenck, H., and Larhed, M. (2004) *J. Org. Chem.*, **69**, 5212; (c) Enquist, P.-A., Lindh, J., Nilsson, P., and Larhed, M. (2006) *Green Chem.*, **8**, 338; (d) Lindh, J., Enquist, P.-A., Pilotti, Å., Nilsson, P., and Larhed, M. (2007) *J. Org. Chem.*, **72**, 7957; (e) Yoo, K.S., Yoon, C.H., and Jung, K.W. (2006) *J. Am. Chem. Soc.*, **128**, 16384.
103. (a) Martin, R. and Buchwald, S.L. (2008) *Acc. Chem. Res.*, **41**, 1461; (b) Surry, D.S. and Buchwald, S.L. (2008) *Angew. Chem. Int. Ed.*, **47**, 6338; (c) Hartwig, J.F. (2008) *Acc. Chem. Res.*, **41**, 1534.
104. Stahl, S.S. (2004) *Angew. Chem. Int. Ed.*, **43**, 3400.
105. (a) McDonald, R.I. and Stahl, S.S. (2010) *Angew. Chem. Int. Ed.*, **49**, 5529; (b) Diao, T. and Stahl, S.S. (2011) *J. Am. Chem. Soc.*, **133**, 14566; (c) Diao, T., White, P., Guzei, I., and Stahl, S.S. (2012) *Inorg. Chem.*, **51**, 11898; (d) Diao, T., Pun, D., and Stahl, S.S. (2013) *J. Am. Chem. Soc.*, **135**, 8205.
106. (a) Nishimura, T., Onoue, T., Ohe, K., and Uemura, S. (1998) *Tetrahedron Lett.*, **39**, 6011; (b) Nishimura, T., Onoue, T., Ohe, K., and Uemura, S. (1999) *J. Org. Chem.*, **64**, 6750; (c) Nishi-mura, T., Maeda, Y., Kakiuchi, N., and Uemura, S. (2000) *J. Chem. Soc., Perkin Trans. 1*, 4301.
107. Brink, G.-J.T., Arends, I.W.C.E., and Sheldon, R.A. (2000) *Science*, **287**, 1636.
108. (a) Ferreira, E.M. and Stoltz, B.M. (2003) *J. Am. Chem. Soc.*, **125**, 9578; (b) Ferreira, E.M., Zhang, H., and Stoltz, B.M. (2008) *Tetrahedron*, **64**, 5987; (c) Zhu, R. and Buchwald, S.L. (2012) *Angew. Chem. Int. Ed.*, **51**, 1926.
109. Zhang, Y.-H., Shi, B.-F., and Yu, J.-Q. (2009) *J. Am. Chem. Soc.*, **131**, 5072.
110. (a) Steinhoff, B.A., Guzei, I.A., and Stahl, S.S. (2004) *J. Am. Chem. Soc.*, **126**, 11268; (b) Steinhoff, B.A. and Stahl, S.S. (2002) *Org. Lett.*, **4**, 4179; (c) Emmert, M.H., Cook, A.K., Xie, Y.J., and Sanford, M.S. (2011) *Angew. Chem. Int. Ed.*, **50**, 9409.
111. (a) Jensen, D.R., Schultz, M.J., Mueller, J.A., and Sigman, M.S. (2003) *Angew. Chem. Int. Ed.*, **42**, 3810; (b) Mueller, J.A., Goller, C.P., and Sigman, M.S. (2004) *J. Am. Chem. Soc.*, **126**, 9724; (c) Muniz, K. (2004) *Adv. Synth. Catal.*, **346**, 1425; (d) Nielsen, R.J. and Goddard, W.A. III, (2006) *J. Am. Chem. Soc.*, **128**, 9651; (e) Rogers, M.M., Wendlandt, J.E., Guzei, I.A., and Stahl, S.S. (2006) *Org. Lett.*, **8**, 2257; (f) Rogers, M.M. and Stahl, S.S. (2007) *Top. Organomet. Chem.*, **21**, 21–46; (g) Xiao, B., Gong, T.J., Liu, Z.J., Liu, J.H., Luo, D.F., Xu, J., and Liu, L. (2011) *J. Am. Chem. Soc.*, **133**, 9250; (h) Ye, X., White, P.B., and Stahl, S.S. (2013) *J. Org. Chem.*, **78**, 2083.
112. Campbell, A.N., White, P.B., Guzei, I.A., and Stahl, S.S. (2010) *J. Am. Chem. Soc.*, **132**, 1511.
113. Piera, J. and Bäckvall, J.-E. (2008) *Angew. Chem. Int. Ed.*, **47** (19), 3506–3523.
114. Campbell, A.N. and Stahl, S.S. (2012) *Acc. Chem. Res.*, **45**, 851 and references therein.
115. Diao, T. and Stahl, S.S. (2014) *Polyhedron*, **84**, 96.
116. Musaev, D.G., Kaledinm, A.L., Shi, B.-F., and Yu, J.-Q. (2012) *J. Am. Chem. Soc.*, **134**, 1690.
117. Eagel, K.M., Wang, D.-H., and Yu, J.-Q. (2010) *J. Am. Chem. Soc.*, **132**, 14137.
118. Yoon, C.H., Yoo, K.S., Yi, S.W., Mishra, R.K., and Jung, K.W. (2004) *Org. Lett.*, **6**, 4037.
119. Zheng, C., Wang, D., and Stahl, S.S. (2012) *J. Am. Chem. Soc.*, **134**, 16496.
120. Thiemann, M., Scheibler, E., and Wiegand, K.W. (1991) Nitric acid, nitrous acid, and nitrogen oxides, in *Ullmann's Encyclopedia of Industrial Chemistry*, (eds B. Elvers, S. Hawkins, and G. Schulz), VCH Verlagsgesellschaft mbH, Weinheim.
121. Solar, J.P., Mares, F., and Diamond, S.E. (1985) *Catal. Rev.-Sci. Eng.*, **27**, 1.
122. Feringa, B.L. (1986) *J. Chem. Soc., Chem. Commun.*, 909.
123. (a) Wickens, Z.K., Morandi, B., and Grubbs, R.H. (2013) *Angew. Chem. Int. Ed.*, **52**, 11257; (b) Wickens, Z.K., Skakuj, K., Morandi, B., and Grubbs,

R.H. (2014) *J. Am. Chem. Soc.*, **136**, 890.
124. Jiang, Y.-Y., Zhang, Q., and Fu, Y. (2015) *ACS Catal.*, **5**, 1414.
125. Cámpora, J., Palma, P., del Rio, D., Carmona, E., Graiff, C., and Tiripicchio, A. (2003) *Organometallics*, **22**, 3345.
126. Wickens, Z.K., Guzman, P.E., and Grubbs, R.H. (2015) *Angew. Chem. Int. Ed.*, **54**, 236.
127. Stowers, K.J., Kubota, A., and Sanford, M.S. (2012) *Chem. Sci.*, **3**, 3192.
128. Goldschleger, N.F., Eskova, V.V., Shilov, A.E., and Shteinman, A.A. (1972) *Russ. J. Phys. Chem.*, **46**, 785.
129. Lin, M. and Sen, A. (1999) *J. Am. Chem. Soc.*, **114**, 7307.
130. Periana, R.A., Mironov, O., Taube, D., Bhalla, G., and Jones, C. (2003) *Science*, **301**, 814.
131. An, Z., Pan, X., Liu, X., Han, X., and Bao, X. (2006) *J. Am. Chem. Soc.*, **128**, 16028.
132. Yuan, J., Wang, L., and Wang, Y. (2011) *Ind. Eng. Chem. Res.*, **50**, 6513.
133. (a) Muehlhofer, M., Strassner, T., and Herrmann, W.A. (2002) *Angew. Chem. Int. Ed.*, **41**, 1745; (b) Munz, D., Meyer, D., and Strassner, T. (2013) *Organometallics*, **32**, 3469; (c) Munz, D. and Strassner, T. (2014) *Angew. Chem. Int. Ed.*, **53**, 2485.

9
Acetaldehyde from Ethylene and Related Wacker-Type Reactions

Reinhard Jira

9.1 Introduction

After World War II, when the big oil companies installed a large number of refineries in industrialized countries, lower olefins, particularly ethylene, became available in large quantities. Chemical companies replaced successively their acetylene-based processes for the production of aliphatic C_2, C_4, and C_8 compounds by much cheaper processes with ethylene as the feedstock. Researchers of the Consortium für elektrochemische Industrie GmbH, the research organization of Wacker Chemie GmbH, succeeded in finding a new process for the manufacture of the important industrial intermediate acetaldehyde from ethylene [1, 2].

$$H_2C=CH_2 + 1/2 O_2 \xrightarrow{\text{cat.}} \text{CH}_3\text{CHO} \tag{9.1}$$

At first, acetaldehyde was observed when ethylene and oxygen with a trace of hydrogen were passed over a palladium hydrogenation catalyst. It was suggested that palladium plays an essential role in this reaction. In fact, an old publication [3] describes that acetaldehyde and a black precipitate of palladium metal are formed when ethylene is passed through an aqueous solution of palladium chloride. Furthermore, a binuclear ethylene–palladium complex previously synthesized by Kharasch *et al.* (**1**) [4] is immediately decomposed by water into palladium metal, hydrochloric acid, and acetaldehyde, and similarly the so-called Zeise's salt $K[PtCl_3(\pi\text{-}C_2H_4)]$ is decomposed at elevated temperature to give acetaldehyde and metallic platinum [5]. It was shown [2] that the reaction of ethylene with aqueous palladium chloride proceeds virtually stoichiometrically (Eq. (9.2)) [2, 6].

1

Liquid Phase Aerobic Oxidation Catalysis: Industrial Applications and Academic Perspectives,
First Edition. Edited by Shannon S. Stahl and Paul L. Alsters.
© 2016 Wiley-VCH Verlag GmbH & Co. KGaA. Published 2016 by Wiley-VCH Verlag GmbH & Co. KGaA.

$$PdCl_2 + H_2C=CH_2 + H_2O \longrightarrow CH_3CHO + Pd + 2HCl \qquad (9.2)$$

Other palladium salts, such as sulfate, nitrate, and acetate, react similarly. Also, other noble metal salts such as those of platinum, rhodium, or ruthenium give acetaldehyde in the same way but at a distinctly reduced rate.

9.2
Chemistry and Catalysis

9.2.1
Oxidation of Olefinic Compounds to Carbonyl Compounds

Oxidation of ethylene according to Eq. (9.2) can be applied to higher olefins and olefins with functional groups [2, 7]. In this regiospecific reaction, the carbonyl group is formed at that carbon atom of the double bond where the nucleophile enters in a Markovnikov-like addition (Eq. (9.3)). Thus, primary olefins give methylketones. The corresponding aldehydes are formed only to a lower percentage.

$$R\text{-}CH=CH_2 + PdCl_2 + H_2O \longrightarrow R\text{-}CO\text{-}CH_3 + Pd + 2HCl \qquad (9.3)$$

R = alkyl

Electron-withdrawing functional groups, such as carbonyl, carboxy, hydroxy, nitro, cyano, and olefinic groups, direct the carbonyl group into the β-position (Eq. (9.4)).

$$R\text{-}CH=CH\text{-}X + PdCl_2 + H_2O \longrightarrow R\text{-}CO\text{-}CH_2\text{-}X + Pd + 2HCl \qquad (9.4)$$

R = H, alkyl, X = NO_2, CN, COOH, –CH=CH–

In the case of the reaction of α,β-unsaturated carboxylic acids, oxidation is followed by decarboxylation. Examples are given in Table 9.1.

9.2.2
Kinetics and Mechanism

Two research groups, a Russian one with Moiseev et al. [11–13] and an American one with Henry [14, 15], published an identical rate equation for reaction (9.2)

$$-\frac{d[C_2H_4]}{dt} = \frac{K \cdot k[C_2H_4][(PdCl_4)^{2-}]}{[Cl^-]^2[H^+]} \qquad (9.5)$$

with an inhibition of second order for Cl^- ions and first order for H^+ ions. There is no dissent that the first step in the reaction is the formation of a complex between

Table 9.1 Some examples of the oxidation of olefinic compounds with aqueous palladium chloride solution [2, 7, 8].

Substrate	Intermediate assumed	Product
Ethylene		Acetaldehyde
Propene		Acetone (propionaldehyde)
1- and 2-Butene		Butanone (butyraldehyde)
1-Olefins		Methyl ketones (aldehydes)
Cyclopentene		Cyclopentanone
Cyclohexene		Cyclohexanone
Styrene		Acetophenone (phenylacetaldehyde)
Acrylic acid	$OHC-CH_2-COOH$	Acetaldehyde
Crotonic acid	$CH_3-CO-CH_2-COOH$	Acetone
Maleic acid	$HOOC-CO-CH_2-COOH$	Pyruvic acid
1,3-Butadiene		Crotonaldehyde [9]
Nitroethylene		Nitroacetaldehyde
Acrylonitrile		Cyanoacetaldehyde
Allyl alcohol	$OHC-CH_2-CH_2OH$	Acrolein [10]
Crotonaldehyde	$CH_3-CO-CH_2-CHO$	1,3,5-Triacetylbenzene

ethylene and palladium according to Eq. (9.6) where one chloro ligand of the tetrachloropalladium complex $[PdCl_4]^{2-}$ will be replaced by ethylene, responsible for the first inhibiting Cl^- ion.

$$[PdCl_4]^{2-} + H_2C=CH_2 \underset{2}{\overset{K}{\rightleftharpoons}} [(C_2H_4)PdCl_3]^- + Cl^- \qquad (9.6)$$

The formation of the next intermediate, **5**, in a hydroxypalladation step has been the subject of controversy over many years. According to his kinetic investigations, Henry postulated a *syn*-attack (*cis*-ligand insertion reaction) of a complexed OH^- ion to the complexed olefin (Eq. (9.7)).

$$\begin{bmatrix} H_2C\overset{CH_2}{\diagup}\overset{Cl}{\underset{Pd}{\diagdown}} \\ HO \quad Cl \end{bmatrix}^- \longrightarrow \begin{bmatrix} \overset{Cl}{\underset{Pd}{\diagdown}} \\ HO \quad Cl \end{bmatrix}^- \qquad (9.7)$$

$$\quad\quad 4 \qquad\qquad\qquad\quad 5$$

Intermediate **4** should be formed by replacement of a chloro ligand of complex **2** by a solvent water molecule, followed by dissociation of an H^+ ion (Eqs. (9.8) and (9.9)), thus explaining the second inhibiting Cl^- ion and the inhibiting H^+ ion in rate Eq. (9.5).

$$[(C_2H_4)PdCl_3]^- + H_2O \longrightarrow \begin{bmatrix} H_2C\overset{CH_2}{\diagup}\overset{Cl}{\underset{Pd}{\diagdown}} \\ H_2O \quad Cl \end{bmatrix} + Cl^- \qquad (9.8)$$

$$\quad 2 \qquad\qquad\qquad\qquad\quad 3$$

$$[\text{H}_2\text{C}\overset{\text{CH}_2}{\underset{\text{H}_2\text{O}}{\overset{\diagup}{\underset{\diagdown}{\text{Pd}}}}}\overset{\text{Cl}}{\underset{\text{Cl}}{}}] \rightleftharpoons [\text{H}_2\text{C}\overset{\text{CH}_2}{\underset{\text{HO}}{\overset{\diagup}{\underset{\diagdown}{\text{Pd}}}}}\overset{\text{Cl}}{\underset{\text{Cl}}{}}] + \text{H}^+ \qquad (9.9)$$

$$\mathbf{3}\mathbf{4}$$

Other authors have claimed an *anti*-attack of a OH⁻ ion or a water molecule from model reactions carried out with functional olefinic compounds and/or in media different from Wacker conditions. So, in strong complexing agents such as CO or solvents such as acetonitrile, even high chloride ion or cupric chloride concentrations (see Section 4.1) may prevent an OH⁻ or OH$_2$ ligand on the Pd central atom and therefore the interaction of the complexed OH⁻ with the olefin. A reaction with diolefins where both olefinic groups are complexed, thus hindering a rotation of the reacting olefin, also prevents the interaction of the olefin with OH⁻ within the complex sphere. Some examples of reaction with such features are discussed at the end of this section. Keith and Henry [16] gave a profound view of this controversial problem and came to the conclusion that the path in which the oxypalladation occurs is strongly dependent on the reaction conditions, the model compounds, and even the chloride ion concentration. At low Cl⁻ concentration, the *syn*-path takes place, while at high Cl⁻ concentration, the *anti*-path is preferred.

In these considerations, one fact has not been taken into account. In Henry's mechanism, the water molecule replaces the chloro ligand in *cis*-position to the olefin which is a prerequisite for the oxypalladation. However, for the analog platinum complex, it could be shown that due to the "*trans*-effect," water replaces at first the chloro ligand in the *trans*-position [17], and so the corresponding Pd complex Eq. (9.10) would be formed.

$$[(\text{C}_2\text{H}_4)\text{PdCl}_3]^- + \text{H}_2\text{O} \rightleftharpoons [\text{H}_2\text{C}\overset{\text{CH}_2}{\underset{\text{Cl}}{\overset{\diagup}{\underset{\diagdown}{\text{Pd}}}}}\overset{\text{Cl}}{\underset{\text{OH}}{}}]^- + \text{H}^+ + \text{Cl}^- \qquad (9.10)$$

$$\mathbf{2}\mathbf{6}$$

Evidence for a *trans*–*cis* isomerization of the OH ligand in complex **6** was revealed by a more detailed kinetic study of the reaction of ethylene with aqueous palladium chloride at low chloride ion concentration. It was demonstrated that H⁺ and Cl⁻ ions were found to both impede and accelerate the reaction, depending on their concentrations. From these studies, empirical rate Eq. (9.11) for constant ethylene concentration was derived [18], which transmutes into Henry's rate Eq. (9.5) at higher H⁺ and Cl⁻ concentrations.

$$-\frac{d[\text{C}_2\text{H}_4]}{dt} = \frac{a[\text{H}^+][\text{Cl}^-]}{b + [\text{H}^+]^2[\text{Cl}^-]^3} \qquad (9.11)$$

This kinetic behavior can be explained through another replacement of a chloro ligand by a water molecule followed by dissociation of an H⁺ ion, while the accelerating effect is due to the replacement of the OH⁻ ligand in *trans*-position responsible for the *trans–cis* isomerization of the OH⁻ ligand (Eqs. (9.12) and (9.13)).

$$\left[\begin{array}{c} H_2C\overset{CH_2}{\diagup}Cl \\ Pd \\ ClOH \end{array}\right]^{-} + H_2O \rightleftharpoons \left[\begin{array}{c} H_2C\overset{CH_2}{\diagup}Cl \\ Pd \\ HOOH \end{array}\right]^{-} + H^+ + Cl^- \quad (9.12)$$

$$67$$

$$\left[\begin{array}{c} H_2C\overset{CH_2}{\diagup}Cl \\ Pd \\ HOOH \end{array}\right]^{-} + H^+ + Cl^- \rightleftharpoons \left[\begin{array}{c} H_2C\overset{CH_2}{\diagup}Cl \\ Pd \\ HOCl \end{array}\right]^{-} + H_2O \quad (9.13)$$

$$74$$

Complex **4** possesses now the prerequisites for a *cis*-ligand insertion reaction: a close (*cis*) position of the interacting ligands and activation of the π-bonded olefin by weakening the olefin–metal π-bond and lowering the olefin rotation barrier through the *trans*-effect of the *trans*-chloro ligand.

A weakening of the ethylene–platinum bond in an aqueous solution of Zeise's salt with increasing Cl⁻ concentration could be shown by NMR spectroscopy. The proton peaks of the complexed olefin were shifted to lower fields under the influence of increasing Cl⁻ concentration [18].

Under the premise of this sequence of reactions (Eq. (9.14) = (9.6) + (9.10) + (9.12)), a rate equation could be derived (see [19], pp. 18–19).

$$[PdCl_4]^{2-} + 2H_2O + C_2H_4 \overset{K_1}{\rightleftharpoons} [(\pi\text{-}C_2H_4)ClPd(OH)_2]^- + 3Cl^- + 2H^+ \quad (9.14)$$

$$[(\pi\text{-}C_2H_4)ClPd(OH)_2]^- + Cl^- + H^+ \overset{K_2}{\rightleftharpoons} cis\text{-}[(\pi\text{-}C_2H_4)(OH)PdCl_2]^- + H_2O$$

$$(9.13a)$$

$$cis\text{-}[(\pi\text{-}C_2H_4)(OH)PdCl_2]^- \overset{k}{\longrightarrow} \text{CH}_3\text{CHO} + Pd + 2Cl^- + H^+ \quad (9.15)$$

The rate Eq. (9.16) adopts the form:

$$-\frac{d[C_2H_4]}{dt} = \frac{k \cdot K_1 \cdot K_2 [(PdCl_4)^{2-}][C_2H_4][H^+][Cl^-]}{K_1[C_2H_4] + K_1 \cdot K_2[C_2H_4][H^+][Cl^-] + [H^+]^2[Cl^-]^3} \quad (9.16)$$

It is comparable with the empirical rate Eq. (9.11). In this sense, these kinetics support Henry's mechanistic ideas of a *syn*-oxypalladation step, at least at low chloride ion concentration.

Further mechanistic studies suggest now an internal hydride transfer. The reaction of Kharasch's ethylene–Pd complex with D_2O gives only H_3CCHO [18], and the oxidation of C_2D_4 with $PdCl_2$ in normal water results only in D_3CCDO

[14]. This hydride transfer, which follows the oxypalladation step (Eq. (9.7)), is represented by Eqs. (9.17) and (9.18).

$$\left[\text{HO}\diagup\diagdown\text{PdCl}_2 \right]^- \rightleftharpoons \left[\text{HO}\diagup\diagdown\| \right] \longrightarrow \text{PdHCl}_2 \qquad (9.17)$$

5   8

$$\left[\text{HO}\diagdown\| \longrightarrow \text{PdHCl}_2 \right] \rightleftharpoons \left[\overset{\text{OH}}{\diagup\diagdown}\text{PdCl}_2 \right] \qquad (9.18)$$

8   9

The last step, the release of acetaldehyde, can be interpreted as a reductive elimination to give the hydrate of acetaldehyde (Eqs. (9.19) and (9.20)) where Eq. (9.19) represents merely the completion of the complex ligation sphere at the central Pd by a solvent molecule. A reductive elimination is a common reaction of group 8 metal compounds. For this case, it has been first proposed in [20]. Keith et al. [21] derived such a pathway but chloride assisted from theoretical considerations. The barrier heights of the transition states for other pathways, for example, β-hydride elimination, were found to be too high. The route according to Eq. (9.20) would also be valid for chloride-free Pd compounds.

$$\left[\overset{\text{OH}}{\diagup\diagdown}\text{PdCl}_2 \right]^- + \text{H}_2\text{O} \rightleftharpoons \left[\overset{\text{OH}}{\diagup\diagdown}\text{PdCl}_2(\text{OH}) \right]^{2-} + \text{H}^+ \qquad (9.19)$$

9   9a

$$\left[\overset{\text{OH}}{\diagup\diagdown}\text{PdCl}_2(\text{OH}) \right]^{2-} \longrightarrow \overset{\text{OH}}{\diagup\diagdown}\text{OH} + \text{Pd} + 2\text{Cl}^- \qquad (9.20)$$

9a

$\downarrow$

$\diagup\diagdown_{\text{O}} + \text{H}_2\text{O}$

As mentioned earlier, the oxypalladation step is discussed controversially. Thus, Bäckvall et al. [22] reacted E-(ethylene)-d_2 ($C_2H_2D_2$) with palladium chloride and cupric chloride under extreme conditions, that is, extremely high chloride ion concentration as cupric and lithium chlorides. Under such conditions, 2-chloroethanol is formed as the main product besides some acetaldehyde [23]; this is not the normal product of the Wacker reaction. In this study, the formation of cis-1,2-dideuterioethylene oxide, evidently via threo-1,2-dideuteriochloroethanol, suggests trans-addition of water (anti-hydroxypalladation).

In another study [24], bis-[(cis-1,2-dideuterioethylene)dichloropalladium] was reacted with carbon monoxide in aqueous acetonitrile to form trans-2,3-dideuterio-β-propiolactone according to Eq. (9.21) via a suggested trans-hydroxypalladation intermediate.

$$1/2\left[\left(\overset{D}{\underset{H}{\rangle}}=\overset{D}{\underset{H}{\langle}}\right)PdCl_2\right]_2 + H_2O + CO \xrightarrow{CH_3CN} \overset{D}{\underset{O}{\overset{H}{\underset{\|}{\square}}}}\overset{H}{\underset{D}{\|}} + Pd + 2HCl \quad (9.21)$$

In a further study [25], 1,2-dimethylcyclohexa-1,4-diene was reacted with bis(acetonitrile)palladium dichloride in aqueous acetone to form predominantly a *trans*-π-allylpalladium complex as well as some of the Wacker oxidation product 3,4-dimethylcyclohex-3-en-1-one (Eq. (9.22)).

$$\text{[structures]} \quad (9.22)$$

PdCl/2

Assuming the same intermediate for both products, the oxypalladation step should be *trans*-orientated. The authors of these studies claim the "mechanism" of the Wacker reaction, although the conditions and the products are different from it. This is also the case with the hydroxypalladation of cyclooctadiene [26] where the chelated Pd complex is rigid and the rotation of the complexed olefin is prevented to enable the *cis*-interaction.

Molecular orbital calculation investigations have been carried out. In one of them [27], it is concluded that the OH nucleophile would attack the complexed olefin within the coordination sphere of the complex, that is, *syn* (*cis*). Two others [28, 29] claim the attack from the outer side, that is, *anti* (*trans*).

A more extensive discussion of these facts can be found in [30, p. 395]. Further examples for the discussion of the oxypalladation step in the Wacker reaction can be found in Keith's paper [16].

9.2.3
Catalytic Oxidation of Ethylene

9.2.3.1 Oxidation of Ethylene to Acetaldehyde in the Presence of $CuCl_2$

Oxidation of ethylene according to Eq. (9.2) occurs stoichiometrically. A catalytic oxidation is possible if palladium(0) is reoxidized immediately. This happens in the presence of oxidizing agents such as cupric and ferric chlorides; ferric sulfate; chromates; heteropoly acids of phosphoric, molybdic, and vanadic acids; peroxides; and others. Benzoquinone is used by Moiseev et al. for their kinetic investigations [11]. Gaseous oxygen does not oxidize palladium black in a sufficiently short time.

Cupric chloride proved to be the most suitable oxidant since cuprous chloride can easily be reoxidized by gaseous oxygen. The basic process according to Eq. (9.1a) (Eq. (9.1a)=(9.2) + (9.23) + (9.24)); equal to Eq. (9.1)) becomes catalytic (Eqs. (9.23), (9.24), and (9.1a)).

$$2CuCl + 2HCl + 1/2O_2 \rightarrow 2CuCl_2 + H_2O \quad (9.23)$$

$$Pd + 2CuCl_2 \rightarrow PdCl_2 + 2CuCl \quad (9.24)$$

$$H_2C=CH_2 + 1/2 O_2 \xrightarrow{PdCl_2/CuCl_2} \triangle\!\!\!\!\diagdown_O \quad (9.1a)$$
$$\Delta H = -58.2 \text{ kcal } (-221.1 \text{ kJ})$$

At first glance, it seems slightly strange that cupric ions should oxidize palladium in the zero oxidation state according to their oxidation potentials [31]. Evidently, chloride ions play an essential role through stabilization of Pd^{2+} and Cu^+ by complexing. Respective thermodynamic considerations are given in [19].

An aqueous solution of palladium chloride and cupric chloride represents the catalyst of the commercial process to produce acetaldehyde from ethylene. For the process, two versions have been developed (see Section 4.3). In the single-stage process, represented by Eq. (9.1a), a mixture of ethylene and oxygen is contacted with the catalyst solution and acetaldehyde separated from it. In the two-stage process, ethylene and oxygen – air can be used instead of pure oxygen – are reacted in separate reactors. In the first reactor, ethylene is reacted with the catalyst solution stoichiometrically with respect to cupric chloride according to Eq. (9.25) (Eq. (9.25) = (9.2) + (9.23)). In the second reactor, the cuprous chloride is reoxidized by air according to Eq. (9.23).

$$H_2C=CH_2 + 2CuCl_2 + H_2O \xrightarrow{PdCl_2} \triangle\!\!\!\!\diagdown_O + 2CuCl + 2HCl \quad (9.25)$$

Kinetic investigations [32, 33] show dependencies, which are represented by Henry and Moiseev's equation (see Eq. (9.5)). Optimal technical catalysts have a deficit of the chloride ion concentration with respect to the Cu^{2+} concentration. In a completely oxidized catalyst, the Cu^{2+} exists partly as copper oxychloride – its insoluble form has the composition $Cu_2(OH)_3Cl$. The chloride ion concentration controls the pH value of the catalyst and has, therefore, a great influence on the reaction rate. Figure 9.1 shows this correlation [6, 34].

Comparing the reaction rate, that is, ethylene consumption with the pH value, both plotted against the mole ratio $[Cu^{2+}]/([Cu^+] + [Cu^{2+}]) = [Cu^{2+}]/[Cu_{total}]$, that is, the degree of oxidation of the catalyst solution, it can be demonstrated that a decline of the reaction rate corresponds to a decline of the pH value in a neutralization curve depending on the ratio Cl/Cu of the catalyst. In fact, this is consistent with the neutralization of the copper oxychloride.

In the single-stage process, the degree of oxidation of the catalyst remains constant. An optimal one has to be chosen. It depends on the ethylene/oxygen and Cl/Cu ratios. In the two-stage process, the degree of oxidation changes. The catalyst solution is reduced in the "reaction phase" with ethylene and reoxidized in the "oxidation phase" with air. An optimal area of reduction/oxidation has to be chosen, adjusted by the Cl/Cu ratio while preventing the precipitation of catalyst components, such as copper oxychloride in the oxidized state and copper(I)chloride in the reduced state of the catalyst.

Cupric chloride is a very aggressive agent. In the Wacker reaction, it acts as a chlorinating agent (Eq. (9.26)). Thus, chloroacetaldehyde is the main by-product

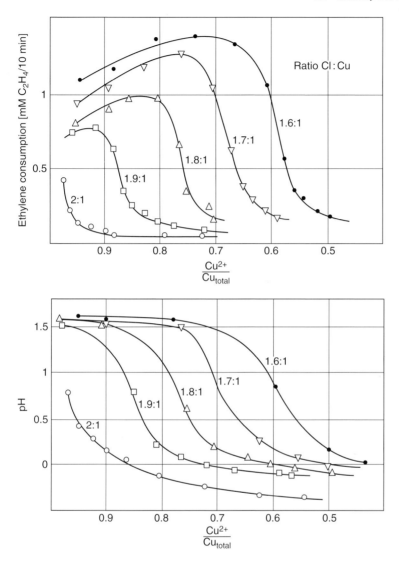

Figure 9.1 Relation between pH value and the rate of the reaction of ethylene with aqueous $PdCl_2$ solutions containing $CuCl_2$.

of the Wacker acetaldehyde process (also see Section 4.3).

$$\text{CH}_2\text{=CH}_2 + 2\text{CuCl}_2 \longrightarrow \text{ClCH}_2\text{CHO} + 2\text{CuCl} + \text{HCl} \tag{9.26}$$

9.2.3.2 Oxidation of Ethylene to 2-Chloroethanol

With a very high concentration of cupric chloride (about $5\,\text{mol·L}^{-1}$) and high temperature and pressure, 2-chloroethanol is the main product of ethylene oxidation besides some acetaldehyde [23]. Cupric chloride is essential. In its absence, despite high chloride ion concentration, absolutely no 2-chloroethanol is obtained. It is assumed that analogous to acetaldehyde formation, a β-hydroxyethyl species bonded to a bi- or oligo-Pd–Cu cluster is an intermediate from which 2-chloroethanol is liberated by reductive elimination.

$$H_2C=CH_2 + 2CuCl_2 + H_2O \xrightarrow{PdCl_2} Cl\diagup\!\!\!\diagdown OH + 2CuCl + HCl \qquad (9.27)$$

2-Chloroethanol is also formed from ethylene in the presence of pyridine and cupric chloride even with low chloride ion concentration [35].

9.3
Process Technology (Wacker Process)

9.3.1
Single-Stage Acetaldehyde Process from Ethylene

The single-stage process was developed by a research group of Hoechst AG (also see [36]). At the time, Hoechst AG owned 50% of the shares of Wacker Chemie. Via the board, they learned at an early time about Wacker's activity on ethylene oxidation and began research on this field. Later on, both companies cooperated and combined their results.

In the single-stage process (Figure 9.2), a mixture of ethylene and oxygen is passed through an aqueous solution of copper chloride and palladium chloride in a towerlike reactor (a). Acetaldehyde is formed according to Eq. (9.1a). In order to avoid an explosive mixture of ethylene and oxygen, ethylene is used in

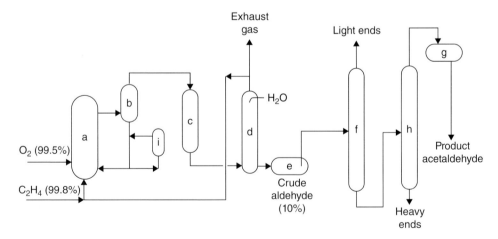

Figure 9.2 Acetaldehyde from ethylene, single-stage process: (a) reactor, (b) separating vessel, (c) cooler, (d) scrubber, (e) crude aldehyde tank, (f) light ends distillation, (g) condenser, (h) purification column, and (i) regeneration unit.

stoichiometric excess over oxygen over the upper explosion limit, so unreacted ethylene leaves the reactor together with acetaldehyde formed. (b) is a separator for the gaseous and liquid parts carried away from the reactor. The gaseous mixture is cooled in (c), and acetaldehyde is separated by scrubbing with water in (d). The gaseous portion is fed back to the reactor and completed with fresh ethylene and oxygen. From the scrubber (d), a small amount of the cycle gas has to be withdrawn in order to avoid an accumulation of inert gases and gaseous by-products. Therefore, high-purity starting material has to be used in order to avoid the loss of a high amount of the starting gas with the exhaust gas. Crude acetaldehyde is collected in container (e).

The purification (h) and regeneration (i) steps have the same function as the analogous steps in the two-stage process described in the following section.

9.3.2
Two-Stage Acetaldehyde Process from Ethylene

The two-stage process has been developed by the research group of Wacker's Consortium für elektrochemische Industrie. While this group worked on the development of a single-stage process, Wacker had to realize that at the time and the site of the scheduled first commercial plant, oxygen was not available at a reasonable price. So, Wacker was forced to develop another version using air as the oxidant.

In the two-stage process (Figure 9.3), ethylene and oxygen are reacted separately in separate meander-like pipe reactors (a, d). First, ethylene is reacted stoichiometrically in (a) with the catalyst solution, reducing cupric chloride to cuprous chloride according to Eq. (9.25). In the second stage (d), cuprous chloride is reoxidized by oxygen to cupric chloride according to Eq. (9.23).

As both gases react separately, their purity need not be high. So, air is usually used for the reoxidation of the catalyst instead of pure oxygen. The reaction is carried out at about 10 bar, a pressure higher than in the single-stage process. Acetaldehyde is separated from the catalyst by expanding it to normal pressure in tower (b). An acetaldehyde–water mixture is evaporated, led into the crude aldehyde column (f) where aldehyde is concentrated up to about 90% using the reaction heat, and collected in the crude aldehyde container (h). The catalyst is cycled from tower (b) to oxidation reactor (d) by pump (c). The exhaust gas (inert gas and gaseous by-products) and air are scrubbed with process water (j) in order to recover acetaldehyde brought with and fed back to the crude aldehyde column. Process water is also recycled to the catalyst in tower (b).

It has been mentioned in Section 4.1 that cupric chloride causes chlorination reactions. Thus, chlorinated aldehydes (see Eq. (9.26)) are the main by-products. Others include acetic acid and chlorinated acetic acids. Light ends are carbon dioxide, methyl, and ethyl chloride. By chlorination, oxidative, and hydrolytic reactions, oxalic acid is formed causing insoluble copper oxalate. In order to avoid an accumulation in the catalyst, it is continuously thermally decomposed by heating a small side stream in the regeneration step (reactor (i) in the one-stage, reactor (m)

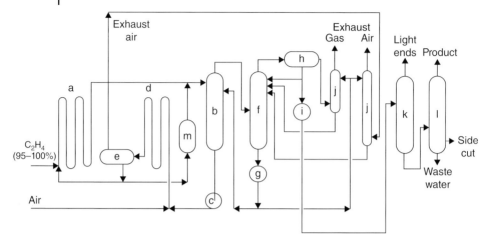

Figure 9.3 Acetaldehyde from ethylene, two-stage process: (a) reactor, (b) expansion tower, (c) catalyst cycle pump, (d) oxidation reactor, (e) exhaust air separator, (f) crude aldehyde column, (g) process water tank, (h) condenser, (i) crude aldehyde container, (j) scrubbers for exhaust gas and air, (k) light ends distillation, (l) purification column, and (m) regeneration.

in the two-stage version). Copper oxalate is decomposed according to Eq. (9.28).

$$CuC_2O_4 + CuCl_2 \xrightarrow{\Delta T} 2CuCl + 2CO_2 \qquad (9.28)$$

Through this, the content of copper oxalate in the catalyst can be kept constantly low. Thus, the catalyst shows a kind of self-cleaning behavior.

As chlorinated by-products are formed during the process and subsequently removed from the catalyst solution, the deficit of chloride in the catalyst must be compensated by feeding the equivalent amount of hydrochloric acid into the catalyst solution.

The subsequent purification of the product is carried out by a two-step distillation. At first (column **k**), the light ends, mainly carbon dioxide, methyl, and ethyl chloride, are removed overhead. In the second step (**l**), acetaldehyde goes overhead; water, acetic acid, and some chlorinated products are removed as bottoms and chlorinated acetaldehydes as a side cut.

Due to environmental requirements, chlorinated and other by-products have to be decomposed by heating and/or alkali treatment and/or biologically. Monochloroacetaldehyde is also a valuable starting material for organic syntheses.

The by-products are the same in both versions with a slight difference in the amounts. Under optimal conditions, the yield of acetaldehyde is about 95% also in both versions.

Due to the high corrosive ability of the catalyst solution, titanium is used as a construction material for all equipment containing catalyst. The reactor of the single-stage process is resin or ceramic lined.

9.4
Other Developments

Propene gives acetone and a small amount of propionaldehyde on oxidation under Wacker conditions, that is, in the presence of $PdCl_2$ and $CuCl_2$. A commercial process has been developed by Hoechst AG, and two plants have been erected in Japan. The process is carried out by the two-stage technology. A regeneration step is also used to decompose copper oxalate thermally. Propionaldehyde is removed by distillation together with the light ends and isolated by a separate distillation step [37, 38].

Butene-1 and *butene-2* give methyl ethyl ketone (butanone) on oxidation with $PdCl_2$ and $CuCl_2$. But here, a particularly high amount of chlorinated butanone, that is, 3-chlorobutanone-2, is formed. Although this product proved to be a valuable starting material for organic syntheses, the process did not gain industrial interest. Wacker ran a pilot plant [38].

Wacker oxidation of the lower olefins follows the same rate law as that of ethylene (Eq. (9.29) [15, 32], compare Eq. (9.5)):

$$-\frac{d[C_3H_6]}{dt} = \frac{k[PdCl_2] \cdot pC_3H_6}{[Cl^-]^{2.23} \cdot [H^+]} \quad (9.29)$$

The rates decrease in the order ethylene > propene > 1-butene > *cis*-2-butene > *trans*-2-butene.

The Wacker process, the oxidation of ethylene to acetaldehyde, lost its original importance over the past 30 years. While at the beginning more than 40 factories with a total capacity of more than 2 million tons of acetaldehyde per year were installed, acetaldehyde as an industrial intermediate was replaced successively by other processes. For example, C_4 compounds such as butyraldehyde/butanol are produced by the oxo process from propylene, and acetic acid by the Monsanto process from methanol and CO or by direct oxidation of ethane. The way via acetaldehyde to these products is dependent on the price of ethylene. Petrochemical ethylene from cracking processes became considerably more expensive during these years. Thus, only few factories would be necessary to meet the demand for other derivatives of acetaldehyde such as alkyl amines, pyridines, glyoxal, and pentaerythritol.

But the regiospecific oxidation of olefinic compounds by palladium catalysts is enjoying increasing importance as a step in the multistep syntheses of complex organic compounds such as pharmaceuticals, pesticides, and fragrances. The carbonyl compounds are reactive intermediates in such syntheses.

It has already been mentioned (Section 9.2) that the Wacker reaction occurs regiospecifically. Thus, 1-olefins give oxidation with palladium chloride to methylketones according to Markovnikov's rule. The synthetic reactions were carried out stoichiometrically as well as catalytically.

There has been much effort spent finding a way to produce aldehydes instead of ketones with *anti*-Markovnikov regioselectivity. Thus, the formation of aldehydes can be increased by higher temperature and high H^+ and Cl^- concentrations [8].

Muzart [39] reviewed examples of such "anti-Markovnikov" reactions forming aldehydes. But many of the examples can be explained by the influence of electron-withdrawing groups, which then are Markovnikov conforming, like the formation of nitroacetaldehyde and cyanoacetaldehyde from nitroethylene and acrylonitrile, respectively (see Eq. (9.4) and Table 9.1).

Feringa succeeded to increase the product ratio in favor of the aldehydes with a palladium catalyst containing acetonitrile and a nitro group as ligands together with cupric chloride [40], and R. H. Grubbs *et al.* found a way to synthesize aldehydes in high yields and selectivities from unsubstituted [41] and substituted [42] terminal olefins using a Wacker catalyst consisting of [$PdCl_2(PhCN)_2$]/$CuCl_2$/O_2 modified with $AgNO_2$ for the first and $NaNO_2$ for the second case. The substituents in the second case were oxygen functions such as aliphatic and aromatic ether and acetoxy groups. For the oxidation of olefinics with oxygen, they used as solvent a mixture of polar organic styrene and substituted styrenes to the respective phenylacetaldehydes with oxygen; Grubbs *et al.* [43] used [$PdCl_2(MeCN)_2$]/benzoquinone as catalyst and *t*-butanol/water as solvent. A formal *anti*-Markovnikov hydration of styrene and substituted styrenes with postulated Wacker oxidation and vinylether intermediates simultaneously catalyzed by a Wacker palladium and a ruthenium hydrogenation catalyst to give the respective primary alcohols has also been published by Grubbs *et al.* [44].

Allylic amines (protected by phthalimide) are also oxidized by oxygen in the presence of a catalyst consisting of $Pd(CH_3CN)Cl(NO_2)$/$CuCl_2$ in DMF/H_2O or *t*-butanol to give predominantly the aldehyde function as an intermediate for a β-amino acid synthesis [45]. An exclusive oxidation to aldehydes was achieved with *p*-benzoquinone as oxidant in *t*-butanol [46]. Protected (as ethers or esters) allylic alcohols are oxidized in a similar way with *p*-benzoquinone as oxidant and diaceto- or dibenzonitrile-$PdCl_2$ in *t*-butanol/acetone to the respective aldehydes [47]. The direction of the entrance of the carbonyl group in these reactions claimed as an *anti*-Markovnikov reaction can also be explained with the electron-withdrawing influence of the amino group or the oxygen function, respectively (compare Eq. (9.4) and Table 9.1) and hence should conform to Markovnikov selectivity.

The oxidation of olefins to carbonyl compounds in the presence of $PdCl_2$ and $CuCl_2$, which represents the catalysts of the commercial acetaldehyde process, is accompanied by chlorinating reactions. These side reactions can reduce the yield of the desired products considerably. Different ways have been suggested to reoxidize Pd^0, avoiding the chlorinating behavior of $CuCl_2$.

As a possible strategy to avoid chlorinating side reactions by cupric chloride, electrochemical oxidation of the Pd(0) has been claimed [48, 49].

The Japanese company Maruzen Oil Co. [50] and Catalytica Associates Co. [51] use ferric sulfate as oxidant (cocatalyst) [52] (also compare [2]); Catalytica claims a commercial process in existing Wacker two-stage plants.

Heteropolyacids of P, Mo, and V such as $H_3PMo_6V_6O_{40}$ are also chloride-free cocatalysts for Pd-catalyzed oxidation of ethylene, propene [53–55], and butene [56]. Numerous patents on usage of such catalysts for commercial olefin oxidation have been filed by the Japanese companies Maruzen Oil Co. and Idemitsu Kosan Co. But, to the knowledge of the author, neither Maruzen nor Idemitsu nor Catalytica is operating their respective plants commercially.

A heterogenized $PdSO_4-VOSO_4-H_2SO_4$-on-coal catalyst has been described [57] for the oxidation of 1-butene to butanone, and another one consisting of $PdCl_2-V_2O_5$ on γ-Al_2O_3 for the oxidation again of 1-butene [58] and of styrene to benzaldehyde, acetophenone, and *trans*-cinnamaldehyde [59]. The oxidation of 1-butene over a heterogeneous catalyst consisting of Pd salt (sulfate or acetate)/V_2O_5 on TiO_2 has also been reported [60]. Also, a heterogenized $PdCl_2/CuCl_2$ catalyst in polyethyleneglycol on silica gel for the oxidation of ethylene to acetaldehyde has been used, but nothing about chlorinating side reactions has been reported [61].

High ethylene conversion and acetaldehyde yield were claimed with a heterogeneous catalyst consisting of Pd^0 and Cu^{2+} or VO_x as cocatalysts on VO_x nanotubes [62].

Benzoquinone was earlier used as an oxidant in the kinetic investigations by Moiseev *et al.* [11–13]. For the oxidation of 1-decene to 2-decanone in the presence of benzoquinone under chloride-free conditions, Tsuji and Minato used electro-reoxidation of the hydroquinone [63]. On the other hand, Bäckvall and Hopkins oxidized 1-decene, 1-dodecene, and some aromatic 1-olefins in DMF by oxygen to the corresponding methylketones in the presence of benzoquinone and iron phthalocyanine as oxygen activator [64]. Grubbs *et al.* [65] reacted a broad range of internal aliphatic, cycloaliphatic, and aromatic olefins, partly substituted with different functional groups (alcohol, acid, ester, amide, and others) in a solvent mixture consisting of *N,N*-dimethylacetamide (DMA), acetonitrile, water, and HBF_4 with $Pd(OAc)_2$ or $[Pd(MeCN)_4](BF_4)_2$ and again benzoquinone as oxidant forming the respective ketones with excellent yields. DMA suppresses isomerization, another side reaction, taking place with higher olefins (discussed later).

Another way to avoid chlorinating side reactions has been published by Tsuji [66, 67]. For the oxidation of olefinics with oxygen he uses as solvent a mixture of a polar organic solvent such as dimethylformamide with water and $PdCl_2/CuCl$ as catalyst. Of course, the reoxidation of Pd(0) runs via $CuCl_2$, but evidently its concentration during the reaction is too low to cause a noticeable chlorination. With this catalyst and $PdCl_2/p$-benzoquinone, Tsuji *et al.* [68] oxidized with oxygen allylic and homoallylic ethers and acetates regiospecifically to β- and γ-alkoxy and acetoxy ketones, respectively. In his paper quoted earlier [66], Tsuji gave a comprehensive review about the application of Wacker olefin oxidation in organic syntheses. He *et al.* [69] used a $PdCl_2/CuCl$ catalyst in supercritical CO_2/poly(ethylene glycol) for the oxidation of styrene and substituted styrenes to the corresponding acetophenones with oxygen.

R. A. Sheldon *et al.* oxidized terminal and cyclic olefins with air and a palladium catalyst with a chelating diamine ligand such as phenanthroline or bathophenanthroline disulfonate in order to prevent the chlorinating $CuCl_2$ [70].

t-Butyl hydroperoxide or hydrogen peroxide together with palladium carboxylate is used by Mimoun *et al.* [71, 72] for the oxidation of terminal olefins (Eq. (9.30)) and by Tsuji *et al.* [73] for the oxidation of α,β-unsaturated esters and ketones to β-ketoesters and 1,3.diketones, respectively.

$$F_3CCO_2PdOOtBu + \text{CH}_2=\text{CHCH}_2\text{CH}_3 \longrightarrow F_3CCO_2PdOtBu + \text{CH}_3COCH}_2\text{CH}_2\text{CH}_3 \tag{9.30}$$

t-Butyl hydroperoxide as oxidant is also used in Wacker-type oxidation of styrene [74] and substituted 1-olefins [75] with palladium chloride complexed with bidentate *N*-heterocyclic ligands. In the latter case, enantioenriched substrates are oxidized with retention of the enantiomeric excess. High conversion and selectivity in direct oxidation of substituted 1-olefins with oxygen catalyzed by palladium chloride complexed with a bidentate heterocyclic diamine (sparteine) is described by Sigman *et al.* [76]. A survey of such reactions is given in [77]. Another method for the oxidation of internal olefins with *t*-butylhydroperoxide and a $PdCl_2$ catalyst complexed with the bidentate *N,N*-heterocycle (4,5-dihydro-2-oxazolyl)quinoline forming intermediates for natural product synthesis has recently been reported by Sigman *et al.* [78].

Oxidation of monoterpenes with hydrogen peroxide and $PdCl_2$ as catalyst in acetonitrile has been reported [79]. Besides the normal Wacker oxidation products, aldehydes, ketones, and other oxidation products such as epoxides and glycols are formed.

As already mentioned, isomerization (double bond shift) is also a side reaction, which reduces the selectivity of the oxidation of higher olefins. High selectivity with respect to the formation of methylketones is obtained with a system consisting of *t*-butyl hydroperoxide as the oxidant, acetonitrile as the solvent, and β-cyclodextrine as a phase transfer catalyst in the oxidation of 1-dodecene [80] and of higher olefins C_{12}–C_{20} [81]. Also, modified β-cyclodextrines increase the selectivity and activity of a $PdSO_4$/$CuSO_4$/phosphomolybdovanadic acid catalyst in biphasic oxidation of 1-decene to 2-decanone [82, 83].

Oxidation of terminal olefins with $PdCl_2$ and oxygen without any cocatalyst has been carried out in polar organic solvents. The best results were obtained with DMA/water [84]. Regiospecific oxidation of internal olefins and olefins substituted in the allylic position with functional groups to the respective ketones with oxygen under slight pressure in the presence of $PdCl_2$ in DMA/water without other oxidants has been described [85–87]. The presence of copper chloride under Tsuji's conditions [66, 67] reduces the rate drastically. It should be noted that in one of these papers [85], the reaction is described as being carried out with $^{18}OH_2$. The carbonyl group formed exclusively contains the heavy oxygen isotope, which proves that the carbonyl oxygen comes from the water and not from the gaseous oxygen.

In DMSO/water/trifluoroacetic acid, styrene, substituted styrenes, and some others are oxidized to the corresponding ketones with sole oxygen and

4-phenyl-1-butene and substituted ones are oxidized and simultaneously dehydrogenated under similar conditions to give α,β-unsaturated ketones [88]. α,β-Unsaturated ketones are synthesized directly from terminal olefins by a simultaneous Wacker oxidation–dehydrogenation with a catalyst consisting of $Pd(CH_3CN)_4(BF_4)_2$ and hypervalent iodonium compound $PhI(OAc)_2$ [89].

Higher aliphatic and aromatic olefins as well as substituted aromatic olefins with oxygen functions are oxidized in THF/water and a solid catalyst Pd^0/C together with $KBrO_3$ to give the corresponding carbonyl compounds, thus also avoiding the use of copper chloride [90].

Polypyrrole, a conducting polymer, is an effective redox system for a palladium catalyst for the oxidation of olefinic compounds with oxygen in acetonitrile/water to the respective carbonyl compounds [91]. 1-Decene and terminal olefins with oxygen functions have been used in these experiments.

A surprising reaction is the formation of acrylic acid, probably via allyl alcohol in an allylic-type reaction by oxidizing propene with a palladium-on-coal-catalyst in aqueous phase, while in the presence of chloride or oxidants, the normal Wacker-type reaction product is formed [92].

References

1. Smidt, J., Hafner, W., Sedlmeier, J., Jira, R., and Rüttinger, R. (1959) Consortium für Elektrochemische Industrie. DE Patent 1049845.
2. Smidt, J., Hafner, W., Jira, R., Sedlmeier, J., Sieber, R., Rüttinger, R., and Kojer, H. (1959) *Angew. Chem.*, **71**, 176–182.
3. Phillips, F.C. (1894) *Am. Chem. J.*, **16**, 255–277.
4. Kharasch, M.S., Seyler, R.C., and Mayo, F.R. (1938) *J. Am. Chem. Soc.*, **60**, 882.
5. Anderson, J.S. (1934) *J. Chem. Soc. II*, 971.
6. Smidt, J., Hafner, W., Jira, R., Sieber, R., Sedlmeier, J., and Sabel, A. (1962) *Angew. Chem.*, **74**, 93–102; *Angew. Chem. Int. Ed. Engl.* (1962) **1**, 80–88.
7. Smidt, J. and Sieber, R. (1959) *Angew. Chem.*, **71**, 626.
8. Hafner, W., Jira, R., Sedlmeier, J., and Smidt, J. (1962) *Chem. Ber.*, **95**, 1575–1581.
9. A lot of Pd(II)-catalysed 1,4-addition reactions to conjugated dienes used for organic syntheses are described by Bäckvall, J.-E. et al. e.g., Castaño, A.M., Person, B.A., and Bäckvall, J.-E. (1997) *Chem. Eur. J.*, **3**, 482 and quoted literature therein.
10. Jira, R. (1971) *Tetrahedron Lett.*, **17**, 1225–1226.
11. Vargaftik, M.N., Moiseev, I.I., and Syrkin, Y.K. (1962) *Dokl. Akad. Nauk SSSR*, **147**, 399.
12. Moiseev, I.I., Vargaftik, M.N., and Syrkin, Y.K. (1963) *Dokl. Akad. Nauk SSSR*, **153**, 140.
13. Vargaftik, M.N., Moiseev, I.I., and Syrkin, Y.K. (1963) *Izv. Akad. Nauk. Otd. Khim. Nauk SSSR*, 1147.
14. Henry, P.M. (1964) *J. Am. Chem. Soc.*, **86**, 3246–3250.
15. Henry, P.M. (1966) *J. Am. Chem. Soc.*, **88**, 1595–1597.
16. Keith, J.A. and Henry, P.M. (2009) *Angew. Chem.*, **121**, 9200–9212; *Angew. Chem. Int. Ed.* (2009) **48**, 9038–9049.
17. Leden, I. and Chatt, J. (1955) *J. Chem. Soc.*, 2936.
18. Jira, R., Sedlmeier, J., and Smidt, J. (1966) *Liebigs Ann. Chem.*, **693**, 99.
19. Jira, R. and Freiesleben, W. (1972) Olefin oxidation and related reactions with group VIII noble metal compounds, in *Organometallic Reactions*, vol. **3** (eds E. Becker and M. Tsutsui), John Wiley & Sons, Inc., New York.

20. Jira, R. (1975) in *Methodicum Chimicum*, vol. **5** (ed F. Korte), Georg Thieme, Stuttgart, p. 286.
21. Keith, J.A., Oxgaard, J., and Goddard, W.A. (2006) *J. Am. Chem. Soc.*, **128**, 3132–3133.
22. Bäckvall, J.E., Åkermark, B., and Ljunggren, S.O. (1979) *J. Am. Chem. Soc.*, **101**, 2411–2416.
23. Stangl, H. and Jira, R. (1970) *Tetrahedron Lett.*, **41**, 3589–3592.
24. Stille, J.K. and Divakaruni, R. (1979) *J. Organomet. Chem.*, **169**, 239–248.
25. Åkermark, B., Söderberg, B.C., and Hall, S.S. (1987) *Organometallics*, **6**, 2608–2610.
26. Stille, J.K. and James, D.E. (1975) *J. Am. Chem. Soc.*, **97**, 674–676.
27. Armstrong, D.R., Fortune, R., and Perkins, P.G. (1976) *J. Catal.*, **45**, 339–348.
28. Bäckvall, J.-E., Björkman, E.E., Pettersson, L., and Siegbahn, P. (1984) *J. Am. Chem. Soc.*, **106**, 4369–4373.
29. Fujimoto, H. and Yamasaki, T. (1986) *J. Am. Chem. Soc.*, **108**, 578–581.
30. Jira, R. (2002) Oxidation of olefins to carbonyl compounds, in *Applied Homogeneous Catalysis with Oranometallic Compounds*, 2nd edn (eds B. Cornils and W. Herrmann), Wiley-VCH Verlag GmbH, Weinheim, 3rd edn, in Press.
31. Latimer, W.M. (1959) *Oxidation Potentials*, 2nd edn, Prentice Hall, Englewood Cliffs, NJ.
32. Kiryu, S. and Shiba T. (1961) Lecture Annual Meeting Japan Petroleum Institute, Tokyo, September, 1961 (cited in Ref. [33]).
33. Dozono, T. and Shiba, T. (1963) *Bull. Japan Petr. Inst.*, **5**, 8 (in English).
34. Jira, R. (1969) Acetaldehyde, in *Ethylene and Its Industrial Derivatives*, Chapter 8 (ed S.A. Miller), Ernest Benn, London.
35. Francis, J.W. and Henry, P.M. (1995) *J. Mol. Catal. A: Chem.*, **99**, 77.
36. Jira, R. (1985) Acetaldehyde, in *Ullmann's Encyclopedia of Industrial Chemistry*, 5th edn, VCH Verlagsgesellschaft mbH, D-6940 Weinheim, vol. **A1**, p. 31.
37. Wöllner, J. and Weber, E. (1974) Aceton in *Ullmanns Encyklopädie der technischen Chemie*, 4th edn., vol. 7, p. 31. Verlag Chemie GmbH, Weinheim/Bergstr.
38. Smidt, J. and Krekeler, H. (1963) *Hydrocarbon Process. Pet. Refiner*, **42**, 149.
39. Muzart, J. (2007) *Tetrahedron*, **63**, 7505–7521.
40. Feringa, B.L. (1986) *J. Chem. Soc., Chem. Commun.*, 909–910.
41. Wickens, Z.K., Morandi, B., and Grubbs, R.H. (2013) *Angew. Chem.*, **125**, 11467–11470; *Angew. Chem. Int. Ed.* (2013) **52**, 11257–11260.
42. Wickens, Z.K., Skakuj, K., Morandi, B., and Grubbs, R.H. (2014) *J. Am. Chem. Soc.*, **136**, 890–893.
43. Teo, P., Wickens, Z.K., Dong, G., and Grubbs, R.H. (2012) *Org. Lett.*, **14**, 3237–3239.
44. Dong, G., Teo, P., Wickens, Z.K., and Grubbs, R.H. (2011) *Science*, **333**, 1609–1612.
45. Weiner, B., Baeza, A., Jerphangnon, T., and Feringa, B.L. (2009) *J. Am. Chem. Soc.*, **131**, 9473–9474.
46. Dong, J.J., Harvey, E.C., Fañanás-Mastral, M., Browne, W.R., and Feringa, B.L. (2014) *J. Am. Chem. Soc.*, **136**, 17302–17307.
47. Dong, J.J., Fañadás-Mastral, M., Alsters, P.L., Browne, W.R., and Feringa, B.L. (2013) *Angew. Chem. Int. Ed.*, **52**, 5561–5565.
48. Tsuji, J. (1987) Production of Carbonyl Compounds. Japan Patent 63192736, Feb. 06, 1987 to Nippon Zeon Co., Ltd., Japan.
49. Inokuchi, T., Ping, L., Hamaue, F., Izawa, M., and Torii, S. (1994) *Chem. Lett.*, 121.
50. Hasegawa, H. and Triuchijima, M. (1971) Process for the Production of Methyl Ethyl Ketone from n-Butene, Maruzen Oil Co.; (1971) Patent GB 1.240.889.
51. Grate, J.H., Hamm, D.R., and Mahayan, S. (1993) *Mol. Eng.*, **3**, 205–229; *Chem. Ind. (Dekker)* (1994) **53**, 213–264.
52. Matveev, K.I., Bukhtoyarov, I.F., Shul'ts, I.I., and Emel'yanova, O.A. (1964) *Kinet. Katal.*, **5**, 649; *Kinet. Katal., Engl. Transl.* (1964) **5**, 572.
53. Matveev, K.I., Zhizhina, E.G., Shitova, N.B., and Kuznetsova, L.I. (1977) *Kinet. Katal.*, **18**, 380–386; *Kinet. Katal., Engl. Transl.*, 320–326.

54. Matveev, K.I. (1977) *Kinet. Katal.*, **18**, 862–877; *Kinet. Katal., Engl. Transl.*, 716–728.
55. Ogawa, H., Fujinami, H., Taya, K., and Teratani, S. (1981) *J. Chem. Soc., Chem. Commun.*, 1274.
56. Davison, S.F., Mann, B.E., and Maitlis, P.M. (1984) *J. Chem. Soc., Dalton Trans.*, 1223–1228.
57. Izumi, Y., Fujii, Y., and Urabe, K. (1984) *J. Catal.*, **85**, 284.
58. Van der Heide, E., Ammerlaan, J.A.M., Gerritsen, A.W., and Scholten, J.J.F. (1989) *J. Mol. Catal.*, **55**, 320.
59. Van der Heide, E., Schenk, J., Gerritsen, A.W., and Scholten, J.J.F. (1990) *Recl. Trav. Chim. Pays-Bas*, **109**, 93.
60. Stobbe-Kreemers, A.W., Makkee, M., and Scholten, J.J.F. (1997) *Appl. Catal., A*, **156**, 219–238.
61. Okamoto, M. and Taniguchi, Y. (2009) *J. Catal.*, **261**, 195–200.
62. Barthos, R., Drotár, E., Szegedi, Á., and Valyon, J. (2012) *Mater. Res. Bull.*, **47**, 4452–4456.
63. Tsuji, J. and Minato, M. (1987) *Tetrahedron Lett.*, **28**, 3683.
64. Bäckvall, J.-E. and Hopkins, R.B. (1988) *Tetrahedron Lett.*, **29**, 2885–2888.
65. Morandi, B., Wickens, Z.K., and Grubbs, R.H. (2013) *Angew. Chem. Int. Ed.*, **52**, 2944–2948.
66. Tsuji, J. (1984) *Synthesis*, **5**, 369–384.
67. Tsuji, J., Nagashima, H., and Nemoto, H. (1990) *Org. Synth., Coll.*, **7**, 137; (1984) **62**, 9.
68. Tsuji, J., Nagashima, H., and Hori, K. (1982) *Tetrahedron Lett.*, **23** (26), 2679–2682.
69. Wang, J.-Q., Cai, F., Wang, E., and He, L.-N. (2007) *Green Chem.*, **9**, 882–887.
70. Ten Brink, G.-J., Arends, I.W.C.E., Papadokianakis, G., and Sheldon, R.A. (1998) *Chem. Commun.*, 2359–2360.
71. Roussel, M. and Mimoun, H. (1980) *J. Org. Chem.*, **45**, 5387–5390.
72. Mimoun, H. (1981) *Pure Appl. Chem.*, **53**, 2389.
73. Tsuji, J., Nagashima, H., and Hori, K. (1980) *Chem. Lett.*, 257.
74. Cornell, C.N. and Sigman, M.S. (2005) *J. Am. Chem. Soc.*, **127**, 2796.
75. Michel, B.W., Camelio, A.M., Cornell, C.N., and Sigman, M.S. (2009) *J. Am. Chem. Soc.*, **131**, 6076.
76. Cornell, C.N. and Sigman, M.S. (2006) *Org. Lett.*, **8** (18), 4117.
77. Cornell, C.N. and Sigman, M.S. (2007) *Inorg. Chem.*, **46**, 1903.
78. DeLuca, R.J., Edwards, J.L., Steffens, L.D., Michel, B.W., Qiao, X., Zhu, C., Cook, S.P., and Sigman, M.S. (2013) *J. Org. Chem.*, **78**, 1682–1686.
79. Viera, L.M.M., da Silva, M.L., and da Silva, M.J. (2013) in *Hydrogen Peroxide*, Chapter 9 (eds G. Aguilar and R.A. Guzman), Nova Science Publishers, Inc., pp. 219–231.
80. Escola, J.M., Botas, J.A., Aquado, J., Sessano, D.P., Vargas, C., and Bravo, M. (2008) *Appl. Catal., A*, **335**, 137.
81. Escola, J.M., Bota, J.A., Vargas, C., and Bravo, M. (2010) *J. Catal.*, **270**, 34.
82. Montflier, E., Tilloy, S., Blouet, E., Barbaux, Y., and Mortreux, A. (1996) *J. Mol. Catal. A: Chem.*, **109**, 27–35.
83. Tilloy, S., Bertoux, F., Mortreux, A., and Monflier, E. (1999) *Catal. Today*, **48**, 245–253.
84. Mitsudome, T., Umetani, T., Nosaka, N., Mori, K., Mizugaki, T., Ebitani, K., and Kaneda, K. (2006) *Angew. Chem.*, **118**, 495; *Angew. Chem. Int. Ed.* (2006) **45**, 481–485.
85. Mitsudome, T., Mizumoto, K., Mizugaki, T., Jitsukawa, K., and Kaneda, K. (2010) *Angew. Chem.*, **122**, 1260–1262.
86. Mitsudome, T., Yoshida, S., Tsubomoto, Y., Mizugaki, T., Jitsukawa, K., and Kaneda, K. (2013) *Tetrahedron Lett.*, **54**, 1596–1598.
87. Mitsudome, T., Yoshida, S., Mizugaki, T., Jitsukawa, K., and Kaneda, K. (2013) *Angew. Chem. Int. Ed.*, **52**, 5961–5964.
88. Wang, Y.-F., Gao, Y.-R., Mao, S., Zhang, Y.-L., Guo, D.-D., Yan, Z.-L., Guo, S.-H., and Wang, Y.-Q. (2014) *Org. Lett.*, **16**, 1610–1613.
89. Bigi, M. and White, M.C. (2013) *J. Am. Chem. Soc.*, **135**, 7831–7834.
90. Kulkarni, M.G., Shaik, Y.B., Borhade, A.S., Chavhan, S.W., Dhondge, A.P., Gaikwad, D.D., Desai, M.P., Birhade, D.R., and Dhatrak, N.R. (2013) *Tetrahedron Lett.*, **54**, 2293–2295.

91. Higuchi, M., Yamaguchi, S., and Hirao, T. (1996) *Synlett*, 1213–1214.
92. Lyons, J.E., Suld, G., and Hsu, C.-Y. (1988) *Chem. Ind. (Dekker) (Catal. Org. React.)*, **33**, 1.

Further Reading

Henry, P.M. (1980) *Palladium Catalysed Oxidation of Hydrocarbons*, D. Reidel, Boston, MA.

Maitlis, P.M. (1971) *The Organic Chemistry of Palladium*, vol. **1** and **2**, Academic Press, New York.

Negishi, E. (2002) *Handbook of Organopalladium Chemistry for Organic Syntheses*, John Wiley & Sons, Inc.

Tsuji, J. (1975) *Organic Synthesis by Means of Tranition Metal Complexes*, Springer-Verlag, Berlin.

Tsuji, J. (1980) *Organic Synthesis with Palladium Compounds*, Topics in Current Chemistry, vol. **90**, Springer-Verlag, Berlin, p. 30.

Tsuji, J. (1995) *Palladium Reagents and Catalysts: Innovation in Organic Synthesis*, John Wiley & Sons, Ltd., Chichester.

Tsuji, J. (2004) *Palladium Reagents and Catalysts: New Perspectives for the 21st Century*, John Wiley & Sons, Ltd.

10
1,4-Butanediol from 1,3-Butadiene

Yusuke Izawa and Toshiharu Yokoyama

10.1
Introduction

Linear aliphatic diols are widely used as raw materials for polymers. Polymers synthesized from even-carbon diols tend to show excellent polymer properties. 1,4-Butanediol is very important as raw material for various polymers such as urethanes and polybutylene terephthalate (PBT), which is an engineering plastic. Since Celanese Corporation described a PBT resin in 1970, the demand for PBT resin, which is mainly used for automotive, electrical, and electronic equipment parts, has been expanding rapidly [1]. THF is also a major 1,4-butanediol derivative as a raw material for poly(tetramethylene ether) glycol used for artificial leather and elastic fibers in addition to being a high-performance solvent. Significant growth in demand for these 1,4-butanediol derivatives is expected in Asia, primarily in China.

The Reppe process is based on acetylene as a raw material. These reactions were developed by Reppe *et al.* [2]. In accordance with the rise of the petrochemical industry, most processes switched from acetylene to olefins as raw material. However, only the 1,4-butanediol production process continued to rely on the Reppe process. Mitsubishi Chemical Corporation developed a totally different production method that uses 1,3-butadiene to produce 1,4-butanediol and THF. Commercial production was launched in 1982 and has been continued ever since. This process ended the over-half-century monopoly of the Reppe method. The Mitsubishi Chemical method has an advantage over the Reppe method with respect to the handling of raw materials and production costs, but in recent years, Chinese companies that can take advantage of inexpensive natural gas and coal have built a new production plant by using the Reppe method and international competition is getting more intense.

10.2
Chemistry and Catalysis

Oxidative acetoxylation of 1,3-butadiene is the key reaction of Mitsubishi Chemical's 1,4-butanediol production method by Mitsubishi Chemical Corporation (Eq. (10.1)).

$$\text{CH}_2=\text{CH−CH}=\text{CH}_2 + 2\,\text{AcOH} + 1/2\,\text{O}_2 \xrightarrow{\text{Pd cat.}} \text{AcO−CH}_2\text{−CH}=\text{CH−CH}_2\text{−OAc} + \text{H}_2\text{O} \quad (10.1)$$

Although similar to the synthesis of vinyl acetate from ethylene [3], oxidative acetoxylation of 1,3-butadiene to the 1,4-diacetoxy compound with 2 equiv. of acetic acid is more complex. Vinyl acetate production from ethylene has succeeded in commercialization in the gas-phase reaction using a Pd-supported catalyst. 1,4-Selectivity and catalyst activity was only moderate when acetoxylation of 1,3-butadiene was performed in a similar system because of competing raw material and product polymerization [4]. As a result of the adhesion of compounds with high boiling points on the catalyst surface, the catalytic activity dramatically decreased in a very short time. To reduce the formation of compounds with high boiling points, catalyst development was carried out in a liquid phase at a lower temperature. Finally, a Pd–Te–C catalyst was found to be suitable. The reaction is considered to proceed via a π-allyl-type intermediate. This reaction is an exothermic reaction, and the removal of the reaction heat is carried out by circulating the reactor outlet gas or liquid. Also, air is used as an oxygen source; preventing an explosive composition of oxygen and the 1,3-butadiene is indispensable. The 1,4-butanediol manufacturing process according to this reaction is highly atom efficient.

10.2.1
Short Overview of Non-butadiene-Based Routes to 1,4-Butanediol

10.2.1.1 Acetylene-Based Reppe Process

The Reppe process is a method that was developed in the 1940s and typical manufacturers include BASF, Ashland, and Invista. Cu–Bi catalyst supported on silica is used to prepare the 1,4-butynediol by reacting formaldehyde and acetylene at 0.5 MPa and 90–110 °C (Eq. (10.2)). The copper used in the reaction is converted to copper(I) acetylide, and the copper complex reacts with the additional acetylene to form the active catalyst. The role of bismuth is to inhibit the formation of water-soluble acetylene polymers (i.e., cuprenes) from the oligomeric acetylene complexes on the catalyst [5a]. The hydrogenation of 1,4-butynediol is accomplished through the use of Raney Ni catalyst to produce 1,4-butanediol (Eq. (10.3)). The total yield of 1,4-butanediol production is 91% from acetylene [5b]. Since acetylene is a highly explosive compound, careful process control is necessary.

$$\text{HC}\equiv\text{CH} + 2\,\text{HCHO} \xrightarrow{\text{Cu–Bi cat.}} \text{HOH}_2\text{CC}\equiv\text{CCH}_2\text{OH} \quad (10.2)$$

$$\text{HOH}_2\text{CC}\equiv\text{CCH}_2\text{OH} \xrightarrow[\text{Raney Ni}]{\text{H}_2} \text{HO−CH}_2\text{CH}_2\text{CH}_2\text{CH}_2\text{−OH} \quad (10.3)$$

10.2.1.2 Butane-Based Process; Selective Oxidation of Butane to Maleic Anhydride

1,4-Butanediol can also be manufactured by hydrogenating maleic acid derivatives obtained by oxidizing n-butane. Various methods have been developed, differing in the reaction system or source of maleic anhydride [6]. The selective oxidation of n-butane to form maleic anhydride is accomplished in either a fixed or fluid bed reactor containing vanadium/phosphorus mixed oxide catalyst. Formed maleic anhydride is then converted to the diester via esterification with a lower alcohol such as ethanol (Eq. (10.4)). The diester is hydrogenated in the gas phase in a fixed bed reactor filled with a copper catalyst in the gas phase (Eq. (10.5)). The alcohol is released and recycled. Since γ-butyrolactone is a reaction intermediate of 1,4-butanediol, hydrogenation conditions can control the product ratio of γ-butyrolactone and 1,4-butanediol. The yield of 1,4-butanediol production is 98% (THF 1%, γ-butyrolactone 0.4%) from dimethyl maleate [6c].

$$\text{maleic anhydride} + 2\text{ROH} \longrightarrow \text{RO-CO-CH=CH-CO-OR} \tag{10.4}$$

$$\text{RO-CO-CH=CH-CO-OR} \xrightarrow[\text{Cu cat.}]{H_2} \text{HO-(CH}_2)_4\text{-OH} \tag{10.5}$$

A catalyst with excellent acid resistance can hydrogenate the maleic acid directly. This route is more cost-efficient due to the elimination of the esterification process.

10.2.1.3 Propylene-Based Process: Hydroformylation of Allyl Alcohol

An alternative route to generate 1,4-butanediol is via the hydroformylation of allyl alcohol. This method was commercialized by LyondellBasell Industries and Dairen Chemical Corporation. Allyl alcohol is produced to isomerize propylene oxide derived from propylene over a lithium phosphate catalyst. Hydroformylation is performed to form 4-hydroxybutyraldehyde (Eq. (10.6)) followed by hydrogenation using Raney Ni catalyst to form 1,4-butanediol (Eq. (10.7)). The patent of Dairen Chemical showed 93% allyl alcohol conversion and 67% 1,4-butanediol selectivity and 31% 2-methyl-propanediol selectivity [7a].

$$\text{CH}_2=\text{CH-CH}_2\text{OH} \xrightarrow[\text{Rh cat.}]{H_2/CO} \text{OHC-CH}_2\text{-CH}_2\text{-CH}_2\text{OH} \tag{10.6}$$

$$\text{OHC-CH}_2\text{-CH}_2\text{-CH}_2\text{OH} \xrightarrow[\text{Raney Ni}]{H_2} \text{HO-(CH}_2)_4\text{-OH} \tag{10.7}$$

The Rh(II) bidentate ligand hydroformylation catalyst developed by Kuraray has been licensed [7b]. This catalyst gives high linear selectivity and high activity at low pressure for the olefin having a functional group.

10.2.2
Butadiene-Based Routes to 1,4-Butanediol

10.2.2.1 Oxyhalogenation of 1,3-Butadiene

Synthesis of 1,4-diacetoxy-2-butene by stoichiometric reaction of a halide complex was considered in an early period [8], and then some catalysts were developed. Although there are a series of Phillips patents, which include the $InBr_3$–$LiBr$ catalyst system, the 1,4-diacetoxy-2-butene production rate was low and the 1,4-selectivity did not exceed 80%. The reaction of this system was summarized by Stapp [9]; for example, the reaction using $Cu(OAc)_2$–LiX-based catalysts proceeds by a copper-based redox cycle (Scheme 10.1). In addition, V_2O_5–$CuBr_2$–KBr, $CuBr_2$–$NaBr$, and $Ag(OAc)_2$–$LiOAc$ were known for diacetoxylation, but either 1,4-selectivity or reaction rate was low. Furthermore, 1,4-dichloro-2-butene is obtained in the production of chloroprene from 1,3-butadiene.

Hydrolysis of 1,4-dichloro-2-butene to produce 1,4-butanediol was previously commercialized on a small scale by Toyo Soda in Japan.

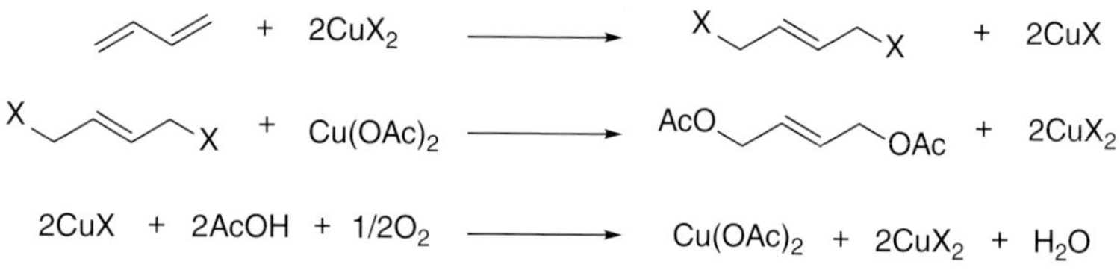

$$\text{CH}_2=\text{CH-CH}=\text{CH}_2 + 2CuX_2 \longrightarrow X\text{-CH}_2\text{-CH}=\text{CH-CH}_2\text{-}X + 2CuX$$

$$X\text{-CH}_2\text{-CH}=\text{CH-CH}_2\text{-}X + Cu(OAc)_2 \longrightarrow AcO\text{-CH}_2\text{-CH}=\text{CH-CH}_2\text{-}OAc + 2CuX_2$$

$$2CuX + 2AcOH + 1/2 O_2 \longrightarrow Cu(OAc)_2 + 2CuX_2 + H_2O$$

Scheme 10.1

10.2.2.2 Oxidative Acetoxylation of 1,3-Butadiene

Exploration of Basic Catalyst Components The study of direct oxidative acetoxylation of 1,3-butadiene began with the use of Wacker-type homogeneous catalyst $Pd(OAc)_2$–$CuCl_2$ [10]. This catalyst system gave low 1,4-diacetoxy-2-butene selectivity, and there was a problem in separating the catalyst. After that, liquid- and vapor-phase methods using a Pd-based catalyst were studied in parallel. Catalyst activity was greatly improved by the addition of Bi or Sb to the Pd catalyst in the gas-phase reaction [11]. However, catalyst activity was reduced by the adhesion of resin by-product derived from unsaturated aldehydes on the catalyst surface. Various improvements have been tried in the gas phase, but catalyst robustness has never met industrial requirements.

Mitsubishi Chemical started solid-state Pd catalyst development in the liquid phase reaction to avoid the polymerization of 1,3-butadiene and by-products. There were some challenges to be overcome, including polymerization, low catalytic activity, and leaching of palladium into the reaction media. At first, the addition of Sb or Bi to Pd alkali metal catalyst on charcoal was found to be beneficial for the catalyst activity because of an inhibition of the polymerization (Table 10.1) [12].

Table 10.1 Ability of Pd–Te–C catalyst to form 1,4-diacetoxy-2-butene.[a]

Catalyst	Amount of metal (mg-atom/g-cat)	Production rate (mol/g-atom Pd*h)	1,4-Diacetoxy-2-butene selectivity (%)
Pd–C[b]	Pd (1.0)	0.458	41.9
Pd–C[c]	Pd (1.0)	0.395	77.1
Pd–Sb–C[d]	Pd (0.5), Sb (0.5)	0.880	72.2
Pd–Bi–C[d]	Pd (0.5), Bi (0.5)	0.600	75.0
Pd–Se–C	Pd (0.5), Se (0.15)	1.08	84.0
Pd–Te–C	Pd (0.5), Te (0.15)	2.58	84.6
Pd–Te–C	Pd (0.5), Te (0.075)	2.76	84.9
Pd–Te–C[e]	Pd (0.5), Te (0.15)	2.67	92.7
Pd–Te–C[e]	Pd (0.5), Te (0.075)	5.92	92.7

a) Reaction conditions: catalyst (10 g), acetic acid (210 ml), 1,3-butadiene (55 mmol/h), and oxygen (55 mmol/h).
b) $CuCl_2$–LiOAc–Ac_2O was added.
c) $SbCl_3$–$CuCl_2$–LiOAc was added.
d) LiOAc and Ac_2O were added.
e) Activated carbon was treated with nitric acid.

Table 10.2 Addition effect of Se and Te to Pd–C catalyst.

Pd content (mg/g-cat)	Sb content (mg/g-cat)	Bi content (mg/g-cat)	Te content (mg/g-cat)	Se content (mg/g-cat)	Production rate (mol/g-atom Pd*h)	1,4-Selectivity (%)	Pd leaching (mg/l)
0.5	—	—	—	—	0.036	75.0	20
0.5	0.5	—	—	—	0.88	72.2	28
0.5	—	0.5	—	—	0.61	74.3	6
0.5	0.5	—	0.15	—	1.18	85.1	<0.5
0.5	0.15	—	0.075	—	1.59	85.2	<0.5
0.5	0.075	—	0.075	—	1.24	85.4	<0.5
0.5	0.5	—	—	0.015	0.98	80.6	<0.5
0.5	—	0.5	0.15	—	0.65	76.7	<0.5

Since the liquid phase method is carried out in acetic acid solvent, the leaching of palladium was the biggest problem. High catalyst activity was finally obtained by the addition of Se or Te, which also prevented the leaching of palladium (Table 10.2) [12]. The addition of the alkali metal to the catalyst is not required in this catalyst system.

Development of Industrial Oxidative Acetoxylation Catalyst When a liquid phase reaction using a solid catalyst is implemented on an industrial scale, a fixed bed or a suspended bed can be considered as a reaction system. Due to the difficulty of the operation for the regeneration of deactivated catalyst and catalyst activation, a suspended bed was not adopted and attention focused on a shaped catalyst for a fixed bed reaction system.

While the addition of Te to the palladium catalyst has a dramatic effect, the preferable range of Te/Pd ratio is narrow. Activating the catalyst, including oxidation and reduction prior to the use of the catalyst, enables us to prepare a Pd_n-Te ($n=4$) alloy catalyst [13, 14a]. This Pd alloy catalyst showed no Pd leaching and enough catalyst activity; therefore, it reached the stage of industrial process development. This catalyst is capable of continuous use over 1 year.

In optimal conditions, Pd–Te–C catalyst shows 27.0 mol/h/g-atom Pd and 89% 1,4-diacetoxy-2-butene selectivity [15]. This method was the first practical application of a liquid phase oxidation reaction with a heterogeneous catalyst.

Plausible Mechanism of Oxidative Acetoxylation of 1,3-Butadiene Several catalysts with different Pd/Te ratios were prepared to understand the origin of catalyst activity. Highest activity was obtained with the catalyst containing Pd species of small positive charge. In the acetoxylation of olefins, metallic palladium generates a π-allyl palladium intermediate. In contrast, a Pd salt such as $PdCl_2$ generates π-bonded palladium olefin complexes as intermediates. Therefore, the industrial acetoxylation catalyst with small positive charge was thought to form a π-allyl palladium intermediate, as shown in Scheme 10.2 [14a–d]. Apart from this mechanism, a radical pathway was also proposed based on a DFT calculation [14e]. The Pd–Te–C system was also suggested as a redox system consisting of a Pd–Te couple. Results of XPS analysis of the Pd–Te–C catalyst prepared from $Pd(NO_3)_2$ indicated that the Pd of the catalyst bears a weak positive charge that contributes to the catalyst activity. Activity of the catalyst when Te is replaced by other elements showed an order of Te ≫ Se > Sb > Bi > As [12a]. Considering that at a catalyst activation temperature of 400 °C, Bi and Se are in the molten state and As_2O_3 sublimes easily, the order of the catalyst activity was relatively consistent with the order of redox potential of Se > Te > Bi > As > Sb [15]. Therefore, one of the roles of Te is assumed to be in the reoxidation of palladium similar to copper in the Wacker reaction.

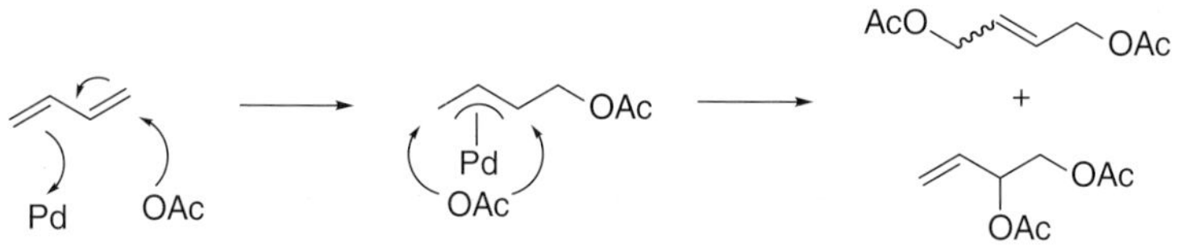

Scheme 10.2

10.3
Process Technology

The oxidative acetoxylation product 1,4-diacetoxy-2-butene can be converted into 1,4-butanediol via hydrogenation and hydrolysis. The order of the hydrolysis and hydrogenation reaction was examined. Although the hydrolysis equilibrium is more favorable for 1,4-diacetoxy-2-butene than for the 1,4-diacetoxybutane hydrogenation product, the thermal stability of the hydrolysis product of 1,4-diacetoxy-2-butene is poor. Therefore, hydrogenation was adopted before hydrolysis [16]. Deacetoxylative cyclization of the hydrolysis intermediate 1-acetoxy-4-hydroxybutane as a precursor of THF proceeds in the presence of cationic ion-exchange resin at a higher temperature (Scheme 10.3, Figure 10.1). Originally, the 1,4-butanediol business was closely related to the THF business; THF has been manufactured from purified 1,4-butanediol via the dehydrative cyclization. Since THF can directly be manufactured from the reaction intermediate, it is a significant improvement in the cost.

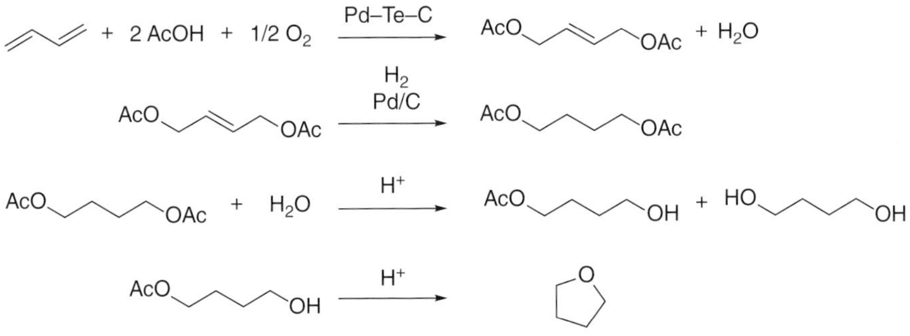

Scheme 10.3

10.3.1
Mitsubishi Chemical's 1,4-Butanediol Manufacturing Process: First-Generation Process

Mitsubishi Chemical's 1,4-butanediol production process is composed of three main steps, namely, diacetoxylation, hydrogenation, and hydrolysis. These steps are discussed in the following sections [17, 18].

10.3.1.1 Oxidative Acetoxylation Step
1,4-Diacetoxy-2-butene is obtained from 1,3-butadiene via catalytic acetoxylation in the liquid phase. This reaction is the most critical production step. 1,3-Butadiene, acetic acid, air, and solid-state Pd catalyst are contacted in the trickle bed reactor at 60–90 °C to obtain 1,4-diacetoxy-2-butene. Unreacted gas is recycled. The oxygen concentration is adjusted by the feed rate of air. After the separation of acetic acid and the formed water in the reaction, the reaction product is fed to the next hydrogenation step. Water and acetic acid are recycled after an acetic acid purification process.

10.3.1.2 Hydrogenation Step
1,4-Diacetoxybutane is produced by hydrogenating the double bond of the 1,4-diacetoxy-2-butene obtained in the diacetoxylation step. Usually, a Pd-based

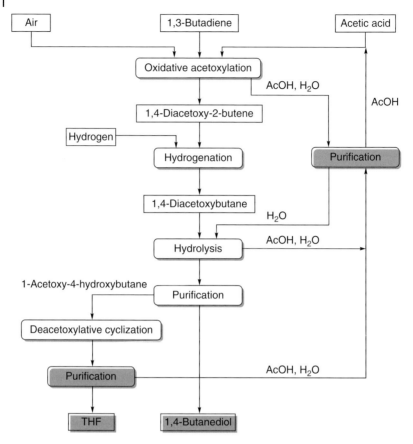

Figure 10.1 Block flow diagram of Mitsubishi Chemical's 1,4-butanediol manufacturing process.

catalyst is used for hydrogenation. High selectivity is observed in mild conditions such as at 40–60 °C and 5 MPaG.

10.3.1.3 Hydrolysis Step

Hydrolysis and alcoholysis have been studied in order to obtain 1,4-butanediol from 1,4-diacetoxybutane. Because of the large energy consumption in the recovery of alcohol and acetic acid via additional hydrolysis of the formed methyl acetate, hydrolysis leading directly to acetic acid as by-product was advantageous. Hydrolysis of 1,4-diacetoxybutane is carried out in the presence of an acidic ion-exchange resin catalyst at 50 °C with the generation of 1,4-butanediol and 1-acetoxy-4-hydroxybutane. The latter compound is converted to THF in the presence of an acidic ion-exchange resin at 70–90 °C. Acetic acid and excess water are recycled after purification. It is possible to vary the production ratio of 1,4-butanediol and THF widely by changing the hydrolysis conditions, which is advantageous in responding to fluctuating market conditions.

10.3.2
Mitsubishi Chemical's 1,4-Butanediol Manufacturing Process: Second-Generation Process

To enhance industrial competitiveness, some improvements were adopted for the latest 1,4-butanediol plant in Mitsubishi Chemical. The previous generation method needed high reaction pressure in the acetoxylation reaction in order to prevent an explosive composition of oxygen and 1,3-butadiene at the reactor inlet. Lower inlet oxygen concentration is achieved by circulating the reactor outlet gas with a low oxygen concentration. Therefore, 9 MPa is required to ensure an adequate partial pressure of oxygen (Figure 10.2). The up-flow type of the liquid phase reactor with fine bubbles enables us to supply a gas with high oxygen concentration while avoiding an explosion. Because the reaction rate increases with the higher concentration of oxygen and 1,3-butadiene, it is possible to carry out the acetoxylation reaction at a lower pressure. As a result of the improvement, the compressor for gas circulation is not required (Figure 10.3). Therefore, electrical costs and construction costs can be reduced [19a]. Furthermore, high dispersion $Pd-Te-SiO_2$ catalyst was developed and catalyst life time was extended [19b]. In addition, in order to reduce the amount of steam used in the hydrolysis process, the number of hydrolysis reactors is reduced to two units from three units by recycling the reaction intermediate 1-acetoxy-4-hydroxybutane in place of water, which requires a large amount of steam during distillation. Energy consumption of the refined process is reduced by 30% in comparison with the previous generation.

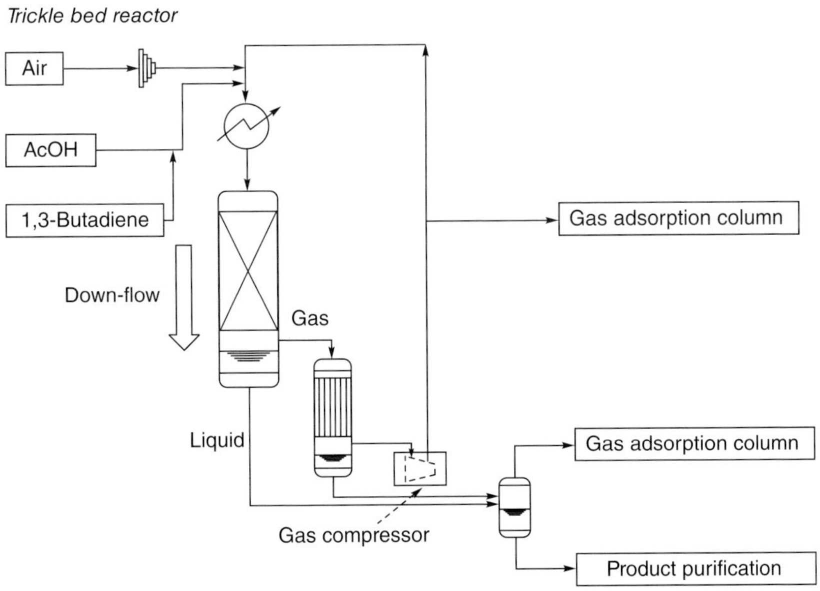

Figure 10.2 Process flow diagram of first generation in oxidative acetoxylation.

Fixed bed up-flow reactor

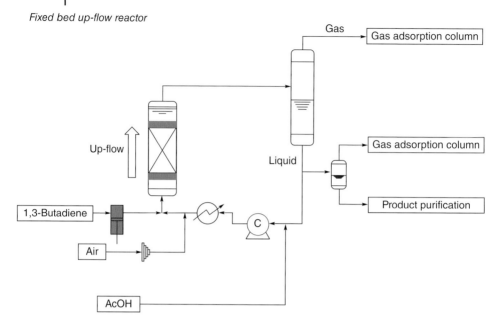

Figure 10.3 Process flow diagram of second generation in oxidative acetoxylation.

This process was applied to the latest 1,4-butanediol plant, which started commercial production with a capacity of 50 000 t/year from 2002 in Mitsubishi Chemical.

10.4
New Developments

10.4.1
Improvement of the Current Process

Rh–Te–C catalyst has been proposed as an improved oxidative acetoxylation catalyst. The advantage of this catalyst is higher 1,4-diacetoxy-2-butene selectivity than the Pd-based catalyst. Although 95% selectivity can be obtained, improvement of durability is desired due to the rhodium leaching [20]. Another feature of the Rh catalyst is the formation of *cis*-isomer. The three-dimensional structure of the π-allyl intermediate might be different in both catalysts.

To increase total yield of 1,4-butanediol, isomerization of 3,4-diacetoxy-1-butene, which is a major by-product in oxidative acetoxylation, has been examined. Isomerization of 3,4-diacetoxy-1-butene to 1,4-diacetoxy-2-butene requires only 1 wt ppm Pd–phosphite catalyst [21]. Therefore, catalyst recovery is not necessary in view of the production cost.

(10.8)

Ideally, direct oxidative hydration of 1,3-butadiene is desired rather than acetoxylation, but the technical difficulty is very high because water is less reactive compared with acetic acid.

10.4.2
Development of Alternative Processes

Eastman Chemical Company proposed a 1,4-butanediol manufacturing process via 3,4-epoxy-1-butene. 1,3-Butadiene is oxidized to 3,4-epoxy-1-butene in the presence of oxygen by using a silver catalyst. 3,4-Epoxy-1-butene is further converted into 2,3-dihydrofuran followed by hydrolysis to form 4-hydroxybutyraldehyde. Finally, hydrogenation of 4-hydroxybutyraldehyde produces 1,4-butanediol [22]. 1,3-Butadiene to furan and propylene to acrolein as an intermediate to produce 1,4-butanediol were also studied, although there are some problems [23]. Genomatica recently developed a bio-1,4-butanediol production method from sugars with their original microorganisms. This method has the potential to produce 1,4-butanediol at lower production cost [24]. The demand for 1,4-butanediol, which is one of the most important monomers, will grow in the future. Since the contribution of raw material cost to the total manufacturing cost is high, competitiveness of various methods is determined by the raw material market. The price of shale oil, shale gas, and biomass resources will become important.

10.5
Summary and Conclusions

The 1,3-butadiene-based 1,4-butanediol process was reviewed in this chapter. This method ended the half-century monopoly of the acetylene-based Reppe method. This development was in line with the switch from coal to petroleum feedstock. Because the acetoxylation of 1,3-butadiene was the first successful commercial technology for a liquid phase oxidation reaction with heterogeneous catalysis, it is an important technology in terms of oxidation catalyst technology. Although 1,3-butadiene derivatives are easy to polymerize, Pd–Te–C catalysts were developed that suppress polymerization in mild conditions by adopting the liquid phase oxidation. This catalyst system also improved the catalytic activity and Pd leaching, thereby allowing commercial production of 1,4-butanediol from 1,3-butadiene. The mechanism of the catalyst system remains unclear but this reaction might proceed via π-allyl Pd intermediates with Te working as a redox cocatalyst. At present, 1,4-butanediol production from propylene and *n*-butane is also commercial. 1,4-Butanediol is a rare example where many routes are competing. Demand for 1,4-butanediol will grow in the future; therefore, further innovation is desired.

References

1. Radusch, H.-J. (2002) *Handbook of Thermoplastic Polymers: Homopolymers, Copolymers, Blends, and Composites*, Wiley-VCH Verlag GmbH, Weinheim, pp. 389–418.
2. Reppe, W., Schlichting, O., Klager, K., and Toepel, T. (1948) *Justus Liebigs Ann. Chem.*, **560**, 1–92.
3. Charles, B.T. (1977) Surface impregnated catalyst, US Patent 4048096.
4. Ono, I., Fukabori, K., and Simomura, S. (1973) Manufacturing method for 1,4-diacyloxy-2-butenes, JP Patent 27290.
5. (a) Haas, T., Jaeger, B., Weber, R., Mitchell, S.F., and King, C.F. (2005) *Appl. Catal., A*, **280**, 83–88; (b) Hara, Y. (2001) The progress of manufacturing technology of 1,4-Butanediol/THF and its polymer application. *Polymer Frontier 21 Proceeding*, 2000, pp. 18–22.
6. (a) Harris, N. and Tuck, M.W. (1990) *Hydrocarbon Process.*, **69**, 79–82; (b) Ohlinger, C. and Kraushaar-Czaranetzki, B. (2003) *Chem. Eng. Sci.*, **58**, 1453–1461; (c) Fischer, R., Kaibel, G., Pinkos, R., and Rahn, R.-T. (2002) Method for producing 1,4-butanediol, US Patent 6350924.
7. (a) Matsumoto, M. and Tamura, M. (1982) *J. Mol. Catal.*, **16**, 195–207; (b) Chen, S.C., Chu, C.C., Lin, F.S., and Chou, J.Y. (1995) Process for preparing 1,4-butanediol, US Patent 5426250.
8. Uemura, S., Hiramoto, T., and Ichikawa, K. (1969) *J. Soc. Chem. Ind., Jpn.*, **72**, 1096–1098.
9. Stapp, P.R. (1979) *J. Org. Chem.*, **44**, 3216–3219.
10. Kohll, C.F., Blaauw, H.J.A., and Van Helden, R. (1969) Process for the preparation of tetrahydrofuran and/or homologues thereof, GB Patent 1170222.
11. (a) Shinohara, H. (1984) *Appl. Catal.*, **10**, 27–42; (b) Shinohara, H. (1985) *Appl. Catal.*, **14**, 145–158; (c) Shinohara, H. (1986) *Appl. Catal.*, **24**, 17–23.
12. (a) Takehira, M. and Ishikawa, T. (1981) *Petrotech*, **4**, 36–46; (b) Onoda, T. and Haji, J. (1973) Process for preparing an unsaturated glycol diester, US Patent 3755423; (c) Onoda, T., Yamura, A., Ohno, A., Haji, J., Toriya, J., Sata, M., and Ishizaki, N. (1975) Process for preparing an unsaturated ester, US Patent 3922300.
13. Weitz, H.-M. and Hartwig, J. (1976) Verfahren zur Herstellung von buten-2-diol-1,4-diacetat, DE Patent 2444004.
14. (a) Takehira, K., Mimoun, H., and Seree De Roch, I. (1979) *J. Catal.*, **58**, 155–169; (b) Robinson, S.D. and Shaw, B.L. (1963) *J. Chem. Soc.*, **919**, 4806–4814; (c) Vekki, A.V. (1993) *Russ. J. Appl. Chem.*, **66**, 2082–2092; (d) Backvall, J.-E., Bystrom, E., and Nordberg, R.E. (1984) *J. Org. Chem.*, **49**, 4619–4631; (e) Shunk, S.A., Baltes, C., and Sundermann, A. (2006) *Catal. Today*, **117**, 304–310.
15. Takehira, M. (1981) *Shokubai*, **23**, 114–123.

16. Ohno, H. (2000) *Shokubai*, **42**, 46–50.
17. Okubo, K. (1998) *Kagakukougaku*, **62**, 596–599.
18. (a) Tanabe, Y. (1981) *Hydrocarbon Process.*, **60**, 187–190; (b) Mitsubishi Kasei Co. (1988) *Chemtech*, **18**, 759–763; (c) Onoda, T. (1988) *Izv. Khim.*, **21**, 340–343.
19. (a) Iwasaka, H. (2006) *Petrotech*, **29**, 92–96; (b) Onoda, T. (2002) *Miraizairyou*, **2**, 58–63.
20. (a) Masuda, T., Matsuzaki, K., and Watanabe, Y. (1978) Method for producing an unsaturated glycol diester, JP Patent 44502; (b) Takehira, K., Trinidad Chena, J.A., Niwa, S., Hayakawa, T., and Ishikawa, T. (1982) *J. Catal.*, **76**, 354–368.
21. (a) Utsunomiya, M., Izawa, Y., and Okubo, M. (2011) *J. Synth. Org. Chem Jpn.*, **69**, 552–561; (b) Utsunomiya, M., Izawa, Y., Mitsuba, Y., Mizoguchi, M., and Tanaka, Y. (2008) Method for isomerizing allyl compound, WO Patent 143036.
22. MacKenzie, P.B., Kanel, J.S., Falling, S.N., and Wilson, A.K. (1999) Process for the preparation of 2-alkene-1,4-diols and 3-alkene-1,2-diols from g,d-epoxyalkenes, US Patent 5959162.
23. (a) Ito, M., Terada, S., Higashino, Y., and Koyasu, Y. (2003) *Stud. Surf. Sci. Catal.*, **145**, 505–506; (b) Ichikawa, S., Ohgomori, Y., Sumitani, N., Hayashi, H., and Imanari, M. (1995) *Ind. Eng. Chem. Res.*, **34**, 971–973.
24. (a) Burgard, A.P., Van Dien, S.J., and Burk, M. (2009) Methods and organisms for the growth-coupled production of 1,4-butanediol, US Patent 0047719; (b) Clark, W., Japs, M., and Burk, M. (2011) Process of separating components of a fermentation broth, US Patent 0003355; (c) Yim, H., Haselbeck, R., Niu, N., Pujol-Baxley, C., Burgard, A., Boldt, J., Khandurina, J., Trawick, J.D., Osterhout, R.E., Stephen, R., Estadilla, J., Teisan, S., Schreyer, H.B., Andrae, S., Yang, T.H., Lee, S.Y., Burk, M., and Van Dien, S. (2011) *Nat. Chem. Biol.*, **7**, 445–452.

11
Mitsubishi Chemicals Liquid Phase Palladium-Catalyzed Oxidation Technology: Oxidation of Cyclohexene, Acrolein, and Methyl Acrylate to Useful Industrial Chemicals

Yoshiyuki Tanaka, Jun P. Takahara, Tohru Setoyama, and Hans E. B. Lempers

11.1
Introduction

The palladium-catalyzed oxidation of ethylene to acetaldehyde, often called the *Wacker process* after the company that developed it, is one of the oldest and best known reactions of palladium (see Chapter 9). The oxidation is usually carried out in water with oxygen as the oxidant in the presence of cupric chloride, which catalyzes the oxidation of Pd(0) formed in the ethylene oxidation back to the active state [1].

The application of the palladium(II)-ethylene oxidation to higher alkenes results in the formation of ketones, even in the case of terminal alkenes. The reaction is more useful for terminal alkenes, as internal alkenes react more slowly and may give mixtures of allylic oxidation as well as ketone products, which originate from alkene isomerization. Moreover, cyclic alkenes react even more sluggishly than corresponding linear internal alkenes [2]. As a result, the pseudo-first-order rate constant of ethylene oxidation is about 1000 times higher than that of cyclohexene oxidation.

We are particularly interested in the Wacker oxidation of cyclohexene as the product, cyclohexanone, is a starting material in the synthesis of caprolactam, which is an intermediate in nylon production. Furthermore, we have strong interest in oxidation of acrolein in particular and acrylic compounds in general. Acrolein oxidation leads to a convenient route to 1,3-propanediol, while methyl acrylate oxidation leads to a starting material for adhesives.

At present, most of the cyclohexanone is produced by aerobic liquid phase oxidation of cyclohexane [3] (also see Chapter 3). Overoxidation of cyclohexanone necessitates a very low conversion level and consequently the recycling of a large amount of unreacted cyclohexane, leading to a low energy efficiency. The cyclohexanol coproduct must be dehydrogenated to cyclohexanone. The disadvantages of this process, such as dangerous nature of the oxidation, overoxidations, and co-formation of cyclohexanol, prompted us to investigate the alternative Wacker oxidation of cyclohexene to cyclohexanone.

Liquid Phase Aerobic Oxidation Catalysis: Industrial Applications and Academic Perspectives,
First Edition. Edited by Shannon S. Stahl and Paul L. Alsters.
© 2016 Wiley-VCH Verlag GmbH & Co. KGaA. Published 2016 by Wiley-VCH Verlag GmbH & Co. KGaA.

Initial experiments were done in water and resulted in low cyclohexene conversions, low product selectivities, and extensive palladium deactivation by Pd black formation. The low cyclohexanone yield originated from overoxidation of cyclohexanone to 2-cyclohexenone, which undergoes further oxidation to a plethora of by-products. The low cyclohexene conversion can be attributed to the aforementioned low reactivity of the internal double bond as well as the low solubility of cyclohexene in water. Several reaction media have been described in which higher alkenes are oxidized to ketones in organic solvent-based systems. Some typical examples are DMF [4], water mixtures with chlorobenzene, dodecane, sulfolane [5], 3-methylsulfolane and N-methylpyrrolidone [6], or alcohols [7]. These solvent systems indeed lead to increased cyclohexene conversions but still suffer from overoxidation and catalyst deactivation by Pd black formation. Hence, the goal of our research was to find a variation to the Wacker oxidation without over-oxidation of the product and deactivation of the palladium catalyst.

11.2
Chemistry and Catalysis

11.2.1
Aerobic Palladium-Catalyzed Oxidation of Cyclohexene to 1,4-Dioxospiro-[4,5]-decane

11.2.1.1 Optimization of the Reaction Conditions

When the oxidation of cyclohexene is carried out in alcoholic solvents, cyclohexanone can undergo subsequent ketal formation with the solvent (Figure 11.1, Table 11.1). Under our mild oxidation conditions, the oxidation of the alcohol solvent is virtually absent.

It appeared that overoxidation decreased with ketal formation selectivity. Thus, whereas the reaction product obtained in the case of ethylene glycol as

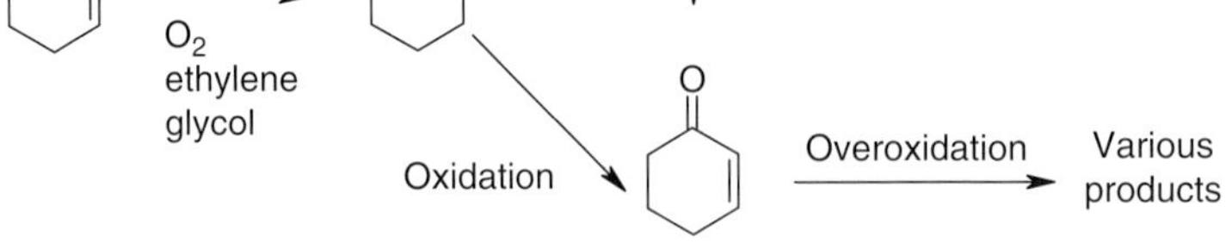

Figure 11.1 Palladium-catalyzed oxidation of cyclohexene to either cyclohexanone or to 1,4-dioxaspiro[4,5]decane. Cyclohexanone is susceptible to overoxidation, while the 1,4-dioxaspiro[4,5]decane is stable.

Table 11.1 Palladium-catalyzed oxidation of cyclohexene in various solvents showing the importance of 1,4-dioxaspiro[4,5]decane formation in the protection of the product against overoxidation.[a]

Solvent	Ketal/ketone	Cyclohexanone/2-cyclohexenone
Ethylene glycol	>1000	No overoxidation
1,3-Propanediol	4.3	19.3
1,4-Butanediol	0.9	5.7
2,3-Butanediol	7.5	25.3
1,2-Cyclohexanedimethanol	4.5	18.7
Ethanol	0	1.3

a) 0.1 mmol $Pd(CH_3CN)_2Cl_2$, 0.1 mmol $FeCl_3$, 0.1 $CuCl_2$, and 20 mmol cyclohexene in 10 ml solvent are stirred at 40 °C and 0.1 MPa oxygen for 5 h.

the solvent consists exclusively of the ketal of cyclohexanone and ethylene glycol, that is, 1,4-dioxaspiro[4,5]decane, in the case of ethanol, the primary product is free cyclohexanone. As the ketal protecting group provides stability toward overoxidation, no by-products are observed with ethylene glycol as the solvent. In ethanol where ketal formation is absent, pronounced overoxidation of cyclohexanone to 2-cyclohexenone (Figure 11.1) to a ratio of cyclohexanone/2-cyclohexenone of 1.3 takes place. Moreover, a wide variety of overoxidized products were observed by GC analysis. The other solvents that were tested were less efficient in ketal formation than ethylene glycol but more efficient than ethanol. Significant overoxidation was observed with solvents that were less efficient in ketal formation.

After having minimized overoxidation by using an appropriate alcoholic medium, attention focused on suppressing catalyst deactivation by Pd black formation. Typically when cyclohexene is oxidized using a $Pd(CH_3CN)_2Cl_2/CuCl_2$ catalyst, ethylene glycol as the solvent at 40 °C and 0.1 MPa oxygen pressure or at 80 °C and 0.7 MPa oxygen pressure (Table 11.2, entries 1 and 4), the reaction solution became black by the formation of colloidal palladium, and a palladium mirror on the inside of the reactor could be observed.

When the $CuCl_2$ was substituted by $FeCl_3$ (entries 2 and 5), no Pd black or Pd mirror was observed anymore. In addition, the activity increased. A disadvantage of the $Pd(CH_3CN)_2Cl_2/FeCl_3$ combination is that the product stream is contaminated with trace amounts of chlorinated product, which is unacceptable for a bulk industrial process. Combining the $CuCl_2$ and the $FeCl_3$ cocatalyst gave superior activity (entries 3 and 6). Furthermore, no Pd black and no chlorinated products were observed [8].

Apparently, the reoxidation of Pd(0) by $CuCl_2$ alone is not sufficient enough to prevent Pd black formation. Reoxidation of Pd(0) by $FeCl_3$ is more efficient, thus increasing the reaction rate. Furthermore, when the $CuCl_2/FeCl_3$ combination is

Table 11.2 Palladium-catalyzed oxidation of cyclohexene with different Cu and Fe cocatalyst compositions showing that the Pd/Cu/Fe combination gives a stable catalyst which gives clean oxidation.

Entry	Pd	Cu	Fe	MPa	T (°C)	Rate (mol/l/h)	Remark
1[a]	0.1	0.1	0	0.1	40	0.0076	Pd black
2[a]	0.1	0	0.1	0.1	40	0.0083	Chlorination
3[a]	0.1	0.1	0.1	0.1	40	0.0083	Clear solution
4[b]	0.1	0.1	0	0.7	80	0.348	Pd black
5[b]	0.1	0	0.1	0.7	80	0.498	Chlorination
6[b]	0.1	0.1	0.1	0.7	80	0.556	Clear solution

a) 0.1 mmol $Pd(CH_3CN)_2Cl_2$ was stirred with either 0.1 mmol $FeCl_3$ or 0.1 mmol $CuCl_2$ or 0.1 mmol of both, 20 mmol cyclohexene in 10 ml ethylene glycol at 40 °C and 0.1 MPa oxygen for 5 h.

b) 0.1 mmol $Pd(CH_3CN)_2Cl_2$ was stirred with either 0.1 mmol $FeCl_3$ or 0.1 mmol $CuCl_2$ or 0.1 mmol of both, 20 mmol cyclohexene in 10 ml ethylene glycol at 80 °C and 0.7 MPa oxygen for 1 h.

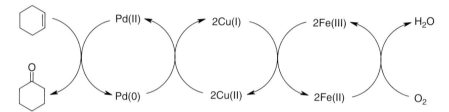

Figure 11.2 The simplified proposed mechanism for the catalytic cycle.

used, rapid oxidation of Cu(I) to Cu(II) by Fe(III) takes place[1], leading to a higher efficiency in the overall reoxidation. The proposed reaction sequence is presented in Figure 11.2.

The activity of the $Pd(CH_3CN)_2Cl_2/CuCl_2/FeCl_3$ combination could easily be increased by either increasing the reaction temperature or by increasing the oxygen pressure, without any deactivation of the catalyst by Pd black formation [9].

11.2.2
Aerobic Palladium-Catalyzed Oxidation of Other Types of Olefins

Oxidation of olefins in appropriate alcohols or diols to the corresponding ketals or ketones using the $Pd(CH_3CN)_2Cl_2/CuCl_2/FeCl_3$ catalyst mixture was also applied to several other aliphatic, aromatic, and acrylic olefins (Table 11.3):

1) CuCl is not soluble in alcoholic solvents. When a solution of $FeCl_3$ is added to CuCl, it is rapidly oxidized to $CuCl_2$ and smoothly dissolves. This effect is also observed with metallic Cu. A solution of $FeCl_3$ smoothly oxidizes and dissolves metallic Cu.

Table 11.3 Various substrates oxidized by the Pd/Cu/Fe-catalyst combination in alcohol solvents using oxygen as the oxidant.

Substrate	Solvent	Product	Yield
1-octene [a]	Ethylene glycol	Ketone/ketal isomers	16
2-octene [a]	Ethylene glycol	Ketone/ketal isomers	17
Styrene [a]	Ethylene glycol	Aldehyde/acetal	70
α-methylstyrene [a]	Ethylene glycol	Aldehyde/acetal	27
Methyl acrylate [b]	Methanol	MeO–CH(OMe)–CH(OMe)–C(=O)OMe	93
Methyl acrylate [b]	Ethanol	EtO–CH(OEt)–CH(OEt)–C(=O)OMe	97
Ethyl acrylate [b]	Methanol	MeO–CH(OMe)–CH(OMe)–C(=O)OEt	95
Ethyl acrylate [b]	Ethanol	EtO–CH(OEt)–CH(OEt)–C(=O)OEt	94
Acrylamide [b]	Methanol	MeO–CH(OMe)–CH(OMe)–C(=O)NH$_2$	90
Acrylonitrile [b]	Methanol	MeO–CH(OMe)–CH$_2$–CN	90
Allyl alcohol (as aldehyde) [b]	Methanol	MeO–CH(OMe)–CH(OMe)–OMe... MeO–CH$_2$–CH(OMe)$_2$	92

[a] 20 mmol substrate, 0.1 mmol Pd(CH$_3$CN)$_2$Cl$_2$, 0.1 mmol CuCl$_2$, 0.1 mmol FeCl$_3$, 10 ml ethylene glycol, 0.7 MPa, 80 °C, GC yield after 1 h.

[b] 200 mmol substrate, 4 mol solvent, 0.4 mmol Pd(CH$_3$CN)$_2$Cl$_2$, 0.4 mmol CuCl$_2$, 0.4 mmol FeCl$_3$ 0.7 MPa, RT, 24 h isolated yield.

Products in both 1- and 2-octene oxidation originate partly from prior double-bond isomerization before olefin oxidation, as in both reactions 2-, 3-, and 4-octanone and the corresponding ethylene ketals were observed. For styrene, where double-bond isomerization is not possible, the product mixture is less complex and only acetophenone, phenylacetaldehyde, and the corresponding ketal and acetal were observed. With α-methylstyrene as substrate, clean oxidation to aldehyde and acetal takes place.

The $Pd(CH_3CN)_2Cl_2/CuCl_2/FeCl_3$-catalyzed oxidation toward acetal products has a high synthetic utility when electron-deficient acrylic-type olefins are used as substrate. For example, methyl acrylate and ethyl acrylate are oxidized at room temperature in methanol and ethanol to the dimethoxy and diethoxy acetals, respectively, in isolated yields of higher than 90%. This oxidation proved to be very mild, as transesterification of the methyl ester functional group in methyl acrylate and the ethyl ester functional group in ethyl acrylate in ethanol and methanol, respectively, did not take place. Also acrylamide, acrylonitrile, and acrolein smoothly afforded the corresponding acetals at room temperature in isolated yields of higher than 90%. In the case of acrolein oxidation, the product is 1,1,3,3-tetramethoxy propane, which originates from acetalization of both aldehyde functions. The products of these acrylic substrates are of particular interest as they contain a protected aldehyde functional group, an activated methylene functional group, and a third functional group of which the nature depends on the starting acrylic substrate. These three functional groups make these compounds very useful building blocks. For example, 3,3-dimethoxy methyl propionate obtained from methyl acrylate in methanol [9] is a useful intermediate in the synthesis of methyl cyanoacetate [10], a starting material for adhesives [11], whereas 1,1,3,3-tetramethoxy propane from acrolein oxidation in methanol can be transformed into 1,3-propanediol [12], a polyester monomer [13].

11.2.3
Aerobic Palladium-Catalyzed Oxidation of Acrolein to Malonaldehyde Bis-(1,3-dioxan-2-yl)-acetal Followed by Hydrolysis/Hydrogenation to 1,3-Propanediol

Given the use of 1,3-propanediol as a polyester monomer, the oxidation of acrolein and the conversion of the product to 1,3-propanediol were investigated in more detail.

Figure 11.3 outlines our three-step route from acrolein to 1,3-propanediol based on the palladium-catalyzed aerobic oxidation as presented in Section 11.2.2.

The sequence consists of a protection step of the aldehyde functional group, an oxidation step of the olefin functional group, and a hydrolysis/hydrogenation step of the two acetal functional groups.

We learned in Section 11.2.2 that it is important to protect the carbonyls as acetals/ketals, in order to diminish overoxidation. By using 1,3-propanediol to that end, no cumbersome diol separation between product diol and protecting diol is needed.

Several catalysts were investigated for this protecting step and the best appeared to be acidic montmorillonite. In a typical reaction sequence (see Figure 11.4 for (by-)products) [12], DAC could be obtained in 83.5% selectivity, which is too low for an industrial production of 1,3-propanediol.

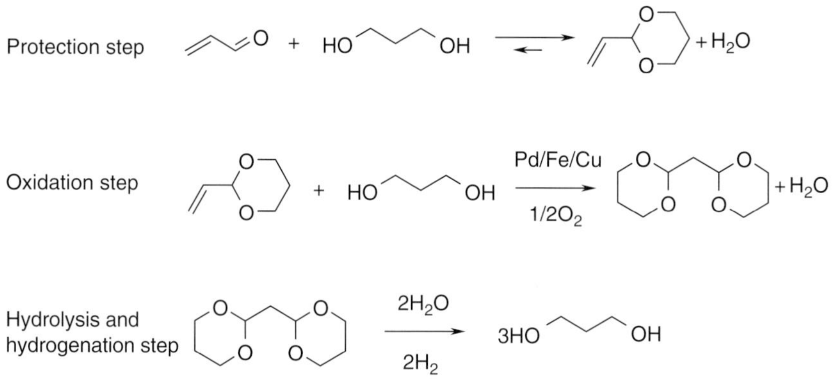

11.2 Chemistry and Catalysis

Protection step

Oxidation step (Pd/Fe/Cu, 1/2 O$_2$)

Hydrolysis and hydrogenation step (2H$_2$O, 2H$_2$) → 3HO~~~OH

Figure 11.3 Three-step synthesis of 1,3-propanediol from acrolein.

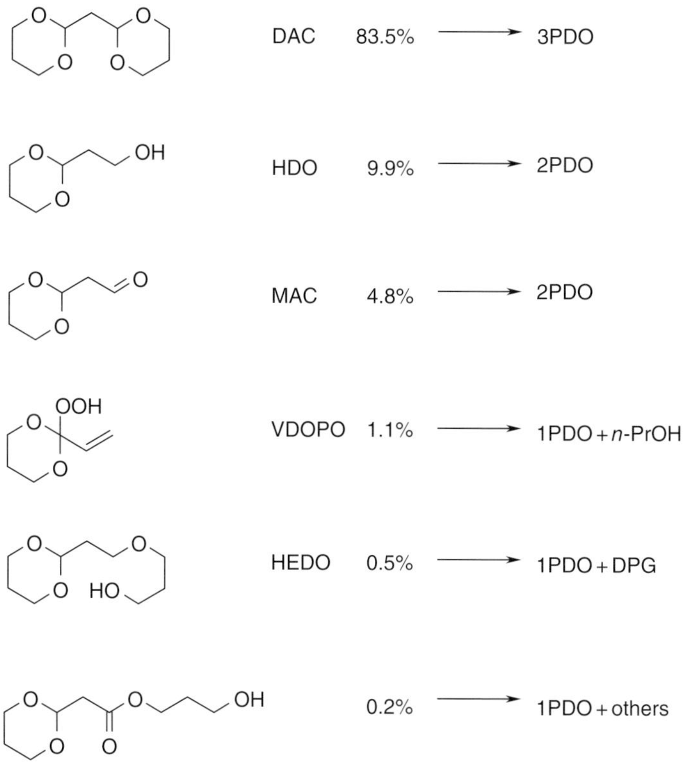

Structure	Name	Yield	Product
	DAC	83.5%	→ 3PDO
	HDO	9.9%	→ 2PDO
	MAC	4.8%	→ 2PDO
	VDOPO	1.1%	→ 1PDO + n-PrOH
	HEDO	0.5%	→ 1PDO + DPG
		0.2%	→ 1PDO + others

Figure 11.4 Hydrolysis and hydrogenation of the by-products (7% Rh/Al$_2$O$_3$, 2 MPa H$_2$, 90 °C, slight excess H$_2$O).

Figure 11.4 also shows the products formed after hydrolysis/hydrogenation of the various (by-)products. The two main by-products (HDO and MAC) are about 99%.

The catalyst phase is very easily separated from the solvent/product phase. Analysis of the solvent phase showed that it contained 93.8% of the formed DAC, free from palladium (by ICP-AES).

In the third step, the reaction mixture obtained from step 2 was subjected to a one-step hydrolysis/hydrogenation. The most promising result was obtained using 7% Rh/Al$_2$O$_3$, 2 MPa hydrogen and slightly excess water at 90 °C. This condition would give 1.48 kg 1,3-propanediol/g catalyst per hour. After a final distillation step, impurities could only be found at parts per million level.

11.3
Prospects for Scale-Up

11.3.1
Aerobic Palladium-Catalyzed Oxidation of Methyl Acrylate (MA) to 3,3-Dimethoxy Methyl Propionate: Process Optimization and Scale-Up

3,3-Dimethoxy methyl propionate is a versatile multifunctional synthetic building block, as illustrated by its use as starting material in our route to methyl cyanoacetate [10] as precursor for cyanomethyl methacrylate [10], which finds wide application in the adhesive industry [11]. We envisaged the synthesis of 3,3-dimethoxy methyl propionate by Wacker-type oxidation of cheap methyl acrylate in methanol as the solvent (Figure 11.5).

This oxidation makes use of the same beneficial aspects that were discussed previously: the stable Pd/Cu/Fe oxidation catalyst and methanol solvent for acetal product stability.

This process makes use of a catalytic reaction using an environmentally friendly oxidant and fits nicely into the context of green chemistry. The main problems of this type of reaction are, however, the danger of an explosion due to oxygen-rich gas phase, the danger of runaway reactions due to the exothermic nature of the reaction, and overoxidation of the product, especially at high conversions.

While studying the palladium-catalyzed oxidation of methyl acrylate in methanol using oxygen as the oxidant to form 3,3-dimethoxy methyl propionate, we encountered all three of these problems. The initial experiments were

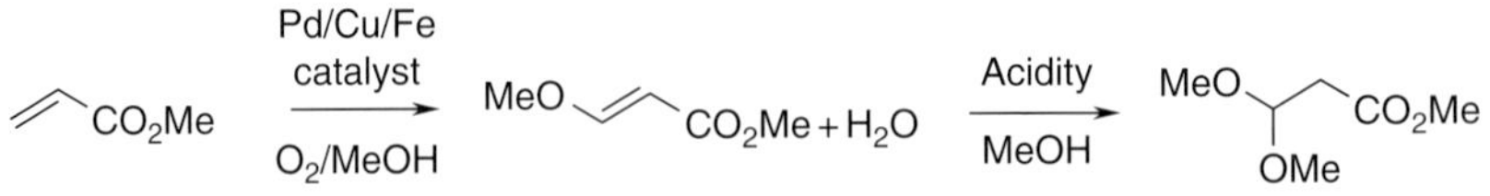

Figure 11.5 Pd/Fe/Cu-catalyzed oxidation of methyl acrylate to 3,3-dimethoxy methyl propionate using oxygen as the oxidant [9].

conducted using pure oxygen, and we were consequently limited in reaction scale using small robust autoclaves equipped with safety valves in case of explosion. Especially with high substrate and catalyst concentrations, the inside reaction mixture temperature rapidly rose considerably beyond heating bath temperature due to the exothermic nature of the oxidation. This effect was enhanced by thermally insulating Teflon inserts to protect the autoclave against the corrosive catalyst solution. Their use easily led to runaway reactions with inside reaction temperatures up to 80 °C beyond heating bath temperatures (Chart 11.1):

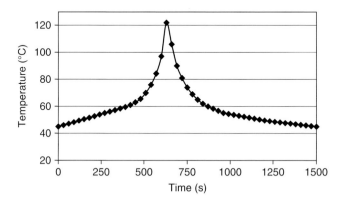

Chart 11.1 Inside reactor temperature increase due to the exothermicity of the oxidation of methyl acrylate to 3,3-dimethoxy methyl propionate [9].

In this particular experiment, methyl acrylate was reacted with the Pd/Cu/Fe catalyst at 0.9 MPa oxygen pressure at 45 °C. In about 10 min, the inside reaction temperature reached 120 °C, and at this point, oxygen deficiency set in, leading to palladium black formation. Furthermore, product selectivity decreased due to overoxidation and other side reactions. Experiments such as this on a large scale could very easily lead to dangerous runaway situations.

First, we focused on controlling exothermicity, catalyst stability, and product selectivity by small-scale optimization of reaction parameters such as the nature and ratio of the catalyst components, oxygen pressure, and temperature. Combined with additional data on explosion safe working conditions, this allowed successful scale-up up to 2 l. By fitting our reaction kinetics with our proposed reaction mechanism, we are now able to simulate different reaction conditions, and we use this tool for further scale-up to 1000 t/year.

11.3.2
Small-Scale Reaction Optimization

The first task was to find a combination of catalyst components with the lowest contribution to the price per kilogram product. Palladium contributes negligibly to the cost price, since the oxidation works well with a very small amount of relatively cheap Na_2PdCl_4 compared with the substrate. The Cu and Fe cocatalysts are, on the other hand, used in larger amounts (Pd/Cu/Fe = 1/230/230) and contribute significantly to product price. An extensive study [9] showed that metallic Cu and $FeCl_3$ gave the lowest contribution to the 3,3-dimethoxy methyl propionate cost price.

The fact that metallic copper is used as catalyst component does not influence the reaction negatively. This is because the catalyst solution is prepared prior to use by mixing $FeCl_3$ and metallic copper in methanol. $FeCl_3$ is a strong enough oxidant to dissolve elemental Cu by oxidation. Therefore, a Na_2PdCl_4, $FeCl_3$, and Cu mixture was used as default catalyst combination.

The aforementioned Pd/Cu/Fe catalyst ratio of 1/230/230 (i.e., (Cu, Fe)/Pd = 230) was arbitrarily chosen. We investigated the effect of changing (Cu, Fe)/Pd ratios (100–800) at different oxygen pressures (0.2–0.4 MPa) but constant temperature of 70 °C (Chart 11.2a). At (Cu, Fe)/Pd ratios lower than 400, the reaction rate is first order in (Cu, Fe)/Pd ratio and oxygen pressure and zero order in palladium and substrate concentration. At (Cu, Fe)/Pd ratios higher than 400, the reaction rate is zero order in (Cu, Fe)/Pd ratio and oxygen pressure, while it is first order in palladium and substrate concentration. The zero-order dependence on oxygen at (Cu, Fe)/Pd ratios beyond 400 positively impacts process development as it assures high reaction rates even at low oxygen pressure. Such a high (Cu, Fe)/Pd ratio catalyst solution at an industrial scale would decrease plant investment costs. Furthermore, process robustness will increase as catalyst performance will be less sensitive to fluctuations of partial oxygen pressure due to, for example, inefficiencies in stirring. In the following experiments, we use Na_2PdCl_4/$FeCl_3$/Cu ratios of 1/500/500 at an oxygen pressure of 0.2 MPa.

The reaction temperature was optimized by experiments with the same conversion curves while using different temperatures and then comparing the product selectivity of 3,3-dimethoxy methyl propionate. This was done by careful adjustment of the methyl acrylate Pd ratio. The same conversion curve is obtained when methyl acrylate Pd is decreased from 40 000 to 25 000 while decreasing the reaction temperature from 80 to 70 °C (Chart 11.2b).

Both at 70 and 80 °C, selectivity starts to decrease above 60% conversion, but more rapidly at 80 °C (product yield at 95% conversion: <60% at 80 °C; >75% yield at 70 °C). Selectivity loss at higher conversions can be explained by competing addition of coproduced water instead of methanol to the 3-methoxyacrylate intermediate. In contrast to the stable 3,3-dimethoxy methyl propionate acetal product formed by methanol addition, the hemiacetal 3-methoxy-3-hydroxy methyl propionate generated by addition of water is prone to overoxidation, especially at higher reaction temperatures. Consequently, 3,3-dimethoxy methyl propionate selectivity benefits from increasing the methanol/methyl acrylate ratio. The results of variation of methanol/methyl acrylate ratio on activity and selectivity are depicted in Chart 11.2c,d, respectively (Pd/Cu/Fe/methyl acrylate: 1/500/500/25000; oxygen pressure: 0.2 MPa). Whereas conversion rates are equal at different methanol/methyl acrylate ratios, 3,3-dimethoxy methyl propionate selectivity erodes to 50% above 90% conversion at a low methanol/methyl acrylate

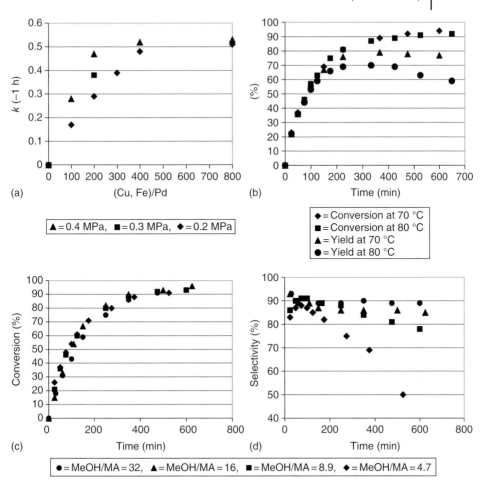

Chart 11.2 Investigation of reaction parameters [9]. (a) Effect of catalyst composition on the reaction rate at different pressures (Pd = 4.4 × 10^{-6} mol, MA = 0.11 mol, MeOH = 0.98 mol, 70 °C, 1000 rpm, pure oxygen, varying Cu and Fe amounts). (b) Product selectivity at different reaction temperatures (Pd = 4.4 × 10^{-6}, Cu = 2.2 × 10^{-3}, Fe = 2.2 × 10^{-3}, MA = 0.11 mol, MeOH = 0.98 mol, 1000 rpm, 0.2 MPa pure oxygen, varying reaction temperatures). (c) Conversion curves at different MeOH/Ma ratios (Pd = 4.4 × 10^{-6}, Cu = 2.2 × 10^{-3}, Fe = 2.2 × 10^{-3}, MA = 0.11 mol, 1000 rpm, 70 °C, 0.2 MPa pure oxygen, varying amounts of methanol). (d) Selectivities at different MeOH/MA ratios (Pd = 4.4 × 10^{-6}, Cu = 2.2 × 10^{-3}, Fe = 2.2 × 10^{-3}, MA = 0.11 mol, 1000 rpm, 70 °C, 0.2 MPa pure oxygen, varying amounts of methanol).

ratio of 4.7, whereas a methanol/methyl acrylate ratio of 32 gives a selectivity of 90% at more than 94% conversion. However, this high ratio of 32 makes the reaction mixture too diluted and consequently increases the energy costs too much. A methanol/methyl acrylate ratio of 16 provides the best compromise between process efficiency and selectivity.

To avoid explosive conditions as in the foregoing experiments with pure oxygen, the use of pressurized N_2/O_2 (range: 3.4–13%) mixtures was investigated, with a Pd/Cu/Fe/methyl acrylate ratio of 1/500/500/25 000 at 70 °C and 0.12 MPa total pressure. Under these conditions where the kinetics was limited by the oxygen supply, a linear decrease in the O_2 partial pressure led to a linear decrease in reaction rate [9]. The linear decrease in reaction rate means that even at very low oxygen concentration of the reaction solution, no deactivation of the palladium catalyst takes place. This is of major importance on a large scale where the N_2/O_2 feed will have a concentration of oxygen just below the limiting oxygen concentration (LOC), but the N_2/O_2 outlet will be much lower in oxygen concentration due to the consumption of oxygen. This outlet oxygen concentration must be higher than the value where palladium deactivation takes place by palladium black formation. Fortunately, even with only 3% oxygen in the gas outlet, there is still no deactivation of the catalyst.

Determination of explosion limits on both the starting reaction mixture and the final product mixture gave LOC values (in nitrogen) of 17.7% and 13.8% for the starting reaction mixture and product mixture, respectively. In our experiments, we used an oxygen/nitrogen gas mixture with 13% oxygen.

11.3.3
Large-Scale Methyl Acrylate Oxidation Reaction and Work-Up

Having optimized the catalyst composition, temperature, pressure, and methanol/methyl acrylate ratio, while knowing how to carry out the reaction safely, the reaction was scaled up in a 3 l autoclave (Pd/Cu/Fe/methyl acrylate ratio of 1/500/500/25 000; methanol/methyl acrylate = 16; 70 °C; 13% O_2 in N_2 at 0.61 MPa). A two-layer system was obtained after neutralization and evaporation. Although analysis of the water layer showed that it contained more than 95% of the catalysts, these were not recycled as calculations for a 1000 t/year scale process showed that the financial benefit would be minimal. On an industrial scale, catalyst recycling would be carried out for environmental reasons. After removal of the water layer, 82% isolated yield was obtained by distillation. Furthermore, 90% of the methanol was recovered.

11.3.4
Reaction Simulation Studies as Aid for Further Scale-Up

Scheme 11.1 shows the reaction mechanism of the palladium-catalyzed oxidation of methyl acrylate to 3,3-dimethoxy methyl propionate.

Palladium oxidizes methyl acrylate in the presence of methanol to form 3-methoxy methyl acrylate and equimolar amounts of water with a reaction rate R3. Reduced Pd is reoxidized by the Fe/Cu cocatalyst with a reaction rate R2 producing a reduced Fe/Cu cocatalyst, which is reoxidized by oxygen. The palladium catalyst is also responsible for the formation of by-products from oxidation of methanol (methyl formate, dimethoxymethane). 3-Methoxy methyl acrylate

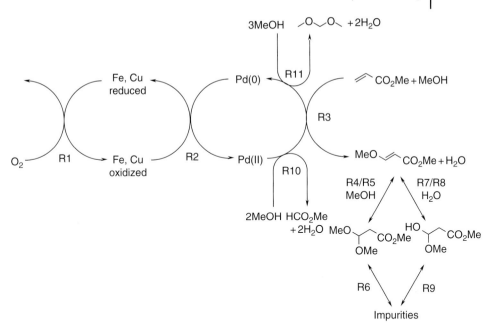

Scheme 11.1 The reaction mechanism of palladium-catalyzed oxidation of methyl acrylate to 3,3-dimethoxy methyl propionate [9].

reacts either with a molecule of methanol to desired 3,3-dimethoxy methyl propionate or with a molecule of water to form 3-hydroxy-3-methoxy methyl propionate. Both product molecules can lead to the formation of impurities, thus decreasing product selectivity. The mechanism in Scheme 11.1 was fitted with experimental data using reaction equations for the different reaction pathways together with fitted values of activation energy and pre-exponential factors, as shown in Scheme 11.2. The reaction is 1.32 order in methyl acrylate and 2.42 order in methanol.

Chart 11.3 compares experimental with simulated substrate conversion and product formation using a Pd/Cu/Fe/methyl acrylate ratio of 1/500/500/25 000 at 70 °C and oxygen pressure of 0.2 MPa at a methanol/methyl acrylate ratio of 8.90. The experimental data fit very well to the proposed reaction mechanism in Scheme 11.1.

Simulations also provided the basis for the experiments described in Section 11.3.2 related to Chart 11.2. The highly relevant result from these simulations that at a high (Cu, Fe)/Pd ratio the rate is independent on the oxygen pressure was fully confirmed experimentally and allowed us to use low partial oxygen pressures without catalyst deactivation while achieving safe processing. This example demonstrates how simulation of reaction conditions led to improvement of the design of the catalytic system. We use this simulation tool for further scale-up to 1000 t/year.

R1	Cu, Fe (reduced)→Cu, Fe (oxidized)	
	$r1 = k1 \times [\text{Cu, Fe (red)}] \times [PO_2]$	$A = 3.14 \times 10^{+11}$ s^{-1}, Ea = 71.8 kJ/mol
R2	Cu, Fe (ox) + Pd(0)→Cu, Fe (red) + Pd(II)	
	$r2 = k2 \times [\text{Cu, Fe (ox)}] \times [Pd(0)]$	$A = 4.95 \times 10^{+5}$ s^{-1}, Ea = 0.0 kJ/mol
R3	Pd(II) + MA + MeOH → 3MAC + Pd(0) + H$_2$O	
	$r3 = k3 \times [Pd(II)] \times [MA]^{1.32} \times [MeOH]^{2.42}$	$A = 1.23 \times 10^{+10}$ s^{-1}, Ea = 74.6 kJ/mol
R4, R5	3MAC + MeOH ↔ 33MP	
	$r4 = k4 \times [3MAC] \times [\text{Cu, Fe}]$	$A = 7.17 \times 10^{+01}$ s^{-1}, Ea = 0.5 kJ/mol
	$r5 = k5 \times [33MP]$	$A = 4.2 \times 10^{+00}$ s^{-1}, Ea = 0.4 kJ/mol
R6	33MP→IMP	
	$R6 = k6 \times [33MP]$	$A = 1.30 \times 10^{-01}$ s^{-1}, Ea = 23.1 kJ/mol
R7, R8	3MAC + H$_2$O ↔ 3H3M	
	$r7 = k7 \times [3MAC] \times [H_2O]$	$A = 8.59 \times 10^{+05}$ s^{-1}, Ea = 0.0 kJ/mol
	$r8 = k8 \times [3H3M]$	$A = 2.55 \times 10^{+06}$ s^{-1}, Ea = 7.0 kJ/mol
R9	3H3M→IMP	
	$r9 = k9 \times [3H3M]$	$A = 8.93 \times 10^{-01}$ s^{-1}, Ea = 10.4 kJ/mol
R10	2MeOH + 2Pd(II)→MF + 2Pd(0) + 2H$_2$O	
	$r10 = k10 \times [Pd(II)] \times [MeOH]$	$A = 5.15 \times 10^{+11}$ s^{-1}, Ea = 78.0 kJ/mol
R11	3MeOH + Pd(II)→DMM + Pd(0) + 2H$_2$O	
	$r11 = k11 \times [Pd(II)] \times [MeOH]$	$A = 1.23 \times 10^{+17}$ s^{-1}, Ea = 120.0 kJ/mol

Scheme 11.2 Reaction equations and the corresponding activation energies and pre-experimental factors (MA = methyl acrylate, 3MAC = 3-methoxy acrylate, 33MP = 3,3-dimethoxy methyl propionate, IMP = impurity, 3H3M = 3-hydroxy-3-methoxy methyl propionate, DMM = dimethoxy methylene, and MF = methyl formate) [9].

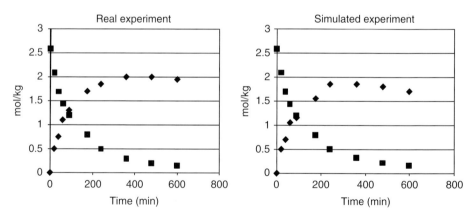

Chart 11.3 Real experiment versus simulated experiment, ■ = substrate conversion and ♦ = product yield.

11.4
Conclusion

In conclusion, we developed an elegant route for the synthesis of cyclohexanone from cyclohexene. Key achievements were product stability due to protection of the ketone by ketal formation and catalyst stability due to the usage of $Na_2PdCl_4/CuCl_2/FeCl_3$ as a catalyst combination.

This strategy could also be applied to acrolein oxidation. Via the sequence of protection, oxidation, hydration, and hydrogenation, acrolein could be converted in high yield into the industrially important 1,3-propanediol.

Finally, we developed a safe synthetic route for the production of 3,3-dimethoxy methyl propionate by a palladium-catalyzed oxidation of methyl acrylate in methanol using oxygen as oxidant. In 2 l scale experiments, we could reach 82% yield of 99% pure 3,3-dimethoxy methyl propionate. This synthetic procedure will aid the development of adhesives, pharmaceutics, agrochemicals, and functionalized polymers.

References

1. (a) Smidt, J., Hafner, W., Jira, R., Sedlmeier, J., Sieber, R., Rüttinger, R., and Kojer, H. (1959) *Angew. Chem.*, **71**, 176–182; (b) Smidt, J., Hafner, W., Jira, R., Sieber, R., Sedlmeier, and Sabel, A., (1962) *Angew. Chem. Int. Ed. Engl.*, **1**, 80–88.
2. Kolb, M., Bratz, E., and Dialer, K. (1977) *J. Catal.*, **2**, 399.
3. Musser, M.T. (2005) Cyclohexanol and cyclohexanone, in *Ullman's Encyclopedia of Industrial Chemistry*, Wiley-VCH Verlag GmbH, Weinheim.
4. Clement, W.H. and Selwitz, C.M. (1964) *J. Org. Chem.*, **29**, 241.
5. Shioyama, T.K. (1985) Patent US4507506, alfa-olefin conversion.
6. Fahey, D.R. and Zuech, E.A. (1974) *J. Org. Chem.*, **39**, 3276.
7. Lloyd, W.G. and Luberoff, B.J. (1969) *J. Org. Chem.*, **34**, 3949.
8. Lempers, H.E.B. and Setoyama, T. (2001) Patent WO0181276, method for producing ketal and/or acetal.
9. Tanaka, Y., Takahara, J.P., and Lempers, H.E.B. (2009) *Org. Process Res. Dev.*, **13** (3), 548–554.
10. Lempers, H.E.B. and Takahara, J.P. (2005) Patent JP2005097233, method for producing nitrile.
11. Cary, R. (2001) Concise International Chemical Assessment, Document 36.
12. Takahara, J.P. and Setoyama, T. (2002) Patent WO0249999, process for producing polyhedric alcohols.
13. Weissermel, K. and Arpe, H.-J. (2007) *Industrial Organic Chemistry*, John Wiley & Sons, Inc.

12
Oxidative Carbonylation: Diphenyl Carbonate
Grigorii L. Soloveichik

12.1
Introduction

12.1.1
Diphenyl Carbonate in the Manufacturing of Polycarbonates

Diphenyl carbonate (DPC) is a carbonate ester of formula $(C_6H_5O)_2CO$. The principal use of DPC is as a critical comonomer in the manufacture of polycarbonate resin, an engineering thermoplastic with high impact resistance and excellent optical properties. It is widely used as a heat-resistant and flame-retardant insulator in electronics, a construction material, and an optical material for data storage (CD and DVD disks) and in manufacturing transparent parts in automotive and aerospace industries [1].

Polycarbonate plastic was independently discovered in 1953 by Daniel Fox of General Electric (GE) and Hermann Schnell of Bayer AG. The base patent was granted to Bayer because its invention was made only a week (!) earlier than GE. Both companies became the first manufacturers of polycarbonate resin under trade names Makrolon® (Bayer) and Lexan® (GE) [2]. Makrolon® and Lexan® (currently produced by SABIC Innovative Plastics (IP) after acquiring GE Plastics in 2007) are still leading trade names in the polycarbonate world. Other major polycarbonate manufacturers include Asahi Kasei Chemicals Corporation (Wonderlite™, Japan), Chi Mei Corporation (WONDERLITE®, China), Teijin Ltd. (PANLITE®, Japan), Styron (CALIBRE®, United States), Mitsubishi Engineering-Plastics (IUPILON®, Japan), Idemitsu Kosan Ltd. (TARFLON®, Japan), and Formosa Chemicals & Fibre Corporation (Taiwan). The annual demand for polycarbonates was 3.7 million metric tons in 2012 and is expected to reach 5 million tons in 2015.

Originally, polycarbonate was produced by interfacial (organic/aqueous) polycondensation of phosgene $(COCl_2)$ with disodium salt of a bisphenol, such as 2,2-bis(4-hydroxyphenyl)propane (bisphenol A or BPA). Initially, the reaction produces an intermediate chloroformate ROC(O)Cl (Eq. (12.1)), which subsequently reacts with another phenoxide molecule, growing a polymer chain

Liquid Phase Aerobic Oxidation Catalysis: Industrial Applications and Academic Perspectives,
First Edition. Edited by Shannon S. Stahl and Paul L. Alsters.
© 2016 Wiley-VCH Verlag GmbH & Co. KGaA. Published 2016 by Wiley-VCH Verlag GmbH & Co. KGaA.

(Eq. (12.2)) in the presence of a phase transfer catalyst. A monohydroxyphenol (e.g., *p*-cumyl phenol) is used as a chain stopper [3].

$$(HOC_6H_4)_2CMe_2 + 2NaOH \rightarrow Na_2(OC_6H_4)_2CMe_2 + 2H_2O \quad (12.1)$$

$$Na_2(OC_6H_4)_2CMe_2 + COCl_2 \rightarrow 1/n\,[OC(OC_6H_4)_2CMe_2]_n + 2NaCl \quad (12.2)$$

However, a phosgene-free route was desired because of environmental hazards and governmental restrictions associated with the production and storage of extremely toxic phosgene (and chlorine used to synthesize it) and the use of chlorinated solvents. An important phosgene-free, so-called *melt* process, for the DPC synthesis involving transesterification of DPC in the absence of solvents was independently developed by GE, Bayer, Asahi/Chi Mei, and Mitsubishi.

Interestingly, early production of polycarbonate was actually based on preparation of DPC from phosgene and phenol followed by reaction with BPA to produce the polymer and regenerate phenol. However, slow reaction rates and process complexity favored the interfacial process that was commercialized. In all cases, DPC is produced by indirect methods, mainly by transesterification of dialkylcarbonates such as dimethyl carbonate produced either from CO, methanol, and oxygen using EniChem technology (SABIC IP) or via carbonylation of methyl nitrite from reaction of NO with methanol (Bayer) or by methanolysis of ethylene carbonate (Asahi/Chi Mei) or di-*n*-butyl carbonate from urea and *n*-butyl alcohol (Mitsubishi Chemical) [4]. Ideally, DPC can be prepared directly from phenol, carbon monoxide, and oxygen in a simpler (one-step) energetically favorable reaction (Scheme 12.1).

$$2\,PhOH + CO + 1/2\,O_2 \xrightarrow{\text{Catalyst}} PhO-C(=O)-OPh + H_2O$$

Scheme 12.1 Direct synthesis of diphenyl carbonate by oxidative carbonylation.

12.1.2
History of Direct Diphenyl Carbonate Process at GE

The history of direct DPC process development at GE can be considered as a case study. In the mid-1970s, Chalk discovered that *para*-substituted phenols can be oxidatively carbonylated using stoichiometric amounts of a Ru, Rh, Os, Ir, or Pd salt, from which the palladium salts were the most active (Eq. (12.3)) [5]. Yields of carbonylation products were high, but the reaction yielded 1 mol of Pd^0 for every mole of product, so it was stoichiometric [6]. In the late 1970s, Hallgren and Matthews assumed that DPC was formed by reductive elimination from the acyl Pd(II) carbonyl complex [7].

$$2\,PhOH + CO + Pd^{II}X_2 \rightarrow (PhO)_2CO + Pd^0 + 2HX \quad (12.3)$$

However, DPC yield and Pd turnover number (Pd TON defined as mol DPC/mol Pd charged) obtained in early works remained low. Unlike aliphatic alcohols, the oxidative carbonylation of aromatic alcohols is difficult due to their increased acidity and oxidative instability. They, therefore, require expensive catalysts [7]. It was the high cost of the optimal Pd-based catalysts, which were superior compared with all others, and low reaction rates that prevented the direct DPC process from being commercially viable.

Up until 1995, the bulk of the patents issued on a direct, one-step DPC process were to GE. These systems were based on palladium salts with cobalt or manganese compounds as the inorganic cocatalyst (ICC). Many systems also used an organic cocatalyst (OCC), usually benzoquinone (BQ). Additional components of the catalytic "package" included a quaternary ammonium or phosphonium bromide and a base. The best systems had TON less than 500 and a rate of less than 1.0 mol DPC/h [8]. High palladium cost demanded higher TONs and a high level of palladium recovery. Because the efficiency of phenol carbonylation had to be increased, it was also necessary to reduce the nonproductive CO conversion to CO_2 and the conversion of phenol into non-DPC products. It is worth noting that all efforts to replace palladium as the main catalyst in this reaction up until then, and even to date, have failed.

Starting in 1995, GE developed a major research program to commercialize the one-step DPC synthesis process. This effort included catalyst development, engineering, and rigorous economic analysis. It was clear that the main problem was to identify an active cocatalyst capable of fast reoxidation of Pd(0), which was generated at DPC formation (Eq. (12.3)). A working hypothesis was that a redox chain consisting of two (or more) redox components, similar to that in natural redox chains, could provide faster reaction rates and deliver higher Pd TON numbers. Typical one-step DPC reactions were run in an autoclave at high pressure (11.5 MPa, 1700 psig) and at elevated temperatures (100 °C) for 1–3 h with vigorous stirring. The setup and tear downtime of the reactor limit the number of reactions that can be run in one reactor to 1–2 per day. At the same time, discoveries of new and effective cocatalysts for this reaction [9] inspired an expanded search for new ICCs. However, an enormous number of potential cocatalysts and conditions to explore made traditional-scale reactions cost prohibitive. As a result, in 1997, a group at GE was put together to tackle the problem of increasing throughput of the one-step DPC reaction.

To accelerate the DPC catalyst development, a high-throughput screening (HTS) method [10] based on the use of a miniaturized reactor system capable of working under high pressure and elevated temperature was developed at GE Global Research [11]. This system allowed for delivering the pressurized mixture of O_2 and CO to all the 25–75 µl samples without cross-contamination at the same temperature and pressure [11, 12]. Catalyst leads (*vide infra*) identified on the basis of Pd TON were optimized for the best formulations using design-of-experiment (DOE) methods, first at the HTS scale and then at a standard batch scale (60 g), after which they were transferred to a continuous 1 Gal benchtop unit (BTU), producing about 1 kg/h of DPC. Thus, GE progressed from discovery of the direct DPC process to the pilot stage [2].

12.2
Chemistry and Catalysis

12.2.1
Mechanism of Oxidative Carbonylation of Phenol

Pd(II) salts stoichiometrically react with phenol in the presence of CO producing 1 mol of diphenyl carbonate per mole of Pd complex (Scheme 12.2) [7]. A key phenoxycarbonyl palladium complex PdCl(CO$_2$Ph)(PPh$_3$)$_2$, stabilized by two phosphine ligands, was isolated and characterized by a single-crystal X-ray analysis [13]. It reacts with sodium phenoxide or phenyl chloroformate to give DPC via formation of an intermediate, Pd(OPh)(CO$_2$Ph)(PPh$_3$)$_2$, which is stable at −20 °C and was characterized by NMR [13]. *ortho*-Metalation and activation of C–H bond in *para*-position give by-products phenyl salicylate (PS) and *para*-hydroxyphenyl benzoate (PHB) (Scheme 12.2), but DPC remains the major product. The ratio of carbonate to PS and PHB is controlled by the nature of the *para*-substituent, the Pd source, and the order of reagent addition [7]. In the presence of oxygen, catalytic formation of DPC is observed due to the oxidation of Pd(0) back to Pd(II). However, the rate of reoxidation of Pd(0) by oxygen is very slow, and the Pd TON (mol DPC/mol Pd charged) is less than 10, which makes the process impractical and requires a cocatalyst to activate O$_2$. Both inorganic (salts of transition metals

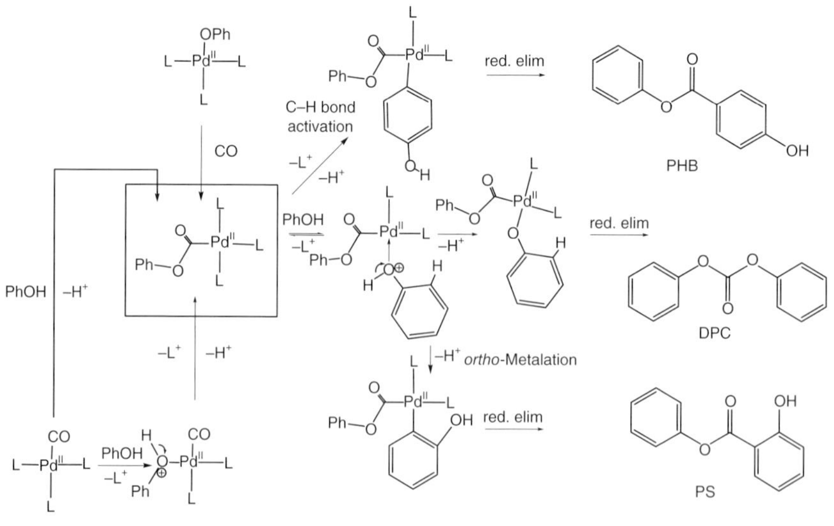

Scheme 12.2 Pathways of phenol oxidative carbonylation by Pd(II).

or lead) and organic (e.g., BQ) cocatalysts (or their combinations) were applied for this purpose.

For the catalyst development, it is very important to elucidate the mechanism of Pd(0) reoxidation. Unfortunately, there is no literature data on redox potentials of potential cocatalysts in PhOH solution. Assuming that the difference in potentials in aqueous and phenolic solutions is similar, all metals used as cocatalyst can be divided into three groups: (i) those with potentials less than the potential of Pd^{2+}/Pd^0 couple (+0.60 V vs. SHE in the presence of bromide anions), for example, Yb^{3+}, Zn^{2+}, and Pb^{2+}; (ii) those with potentials between the potentials of Pd^{2+}/Pd^0 and $(O_2 + H^+)/H_2O$ couples, for example, Fe^{3+} and Mn^{4+}; and (iii) those with potentials higher than that for the $(O_2 + H^+)/H_2O$ couple, for example, Pb^{4+}, Ce^{4+}, and Co^{3+}. In other words, only group 2 metals can oxidize Pd(0) and be reoxidized by oxygen, while metals from group 3 can oxidize Pd(0) but cannot be reoxidized by oxygen, and group 1 metals can be reoxidized by O_2 but cannot oxidize Pd(0).

These data show that there is another mechanism of Pd(0) reoxidation, in addition to the classic mechanism, where electrons from Pd(0) are transferred to the metal in the higher oxidation state. As one explanation, in this case the role of a cocatalyst is to activate oxygen and generate hot oxygen-containing radicals such as HOO• that are capable of oxidizing Pd(0). Most probably, however, the cocatalyst forms a complex with Pd in which the palladium reoxidation rate is greatly increased due to lower oxidation potential or/and symmetry decrease. It was shown that the activity of metal cocatalysts is practically the same for complexes with various ligands, which most probably exchanged for phenoxide or bromide ligands present in a large excess. Therefore, it is likely that such a mixed-metal complex is formed via metal–phenoxide and metal–bromide bridge bonds. In addition, bromide can serve as a redox mediator. The potential of Br_2/Br^- couple allows the oxidation of Br^- by oxygen and oxidation of Pd(0) by generated bromine. This may explain the absolutely crucial role of a bromide salt, which is superior to other halides, in DPC synthesis. Nonadditive effect (positive or negative) of two or more cocatalysts confirms the hypothesis of formation of mixed-metal complexes in DPC catalytic system.

12.2.2
Catalysts for Oxidative Carbonylation of Phenol

Catalyst testing for direct DPC synthesis reported by different research groups was done at elevated pressure (usually 100 bar) using nonexplosive mixtures of O_2 and CO or air and CO (1:2). The low flammability limit in such mixtures is 12.5 vol% for CO concentration [14]. However, the presence of water vapors decreases the flammability limit. The safe oxygen concentration is generally assumed to be below 8 vol% though in many works 10 vol% of O_2 was used. Initially, reactions were carried out in a solvent such as methylene chloride. Later, it was recognized that phenol could serve not only as a reagent but also as a solvent, and reactions were performed in neat PhOH. It should be emphasized that water formed during the course of reaction (Scheme 12.2) is highly detrimental to the direct DPC process because it inhibits Pd catalyst and hydrolyzes DPC in the reaction mixture [15]. It was recognized earlier that the presence of a desiccant (3 or 4 Å molecular sieves) could suppress hydrolysis by sequestering water and increase DPC yield three- to fourfold. The results discussed in later sections were obtained using molecular sieves unless otherwise indicated.

Initially, simple palladium salts, such as $PdBr_2$ [16], $Pd(OAc)_2$ [17], or $Pd(acac)_2$ [12, 18], were used as a Pd source in combination with ICCs and a base. There was a consensus that anions in these salts are eventually replaced by phenolate and do not play a significant role. Indeed, reaction of diaryloxy palladium complex with CO produces DPC when reacted with CO under pressure at 100 °C [19]. However, the testing of Pd complexes with stronger P- and N-containing ligands revealed that both ligands and anions affect catalytic activity and selectivity. Activity of $PdCl_2$ complexes with bidentate phosphines increased in the order 1,2-bis(diphenylphosphino)ethane (dppe) < 1,3-bis(diphenylphosphino)propane (dppp) < 1,4-bis(diphenylphosphino)butane (dppb). Thus, the complex $PdCl_2$(dppe) exhibited lower selectivity and activity an order of magnitude less than $PdCl_2$(dppp) analog [20]. Possibly, stronger chelate ligands block coordination sites at the Pd atom. Indeed, addition of a large excess of chelating ligands completely inhibits DPC formation [21]. In another catalytic package containing lead bromophenoxide and NEt_4Br as cocatalysts, dppp showed better activity than dppb (Pd TON in a large-scale run 6075 vs. 5260 compared with 4340 when no phosphine added) but lower selectivity (77% vs. 82%) [22]. In the system containing $Pd/Co(acac)_3/BQ/TBAB$ in phenol, Pd complexes with bulky bidentate N−N ligands such as 2,9-dimethyl-1,10-phenanthroline (dmphen) showed the highest Pd TON. The performance of Pd complexes with chelating nitrogen ligands was superior to individual Pd salts (2.2×) and complexes of $PdCl_2$ (1.5×) with phosphine ligands, though initial activity was comparable [20]. N,N-Bis(diphenylphosphino)propylamine (dppa) also showed very high TON = 6040.

Binuclear complexes with bridging diphenylphosphinomethane (dppm) [23] and diphenyl-2-pyridylphosphine (Ph_2PPy) [24] ligands were more active than $PdBr_2$ in combination with Ce and Mn ICCs. Addition of a quaternary bromide salt (QBr) increased the catalyst activity but decreased selectivity due to the formation of PS [23]. The highest TON (58 in 3 h) was reached at small scale with $Pd_2(Ph_2PPy)(NO_3)_2$ and Ce ICC in neat PhOH [24]. Tin-containing heterometallic Pd complexes $Pd_2(dppm)(SnCl_3)Cl$ and $Pd_2(dppm)(SnCl_3)_2$ catalyzed DPC synthesis with Mn [25] or Ce [26] cocatalysts without ammonium bromide, but the Pd TON was low (9 in 3 h) [26].

Among Pd complexes with chelating N-heterocyclic carbene ligands, 1,1′-dialkyl(aryl)-3,3′-methylenediimidazolin-2,2′-diylidene, in a system including $Ce(tmhd)_4$ (tmhd = 2,2,6,6,-tetramethyl-3,5-heptanedionate), Bu_4NBr, and HQ (hydroquinone) in CH_2Cl_2, only the $Pd(carbene)_2Br_2$ complex with *tert*-butyl substituent showed activity exceeding (by 80%) the activity of $PdBr_2$, while others were much less active [27]. This result was explained by steric properties of the carbene substituents that could either ease reductive elimination of DPC (R = Bu^t) or interfere with the coordination of phenolate (R = Bu^n) [27]. A Pd giant cluster Pd$_{561}$phen$_{60}$(OAc)$_{180}$ (phen = 1,10-phenanthroline), in contrast to Pd(OAc)$_2$ and [Pd(CO)Cl]$_n$, produces DPC with nitrobenzene used as an oxidant but not with O$_2$ even at 150 bar [28].

Heterogeneous Pd/C catalysts showed superior activity compared with homogeneous Pd(OAc)$_2$ with Cu$_2$O but least activity with Ce(OAc)$_3$ cocatalyst [29]. It was shown that CuO reacted with PdO on the support surface to form PdCuO$_2$ compound that transfers electrons in the crystal lattice for Pd reactivation [30]. Takagi et al. [31] concluded that Pd/C actually worked as a homogeneous catalyst in combination with PbO and NMe$_4$Br. However, later it was found that the activity of Pd heterogeneous catalysts exceed the activity of homogeneous Pd(OAc)$_2$ catalyst and depends on the support, with increasing activity in the order MgO < SiO$_2$ < Al$_2$O$_3$ < C. No soluble Pd in the reaction liquid phase was detected by the ICP analysis. The higher DPC yield for the Pd/C catalyst was attributed to the hydrophobicity of carbon surface, which repels water from poisoning the catalyst [32]. Selectivity of all heterogeneous Pd catalysts was lower (50–70%) [32]. Heterogeneous Pd catalysts on carbon and Amberlyst supports showed faster deactivation at initial activity comparable to homogeneous ones [20]. TBAB is critical for high TON, presumably due to stabilization of easily oxidizable Pd nanoparticles [20]. Activity of Pd(MeCN)$_2$Cl$_2$ with a Mn ICC and a quaternary ammonium bromide in the presence of polyvinylpyrrolidone was increased by about 30% compared with a purely homogeneous catalyst system [33]. A catalytic package consisting of Pd(OAc)$_2$, TBAB, and polyaniline with a molecular weight of about 3380 demonstrated good DPC yield but very poor selectivity (40–52%) [34].

Several attempts to use a redox-active support for Pd catalyst have been made. Pd catalyst on the La$_x$Pb$_y$Mn$_z$O support prepared by the sol–gel method produced DPC with a yield of 26.8% and selectivity of 99% [35]. Optimization of Cu, Mn, and Co oxide mixture showed that the most active was the composition with the ratio 1 : 1 : 1 that reached a DPC yield of 38% at 99% selectivity [36]. The use of Pd–Co catalyst on a Hopcalite (a mixture of copper and manganese oxides) support increased DPC yield to 43.5% with exceptional selectivity (99.6%) [37]. Anchoring of Pd via a carbene linkage to a polymer resin substantially increased Pd TON (up to 5100) compared with homogeneous carbene systems (80–150) [38].

Motivated by the high and volatile price of Pd, an intensive search for a Pd replacement was undertaken using HTS. Iron compounds such as FeBr$_3$ in combination with various ICCs (e.g., Pb, Cu) and a base showed 2–6 TON [39]. Other nonplatinum group metals, such as manganese [40], cobalt [41], and nickel [42], also catalyze DPC synthesis, but TON values were less than 10–15 (though well above stoichiometric). Even taking into account the low cost of these metals compared with Pd, their catalytic activity was too low to be practical.

12.2.3
Cocatalysts for Oxidative Carbonylation of Phenol

12.2.3.1 Organic Cocatalysts

Redox-active quinones/hydroquinones were proposed to assist in the reoxidation of Pd(0) in addition to Pd–ICC–NR_4Br systems [43]. Formation of HQ carbonate as a side product was not detected [27]. A similar system, $Pd(OAc)_2$/BQ/$Co(acac)_3$/Bu_4NBr, in the presence of dmphen delivered a Pd TON of 700 [21]. The ratio BQ:Pd affects the catalytic activity until it reaches 30:1, and the addition of dmphen just shifts the curve to higher TON [21]. BQ and anthraquinone were also used with copper ICCs [44]. Addition of BQ increased Pd TON two to four times in the systems Pd–Co–TBAB and Pd–Cu–TBAB [45].

Another class of OCCs is aromatic N-heterocycles, which are most effective at low OCC:Pd ratios and probably serve as Pd or ICC ligands rather than redox mediators. Terpyridine (terpy) improved the performance of a Pd–Co catalyst compared with BQ [8c, 46]. An optimal terpy:Pd ratio was 0.5, and its increase led to the deactivation of the catalyst, probably due to coordination to Pd [47]. Addition of carbazoles to a system containing $Mn(acac)_2$ and a base increased DPC yield by 25–80% [48].

At early development stages, OCCs were deemed as an essential part of catalytic packages for direct DPC synthesis, but difficulties in recovery, cost, and, most importantly, development of very efficient ICCs forced them out of consideration.

12.2.3.2 Inorganic Cocatalysts

Cobalt compounds were present in the first catalytic packages for DPC direct synthesis [16]. Co(salen) showed the best performance (Pd TON = 25) among other cobalt salts tested with $Pd(OAc)_2$ and TBAB, which was comparable with performance of copper salts in CH_2Cl_2 [49] and in neat PhOH [8c]. $Co(acac)_2$ was more active than other Co salts (Pd TON = 69) in the package Pd:Co:BQ:TBAB (1 : 2 : 30 : 60) in neat PhOH [45].

Copper(II) acetate in combination with $PdCl_2$, TBAB, and HQ was the most active of a number of metal salts when the reaction was performed in CH_2Cl_2 [50]. $Cu(acac)_2$ was the most active in the case of the Pd:Cu:BQ:TBAB system with a Pd TON of 89 [45]. With the addition of HQ, a Pd TON of 250 was reached [51]. With heterogeneous Pd/C catalyst, Cu_2O showed superior performance in oxidative carbonylation of phenols [29]. However, copper ICCs promote side reactions such as phenol oxidation by oxygen to form 2,2-biphenol [16] or formation of o-phenylene carbonate [51].

Cerium salts turned out to be the most active (Ce ~ Cu > Mn > Co > Cr > Mo ≫ Fe) in DPC synthesis in CH_2Cl_2 solution using $PdCl_2$, TBAB, HQ, and an ICC [50]. Though the activity of Ce and Cu acetates was comparable (yield 76%, Pd TON 250), cerium cocatalysts delivered better selectivity [50]. In the catalytic system including a Pd carbene complex, a Ce salt, NBu_4Br, and HQ in CH_2Cl_2, the activity of cerium cocatalysts increased in the order $Ce(OAc)_3 < Ce(trop)_4 <$

Ce(acac)$_3$ < Ce(tmhd)$_4$ (trop = tropolonate) [27]. This order correlates with the increase of electron donor ability of ligands that should ease Ce(III) reoxidation with oxygen [27].

A breakthrough in the catalyst productivity was made by Mitsubishi in 1996 using lead compounds as cocatalysts [9, 52]. The use of PbO as cocatalyst with 5% Pd/C catalyst and NMe$_4$Br (PbO:Br:Pd = 10:20:1) in neat PhOH with a mixture of CO and air (2:1) at 90 atm and 100 °C allowed a Pd TON of 1273 in 3 h [31]. This value was substantially higher than the Pd TON obtained under similar conditions with other cocatalysts such as Ce(OAc)$_3$ (577) or Co(OAc)$_2$ (150). In contrast to cobalt ICCs, the system with PbO ICC produced less PS but more bromophenols, resulting in better overall DPC selectivity. Addition of a second ICC such as CuO or cobalt complexes did not affect Pd TON, but suppressed the formation of bromophenols. However, it also increased the production of PS, thus reducing DPC selectivity [31]. The initial rate of DPC formation is first order in [Pb] and [O$_2$] and independent of other reactants including Pd, which indicated that O$_2$ activation by Pb species was the rate-determining step [31]. Increasing the Br:Pd ratio to 1000–1500:1 resulted in higher Pd TON (3300–4100) [9]. Replacement of NBu$_4$Br with hexaethylguanidinium bromide (HEGBr) further increased the Pd TON to greater than 11 500 [53]. Increasing the Pb:Pd ratio well above the initial findings (from 13 to 60) not only increased Pd TON in batch runs but also prevented catalyst deactivation during continuous BTU runs for at least 22 h [18]. In the continuous process, solutions of a Pd source in PbO phenolic solution and of QBr in PhOH were simultaneously fed to a pressurized reactor, while a part of reaction mixture was removed [54].

Due to heterogeneity of reaction mixtures containing PbO, the catalyst performance was affected by the order of mixing, temperature, and time of the catalyst preparation [55]. PbO is very soluble in molten phenol (up to 20 wt%), and evaporation of phenol from such solution produces lead phenolate Pb(OPh)$_2$. Study of the interactions between catalyst components in molten phenol showed that the phase composition of resulting mixtures depends on the temperature and the QBr:PbO ratio (Scheme 12.3). Below 140 °C, a precipitate formed in this reaction contains phases of PbBr$_2$ and Pb(OH)Br (minor component). Pd(acac)$_2$ is unstable in PhOH solution in the presence of a bromide salt and decomposes with the formation of bromide Pd clusters (Pd:Br = 6:1) and Pd···Pd distance

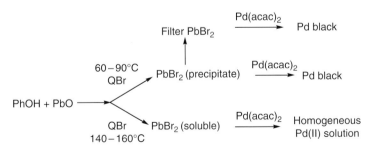

Scheme 12.3 Reactions between the main components of lead-based catalytic packages.

2.81 Å (2.76 Å in Pd metal) [55]. After the reaction, Pd had a strong metallic character [29]. A large excess of Br$^-$ or stronger chelates keeps Pd in solution. Interaction of PbO with bromides in phenol at the ratio Br: PbO < 2 results in the formation of water and precipitation of white sediments of the general formula $Pb_nO_m(OPh)_{(2-z)(n-m)}Br_{z(n-m)}$, where $n=4-10$, $m=1-4$, $n/m=3-4$, $z=0-1$ with short Pb$\cdots$Pb nonbonding distances (3.63–4.01 Å) [55]. It was suggested that the structure of these phases is similar to the structures of known complexes $Pb_6O_4(O^iPr)_4$ [56] and $Pb_4O(OSiMe_3)_6$ [57]. X-ray analysis of isolated $Pb_{10}O_4(OPh)_{12}\cdot 2MeCN$ complex revealed very short Pb$\cdots$Pb distances (average 2.67 Å), the same as determined by EXAFS [55].

Similar polynuclear lead species with short Pb$\cdots$Pb bond distances (3.83 Å) were found by EXAFS in frozen active DPC reaction mixtures. However, in the final reaction mixtures containing inactive catalyst, the short Pb$\cdots$Pb distances were not detected. Compared with PbO in the same conditions, these polynuclear lead bromophenoxides exhibited both higher Pd TON (6500 vs. 5000) and reaction selectivity (75–84% vs. 65–67%), with the highest selectivity observed for the complex $Pb_6O_2(OPh)_6Br_2$ [55]. Applying palladium complexes with dppp and dppa in this system increased Pd TON by 40% [22, 58]. Though the use of lead bromophenoxides instead of PbO requires an additional preparation step thus increasing the process cost, the gain in catalyst productivity may justify the added complexity. In the catalytic package including $PdCl_2$, TBAB, HQ, and an ICC, cerium salts turned out to be the most active (Ce ~ Cu > Mn > Co > Cr > Mo $\gg$ Fe) in DPC synthesis in CH_2Cl_2 solution [50]. Though Ce and Cu acetates were comparable (yield 76%, Pd TON 250) in activity, cerium ICCs delivered better selectivity [50]. In the catalytic system including a Pd carbene complex, a Ce salt, HQ, and NBu_4Br in CH_2Cl_2, the activity was increased in the order $Ce(OAc)_3 < Ce(trop)_4 < Ce(acac)_3 < Ce(tmhd)_4$ [27]. This order correlates with the increase of electron donor ability of ligands that should ease Ce(III) reoxidation with O_2 [27].

Successful use of Pb, Co, and Mn ICCs in the Pd/Br system stimulated a broad search for other nonlead cocatalysts [11]. For individual metal cocatalysts screened by the HTS method, Pd TON decreased in the following order: Ce > Zn > Cu > Yb > Mn > Pb > Eu > Co > Ni > Bi [12]. Surprisingly, in contrast to the common belief that the ICC should be able to reoxidize Pd(0) [12], it was discovered that some ICCs such as Zn [59], Yb [60], and Eu [60] with redox potentials too negative to oxidize Pd(0) were active. For example, a catalyst containing $Yb(acac)_2$ (Pd:Yb:Br = 1 : 14 : 120) demonstrated a Pd TON of 790 [60]. Addition of Zn, Bi, and Cu cocatalysts to a catalytic package containing $Pd(OAc)_2-TiO(acac)_2-NR_4Br$ improved Pd TONs from 150 to about 900, 1000, and 2000, respectively [61]. Therefore, these results may indicate that electron transfer to a transition metal may be not the only mechanism for palladium reoxidation.

12.2.4
Multicomponent Catalytic Packages

The catalyst packages containing only PbO as an ICC showed high TONs but low reaction rates and significant induction periods. To improve catalyst performance, additional redox cocatalysts have been screened. A synergistic effect was found for titanium [62], cerium, manganese, iron, and zirconium ICCs [12]. The Pb–Ti and Pb–Ce combinations showed the highest rate [55]. QBr may be replaced with quaternary chloride, but DPC yield decreased three- to fourfold [62].

To find nonlead ICCs, binary combinations of cocatalysts have been studied. Though the majority of the combinations were no better or worse (V, W, Sb, and Sn salts) than the individual runs, a number of synergies (Pd TON higher than the sum of the TON values for single-metal ICCs) were discovered [11]. The most interesting systems involved cocatalysts such as Ti, Fe, and Zr, which alone had low TONs but significantly improved the performance of other ICCs such as Pb, Cu, and Mn. Combination of Eu salts with Mn and Fe acetylacetonates as ICC with Pd–HegBr catalyst showed good activity (Pd TON up to 1100 and 940, respectively, in small-scale reactions) [59a]. Even higher Pd TONs were found in the system with Cu and Fe ICCs (Pd TON = 1470 at the ratio Pd:Cu:Fe:HegBr = 1:28:5.6:120) and copper and manganese (Pd TON = 2140 at the ratio Pd:Cu:Mn:HegBr = 1:28:5.6:120) [63]. Addition of cerium, nickel, or iron compounds to a package containing a Pd salt, QBr, and manganese ICC also improved the Pd TONs [64].

The performance was further improved by the addition of a third cocatalyst, often unpredictable. For instance, the addition of Fe (TON = 9 for single metal) to the Pb/Ti system increased the TON from 1068 to 1631 but reduced it for another active Cu/Ti system [12]. By examining the frequency of cocatalysts that appear in the best ternary system after testing over 10 000 unique catalyst combinations, it can be seen that the top three metals are Pb, Ti, and Mn. Despite the success of the ternary systems, the use of three metal cocatalysts in addition to palladium made the economics of metal separation of recovery unfavorable. Therefore, the R&D was focused on optimization of the best binary systems.

12.2.5
Role of Bromide in Direct Synthesis of Diphenyl Carbonate

At early stages of the process development, it was discovered that all effective catalytic systems for DPC synthesis contained a bromide quaternary ammonium salt. Other bromide salts were practically inactive, while quaternary ammonium iodides and chlorides showed poor performance [51]. The DPC yield grows with bromide concentration, for example, in the catalytic package $Pd(OAc)_2/Co(acac)_3/NBu_4Br$ up to the ratio Br:Pd = 20 [21].

The role of bromide in DPC catalysis is still unclear. It was suggested that NR_4Br works as a base to ionize phenol to phenoxide anion [51] or as a surfactant to stabilize Pd(0) nanoparticles [65]. However, no other bases or surfactants have even come close to its activity in DPC synthesis. Quaternary ammonium and phosphonium bromides are expensive and difficult to recover in downstream processing and, therefore, have a substantial impact on the DPC cost structure. For the process economics, it was critical to replace these compounds with less expensive equivalents, which could be done by two ways: either replacing QBrs with inexpensive NaBr or nonbromide nonquaternary salts.

Early attempts to simply replace QBrs with inexpensive alkali metal bromides were unsuccessful. CsBr showed the best, but still unsatisfactory, activity. For example, in combination with Pd and Co acetates, it produced just 0.73% DPC [66]. However, the GE team found that alkali metal bromides can be used in DPC catalytic packages in combination with activating solvents that promote dissociation of MBr and achieve the catalyst performance comparable with the performance of QBrs. The least expensive NaBr turned out to be the best choice. The most effective activating solvents were nitriles (acetonitrile) [67], aliphatic (tetraglyme (TG)) and cyclic (15-crown-5) ethers [68], sulfones (sulfolane) [69], and amides (N-methylpyrrolidone) [70]. The activity of catalytic packages containing NaBr–polyether increased in the order diglyme < triglyme < tetraglyme (TG) < polyethylene glycol dimethyl ether (PEGDME) < crown ethers [68a]. TG has the best practical combination of properties, as crown ethers are too expensive and PEGDME is difficult to recover. In large-scale runs, a Pd TON of 6145 was reached using the $Pd(acac)_2$–PbO–$TiO(acac)_2$–NaBr (1:50:11:650) system containing 7.2 vol% TG. Economic analysis showed that systems containing NaBr and MeCN or glymes delivered an acceptable cost. A combination of two solvents, one solvating Na^+ such as TG and second with high dielectric constant such as ethylene carbonate or dimethylacetamide used in a catalytic package containing $Pd(acac)_2$, NaBr, $TiO(acac)_2$, and $Cu(acac)_2$ or PbO, delivered 15–35% higher Pd TON compared with the same package with a single activating solvent [71].

A nonbromide catalytic package including Pd, Pb, Ti, and Bu_4NCl delivered Pd TON only about two times less compared with a bromide salt [62a]. Bu_4NCl can also be used with copper cocatalyst and Cu–Ti and Cu–Zr ICCs [72]. The presence of a base (NaOH or NaOPh) improves the performance of all catalytic packages with onium chlorides [72]. Bimetallic cocatalysts with Bu_4NCl demonstrated a twofold increase in the Pd TON compared with lead cocatalyst alone [55]. In contrast to bromide-based systems, the activity of chloride-based systems is not dependent on halide concentration. However, the increase of lead concentration from 12 to 50 equiv. versus Pd increases Pd TON from 1100 to 2800 with Ce cocatalyst [55]. A package including Pd/C, $Ce(trop)_4$, and CsCl gave 20.4% DPC in neat PhOH at 100° in 3 h [52]. NaCl activated by a polyether such as TG can be used with Cu–Ti, Cu–Zr, Pb–Ce, and Pb–Ti ICCs with Pd TON up to 5925 in large-scale runs (for the Pd–Cu–Ti–NaOH system) [73]. The activity order for alkali metals depends on the catalytic packages, for example, Li > Na > K > Cs for the Pb–Ti system but Cs ~ K > Na ~ Li for the Cu–Zr system [73].

A catalyst containing Pd(acac)$_2$ and Bu$_4$NI produced 26.7% DPC in 4 h under 100 bar with 99.1% selectivity [74]. Pd(acac)$_2$ in combination with Ce(acac)$_3$ or PbO catalyzed oxidative carbonylation of phenol in the presence of sodium iodide with Pd TON above 1000 [75]. In this case, bisphosphine ligands are necessary to maintain high regioselectivity of carbonylation reaction (96% vs. 72% for Pd(acac)$_2$) [55].

Several nonhalide catalytic packages have been developed using tetrabutylammonium benzoate. The Pd–PbO–TiO(acac)$_2$ and Pd–PbO–Cu(acac)$_2$ combinations in the presence of NaOH delivered Pd TONs of 1050 [76] and 1640 [77], respectively. The use of tetraalkylammonium alkyl- or arylsulfonates with Pb–Ti, Pb–Ce, and Pb–Mn ICCs increased Pd TON to 1500–2000, while Pb–Cu cocatalyst was less active [78]. Tetrabutylammonium salts can be replaced with NaNO$_3$ plus TG as an activating solvent. However, a base is essential for good catalyst performance increasing the Pd TON twofold [77, 79].

12.3 Prospects for Scale-Up

12.3.1 Catalyst Optimization

Optimization of lead-based DPC catalysts was performed using the HTS technique [12], enabling to carry out a multifactorial and multilevel DOE in a single run. The HTS results exhibited an excellent correlation with large-scale experiments [55]. The Pd TON reaches a maximum at a bromide concentration of 715 equiv. versus Pd and drops at high bromide concentrations, probably due to the formation of coordinatively saturated [PdBr$_4$]$^{2-}$ anion. The Pd TON grows with the lead concentration, but the practical limit of lead cocatalyst level is approximately 100 equiv. versus Pd due to growing heterogeneity of the reaction mixture. NEt$_4$Br and NBu$_4$Br (TBAB) as a bromide source showed similar behavior though TBAB-based formulations were approximately 15% more active [55]. Pd TON increases with lead concentration also in two-metal Pd–Pb–bromide catalytic packages but decreases with the concentration of a second cocatalyst. However, addition of a small amount of Ti or Ce is well justified because it improves reaction rate and eliminates the induction period characteristic of the lead-only package. The best-optimized catalytic packages contain about 100 equiv. (vs. Pd) lead, 2–4 equiv. of a second cocatalyst (Ti or Ce), and 600–800 equiv. of a QBr [55]. Addition of copper compounds and NaOH to the Pd–Pb catalyst (Pd:Pb:Cu:NaOH = 1 : 60 : 10 : 300) does not affect Pd TON significantly (7000–7760) but substantially improves the DPC/bromophenol ratio and overall DPC selectivity [80].

Addition of titanium salt to copper-containing catalytic packages allows for substantial increase in Pd TON [81] but requires a base for good performance [82]. Tetrabutylammonium chloride can be used instead of the bromide, although the catalyst activity is much less [83]. Combination of $Cu(acac)_2$ and $TiO(acac)_2$ demonstrated the highest activity [82]. The best results were obtained using low concentration of Pd (13 ppm) with the ratio Pd:Cu:Ti:NaOH:NaBr = 1 : 21 : 21 : 1127 : 477 in the presence of 7 wt% TG achieving 10 570 Pd TON and the lowest concentration of bromophenols (0.12 wt%) [84].

12.3.2
Water Removal in Direct Diphenyl Carbonate Process

As already mentioned, water removal in the oxidative carbonylation of phenols to make diaryl carbonates is a necessary process since it greatly enhances the productivity of the reaction and thus reduces reactor cost per unit mass of product. The use of 3 or 4 Å molecular sieves, effectively removing water in laboratory-scale runs [8d], is impractical at an industrial scale.

The use of an inert stripping agent such as pentane vapor for the removal of water from a direct DPC process was claimed [85]. Pentane was condensed to separate from the water stripping step and recycled back to the water stripping step [85].

About 2.5 times improvement in DPC yield for a batch process catalyzed by $PdBr_2$ in combination with $Mn(acac)_2$, TBAB, and NaOPh was achieved if the generated water was removed continuously from a partial flow of the reaction under reduced pressure at isothermal conditions, and the dehydrated partial flow is returned to the reaction [86]. It was proposed to perform water removal during continuous process by removing a portion of liquid stream from a reactor to a flash vessel subjecting to reduced pressure and returning a portion of dried liquid stream to the reactor [87]. The DPC yield obtained with this procedure was comparable with that obtained with the use of 3 Å molecular sieves as a desiccant.

Addition of a water removal train to the direct DPC process affects the total process capital and operational expenses. It is a challenging engineering problem that includes depressurization/repressurization cycles. It was shown that depressurization adversely affects the catalyst performance (about two times loss of Pd TON). Therefore, an intermediate disengaging vessel kept at the same pressure as the reactor was proposed that will increase the capital cost even more [87b, 88]. An ideal solution to this problem would be a water-resistant catalyst for phenol oxidative carbonylation with the water resistivity defined as the ratio of Pd TON without and with a desiccant equal to 1. Eventually, such a catalyst was developed at GE [89]. It was discovered that some metal halides including LiBr and $MgBr_2$ greatly improve the water resistivity of the DPC catalyst defined as the ratio of Pd TON without and with a desiccant [89]. For the catalytic package Pd–Cu–Ti–NaOH–NaBr–TG, the water resistivity was 0.87 for $MgBr_2$ ($Pd:MgBr_2$ = 1 : 100) and 0.98 for LiBr (Pd:LiBr = 1 : 350). Lithium ions can be introduced in the form of LiOH instead of NaOH to give water resistivity coefficient above 1 [89]. Replacement of NaOH with an organic base, NMe_4OH, or NEt_3 results in better Pd TON and selectivity in phenol carbonylation reactions without a desiccant or even with intentionally added water.

12.3.3
Downstream Processing and Catalyst Recovery

Economics dictates the necessity of effective recovery of an expensive catalyst and downstream processing. Basic downstream processing steps include the isolation of the target product, DPC, recovery of unreacted phenol and organic activating solvent, isolation, and reconstitution of Pd, metal cocatalyst, and bromide components, as well as cracking of organic by-products such as bromophenols and oligomeric compounds. Addition of an acid such as HBr, EDTA, or oxalic acid suppresses the decomposition of DPC stored in the reaction mixture [90]. Selection of the acid depends on the catalytic package used: HBr worked well for Pd–Pb–TEAB and Pd–Pb–Ti–NaOH–NaBr–TG packages but did not prevent the decomposition in the case of Pd–Pb–Cu–NaOH–NaBr–TG and Pd–Cu–Ti–NaOH–NaBr–TG. For the latter, citric acid was the best stabilizing agent [90]. DPC can be separated from the carbonylation reaction mixture as a 1:1 M DPC–PhOH adduct. An aqueous stream was used to precipitate and separate metal contaminants and produce purified DPC [91]. Phenol from the reaction mixture can be extracted with water and then with anisole at room temperature or with DPC at 85 °C [92]. The main metal removal steps from reaction mixtures using the Pd–Co–QBr package included palladium extraction with an aqueous acid such as HCl, cobalt extraction with an aqueous chelating agent such as trisodium nitrilotriacetate, and extraction with water to remove the bromide source [93]. Palladium was separated as Pd black after reduction with formic acid. Lead can be recovered from reaction mixtures with an efficiency of more than 99.5% by precipitation with oxalic acid followed by calcination of resulting lead oxalate at 600 °C back to PbO [94]. Tetraalkylammonium salts can be extracted with aqueous HBr with an efficiency of 79% either from reaction mixture or from a solid residue after the evaporation of organic products [95].

A set of consecutive extractions with aqueous HBr and TG allowed for the isolation of all catalyst components from the reaction mixture after the use of the Pd–Pb–Ti–NaOH–NaBr–TG catalytic package [96]. An aqueous extract of Pd and Co salts was treated with Na(acac) to form the $Pd(acac)_2$ precipitate, thus effectively separating these metals [97]. It was found that at PhOH concentration above 40%, all TG remains in an organic phase during the extraction with aqueous HCl or HBr. At the same time, all metal components are concentrated in an aqueous phase, which simplifies the recovery of catalyst components [98]. Extraction with a mixture of HBr and NaBr in water was the most effective [99].

12.4
Conclusions and Outlook

Recently, the research focus was shifted to heterogeneous Pd catalysts due to ease of separation from the reaction mixture. Magnetic perovskites such as $La_{1-x}Pb_xMnO_3$ ($x = 0.4$–0.5) with a higher concentration of oxygen vacancies were proposed as a support for heterogeneous palladium catalysts for DPC synthesis to operate in the magnetically stabilized bed reactor [100]. The use of ionic liquids such as 1-butyl-3-methylpyrrolidinium trifluoroacetate as an additive to a Pd–Mn base catalytic package in chlorobenzene resulted in a homogeneous reaction mixture without precipitation [101].

Recently, electrocarbonylation of phenol with CO to DPC was achieved in MeCN solution at ambient CO pressure using a heterogeneous Pd/C [102] and a homogeneous N-heterocyclic carbene Pd electrocatalyst, the activity of which increased with the electron-donating ability of the ligand (Eq. (12.4)) [103]:

$$2PhOH + CO \rightarrow (PhO)_2CO + H_2 \qquad (12.4)$$

Another trend is the application of the direct DPC process development findings to a direct polycarbonate process. Replacing PhOH with a bifunctional phenol in the direct DPC process opens a way for direct production of polycarbonates, thus eliminating an intermediate transesterification step. $PdBr_2$ and its complexes with substituted bipyridyls and phenanthrolines as well as dinuclear Pd complexes with pyridylphosphine ligand combined with $Mn(tmhd)_3$ or $Ce(tmhd)_4$, $(Ph_2P=)_2NBr$, and HQ produced PC with M_w in the range 2500–9500 and yield 46–86% (Pd TON up to 300) [104]. Interestingly, Pd catalysts with unsubstituted bipyridyl and phenanthroline ligands produced only dimers or trimers but not high-molecular-weight products. Reaction of BPA with CO and O_2 catalyzed by Pd carbene complexes with $Ce(tmhd)_4/HQ/^nBu_4PBr$ cocatalyst in CH_2Cl_2 under optimum conditions developed for DPC synthesis generated PC with 80% yield (Pd TON = 133). These polymer properties ($M_w = 24\,000$, $M_n = 9400$) are comparable with industrial requirements ($M_w > 10\,000$) [105]. Oxidative carbonylation of BPA using a palladium catalyst anchored to a resin via a heterocyclic carbene ligand resulted in the preparation of PC with high molecular weight ($M_w = 22\,400$) at 95% yield [38].

The patent activity in direct DPC process decreased in the last several years. Competitive processes such as two-step process via the intermediate formation of cyclic carbonates by reaction with CO_2 are under evaluation [106]. Nevertheless, owing to the excellent progress made, the direct DPC process is a viable cost-competitive option, especially in cases when the use of phosgene is prohibited. A continuous process developed by GE [18, 54] and Bayer [107] seems to be more cost effective than a traditional batch process. High achieved Pd TONs (>10 000) and efficient recovery of Pd greatly reduced its impact on the DPC cost, though the high cost of Pd still requires a substantial capital investment in Pd metal. All key steps in downstream processing have been tested and verified. Still, process optimization and verification at larger scale at a pilot plant facility are necessary for the practical implementation of the direct DPC process.

Acknowledgments

The author thanks the members of the GE team who worked with him on the direct DPC process: Dick Battista, Peter Bonitatibus Jr., Michael Brennan, James

Cawse, Timothy Chuck, Yan Gao, Marsha Grade, Bruce Johnson, Tracey Jordan, Richard Kilmer, Jonathan Male, Bahram Moasser, Phil Moreno, Ben Patel, Eric Pressman, John Ofori, Kirill Shalyaev, David Smith, James Spivack, Ignacio Vic, Donald Whisenhunt Jr., and Eric Williams. The author also thanks Thomas Miebach and Gary Yeager for their help with the manuscript preparation.

References

1. Bendler, J.T. (1999) *Handbook of Polycarbonate Science and Technology*, Taylor & Francis.
2. Coe, J.T. (2010) *Unlikely Victory: How General Electric Succeeded in the Chemical Industry*, John Wiley & Sons, Inc.
3. LeGrand, D.G. and Bendler, J.T. (2000) *Handbook of Polycarbonate Science and Technology, Plastics Engineering*, Marcel Dekker, Inc., New York.
4. Gong, J., Ma, X., and Wang, S. (2007) *Appl. Catal., A*, **316**, 1–21.
5. Chalk, A.J. (1978) US Patent 4096169, General Electric Co.
6. Chalk, A.J. (1980) US Patent 4187242, General Electric Co.
7. Hallgren, J.E. and Matthews, R.O. (1979) *J. Organomet. Chem.*, **175**, 135–142.
8. (a) Hallgren, J.E. (1980) US Patent 4201721, General Electric Co.; (b) Hallgren, J.E. (1981) US Patent 4260802, General Electric Co.; (c) Joyce, R.P., King, J.A. Jr., and Pressman, E.J. (1993) US Patent 5231210, General Electric Co.; (d) King, J.A., MacKenzie, P.D., and Pressman, E.J. (1995) US Patent 5399734, General Electric Co.; (e) Pressman, E.J. and King, J.A. Jr. (1994) US Patent 5284964, General Electric Co.
9. Takagi, M., Miyagi, H., Ohgomori, Y., and Iwane, H. (1996) US Patent 5498789, Mitsubishi Chemical Corp.
10. Hanak, J.J. (1970) *J. Mater. Sci.*, **5**, 964–971.
11. Whisenhunt, D.W. Jr. and Soloveichik, G. (2007) *Combinatorial and High-Throughput Discovery and Optimization of Catalysts and Materials*, CRC Press LLC, pp. 129–148.
12. Spivack, J.L., Cawse, J.N., Whisenhunt, D.W., Johnson, B.F., Shalyaev, K.V., Male, J., Pressman, E.J., Ofori, J.Y., Soloveichik, G.L., Patel, B.P., Chuck, T.L., Smith, D.J., Jordan, T.M., Brennan, M.R., Kilmer, R.J., and Williams, E.D. (2003) *Appl. Catal., A*, **254**, 5–25.
13. Yasuda, H., Maki, N., Choi, J.-C., and Sakakura, T. (2003) *J. Organomet. Chem.*, **682**, 66–72.
14. Zlochower, I. and Green, G. (2009) *J. Loss Prev. Process Ind.*, **22**, 499–505.
15. Yasuda, H., Watarai, K., Choi, J.-C., and Sakakura, T. (2005) *J. Mol. Catal. A: Chem.*, **236**, 149–155.
16. Hallgren, J.E., Lucas, G.M., and Matthews, R.O. (1981) *J. Organomet. Chem.*, **204**, 135–138.
17. Mizukami, M., Hayashi, K., Iura, K., and Kawaki, T. (1955) US Patent 5380907 Mitsubishi Gas Chemical Co.
18. Pressman, E.J., Johnson, B.F., Moreno, P.O., and Battista, R.A. (2001) US Patent 6191299, General Electric Co.
19. Yasuda, H., Choi, J.-C., Lee, S.-C., and Sakakura, T. (2002) *Organometallics*, **21**, 1216–1220.
20. Ronchin, L., Vavasori, A., Amadio, E., Cavinato, G., and Toniolo, L. (2009) *J. Mol. Catal. A: Chem.*, **298**, 23–30.
21. Vavasori, A. and Toniolo, L. (2000) *J. Mol. Catal. A: Chem.*, **151**, 37–45.
22. Soloveichik, G.L., Patel, B.P., Ofori, J.Y., and Shalyaev, K.V. (2002) US Patent 6407027, General Electric Co.
23. Ishii, H., Ueda, M., Takeuchi, K., and Asai, M. (1999) *J. Mol. Catal. A: Chem.*, **144**, 477–480.
24. Ishii, H., Goyal, M., Ueda, M., Takeuchi, K., and Asai, M. (1999) *J. Mol. Catal. A: Chem.*, **148**, 289–293.
25. Ishii, H., Ueda, M., Takeuchi, K., and Asai, M. (1999) *J. Mol. Catal. A: Chem.*, **144**, 369–372.

26. Ishii, H., Ueda, M., Takeuchi, K., and Asai, M. (1999) *J. Mol. Catal. A: Chem.*, **138**, 311–313.
27. Okuyama, K.-I., Sugiyama, J.-I., Nagahata, R., Asai, M., Ueda, M., and Takeuchi, K. (2003) *J. Mol. Catal. A: Chem.*, **203**, 21–27.
28. Moiseev, I.I., Vargaftik, M.N., Chernysheva, T.V., Stromnova, T.A., Gekhman, A.E., Tsirkov, G.A., and Makhlina, A.M. (1996) *J. Mol. Catal. A: Chem.*, **108**, 77–85.
29. Kim, W.B., Park, E.D., and Lee, J.S. (2003) *Appl. Catal., A*, **242**, 335–345.
30. Xue, W., Zhang, J., Wang, Y., Zhao, X., and Zhao, Q. (2005) *Catal. Commun.*, **6**, 431–436.
31. Takagi, M., Miyagi, H., Yoneyama, T., and Ohgomori, Y. (1998) *J. Mol. Catal. A: Chem.*, **129**, L1–L3.
32. Song, H.Y., Park, E.D., and Lee, J.S. (2000) *J. Mol. Catal. A: Chem.*, **154**, 243–250.
33. Ishii, H., Takeuchi, K., Asai, M., and Ueda, M. (2001) *Catal. Commun.*, **2**, 145–150.
34. Chaudhari, R.V., Gupte, S.P., Kanagasabapathy, S., Kelkar, A.A., and Radhakrishnan, S. (2003) US Patent 5917077, General Electric Co.
35. Zhang, G., Wu, Y., Ma, P., Tian, Q., Wu, G., and Li, D. (2004) *Chin. J. Chem. Eng.*, **12**, 191–195.
36. Guo, H., Chen, H., Liang, Y., Rui, Y., Lü, J., and Fu, Z. (2008) *Chin. J. Chem. Eng.*, **16**, 223–227.
37. Guo, H.X., Lü, J.D., Wu, H.Q., Xiao, S.J., and Han, J. (2013) *Adv. Mater. Res.*, **750–752**, 1292–1295.
38. Okuyama, K.-I., Sugiyama, J.-I., Nagahata, R., Asai, M., Ueda, M., and Takeuchi, K. (2003) *Green Chem.*, **5**, 563–566.
39. (a) Patel, B.P., Soloveichik, G.L., Whisenhunt, D.W., and Shalyaev, K.V. (2001) US Patent 6187942, General Electric Co.; (b) Patel, B.P., Soloveichik, G.L., Whisenhunt, D.W., and Shalyaev, K.V. (2002) US Patent 6355824, General Electric Co.
40. (a) Patel, B.P., Soloveichik, G.L., Whisenhunt, D.W. Jr., and Shalyaev, K.V. (2001) US Patent 6175033, General Electric Co.; (b) Patel, B.P., Soloveichik, G.L., Whisenhunt, D.W., and Shalyaev, K.V. (2002) US Patent 6380418, General Electric Co.
41. (a) Patel, B.P., Soloveichik, G.L., Whisenhunt, D.W. Jr., and Shalyaev, K.V. (2001) US Patent 6175032, General Electric Co.; (b) Patel, B.P., Soloveichik, G.L., Whisenhunt, D.W., and Shalyaev, K.V. (2001) US Patent 6323358, General Electric Co.
42. (a) Patel, B.P., Soloveichik, G.L., Whisenhunt, D.W. Jr., and Shalyaev, K.V. (2001) US Patent 6184409, General Electric Co.; (b) Patel, B.P., Soloveichik, G.L., Whisenhunt, D.W., and Shalyaev, K.V. (2003) US Patent 6509489, General Electric Co.
43. (a) Chang, T.C.T. (1990) EP Patent 350700, General Electric Co.; (b) Chang, T.C.T. (1990) EP Patent 350697, General Electric Co.
44. Wang, Y., Xue, W., and Zhao, X. (2006) Hebei University of Technology, People's Republic of China, CN Patent 1775734.
45. Vavasori, A. and Toniolo, L. (1999) *J. Mol. Catal. A: Chem.*, **149**, 321.
46. Lapidus, A.L., Sukhov, V.V., and Eliseev, O.L. (2000) DGMK Tagungsbericht 2000–3, pp. 221–226.
47. Pressman, E.J. and King, J.A. (1994) US Patent 5284964, General Electric Co.
48. Huang, Z.W., Chang, C.W., and Tsai, C.J. (2012) US Patent 8212066, China Petrochemical Development Corporation.
49. Lapidus, A.L., Pirozhkov, S.D., and Sukhov, V.V. (1999) *Russ. Chem. Bull.*, **48**, 1113–1117.
50. Goyal, M., Nagahata, R., Sugiyama, J.-I., Asai, M., Ueda, M., and Takeuchi, K. (1998) *Catal. Lett.*, **54**, 29–31.
51. Goyal, M., Nagahata, R., Sugiyama, J.-I., Asai, M., Ueda, M., and Takeuchi, K. (1999) *J. Mol. Catal. A: Chem.*, **137**, 147–154.
52. Ookago, J., Hayashi, H., Myagi, H., Kujira, K., Takagi, M., and Suzuki, N. (1996) JP Patent 08193056, Mitsubishi Chemical Corp.
53. Pressman, E.J. and Shafer, S.J. (1999) US Patent 5898079, General Electric Co.

54. Moreno, P. (2000) US Patent 6034262, General Electric Co.
55. Soloveichik, G.L., Shalyaev, K.V., Patel, B.P., Gao, Y., and Pressman, E.J. (2005) *Chem. Ind. (Boca Raton, FL)*, **104**, 185–194.
56. Yanovsky, A.I., Turova, N.Y., Turevskaya, E.P., and Struchkov, Y.T. (1982) *Koord. Khim.*, **8**, 153.
57. Gaffney, C., Harrison, P.G., and King, T.J. (1980) *J. Chem. Soc., Chem. Commun.*, 1251–1252.
58. Soloveichik, G.L., Patel, B.P., Ofori, J.Y., and Shalyaev, K.V. (2001) US Patent 6245929, General Electric Co.
59. (a) Spivack, J.L., Whisenhunt, D.W., Cawse, J.N., Johnson, B.F., Soloveichik, G.L., Ofori, J.Y., and Pressman, E.J. (2000) US Patent 6143914, General Electric Co.; (b) Spivack, J.L., Cawse, J.N., Whisenhunt, D.W. Jr., Johnson, B.F., and Soloveichik, G.L. (2000) WO Patent 2000066530, General Electric Co.
60. Spivack, J.L., Whisenhunt, D.W., Cawse, J.N., and Soloveichik, G.L. (2003) US Patent 6566299, General Electric Co.
61. Spivack, J.L., Cawse, J.N., Whisenhunt, D.W., Johnson, B.F., and Soloveichik, G.L. (2003) US Patent 6514900, General Electric Co.
62. (a) Spivack, J.L., Whisenhunt, D.W., Cawse, J.N., Johnson, B.F., Soloveichik, G.L., Ofori, J.Y., and Pressman, E.J. (2002) US Patent 6420587, General Electric Co.; (b) Spivack, J.L., Whisenhunt, D.W., Cawse, J.N., Johnson, B.F., Soloveichik, G.L., Ofori, J.Y., and Pressman, E.J. (2001) US Patent 6197991, General Electric Co.
63. Spivack, J.L., Whisenhunt, D.W., Cawse, J.N., Johnson, B.F., Grade, M.M., Soloveichik, G.L., Ofori, J.Y., and Pressman, E.J. (2000) US Patent 6160154, General Electric Co.
64. Spivack, J.L., Whisenhunt, D.W., Cawse, J.N., Johnson, B.F., Soloveichik, G.L., Ofori, J.Y., and Pressman, E.J. (2002) US Patent 6380417, General Electric Co.
65. Vavasori, A. and Toniolo, L. (1999) *J. Mol. Catal. A: Chem.*, **139**, 109–119.
66. Kiso, Y., Nagata, T., Fujita, T., and Iwasaki, H. (1993) JP Patent 05058961, Mitsui Petrochemical Industries, Co.
67. (a) Pressman, E.J., Soloveichik, G.L., Johnson, B.F., and Shalyaev, K.V. (2002) US Patent 6403821, General Electric Co.; (b) Pressman, E.J., Soloveichik, G.L., Johnson, B.F., and Shalyaev, K.V. (2001) US Patent 6172254, General Electric Co.
68. (a) Pressman, E.J., Soloveichik, G.L., Shalyaev, K.V., and Johnson, B.F. (2000) US Patent 6114564, General Electric Co.; (b) Pressman, E.J., Soloveichik, G.L., Shalyaev, K.V., and Johnson, B.F. (2002) US Patent 6346500, General Electric Co.
69. Johnson, B.F., Shalyaev, K.V., Soloveichik, G.L., and Pressman, E.J. (2001) US Patent 6265340, General Electric Co.
70. (a) Johnson, B.F., Soloveichik, G.L., Pressman, E.J., and Shalyaev, K.V. (2001) US Patent 6180812, General Electric Co.; (b) Johnson, B.F., Soloveichik, G.L., Pressman, E.J., and Shalyaev, K.V. (2002) US Patent 6346499, General Electric Co.
71. Soloveichik, G.L. (2005) US Patent 6903049, General Electric Co.
72. Shalyaev, K.V., Soloveichik, G.L., Johnson, B.F., and Whisenhunt, D.W. (2002) US Patent 6372683, General Electric Co.
73. Shalyaev, K.V., Soloveichik, G.L., Johnson, B.F., and Whisenhunt, D.W. (2002) US Patent 6365538, General Electric Co.
74. Fukuoka, S., Ogawa, H., and Watanabe, T. (1989) JP Patent 01165551, Asahi Chemical Industry Co.
75. Patel, B.P., Soloveichik, G.L., and Ofori, J.Y. (2003) US Patent 6617279, General Electric Co.
76. Shalyaev, K.V., Johnson, B.F., Whisenhunt, D.W., and Soloveichik, G.L. (2004) US Patent 6700008, General Electric Co.
77. Shalyaev, K.V., Soloveichik, G.L., Whisenhunt, D.W., and Johnson, B.F. (2004) US Patent 6706908, General Electric Co.
78. Shalyaev, K.V., Johnson, B.F., Whisenhunt, D.W. Jr., and Soloveichik,

G.L. (2002) US Patent 6440893, General Electric Co.
79. Shalyaev, K.V., Soloveichik, G.L., Whisenhunt, D.W. Jr., and Johnson, B.F. (2002) US Patent 6440892, General Electric Co.
80. Soloveichik, G.L., Shalyaev, K.V., Grade, M.M., and Johnson, B.F. (2004) US Patent 6700009, General Electric Co.
81. Spivack, J.L., Cawse, J.N., Whisenhunt, D.W., Johnson, B.F., and Soloveichik, G.L. (2001) US Patent 6201146, General Electric Co.
82. Shalyaev, K.V., Soloveichik, G.L., and Johnson, B.F. (2003) US Patent 6512134, General Electric Co.
83. Shalyaev, K.V., Soloveichik, G.L., Johnson, B.F., and Whisenhunt, D.W. Jr. (2001) US Patent 6207849, General Electric Co.
84. Shalyaev, K.V., Soloveichik, G.L., and Johnson, B.F. (2003) US Patent 6566295, General Electric Co.
85. Battista, R.A., Lo, F.S., and Tatterson, R.L. (1999) US Patent 5917078, General Electric Co.
86. Buysch, H.J., Hesse, C., and Rechner, J. (1997) US Patent 5625091, Bayer Aktiengesellschaft.
87. (a) Ofori, J.Y., Pressman, E.J., Shalyaev, K.V., Williams, E.D., and Battista, R.A. (2002) US Patent 6384262, General Electric Co.; (b) Ofori, J.Y., Pressman, E.J., Shalyaev, K.V., Williams, E.D., and Battista, R.A. (2002) US Patent 6472551, General Electric Co.
88. Ofori, J.Y., Pressman, E.J., Shalyaev, K.V., Williams, E.D., and Battista, R.A. (2003) US Patent 6521777, General Electric Co.
89. Soloveichik, G.L., Chuck, T.L., Shalyaev, K.V., Pressman, E.J., and Bonitatebus, P.J. (2006) US Patent 7084291, General Electric Co.
90. Shalyaev, K.V., Pressman, E.J., Hallgren, J.E., Ofori, J.Y., and Male, J.L. (2002) US Patent 6441215, General Electric Co.
91. Fernandez, I.V., Lyakhovych, M., and Nadal, S.F. (2014) US Patent 20140275473, Sabic Innovative Plastics.
92. Ofori, J.Y. (2000) US Patent 6143937, General Electric Co.
93. Ofori, J.Y., Shafer, S.J., Pressman, E.J., Kailasam, G., and Lee, J.L. (1999) US Patent 5981788, General Electric Co.
94. Ofori, J.Y. (2000) US Patent 6090737, General Electric Co.
95. Ofori, J.Y., Pressman, E.J., Patel, B.P., Moreno, P.O., and Battista, R.A. (2001) US Patent 6310232, General Electric Co.
96. Ofori, J.Y. and Bonitatebus, P.J. (2004) US Patent 6683015, General Electric Co.
97. Ofori, J.Y. (2001) US Patent 6191060, General Electric Co.
98. Grade, M.M., Ofori, J.Y., and Pressman, E.J. (2002) US Patent 6410774, General Electric Co.
99. Grade, M.M., Ofori, J.Y., and Pressman, E.J. (2003) US Patent 6506924, General Electric Co.
100. Lu, W., Du, Z., Yuan, H., Tian, Q., and Wu, Y. (2013) *Chin. J. Chem. Eng.*, **21**, 8–13.
101. Böwing, A.G., Wolf, A., Tellmann, K., and Mleczko, L. (2013) US Patent 8507712, Bayer Technology Gmbh.
102. Murayama, T., Hayashi, T., Kanega, R., and Yamanaka, I. (2012) *J. Phys. Chem. C*, **116**, 10607–10616.
103. Kanega, R., Hayashi, T., and Yamanaka, I. (2013) *ACS Catal.*, **3**, 389–392.
104. Ishii, H., Goyal, M., Ueda, M., Takeuchi, K., and Asai, M. (2001) *Macromol. Rapid Commun.*, **22**, 376–381.
105. Okuyama, K.-I., Sugiyama, J.-I., Nagahata, R., Asai, M., Ueda, M., and Takeuchi, K. (2003) *Macromolecules*, **36**, 6953–6955.
106. Ryu, J.Y. (2012) US Patent 8110698, Shell Oil Co.
107. Buysch, H.-J., Hesse, C., and Rechner, J. (1997) EP Patent 801051, Bayer A.-G.

13
Aerobic Oxidative Esterification of Aldehydes with Alcohols: The Evolution from Pd–Pb Intermetallic Catalysts to Au–NiO$_x$ Nanoparticle Catalysts for the Production of Methyl Methacrylate

Ken Suzuki and Setsuo Yamamatsu

13.1
Introduction

Esters are very useful chemical intermediates, in terms of atom economy and versatility, and can be helpful in further transformations. Esterification is one of the fundamental transformations in organic synthesis and is widely used in laboratories and industry [1]. Oxidative esterification of aldehydes with alcohols is an attractive method for the synthesis of esters because aldehydes are readily available raw materials on a commercial scale. Although several facile and selective esterification reactions have been reported [2], the development of a catalytic method for the direct oxidative esterification of aldehydes with alcohols under mild and neutral conditions in the presence of molecular oxygen as the terminal oxidant is highly desirable from both economic and environmental aspects.

As an example, the aerobic catalytic esterification of methacrolein with methanol to form methyl methacrylate (MMA) was investigated under neutral conditions. The monomer MMA is mainly used to produce acrylic plastics such as poly(methyl methacrylate) (PMMA) and other polymer dispersions used in paints and coatings. MMA can be manufactured in numerous ways from C_2–C_4 hydrocarbon feedstocks [3]. Currently, MMA is mainly produced via the acetone cyanohydrin (ACH) process, but there are problems in handling the resulting ammonium bisulfate waste and toxic hydrogen cyanide. Some manufacturers use isobutene or *tert*-butyl alcohol (TBA) as the starting material, which is sequentially oxidized first to methacrolein and then to methacrylic acid, which in turn is esterified with methanol. However, the existing synthetic methods still suffer from several disadvantages. Therefore, the development of an efficient and highly selective catalytic system based on the above mentioned reaction remains a challenge. In this chapter, we report a highly selective and efficient catalytic method for the oxidative esterification of methacrolein (**1**) in

---

Liquid Phase Aerobic Oxidation Catalysis: Industrial Applications and Academic Perspectives,
First Edition. Edited by Shannon S. Stahl and Paul L. Alsters.
© 2016 Wiley-VCH Verlag GmbH & Co. KGaA. Published 2016 by Wiley-VCH Verlag GmbH & Co. KGaA.

methanol to MMA (**2**) that employs supported Pd–Pb intermetallic compounds as the catalyst and molecular oxygen as the terminal oxidant (Eq. (13.1)) [4]. This catalytic system was commercialized in 1999 by Asahi Kasei. Thus, isobutene (or its precursor TBA) was oxidized in the gas phase using a Mo–Bi catalyst to synthesize methacrolein. Subsequent liquid phase aerobic oxidative esterification of methacrolein in the presence of methanol using the Pd–Pb catalyst produced MMA. This method results in a high yield of MMA and uses the raw material very efficiently. Furthermore, the present process has evolved greatly with the development of Au–NiO$_x$ nanoparticle catalysts.

$$\underset{1}{\text{CH}_2=\text{C(CH}_3)\text{CHO}} + \text{CH}_3\text{OH} \xrightarrow[\text{O}_2]{\text{Pd–Pb (cat.)}} \underset{2}{\text{CH}_2=\text{C(CH}_3)\text{COOCH}_3} \qquad (13.1)$$

13.2
Chemistry and Catalysis

13.2.1
Discovery of the Pd–Pb Catalyst

In the 1970s, it was already known that oxidative esterification of methacrolein with methanol and hydrogen peroxide, in the presence of a strong acid, and using Pd metal as a catalyst would result in a high yield of MMA [5]. It is presumed that this reaction proceeds via a hemiacetal intermediate. On the other hand, when aerobic oxidation is used, the yield of MMA could not surpass 30%. This is because, unlike oxidation with hydrogen peroxide, decarbonylation tends to occur as a side reaction. It is presumed that, when oxidizing in air, the reaction proceeds via an acyl palladium intermediate. In order to improve the oxidative esterification of methacrolein under aerobic oxidation to a practical level, the acyl palladium intermediate must be stabilized by suppressing the decarbonylation as a side reaction.

When by chance a catalyst consisting of Pd metal was contaminated with Pb (i.e., a Lindlar catalyst) and it was tested in this reaction, it was discovered that decarbonylation was effectively suppressed. Taking advantage of this fortuitous discovery, binary catalysts consisting of a combination of Pd metal with various elements were widely examined, and it was discovered that a catalyst system consisting of a combination of Pd with various sixth period elements of the periodic table – Hg, Tl, Pb, and Bi – was superior. Figure 13.1 shows an example of the reaction performance when metallic Pd is used as a catalyst in combination with Pb(II) acetate as a Pb species.

13.2.2
Pd–Pb Intermetallic Compounds

As the study progressed, it was noted that Pd is capable to form intermetallic compounds such as Pd_3Pb_1 and Pd_3Tl_1 (i.e., Pd_3M_1, M = Hg, Tl, Pb, Bi). In addition,

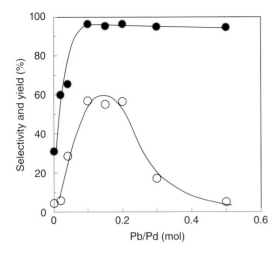

Figure 13.1 Effect of lead on palladium catalyst. Symbols: (●) **2** Selectivity and (○) **2** Yield. Reaction conditions: **1** (47 mmol), catalyst (5% Pd/CaCO$_3$) in methanol (90 ml), and O$_2$ (1 atm, 5 l/h) at 40 °C for 2 h.

it was also noted that these intermetallic compounds, with a strictly controlled atomic ratio, function as high-quality catalysts for the oxidative esterification of methacrolein. In particular, Pd$_3$Pb$_1$ was determined to be the optimum catalyst for the reaction. It is clear from the phase diagram that several chemical species exist among the Pd–Pb intermetallic compounds. However, if the oxidative esterification reaction is conducted with the addition of a lead salt to the Pd metal catalyst (Figure 13.1), or if the Pd and Pb salts are in the liquid phase, reduced with formalin, and so on, Pd$_3$Pb$_1$ is selectively produced. The presumed mechanism is shown in Figure 13.2.

It is important to carefully control and manage the structure of intermetallic compounds, specifically, to leave no unsubstituted Pd in sites of the Pd$_3$Pb$_1$ structure that should be substituted by Pb and not allowing Pb species other than Pd$_3$Pb$_1$ to be present in the catalyst system. The Pb–Pd ratio of approximately 0.15 in Figure 13.1 suggests that about half of Pd has not changed to the Pd$_3$Pb$_1$ structure in these experiments. Based on these two fundamental thoughts, essential

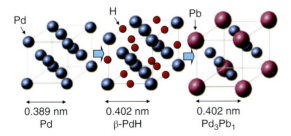

Figure 13.2 Presumed mechanism for synthesis of Pd$_3$Pb$_1$.

technologies for industrialization, such as those for catalyst preparation and control of the optimum conditions for the active species, were developed. The discovery of the Pd–Pb intermetallic catalyst greatly accelerated the industrialization of the oxidative esterification of methacrolein by aerobic oxidation.

13.2.3
Mechanism

Based on the analysis of the reaction kinetics, the mechanism of ester formation from the acyl palladium intermediate and methanol can be assumed (Figure 13.3). Methyl formate (MF) is a by-product of the formation of a formyl palladium intermediate through removal of the α-hydrogen of the methyl group in methanol. Since oxygen is involved in the removal of the α-hydrogen of the methyl group, production of MF – a by-product – can be suppressed by limiting the partial pressure of oxygen. Reducing the conversion rate of methacrolein to a low level is also effective for further suppressing the production of MF as a by-product. This is because the aldehyde adsorbs onto the active sites of the catalyst more strongly than the alcohol (ka > kb; Figure 13.3), and the reaction that produces formaldehyde from methanol is suppressed.

The reaction mechanism that proceeds via the hemiacetal is considered less likely since the Pd–Pb-catalyzed oxidative esterification proceeds in neutral to basic conditions rather than in acidic conditions, and an alkali is also integrated as the third component of the catalyst; thus, a mechanism involving an acyl palladium intermediate is favored. The reaction mechanism in Figure 13.3 rationalizes the observation that the presence of aldehydes suppresses the production of MF. It is presumed that alkali is effective in increasing the adsorption equilibrium of the alcohol.

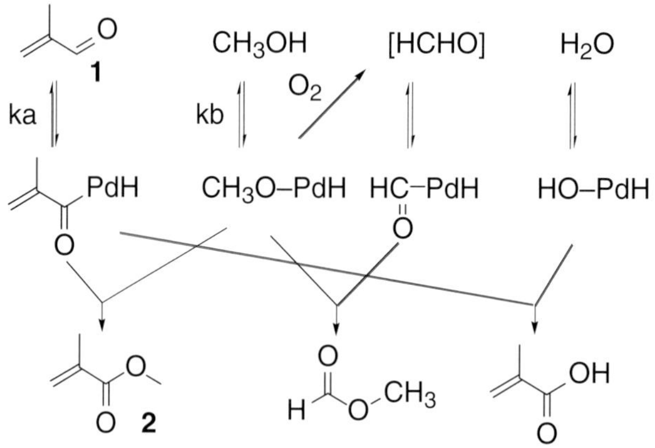

Figure 13.3 Reaction network in oxidative esterification reaction of methacrolein.

13.2.4
The Role of Pb in the Pd–Pb Catalyst

Pb suppresses decarbonylation by stabilizing the acyl palladium intermediate. On the other hand, it was discovered that Pb has an adverse effect that induces production of MF as a by-product, as it speeds up the reaction that produces formaldehyde from methanol. This is the reason that the MMA yield decreases as the ratio of Pb increases (Figure 13.1). Therefore, in order to suppress both decarbonylation and the production of MF as a by-product, it is important to use Pd_3Pb_1 as the catalytic species, with its atomic ratio precisely controlled at 3 : 1 Pd:Pb, and to ensure that there are no Pb species other than Pd_3Pb_1 present in the catalyst system. To use intermetallic compounds correctly as catalysts, development of a control technology to maintain the optimum conditions for formation of the active species during catalyst preparation was necessary.

13.2.5
Industrial Catalyst

Ultimately, a supported catalyst consisting of Pd_3Pb_1 with its support position and thickness accurately controlled was developed as an industrial catalyst (Figure 13.4). By using such a structure, the total consumption of expensive Pd can be reduced. The thickness of the Pd_3Pb_1 support was determined based on in-depth studies. A carrier with a high mechanical strength was developed, which remains sufficiently resistant to vigorous stirring in suspension. In addition, it should be noted that the Pd_3Pb_1 support position was submerged into the carrier at less than 1 μm from the surface. The carrier surface without Pd becomes a protective layer that affords the carrier with high mechanical strength and greatly reduces the loss of the Pd component to the reaction liquid. Even when the catalyst particles collide violently with each other in suspension and the carrier of the catalyst is worn and peeling, there is no concern regarding Pd loss for this catalyst.

This catalyst system has a weakness in that it does not completely suppress the production of MF as a by-product. However, as the first industrial catalyst for

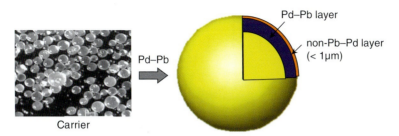

Figure 13.4 Precise control of Pd and Pb distribution in the catalyst.

aerobic oxidative esterification, its technological assets introduced in this chapter are clearly carried over to the Au–NiO$_x$ catalyst discussed in the latter part of this chapter.

13.3
Process Technology

A block diagram of this process is shown in Figure 13.5. Methacrolein synthesis makes use of a fixed-bed reactor to run a gas-phase reaction using a Mo–Bi system catalyst at 300–450 °C. In the reactor, in order to prevent explosions, inert gas is fed in along with the raw materials, TBA/air, in order to lower the concentration of oxygen. By efficiently absorbing methacrolein in the gas with methanol at the reactor outlet in the methacrolein absorption process, the loss is suppressed at the same time as obtaining the feed liquid for MMA synthesis. MMA synthesis is carried out in the presence of a Pd–Pb catalyst as air is passed through methacrolein/methanol (60–80 °C, 1 atm). This reaction is relatively facile and has a high selectivity under an excess of methanol. However, economically, if the utility cost of the recovery system is considered, excess methanol must be cut down. Thus, in order to satisfy both high activity and selectivity under the lowest molar ratio of methanol/methacrolein, a precise synthetic technology for this catalyst is being completed and industrialized. In a methacrolein recovery process, unreacted methacrolein and methanol are effectively separated from rough MMA through azeotropic distillation and are returned to the MMA synthesis reactor. In this process, due to the side reactions of methacrolein and methanol, a small amount of acetal and methoxy compounds is produced as by-products. In the purification process, acetal was decomposed, and the methoxy compound was separated and removed as a high-boiling product.

Since this process is simple, the plant construction cost is low, and the MMA yield is high; thus, the cost/benefit is high, and it can be said that it is a process

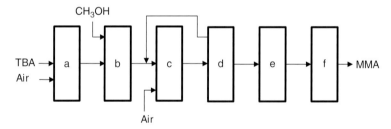

Figure 13.5 The block flow diagram of oxidative esterification process for MMA. (a) TBA oxidation for methacrolein synthesis. (b) Methacrolein absorption by methanol. (c) Oxidative esterification for MMA synthesis. (d) Recovery column for unreacted methacrolein and methanol and removal of low-boiling-point products. (e) High-boiling-point products' separation tower. (f) MMA purification tower.

that can fulfill both resource and energy conservation. Asahi Kasei built the new plant in 1998 and started the production at 60 000 t/year in January 1999.

13.4 New Developments

The discovery of the Pd–Pb catalyst was an important milestone in the aerobic oxidative esterification of aldehydes, as it put forth a clean and efficient method for forming carboxylic esters. However, these methods still suffer from several disadvantages; methods for successful catalytic oxidative esterification are limited as the selective oxidation of methacrolein is extremely difficult because of the instability of α,β-unsaturated aldehydes. Recently, we have developed a highly selective and efficient catalytic method for the aerobic oxidative esterification of methacrolein in methanol to MMA that employs supported gold–nickel oxide (Au–NiO$_x$) nanoparticles as the catalyst [6]. The Au–NiO$_x$ nanoparticles have a core–shell structure, with Au nanoparticles at the core and the surface covered by highly oxidized NiO$_x$. They are supported on a carrier with high dispersion. This novel bimetallic nanoparticulate structure provides superior catalytic performance compared with monometallic nanoparticles. Furthermore, we established the industrial catalytic technology with a long catalyst life by developing a high-strength silica-based carrier, using NiO$_x$ to improve chemical stability, and precisely controlling the distribution of the Au–NiO$_x$ nanoparticles in the catalysts.

Table 13.1 summarizes the activity of various catalysts used in the aerobic esterification of methacrolein **1** with methanol. Molecular oxygen diluted with nitrogen (O$_2$/N$_2$ = 7/93 v/v, 3 MPa), which corresponds to air diluted with molecular nitrogen, was used outside flammability limits with an autoclave. In industry, pure oxygen cannot be used, and air diluted with molecular nitrogen is used. This is such a case. This method is convenient for a large-scale closed system and also for a flow system. Oxidation with molecular oxygen (1 atm, balloon) gave excellent results; therefore, this method is convenient for laboratory organic synthesis.

When Pd/SiO$_2$–Al$_2$O$_3$ was used, MMA **2** could not be obtained in satisfactory yields (entry 1) because the decarbonylation of **1** resulted in the formation of large amounts of propylene and CO$_2$ as by-products. When Pd–Pb/SiO$_2$–Al$_2$O$_3$ was used, decarbonylation was inhibited, and the selectivity for **2** was improved to 84% (entry 2). During oxidative esterification in the presence of methanol, the excess methanol is oxidized to form MF as a by-product (0.2 mol of MF per mole of **2**). The turnover number (TON) was determined to be 61. The reaction of **1** in the presence of Au–NiO$_x$/SiO$_2$–Al$_2$O$_3$–MgO gave **2** with 98% selectivity and 58% conversion (entry 3). The TON of the supported nanoparticle catalyst was determined to be 621, and its activity was approximately 10 times that of the Pd–Pb catalyst. Moreover, a reduced by-product (MF) formation was observed in this case (0.007 mol of MF formed per mole of **2**). The catalyst supported with only Au nanoparticles showed lower activity and selectivity than the supported

Table 13.1 Catalytic activity for aerobic oxidative esterification of methacrolein (1) with methanol to form methyl methacrylate (2).[a]

Entry	Catalyst	Conversion of aldehyde 1 (%)[b]	Selectivity for ester 2 (%)[b]
1[c]	Pd/SiO$_2$–Al$_2$O$_3$	20	40
2[c]	Pd–Pb/SiO$_2$–Al$_2$O$_3$	34	84
3	Au–NiO$_x$/SiO$_2$–Al$_2$O$_3$–MgO	58	98
4	Au/SiO$_2$–Al$_2$O$_3$–MgO	14	91
5	Au–Ni/SiO$_2$–Al$_2$O$_3$–MgO	12	89

a) Reaction conditions: **1** (15 mmol), catalyst (Au: 0.1 mol%) in methanol (10 ml), O$_2$/N$_2$ (7 : 93) (v/v), 3 MPa) at 60 °C for 2 h.
b) Determined by GC analysis using tetradecane as an internal standard.
c) Pd-based catalyst (Pd: 0.5 mol%).

Au–NiO$_x$ catalyst (entry 4). The activity and selectivity of the Au–Ni catalyst, prepared by the reduction of the Au–NiO$_x$ catalyst under a H$_2$ atmosphere at 400 °C for 3 h, was greatly decreased (entry 5). The oxidative esterification activity of the Au–NiO$_x$ catalyst showed a strong dependence on the Au and NiO$_x$ composition in the supported nanoparticle. The maximum activity was observed for 20 mol% of Au.

Spherical particles of the Au–NiO$_x$ catalyst that are uniformly distributed on the carrier can be seen in the transmission electron microscopy (TEM) images. The Au–NiO$_x$ particles have a diameter of 2–3 nm. High-magnification images revealed a lattice of Au (111) particles with a d-spacing of 2.36 Å. Elemental analysis of individual particles by energy-dispersive X-ray (EDX) spectroscopy showed the presence of Ni and Au in the particles. EDX analysis was performed on the scanning transmission electron microscopy (STEM) image of the nanoparticles. The results showed that Ni was distributed on the Au particles as well as around the edges of the particles. A broad diffraction peak attributable to Au^0 was observed in the XRD patterns. The absence of diffraction peaks due to Ni suggested that Ni existed as a noncrystalline phase. The Au 4f and Ni 2p XPS spectra confirmed the oxidation states of Au and Ni to be 0 and +2, respectively. When the variation in the electronically excited state was examined using UV–vis spectroscopy, no surface plasmon absorption peak, as observed in the case of the Au catalyst, originated from the Au nanoparticles (~530 nm) in the case of the Au–NiO$_x$ catalyst. The FTIR spectra of CO adsorbed on the Au catalyst showed an intense band attributed to Au^0–CO. In contrast, the Au–NiO$_x$ catalyst only showed a weak signal attributed to Ni^{2+}–CO. No peak corresponding to Au^0–CO was observed. Based on these results, the Au–NiO$_x$ nanoparticle was assumed to have a core–shell structure with a Au nanoparticle at the core with its surface covered by highly oxidized NiO$_x$ (Figure 13.6). NiO$_x$ exists in a highly oxidized state in the Au–NiO$_x$ catalyst owing to heterometallic bonding interactions with Au (ligand effect). In control experiments, nickel

Figure 13.6 Proposed structure of Au–NiO$_x$ nanoparticles.

peroxide was found to show catalytic activity in oxidative esterification reactions. Therefore, we suggest that the active species in this case is the highly oxidized NiO$_x$ supported on Au.

The practical applicability of this catalytic system was verified in a 100 000 t/year MMA production plant in 2008. This process confirmed the high selectivity, high activity, and long life of the Au–NiO$_x$ catalyst. This catalyst would help in saving energy and resources, in addition to being highly economical. Even more than 6 years after the introduction of this technology, hardly any deactivation in the catalyst is observed, and the operation is stable. This strategy provides an efficient and environmentally benign method for the synthesis of esters.

13.5
Conclusion and Outlook

The Pd–Pb intermetallic catalyst and Au–NiO$_x$ nanoparticle catalyst developed and industrialized by Asahi Kasei are both unprecedented and unique aerobic oxidative esterification catalysts. We believe that three production processes – our oxidative esterification method (TBA to C$_4$ isobutene hydrocarbon feedstock) in addition to the alpha process (C$_2$ ethylene hydrocarbon feedstock) and the improved ACH process (C$_3$ propene hydrocarbon to acetone feedstock; see Chapter 7 in this book on the cumene-based phenol process) – will be competing for the top position in the global market as the MMA manufacturing method.

On the other hand, if the oxidative esterification reaction is considered as a general synthetic method for esters, it is quite significant that the Au–NiO$_x$ catalyst can almost completely suppress production of the ester by-product derived from oxidation of the alcohol reaction partner. As a general organic synthetic method for esters, it is not controlled by the equilibrium theory, and it is expected to develop as a green manufacturing method that does not require acid or alkali and does not proceed via organic acids.

References

1. Otera, J. (2003) *Esterification: Methods, Reaction and Applications*, Wiley-VCH Verlag GmbH, Weinheim.
2. (a) Murahashi, S.-I., Naota, T., Ito, K., Maeda, Y., and Taki, H. (1987) *J. Org. Chem.*, **52**, 4319–4327; (b) Gopinath, R. and Patel, B.K. (2000) *Org. Lett.*, **2**, 577–579; (c) Wu, X.-F. and Darcel, C. (2009) *Eur. J. Org. Chem.*, **2009**, 1144–1147; (d) Hashmi, A.S.K., Lothschuetz, C., Ackermann, M., Doepp, R., Anantharaman, S., Marchetti, B., Bertagnolli, H., and Rominger, F. (2010) *Chem. Eur. J.*, **16**, 8012–8019; (e) Liu, C., Tang, S., Zheng, L., Liu, D., Zhang, H., and Lei, A. (2012) *Angew. Chem. Int. Ed.*, **51**, 5662–5666; (f) Marsden, C., Taarning, E., Hansen, D., Johansen, L., Klitgaard, S.K., Egeblad, K., and Christensen, C.H. (2008) *Green Chem.*, **10**, 168–170; (g) Su, F.-Z., Ni, J., Sun, H., Cao, Y., He, H.-Y., and Fan, K.-N. (2008) *Chem. Eur. J.*, **14**, 7131–7135; (h) Taarning, E., Nielsen, L.S., Egeblad, K., Madsen, R., and Christensen, C.H. (2008) *ChemSusChem*, **1**, 75–78; (i) Xu, B., Liu, X., Haubrich, J., and Friend, C.M. (2009) *Nat. Chem.*, **2**, 61–65.
3. Nagai, K. (2001) *Appl. Catal., A*, **221**, 367–377.
4. Yamamatsu, S., Yamaguchi, T., Yokota, K., Nagano, O., Chono, M., and Aoshima, A. (2010) *Catal. Surv. Asia*, **14**, 124–131.
5. Shiraish, T. and Mouri, Z. (1974) Manufacturing method of α,β-unsaturated acid and its esters. Japanese Patent 49-35322 A1, Apr. 1, 1974, to Sumitomo Chemical Co., Ltd.
6. Suzuki, K., Yamaguchi, T., Matsushita, K., Iitsuka, C., Miura, J., Akaogi, T., and Ishida, H. (2013) *ACS Catal.*, **3**, 1845–1849.

Part IV
Organocatalytic Aerobic Oxidation

14
Quinones in Hydrogen Peroxide Synthesis and Catalytic Aerobic Oxidation Reactions

Alison E. Wendlandt and Shannon S. Stahl

14.1
Introduction

Quinones are important reagents in organic synthesis [1], and common quinones such as 2,3-dichloro-5,6-dicyanobenzoquinone (DDQ), chloranil, and benzoquinone have found important application in dehydrogenation reactions, in oxidative couplings, and as cocatalysts in transition metal-catalyzed reactions. Quinone-mediated transformations of this kind are featured in several prominent process-scale pharmaceutical chemical syntheses, despite the need for (super)stoichiometric quantities of quinone oxidant [2]. Many hydroquinones may be reoxidized by molecular oxygen to regenerate the corresponding quinone (Eq. (14.1)), suggesting the possibility of replacing stoichiometric quinone-mediated reactions with aerobic quinone-catalyzed or cocatalyzed processes.

The reaction of certain hydroquinones with O_2 to afford the corresponding quinone and hydrogen peroxide forms the basis of the anthraquinone oxidation (AO) process for the industrial synthesis of hydrogen peroxide. In 1901, Manchot reported that the autoxidation of hydroquinones, hydrazobenzenes, and other reduced organic compounds resulted in the near-quantitative formation of hydrogen peroxide [3]. A two-step cyclic process for the industrial manufacture of hydrogen peroxide based on sequential oxidation and reduction of hydrazobenzenes was subsequently proposed by Walton and Filson [4].

Later, development by Riedl and Pfleiderer with BASF led to the replacement of hydrazobenzenes with alkylated anthrahydroquinones [5]. The first plants employing the Riedl–Pfleiderer process – the technological basis for all modern AO processes [6] – were developed by BASF between 1935 and 1945 but dismantled at the conclusion of World War II.

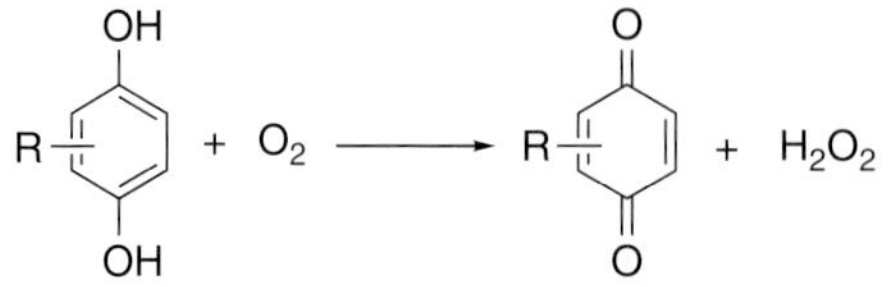

$$R\text{-}C_6H_3(OH)_2 + O_2 \longrightarrow R\text{-}C_6H_3(=O)_2 + H_2O_2 \tag{14.1}$$

In the contemporary AO process, H_2O_2 is formed by the stoichiometric autoxidation of a 2-alkyl-9,10-anthrahydroquinone derivative (Scheme 14.1; autoxidation step). Oxidized anthraquinone coproduct is regenerated in a separate step by catalytic hydrogenation (Scheme 14.1; hydrogenation step). This sequence formally represents the ideal net production of H_2O_2 from molecular oxygen and hydrogen (Scheme 14.1; net reaction). In reality, however, the AO process has suboptimal features, including environmental and economic drawbacks, which motivate considerable contemporary efforts to develop methods for H_2O_2 synthesis via direct catalytic reduction of O_2 by H_2 [7].

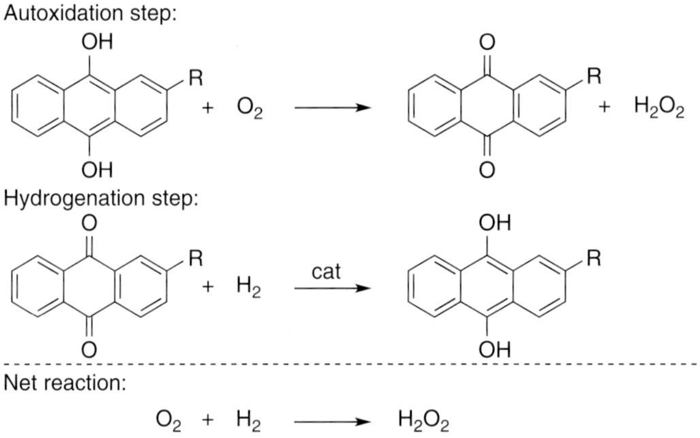

Autoxidation step:

anthrahydroquinone(R) + O_2 ⟶ anthraquinone(R) + H_2O_2

Hydrogenation step:

anthraquinone(R) + H_2 $\xrightarrow{\text{cat}}$ anthrahydroquinone(R)

Net reaction:

$O_2 + H_2 \longrightarrow H_2O_2$

Scheme 14.1 The anthraquinone oxidation (AO) process for industrial synthesis of H_2O_2.

A prominent application of AO technology is the hydrogen peroxide to propylene oxide (HPPO) process, one of the most important commodity-scale oxidation reactions employing hydrogen peroxide as the oxidant (Scheme 14.2) [8]. In the HPPO process, propylene oxide (PO) is formed directly from propylene and aqueous H_2O_2 using TS-1, a Ti silicalite catalyst [9]. H_2O_2 is typically produced via the AO process in a plant situated nearby, or on-site, in an integrated facility [8, 10]. The coupling of these two processes results in an effective net transformation for the synthesis of PO from propylene, O_2, and H_2 (Scheme 14.2; net reaction). Compared with the chlorohydrin route, the HPPO process carried out

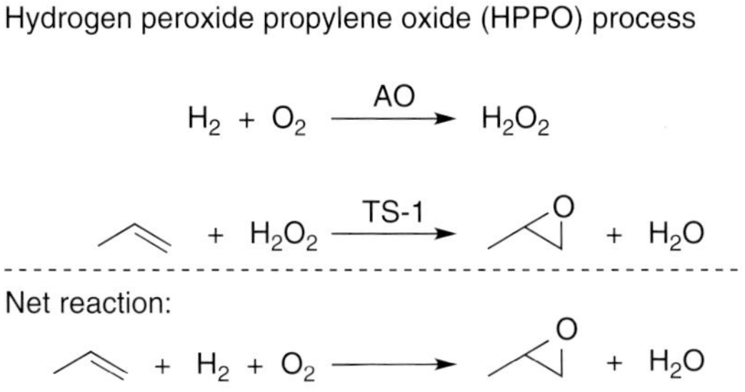

Hydrogen peroxide propylene oxide (HPPO) process

$$H_2 + O_2 \xrightarrow{AO} H_2O_2$$

$$\text{propylene} + H_2O_2 \xrightarrow{TS-1} \text{propylene oxide} + H_2O$$

Net reaction:

$$\text{propylene} + H_2 + O_2 \longrightarrow \text{propylene oxide} + H_2O$$

Scheme 14.2 HPPO process for the synthesis of propylene oxide.

in physically integrated H_2O_2 and PO plants offers numerous economic and environmental benefits, including reduced transportation and H_2O_2 production costs, simpler raw materials, no coproduct formation, 70–80% reduced wastewater generation, and 35% reduced energy usage [11]. Three world-scale plants employing this chemistry are currently in operation.[1] Nonetheless, HPPO is still a relatively minor contributor to worldwide PO production, contributing approximately 8% of the total PO economy (total PO estimated 8 657 000 ta^{-1}, 2010) [12].

The following sections provide more information about the chemistry and process technology related to the AO process. This content is followed by a discussion of recent academic developments highlighting emerging opportunities for the use of quinone catalysts in selective aerobic oxidation of organic molecules.

14.2
Chemistry and Catalysis: Anthraquinone Oxidation (AO) Process

Hydrogen peroxide is an important commodity oxidant, with applications in a variety of industries, including chemical synthesis, paper and pulp bleaching [13], and wastewater remediation [14]. The total global market for H_2O_2 has been projected to reach 4 670 000 ta^{-1} by 2017. Over 95% of all H_2O_2 is produced through the AO process [15, 16].

14.2.1
Autoxidation Process (Hydroquinone to Quinone)

In the autoxidation step of the AO process, the $2H^+/2e^-$ reduction of O_2 to H_2O_2 is coupled to the uncatalyzed oxidation of anthrahydroquinone to anthraquinone. A variety of 2-alkylanthraquinones have been reported, but 2-ethylanthraquinone

---

1) A joint BASF/Dow HPPO plant (300 000 ta^{-1} PO) was constructed in Antwerp in 2006, partnering with Solvay as a supplier of H_2O_2 (230 000 ta^{-1} H_2O_2). Evonik (formerly Degussa) developed an HPPO process with Uhde, opening a plant in Ulsan, South Korea (100 000 ta^{-1} PO), licensed to SKC in 2008. And finally, an HPPO plant operated by Dow/SCG began production in Thailand in 2011 (390 000 ta^{-1} PO) with a companion H_2O_2 plant (330 000 ta^{-1}) nearby.

is most commonly employed. Hydroquinone autoxidation is carried out in solution phase at 30–60 °C under slightly pressurized air (0.5 MPa). Solvent selection is of critical importance, as it is necessary to balance the differential solubilities of the quinone (nonpolar, aromatic solvents) and hydroquinone (polar solvents, i.e., alcohols and esters), which must also show good stability under both oxidative and reductive conditions. Consequently, solvent mixtures are typically employed, including polyalkylbenzenes/alkyl phosphates [17], polyalkylbenzenes/tetraalkyl ureas [18], trimethylbenzene/alkylcyclohexanol esters [19], and methylnaphthalene/nonyl alcohol [20].

The mechanism of hydroquinone autoxidation likely proceeds by a radical chain pathway. Kinetic studies carried out under relevant reaction conditions support a second-order rate law for the reaction, rate $= k[QH_2][O_2]$, with an apparent activation energy of $E_a = 15$ kcal/mol [21]. Based on these kinetic findings, as well as DFT studies [22], anthrahydroquinone autoxidation has been proposed to occur through initial, rate-limiting, direct H-atom abstraction from the hydroquinone species by O_2 (Eq. (14.2)). The semiquinone species then react readily with triplet O_2 (Eq. (14.3)), and hydroperoxy radical, $HO_2^\bullet$, has been proposed to act as a radical chain carrier (Eq. (14.4)).

$$\text{anthrahydroquinone} + O_2 \longrightarrow \text{semiquinone} + HO_2^\bullet \qquad (14.2)$$

$$\text{semiquinone} + O_2 \longrightarrow \text{anthraquinone} + HO_2^\bullet \qquad (14.3)$$

$$\text{anthrahydroquinone} + HO_2^\bullet \longrightarrow \text{semiquinone} + H_2O_2 \qquad (14.4)$$

Depending on the reaction solvent and conditions, the disproportionation of hydroperoxy radical (Eq. (14.5)) may contribute to the reaction (estimated rate constant $10^6 – 10^7$ M^{-1} s^{-1}) [23]. While no autocatalytic effects were observed in the previous kinetic studies, the role of trace quinone on hydroquinone oxidation has been extensively reported [24]. Comproportionation of hydroquinone and quinone leads to the formation of semiquinone species (Eq. (14.6)), which can react readily with O_2. The semiquinone is unstable with respect to disproportionation but has nevertheless been shown to be an important intermediate in hydroquinone autoxidation pathways [24].

$$2HO_2^\bullet \rightarrow O_2 + H_2O_2 \qquad (14.5)$$

$$\text{[anthraquinone]} + \text{[anthrahydroquinone-OH]} \rightleftharpoons 2\,\text{[semiquinone radical]} \quad (14.6)$$

14.2.2
Hydrogenation Process (Quinone to Hydroquinone)

Most of the quinone/hydroquinone decomposition products are formed in the hydrogenation step of the AO process, which is usually carried out over Pd catalysts supported on alumina, silica, or sodium aluminum silicate, or as Pd black. Depending on the specific process, hydrogenation reaction temperatures range from 25 to 75 °C, and hydrogen pressures up to 0.3 MPa are used. The reaction rate is apparently not affected by increasing pressures above 0.4 MPa. To minimize secondary reduction products (discussed later), conversions of quinone to hydroquinone are typically kept around 45–50% but may be performed up to 80%.

In addition to the desired reduction of quinone to hydroquinone, a number of secondary reactions can occur during the hydrogenation process. Of particular significance is the reduction of the unsubstituted anthraquinone ring, leading to the formation of 2-alkyl-5,6,7,8-tetrahydro-9,10-dihydroxyanthracene, **A** (Eq. (14.7)). Ring-reduced product **A** can also be oxidized by O_2 to generate stoichiometric H_2O_2, albeit at a rate 5–10× slower than the parent anthrahydroquinone. The corresponding oxidized 2-alkyl-5,6,7,8-tetrahydroanthraquinone is referred to as "tetra" (Eq. (14.8)).

$$\text{anthraquinone} + 3H_2 \xrightarrow{\text{cat}} \textbf{A} \quad (14.7)$$

$$\textbf{A} + O_2 \longrightarrow \text{"Tetra"} + H_2O_2 \quad (14.8)$$

An important redox equilibrium exists between anthrahydroquinone **A** and tetra (Eq. (14.9)). Because tetra is more readily reduced than the parent anthraquinone, this equilibrium lies almost entirely to the right. Thus, if tetra formation is not suppressed, eventually all the hydroquinone species present in the oxidation process will consist of the reduced form of tetra. Because the analogous "tetra" hydroquinone is oxidized more slowly, the autoxidation step will slow down significantly.

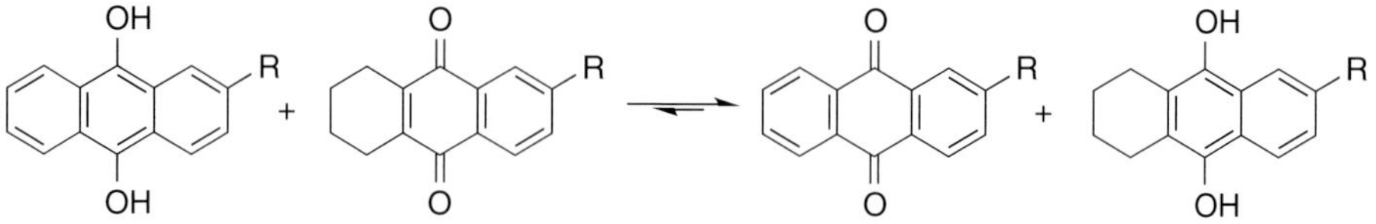

226 | 14 Quinones in Hydrogen Peroxide Synthesis and Catalytic Aerobic Oxidation Reactions

$$\text{(14.9)}$$

If tetra concentration becomes high, further quinone degradation products are observed. Additional ring hydrogenation can occur to give an "octa"-hydroquinone (Eq. (14.10)). Autoxidation of the octa-hydroquinone is too slow to be relevant and thus constitutes a decomposition product from which quinone cannot be regenerated.

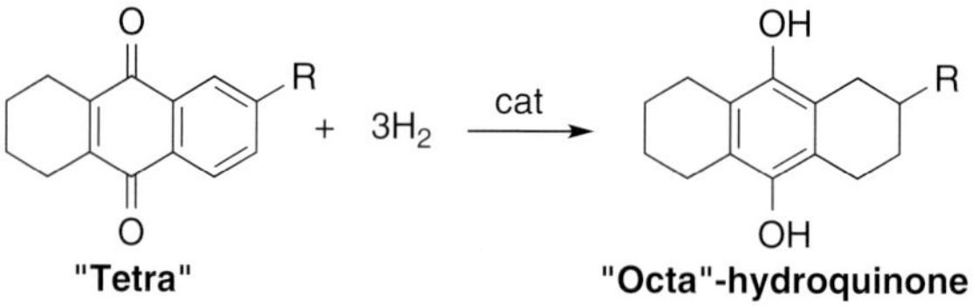

$$\text{"Tetra"} + 3H_2 \xrightarrow{\text{cat}} \text{"Octa"-hydroquinone} \quad (14.10)$$

Under certain conditions, oxanthrone can also be formed during the hydrogenation process (Scheme 14.3). Oxanthrone is not reoxidized in the autoxidation process, and further hydrogenation of oxanthrone leads to anthrone, tetrahydroanthrones, and dianthrones (the latter via oxidative dimerization of anthrone) as decomposition products.

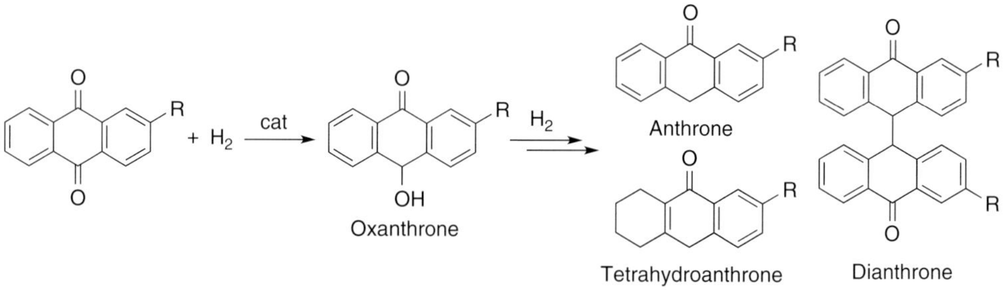

Scheme 14.3 Formation of anthrone by-products in the "anthra" system.

While the chemical yield of hydrogen peroxide in the AO process is very high, the loss of quinone/hydroquinone via the formation of these by-products necessitates the regeneration of the reaction mediators, hydrogenation catalyst, and removal of organic by-products. Periodically, fresh anthraquinone and solvent are added to compensate for losses.

14.3 Process Technology

The specific reaction conditions and process technologies vary significantly across the major H_2O_2 producers, but a generalized process scheme is shown in Scheme 14.4.

The most extensive development efforts have focused on hydrogenation processes, and the hydrogenator design varies considerably from producer to producer. Reactors include continuous-stirred tank, bubble-column, loop, and fixed-bed reactors. Depending on the type of hydrogenation catalyst employed (e.g., Raney Ni), the solution may require a pretreatment column to remove H_2O from the extraction step. Any residual H_2O_2 must also be decomposed prior to contact with the hydrogenation catalyst. This step is typically accomplished by treating the oxidized solution with a portion of the reduced solution (containing hydroquinone) over a supported Ag/Ni catalyst (Eq. (14.11)). The use of Pd black (Degussa) and supported Pd (Laporte) catalysts alleviates the stringent handling precautions necessitated by Raney Ni, as these catalysts are nonpyrophoric. In addition, Pd catalysts are more selective and more easily recycled, particularly in the case of supported catalysts.

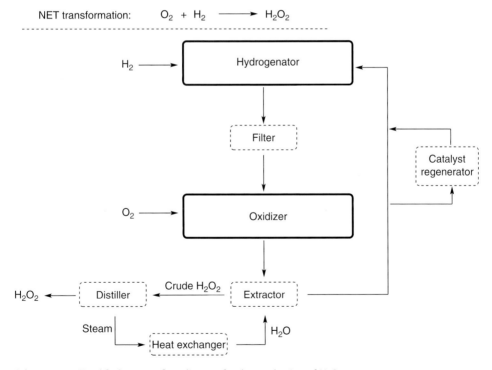

Scheme 14.4 Simplified process flow diagram for the production of H_2O_2.

$$\text{(anthrahydroquinone-R)} + H_2O_2 \xrightarrow{Ni-Ag} \text{(anthraquinone-R)} + 2H_2O \quad (14.11)$$

After leaving the hydrogenator, the solution must pass through a safety filter prior to entering the oxidizer. Failure to remove all of the hydrogenation catalyst could lead to catastrophic H_2O_2 disproportionation in the oxidizer. Fixed-bed reactor designs alleviate the need for independent catalyst removal steps.

As in the hydrogenation step, several unique oxidizer designs are employed commercially, including cocurrent and countercurrent flow oxidations in single or serial column arrangements. Oxidizer off-gas is passed over beds of activated carbon to recover solvent vapor and for purification.

Following oxidation, hydrogen peroxide is extracted into an aqueous phase at a yield of approximately 95% (based on maximum theoretical H_2O_2). The crude aqueous hydrogen peroxide (15–40 wt%) is distilled to remove any organic impurities and concentrated to the desired commercial level. Because H_2O_2 does not form an azeotrope with H_2O, high concentrations can be obtained; typically 35–70 wt% solutions are used commercially. For safety reasons, concentrations above 90 wt% are typically obtained by fractional crystallization from aqueous solutions.

The oxidized reaction medium containing quinone is passed through a drier to adjust the water content of the solution and sent back to the hydrogenator. Because of the buildup of quinone degradation products (discussed earlier), a portion of the solution is passed through a catalyst regeneration/purification step.

Catalyst decomposition depends heavily on the specific process conditions employed. Producers operate under two different regimes: the "all-tetra" system, in which no specific actions are taken to either suppress the formation of tetra or to dehydrogenate it back to anthraquinone, and the "anthra" system, in which efforts are made to minimize the tetra content. Tetra formation can be reduced by the use of selective hydrogenation catalysts and specific operating conditions (i.e., solvent choice and specialized quinones). In addition, tetra can be dehydrogenated in the presence of activated alumina (Al_2O_3) in the catalyst regenerator (Eq. (14.12)).

$$3 \, \text{(tetrahydroanthraquinone-R)} \xrightarrow{Al_2O_3} \text{(anthraquinone-R)} + 2 \, \text{(anthrahydroquinone-R)} \quad (14.12)$$

Most producers operate within the "all-tetra" regime. Dianthrone formation is less problematic in the all-tetra system relative to the anthra system. On the other hand, the anthra system benefits the oxidation process, as formation of more difficult-to-oxidize ring-reduced hydroquinone is suppressed.

14.4 Future Developments: Selective Aerobic Oxidation Reactions Catalyzed by Quinones

Quinone-catalyzed aerobic oxidation of organic substrates bears important conceptual similarity to the AO process (Scheme 14.5). In both instances, aerobic oxidation of hydroquinone forms quinone and H_2O_2 (the desired product in the case of AO). However, instead of reduction of the quinone by catalytic hydrogenation, the quinone may be used to oxidize an organic molecule, $SubH_2$, to give the desired product, Sub^{ox}. Several classes of quinone catalysts have been developed for the aerobic oxidation of organic substrates. In some instances, catalyst regeneration is achieved through direct autoxidation of the reduced hydroquinone (as with anthraquinone); however, hydroquinone reoxidation is more commonly promoted by the use of simple transition metal or inorganic cocatalysts.

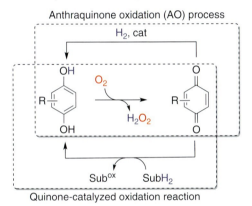

Scheme 14.5 Aerobic oxidation of hydroquinone to quinones, a common step in the AO process and in quinone-catalyzed oxidation reactions.

14.4.1 Aerobic DDQ-Catalyzed Reactions Using NO$_x$ Cocatalysts

High-potential quinones, such as DDQ and chloranil, are important stoichiometric reagents in organic synthesis with notable applications in the oxidative functionalization of activated C–H bonds and dehydrogenation to form (hetero)arenes [1]. These reactions can be rendered catalytic in quinone by using certain transition metal salts, anodic oxidation, or molecular oxygen to regenerate the quinone *in situ*.

The direct aerobic reoxidation of $DDQH_2$ to DDQ is not thermodynamically feasible due to the high potential of DDQ versus the 2e-reduction of O_2 to H_2O_2.

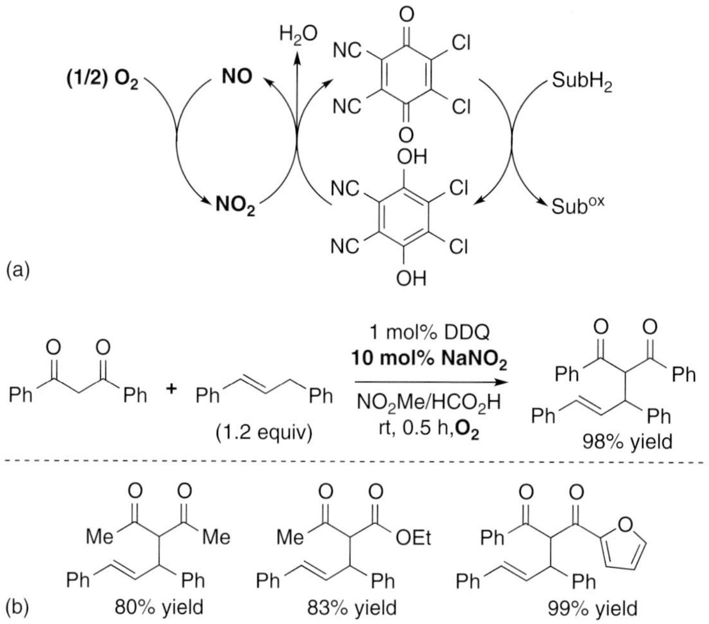

Scheme 14.6 (a) Catalytic oxidation of organic substrates employing catalytic DDQ with molecular oxygen and cocatalytic NO_x and (b) DDQ-catalyzed and $NaNO_2$-cocatalyzed aerobic cross-coupling of diarylpropenes and 1,3-dicarbonyls.

This limitation can be overcome by employing NO_x-based cocatalysts, which enable 4e-reduction of O_2 to $2H_2O$, thereby accessing a larger thermodynamic driving force (Scheme 14.6a) [25, 26]. For example, Yan and coworkers described an aerobic DDQ-catalyzed oxidative cross-coupling of diarylpropenes and 1,3-diketones [27]. New C–C coupled products were obtained in good to excellent yields using 1 mol% DDQ and 10 mol% $NaNO_2$ in $MeNO_2/HCO_2H$ at room temperature under 1 atm O_2 (Scheme 14.6b).

14.4.2
Aerobic Quinone-Catalyzed Reactions Using Other Cocatalysts

Alternative cocatalysts, such as $[Fe(pc)]_2O$ (pc = phthalocyanine) and AIBN, have also been reported for catalytic aerobic DDQ oxidation reactions, though these have seen less widespread application than NO_x [28]. Aerobic quinone-catalyzed dehydrogenation reactions using less oxidizing quinones, such as chloranil, have been reported using polymer-incarcerated noble metal catalysts reported by Kobayashi and colleagues [29, 30].

In addition to their value as synthetic reagents, quinones also play an essential redox role within biological systems, such as in photosynthetic and mitochondrial electron transport chains and as aerobic enzyme cofactors [31]. This concept was exploited for synthetic applications as early as 1960, when Moiseev found benzoquinone to be an efficient terminal oxidant for Pd-catalyzed acetoxylation reactions [32]. Subsequent development of quinone cocatalysts by Backvall and others has led to their use in numerous transition metal-catalyzed aerobic oxidative transformations, such as the aerobic Pd-catalyzed allylic acetoxylation of cyclohexene shown in Scheme 14.7 [33, 34].

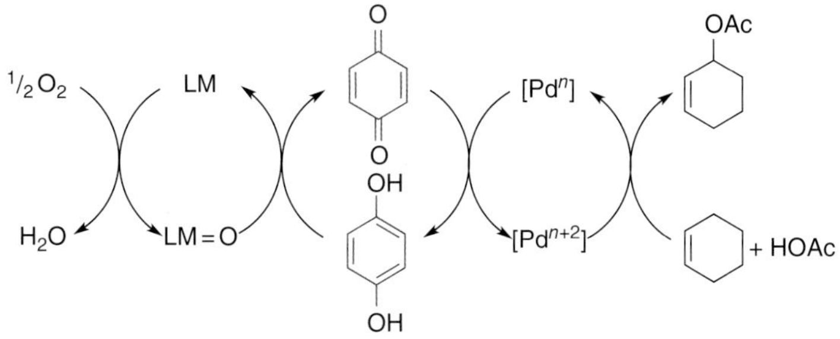

Scheme 14.7 Pd-catalyzed allylic acetoxylation reaction featuring Pd-, quinone-, and metal macrocycle (LM)-coupled catalytic cycles.

14.4.3
CAO Mimics and Selective Oxidation of Amines

Copper amine oxidase (CAO) enzymes carry out the aerobic oxidation of primary amines to aldehydes (Scheme 14.8a). While copper is present in the active site, substrate oxidation proceeds by an organocatalytic pathway involving an o-quinone cofactor via a transamination mechanism (Scheme 14.8b).

Recent studies have shown that o-quinones inspired from CAO cofactors are efficient and highly selective catalysts for synthetic aerobic C–N bond dehydrogenation reactions [35]. For example, o-quinones such as **Q1^{red}**, **TBHBQ**, and **Q2** have been shown to carry out selective aerobic primary amine oxidation to give the corresponding dimeric and cross-coupled secondary imine products (Scheme 14.9) [36–39]. Additional studies have demonstrated the efficiency of similar CAO mimics in the stoichiometric [40] as well as electrochemical [41] oxidation of primary amines. Substrate oxidation by these catalysts is typically thought to involve a transamination-type mechanism similar to that observed in CAOs and model cofactors (cf. Scheme 14.8b) [42, 43]. Exquisite selectivity for primary amine substrates is observed as a result. Alcohols and tertiary amines are not oxidized using these catalysts, and secondary amines have been observed to be mechanism-based inhibitors [44].

Through subtle modification of the reaction mechanism, the scope of o-quinone-catalyzed aerobic oxidations has been expanded to secondary amines and N-heterocyclic compounds as well [45–47]. For example, aerobic oxidation of secondary amines and N-heterocycles has been achieved by using 10-phenanthroline-5,6-dione (phd) as a catalyst (Scheme 14.10a). Using phd and cocatalytic ZnI_2, a diverse range of N-heterocyclic compounds undergo

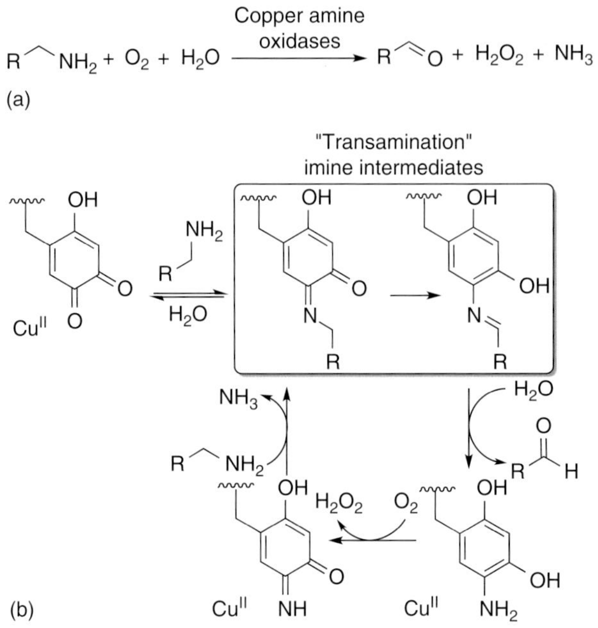

Scheme 14.8 Copper amine oxidases carry out (a) the aerobic oxidation of primary amines *in vivo* via a (b) transamination mechanism.

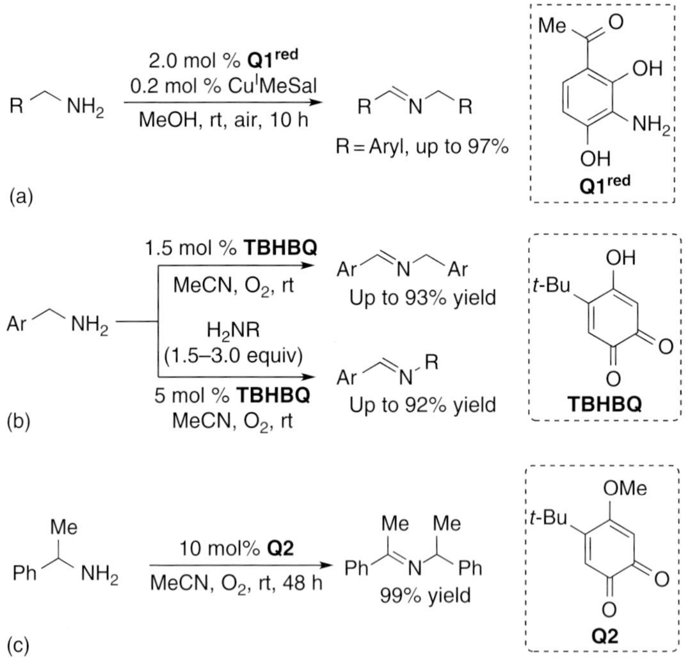

Scheme 14.9 Copper amine oxidase mimics promote the aerobic oxidation of primary amines to imines.

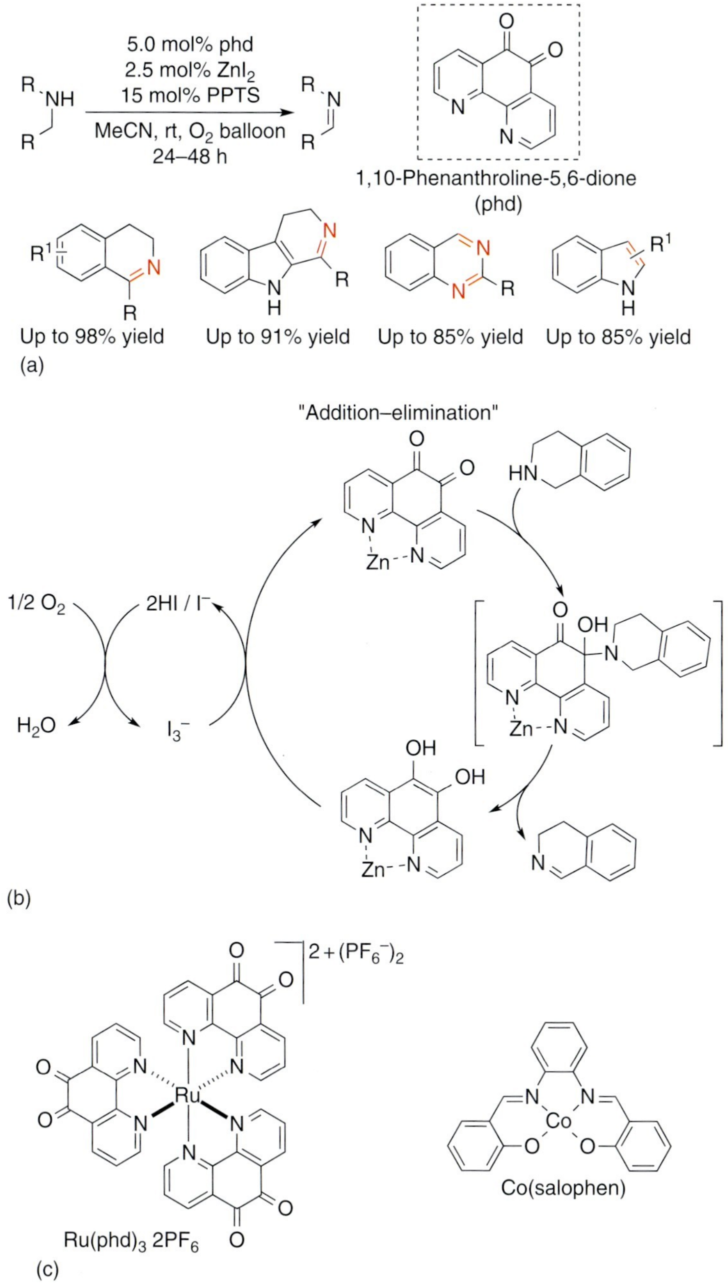

Scheme 14.10 (a) Aerobic phd-catalyzed oxidation of diverse classes of secondary amines through (b) an "addition–elimination" mechanism. (c) Improved reaction efficiency is obtained by replacing Zn^{2+} with Ru^{2+} and I^- cocatalyst with Co(salophen).

dehydrogenation under mild conditions (Scheme 14.10a). The mechanism was shown to proceed by an addition–elimination, rather than transamination, mechanism (Scheme 14.10b) [46a]. Cocatalytic Zn^{2+} enhances the rate of substrate oxidation, and I^- promotes aerobic catalyst turnover via I^-/I_3^- redox couple. In a subsequent study, dramatically enhanced reaction rates and improved substrate scope were achieved by replacing phd/Zn^{2+} with a well-defined homoleptic $[Ru(phd)_3](PF_6)_2$ complex (Scheme 14.10c) and replacing the iodide cocatalyst with Co(salophen) as a redox mediator [46b]. Collectively, the results summarized in this section highlight new opportunities for the development of quinone catalysts to achieve selective oxidation of organic molecules.

References

1. (a) Patai, S. and Rappoport, Z. (Eds.), (1974 and 1988) *The Chemistry of the Quinonoid Compounds*, Wiley Interscience, New York; (b) Buckle, D.R., Collier, S.J., and McLaws, M.D. (2001) *Encyclopedia of Reagents for Organic Synthesis*, John Wiley & Sons, Ltd; (c) Walker, D. and Hiebert, J.D. (1967) *Chem. Rev.*, **67**, 153; (d) Fu, P.P. and Harvey, R.G. (1978) *Chem. Rev.*, **78**, 317; (e) Bharate, S.B. (2006) *Synlett*, **2006**, 0496.

2. (a) Merschaert, A., Boquel, P., Van Hoeck, J.-P., Gorissen, H., Borghese, A., Bonnier, B., Mockel, A., and Napora, F. (2006) *Org. Process Res. Dev.*, **10**, 776; (b) Armitage, M., Bret, G., Choudary, B.M., Kingswood, M., Loft, M., Moore, S., Smith, S., and Urquhart, M.W.J. (2012) *Org. Process Res. Dev.*, **16**, 1626; (c) Williams, J.M. (2010) *The Art of Process Chemistry*, Wiley-VCH Verlag GmbH & Co. KGaA, p. 77; (d) Mickel, S.J., Sedelmeier, G.H., Niederer, D., Schuerch, F., Seger, M., Schreiner, K., Daeffler, R., Osmani, A., Bixel, D., Loiseleur, O., Cercus, J., Stettler, H., Schaer, K., Gamboni, R., Bach, A., Chen, G.-P., Chen, W., Geng, P., Lee, G.T., Loeser, E., McKenna, J., Kinder, F.R., Konigsberger, K., Prasad, K., Ramsey, T.M., Reel, N., Repič, O., Rogers, L., Shieh, W.-C., Wang, R.-M., Waykole, L., Xue, S., Florence, G., and Paterson, I. (2003) *Org. Process Res. Dev.*, **8**, 113; (e) Sharma, P.K., Kolchinski, A., Shea, H.A., Nair, J.J., Gou, Y., Romanczyk, L.J., and Schmitz, H.H. (2007) *Org. Process Res. Dev.*, **11**, 422.

3. (a) Manchot, W. (1901) *Justus Liebigs Ann. Chem.*, **314**, 177; See also: (b) Manchot, W. and Herzog, J. (1901) *Justus Liebigs Ann. Chem.*, **316**, 318; (c) Manchot, W. and Herzog, J. (1901) *Justus Liebigs Ann. Chem.*, **316**, 331.

4. Walton, J.H. and Filson, G.W. (1932) *J. Am. Chem. Soc.*, **54**, 3228–3229.

5. (a) Reidl, H.-J. and Pfleiderer, G. (1936) Production of hydrogen peroxide. US Patent 2,158,525, filed Oct. 3, 1936 and issued May 16, 1939 (I. G. Farbenindustrie); (b) Reidl, H.-J. and Pfleiderer, G. (1938) Production of hydrogen peroxide. US Patent 2,215,883, filed Apr. 5, 1938 and issued Sep. 24, 1940 (I. G. Farbenindustrie).

6. Von Schickh, O. (1960) *Chem. Ing. Tech.*, **32**, 462.

7. (a) Campos-Martin, J.M., Blanco-Brieva, G., and Fierro, J.L.G. (2006) *Angew. Chem. Int. Ed.*, **45**, 6962; (b) Centi, G., Perathoner, S., and Abate, S. (2009) *Modern Heterogeneous Oxidation Catalysis*, Wiley-VCH Verlag GmbH & Co. KGaA, p. 253; (c) Edwards, J.K., Solsona, B., Ntainjua, E.N., Carley, A.F., Herzing, A.A., Kiely, C.J., and Hutchings, G.J. (2009) *Science*, **323**, 1037; (d) Lunsford, J.H. (2003) *J. Catal.*, **216**, 455; (e) Van Weynbergh, J., Schoebrechts, J.P., and Colery, J.C. (1992) Direct synthesis of hydrogen peroxide by heterogeneous catalysis, catalyst for the said synthesis and method of preparation of the said catalyst.

US Patent 5,447,706, filed Feb. 25, 1992 and issued Sep. 5, 1995 (Solvay Interox).
8. Cavani, F. and Teles, J.H. (2009) *ChemSusChem*, **2**, 508.
9. (a) Taramasso, M., Perego, G., and Notari, B. (1982) Preparation of porous crystalline synthetic material comprised of silicon and titanium oxides. US Patent 4,410,501, filed June 29, 1982 and issued Oct. 18, 1983 (Snamprogetti S.p.A.); (b) Neri, C., Esposito, A., and Buonomo, F. (1983) Process for the epoxidation of olefinic compounds. EP Patent 0100119B1, filed July 13, 1983 and issued Sep. 3, 1986 (Enichem Anic S.p.A); (c) Notari, B. (1993) *Catal. Today*, **18**, 163; (d) Notari, B. (1988) *Stud. Surf. Sci. Catal.*, **37**, 413; (e) Meiers, R., Dingerdissen, U., and Hölderich, W.F. (1998) *J. Catal.*, **176**, 376; (f) Clerici, M.G., Bellussi, G., and Romano, U. (1991) *J. Catal.*, **129**, 159.
10. (a) Clerici, M.G., De Angelis, A., and Ingallina, P. (1997) Process for the preparation of epoxides from olefins. European Patent 819,683, filed Apr. 7, 1997 and issued Jan. 21, 1998 (Enichem S.p.A.); (b) Rodriguez, C.L. and Zajacek, J.G. (1994) Integrated process for epoxide production. US Patent 5,463,090, filed Oct. 27, 1994 and issued Oct. 31, 1995 (ARCO Chemical Technology).
11. Cavani, F. and Gaffney, A.M. (2009) in *Sustainable Industrial Processes* (eds F. Cavani et al.), Wiley-VCH Verlag GmbH & Co. KGaA, pp. 319–365.
12. The majority of propylene oxide is produced in the chlorohydrin process (CHPO), or co-product routes: Baer, H., Bergamo, M., Forlin, A., Pottenger, L.H., and Lindner, J. (2012) Propylene oxide, in *Ullmann's Encyclopedia of Industrial Chemistry*, Wiley-VCH Verlag GmbH & Co. KGaA, Weinheim.
13. (a) Farr, J.P., Smith, W.L., and Steichen, D.S. (2000) Bleaching agents, in *Kirk-Othmer Encyclopedia of Chemical Technology* (eds A. Seidel and M. Bickford), John Wiley & Sons, Inc; (b) Ni, Y. and Liu, Z. (2000) Pulp bleaching, in *Kirk-Othmer Encyclopedia of Chemical Technology* (eds A. Seidel and M. Bickford), John Wiley & Sons, Inc; (c) Süss, H.U. (2012) Bleaching, in *Ullmann's Encyclopedia of Industrial Chemistry*, Wiley-VCH Verlag GmbH & Co. KGaA, Weinheim; (d) Ragnar, M., Henriksson, G., Lindström, M.E., Wimby, M., Blechschmidt, J., and Heinemann, S. (2012) Pulp, in *Ullmann's Encyclopedia of Industrial Chemistry*, Wiley-VCH Verlag GmbH & Co. KGaA, Weinheim; (e) Hage, R. and Lienke, A. (2006) *Angew. Chem. Int. Ed.*, **45**, 206.
14. (a) Simmler, W. (2012) Wastewater, 4. Chemical treatment, in *Ullmann's Encyclopedia of Industrial Chemistry*, Wiley-VCH Verlag GmbH & Co. KGaA, Weinheim; (b) Glaze, W.H. (2012) Water, 6. Treatment by oxidation processes, in *Ullmann's Encyclopedia of Industrial Chemistry*, Wiley-VCH Verlag GmbH & Co. KGaA, Weinheim.
15. Alternative, albeit minor processes for the production of hydrogen peroxide include: other autoxidation processes, notably the 2-Propanol Process (Shell Process); and electrochemical processes, including the Degussa-Weissenstein, Münchner, and Riedel-Loewenstein Processes. See, (a) Harris, C.R. (1945) Production of hydrogen peroxide by partial oxidation of alcohols. US Patent 2,479,111A, filed Jun. 29, 1945 and issued Aug. 16, 1949 (E. I. du Pont de Nemours); (b) Rust, F.F. (1955) Manufacture of hydrogen peroxide. US Patent 2,871,104A, filed Jan. 31, 1955 and issued Jan. 27, 1959 (Shell Dev); (c)N. V. De Bataafsche Petroleum Maatschappij (1950) Process for the production of hydrogen peroxide and the hydrogen peroxide so produced. GB Patent 708,339, filed Nov. 30, 1950 and issued May 5, 1954; (d) Skinner, J.R and Steinle, S.E. (1954) DE Patent 1,002,295, filed Dec. 21, 1954 (N. V. De Bataafsche Petroleum Maatschappij).
16. (a) Goor, G., Glenneberg, J., and Jacobi, S. (2012) Hydrogen peroxide, in *Ullmann's Encyclopedia of Industrial Chemistry*, Wiley-VCH Verlag GmbH & Co. KGaA, Weinheim; (b) Eul, W., Moeller, A., and Steiner, N. (2000) Hydrogen peroxide, in *Kirk-Othmer Encyclopedia of Chemical Technology* (eds A. Seidel and M. Bickford), John Wiley & Sons, Inc; (c) Jones, C.W. and

Clark, J.H. (eds) (1999) *Applications of Hydrogen Peroxide and Derivatives*, The Royal Society of Chemistry.
17. Kaebisch, G. (1963) A process for the production of hydrogen peroxide. DE Patent 1,261,838, filed Sep. 3, 1963 (Degussa).
18. Giesselmann, G., Schreyer, G., and Weigert, W. (1970) A process for the production of hydrogen peroxide. DE Patent 2,018,686, filed Apr. 18, 1970 and issued Nov. 15, 1973 (Degussa).
19. LeFeuvre, C.W. (1953) A process for the manufacture of hydrogen peroxide. GB Patent 747190, filed July 1, 1953 and issued Mar. 28, 1956 (Laporte Chemicals Ltd.).
20. Harris, C.R. and Sprauer J.W. (1950) A process for the production of hydrogen peroxide. DE Patent 888,840, filed Nov. 3, 1950 and issued Sep. 3, 1953 (E. I. du Pont de Nemours).
21. (a) Santacesaria, E., Ferro, R., Ricci, S., and Carrà, S. (1987) *Ind. Eng. Chem. Res.*, **26**, 155; (b) Santacesaria, E., Di Serio, M., Russo, A., Leone, U., and Velotti, R. (1999) *Chem. Eng. Sci.*, **54**, 2799.
22. Nishimi, T., Kamachi, T., Kato, K., Kato, T., and Yoshizawa, K. (2011) *Eur. J. Org. Chem.*, **2011** (22), 4113.
23. Sawyer, D.T. and Valentine, J.S. (1981) *Acc. Chem. Res.*, **14**, 393.
24. Song, Y. and Buettner, G.R. (2010) *Free Radical Biol. Med.*, **49**, 919.
25. Gerken, J.B. and Stahl, S.S. (2015) *ACS Cent. Sci.* **1**, 234.
26. For specific examples, see: (a) Zhang, W., Ma, H., Zhou, L., Sun, Z., Du, Z., Miao, H., and Xu, J. (2008) *Molecules*, **13**, 3236; (b) Zhang, W., Ma, H., Zhou, L., Miao, H., and Xu, J. (2009) *Chin. J. Catal.*, **30**, 86; (c) Shen, Z., Dai, J., Xiong, J., He, X., Mo, W., Hu, B., Sun, N., and Hu, X. (2011) *Adv. Synth. Catal.*, **353**, 3031; (d) Wang, L., Li, J., Yang, H., Lv, Y., and Gao, S. (2012) *J. Org. Chem.*, **77**, 790; (e) Shen, Z., Sheng, L., Zhang, X., Mo, W., Hu, B., Sun, N., and Hu, X. (2013) *Tetrahedron Lett.*, **54**, 1579; (f) Walsh, K., Sneddon, H.F., and Moody, C.J. (2014) *Tetrahedron*, **70**, 7380.
27. Cheng, D., Yuan, K., Xu, X., and Yan, J. (2015) *Tetrahedron Lett.*, **56**, 1641.
28. For alternative approaches, see: (a) Alagiri, K., Devadig, P., and Prabhu, K.R. (2012) *Chem. Eur. J.*, **18**, 5160; (b) Ravikanth, M., Achim, C., Tyhonas, J.S., Münck, E., and Lindsey, J.S. (1997) *J. Porphyrins Phthalocyanines*, **1**, 385; (c) Lindsey, J.S., MacCrum, K.A., Tyhonas, J.S., and Chuang, Y.Y. (1994) *J. Org. Chem.*, **59**, 579; (d) Zaidi, S.H.H., Fico, R.M., and Lindsey, J.S. (2006) *Org. Process Res. Dev.*, **10**, 118.
29. (a) Miyamura, H., Maehata, K., and Kobayashi, S. (2010) *Chem. Commun.*, **46**, 8052; (b) Miyamura, H., Shiramizu, M., Matsubara, R., and Kobayashi, S. (2008) *Angew. Chem. Int. Ed.*, **47**, 8093; (c) Miyamura, H., Shiramizu, M., Matsubara, R., and Kobayashi, S. (2008) *Chem. Lett.*, **37**, 360.
30. See also: Neumann, R., Khenkin, A.M., and Vigdergauz, I. (2000) *Chem. Eur. J.*, **6**, 875.
31. Klinman, J.P. (1996) *J. Biol. Chem.*, **271**, 27189.
32. (a) Moiseev, I.I., Vargaftik, M.N., and Syrkin, Y.K. (1960) *Dokl. Akad. Nauk SSSR*, **133**, 377; (b) Moiseev, I.I. and Vargaftik, M.N. (2004) *Coord. Chem. Rev.*, **248**, 2381.
33. (a) Bäckvall, J.E., Awasthi, A.K., and Renko, Z.D. (1987) *J. Am. Chem. Soc.*, **109**, 4750; (b) Bäckvall, J.-E., Hopkins, R.B., Grennberg, H., Mader, M., and Awasthi, A.K. (1990) *J. Am. Chem. Soc.*, **112**, 5160.
34. For a recent review in this area, see: Piera, J. and Bäckvall, J.-E. (2008) *Angew. Chem. Int. Ed.*, **47**, 3506.
35. Largeron, M. and Fleury, M.-B. (2013) *Science*, **339**, 43.
36. (a) Ohshiro, Y., Itoh, S., Kurokawa, K., Kato, J., Hirao, T., and Agawa, T. (1983) *Tetrahedron Lett.*, **24**, 3465; (b) Itoh, S., Kato, N., Ohshiro, Y., and Agawa, T. (1984) *Tetrahedron Lett.*, **25**, 4753.
37. Wendlandt, A.E. and Stahl, S.S. (2012) *Org. Lett.*, **14**, 2850.
38. Largeron, M. and Fleury, M.-B. (2012) *Angew. Chem. Int. Ed.*, **51**, 5409.
39. Qin, Y., Zhang, L., Lv, J., Luo, S., and Cheng, J.-P. (2015) *Org. Lett.*, **17**, 1469.

40. Corey, E.J. and Achiwa, K. (1969) *J. Am. Chem. Soc.*, **91**, 1429.
41. (a) Largeron, M., Chiaroni, A., and Fleury, M.-B. (2008) *Chem. Eur. J.*, **14**, 996; (b) Largeron, M., Fleury, M.-B., and Strolin Benedetti, M. (2010) *Org. Biomol. Chem.*, **8**, 3796; (c) Largeron, M. and Fleury, M.-B. (2009) *Org. Lett.*, **11**, 883.
42. (a) Mure, M., Mills, S.A., and Klinman, J.P. (2002) *Biochemistry*, **41**, 9269; (b) Mure, M. (2004) *Acc. Chem. Res.*, **37**, 131.
43. (a) Mure, M. and Klinman, J.P. (1995) *J. Am. Chem. Soc.*, **117**, 8698; (b) Mure, M. and Klinman, J.P. (1995) *J. Am. Chem. Soc.*, **117**, 8707; (c) Lee, Y. and Sayre, L.M. (1995) *J. Am. Chem. Soc.*, **117**, 11823.
44. (a) Lee, Y., Ling, K.-Q., Lu, X., Silverman, R.B., Shepard, E.M., Dooley, D.M., and Sayre, L.M. (2002) *J. Am. Chem. Soc.*, **124**, 12135; (b) Lee, Y., Huang, H., and Sayre, L.M. (1996) *J. Am. Chem. Soc.*, **118**, 7241; (c) Zhang, Y., Ran, C., Zhou, G., and Sayre, L.M. (2007) *Bioorg. Med. Chem.*, **15**, 1868.
45. (a) Yuan, H., Yoo, W.-J., Miyamura, H., and Kobayashi, S. (2012) *J. Am. Chem. Soc.*, **134**, 13970; (b) Yuan, H., Yoo, W.-J., Miyamura, H., and Kobayashi, S. (2012) *Adv. Synth. Catal.*, **354**, 2899.
46. (a) Wendlandt, A.E. and Stahl, S.S. (2014) *J. Am. Chem. Soc.*, **136**, 506; (b) Wendlandt, A.E. and Stahl, S.S. (2014) *J. Am. Chem. Soc.*, **136**, 11910.
47. Jawale, D.V., Gravel, E., Shah, N., Dauvois, V., Li, H., Namboothiri, I.N.N., and Doris, E. (2015) *Chem. Eur. J.*, **21**, 7039.

15
NO$_x$ Cocatalysts for Aerobic Oxidation Reactions: Application to Alcohol Oxidation

Susan L. Zultanski and Shannon S. Stahl

15.1
Introduction

Many homogeneous catalytic aerobic oxidation reactions employ redox-active cocatalysts that facilitate catalyst re-oxidation by O_2 [1]. A prominent example is the use of Cu salts in Pd-catalyzed oxidation of ethylene to acetaldehyde (the Wacker process; see Chapter 9). Nitrogen oxides, collectively abbreviated as "NO$_x$," have proved to be another highly effective class of cocatalysts in applications ranging from Pd-catalyzed organometallic oxidation reactions to quinone- and organic nitroxyl-catalyzed oxidation of organic molecules [2]. Diverse NO$_x$ species convert into nitric oxide (NO) and/or nitrogen dioxide (NO$_2$) [3], and the aerobic NO/NO$_2$ redox cycle provides an efficient means to couple O_2 reduction to the oxidation of organic molecules. This chapter highlights the advances in aerobic alcohol oxidation chemistry (co-)catalyzed by NO$_x$, including (i) direct NO$_x$-catalyzed aerobic alcohol oxidation in the absence of other redox cocatalysts and (ii) NO$_x$-catalyzed aerobic alcohol oxidation with organic nitroxyl cocatalysts, such as TEMPO (2,2,6,6-tetramethylpiperidine-N-oxyl).

The stepwise reduction of NO$_2$ to NO occurs at standard reduction potentials very close to the standard potential for the reduction of O_2 to H_2O (Table 15.1, Eqs. (15.1–15.3)). This relationship implies that NO$_x$ cocatalysts are able to capture nearly the full thermodynamic driving force of O_2 as a terminal oxidant [4]. This favorable feature, together with the kinetically facile oxidation of NO to NO$_2$ (Table 15.1, Eq. (15.4)), contributes to the effectiveness of NO$_x$-based cocatalysts in aerobic oxidation reactions. Depending on the reaction conditions, NO$_2$ can equilibrate with other nitrogen oxide species, such as N_2O_4 and NO$^+$, which could also serve as catalytically relevant oxidants in NO$_x$-catalyzed aerobic alcohol oxidation reactions (Eq. (15.5)) [5].

$$2NO_2 \rightleftharpoons N_2O_4 \rightleftharpoons NO^+ + NO_3^- \tag{15.5}$$

Table 15.1 Thermodynamic values associated with O_2 reduction and NO_x-based redox reactions.

Equations	Reaction	$E°$ or $\Delta G°$
(15.1)[a]	$NO_2 + H^+ + e^- \rightleftharpoons HNO_2$	1.04 V
(15.2)[b]	$HNO_2 + H^+ + e^- \rightleftharpoons NO + H_2O$	1.06 V
(15.3)	$O_2 + 4H^+ + 4e^- \rightleftharpoons 2H_2O$	1.23 V
(15.4)[b]	$2NO + O_2 \rightleftharpoons 2NO_2$	−8.4 kcal/mol

a) Aqueous phase, 1 atm.
b) Gas phase value.

NO and NO_2 are stable radical species that exist as gases, and they may be used directly in oxidation reactions. It is often more convenient, however, to use solid or liquid nitrate or nitrite sources that convert into NO/NO_2 under the reaction conditions. For example, nitric acid (HNO_3) decomposes thermally to afford NO_2 (Table 15.2, Eq. (15.6)) [6]. Alternatively, the use of metal nitrates (MNO_3) and a strong acid can generate nitric acid *in situ*, which then converts into NO_2 (Table 15.2, Eqs. (15.7) and (15.6)). Metal nitrites (MNO_2) in the presence of relatively weak acids, such as acetic acid, will form nitrous acid (HNO_2), which decomposes at ambient temperature and pressure to afford NO and NO_2 (Table 15.2, Eqs. (15.8) and (15.9)) [7, 8]. Finally, alkyl nitrites (RONO, e.g., *tert*-butyl nitrite and methyl nitrite) have low RO–NO bond dissociation energies (∼40 kcal/mol) and undergo thermal or photolytic generation of NO (Table 15.2, Eq. (15.10)) [9].

Table 15.2 Common NO_x precursors and pathways to *in situ* NO_x generation.

Equations	NO_x source	Reaction(s) to generate NO_x
(15.6)	HNO_3	$4HNO_3 \rightleftharpoons 4NO_2 + 2H_2O + O_2$
(15.7)	MNO_3	$MNO_3 + H^+ \rightleftharpoons HNO_3$ (then Eq. 15.6)
(15.8)	MNO_2	$MNO_2 + HX \rightleftharpoons HNO_2 + MX$
(15.9)	MNO_2	$2HNO_2 \rightleftharpoons NO_2 + NO + H_2O$
(15.10)[a]	RONO	$RONO \underset{}{\overset{\Delta \text{ or } h\nu}{\rightleftharpoons}} RO^\bullet + NO$

a) R = alkyl.

15.2
Chemistry and Catalysis

15.2.1
Aerobic Alcohol Oxidation with NO_x in the Absence of Other Redox Cocatalysts

There have been several reports on the use of catalytic NO_x sources for aerobic alcohol oxidation in the absence of other redox mediators. In 1981, Tovrog disclosed that catalytic quantities of Lewis acids and Co(III)–nitro complexes could be used for the aerobic oxidation of benzyl alcohol and cycloheptanol in moderate and low yields, respectively (Scheme 15.1, top) [10]. Under anaerobic conditions with stoichiometric pyCo(TPP)NO_2 (py = pyridine, TPP = tetraphenylporphyrin), Co(TPP)NO was detected as the reaction by-product. This observation provided the basis for a proposed mechanism in which alcohol addition to a Lewis acid-coordinated LCoNO_2 complex would generate the adduct 1 (Scheme 15.1, bottom), which then afforded the desired carbonyl product and LCo(NO). The latter species can undergo aerobic oxidation to LCo(NO_2).

In another approach, Beletskaya reported the aerobic oxidation of benzyl alcohol to benzaldehyde using catalytic NH_4NO_3 in trifluoroacetic acid (TFA) (Scheme 15.2a) [11]. This was based on previous reports on the stoichiometric NO_x-mediated oxidation of benzyl alcohol in strongly acidic solutions [12]. In 1994, Levina disclosed a system for aerobic aliphatic alcohol oxidation using catalytic $NaNO_2$ in perchloric acid (Scheme 15.2b) [13].

Subsequent efforts led to more mild reaction conditions for aerobic alcohol oxidation. Kakimoto disclosed the use of nanoshell carbon (NSC) catalysts and 67 mol% HNO_3 for the aerobic oxidation of primary and secondary benzylic

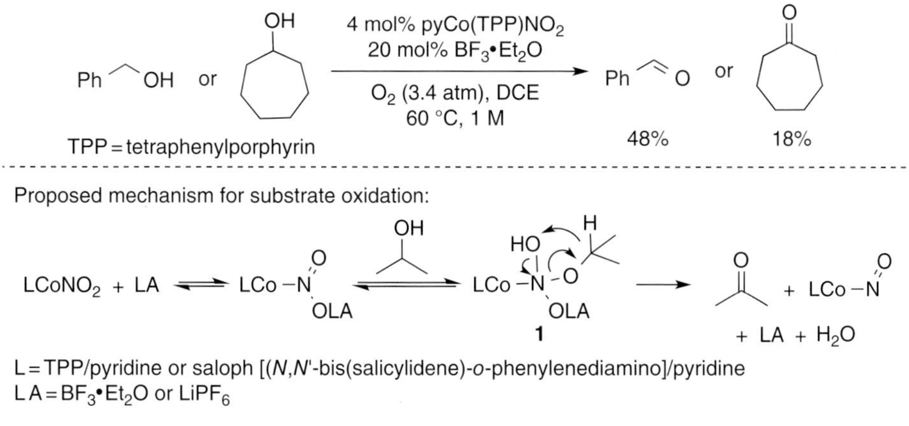

Scheme 15.1 (Top) LCo(III)NO_2- and Lewis acid-catalyzed aerobic oxidation of benzyl alcohol and cycloheptanol. (Bottom) Proposed mechanism for substrate oxidation.

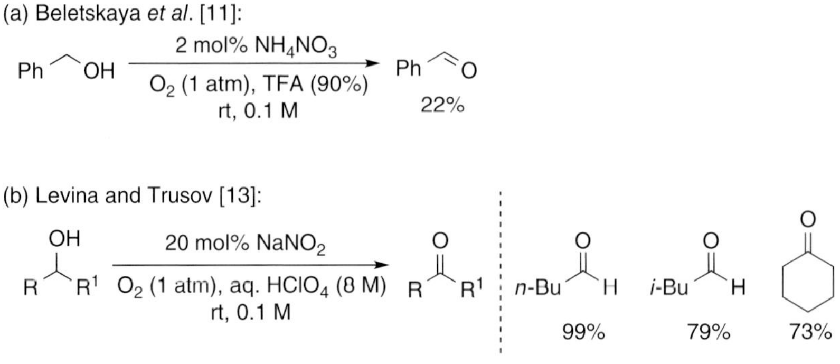

(a) Beletskaya et al. [11]:

Ph–OH → [2 mol% NH$_4$NO$_3$, O$_2$ (1 atm), TFA (90%), rt, 0.1 M] → Ph–CHO
22%

(b) Levina and Trusov [13]:

R–CH(OH)–R^1 → [20 mol% NaNO$_2$, O$_2$ (1 atm), aq. HClO$_4$ (8 M), rt, 0.1 M] → R–C(O)–R^1

n-Bu–C(O)–H 99%
i-Bu–C(O)–H 79%
cyclohexanone 73%

Scheme 15.2 (a) NO$_x$-catalyzed aerobic benzyl alcohol oxidation in TFA. (b) NO$_x$-catalyzed aerobic aliphatic alcohol oxidation in perchloric acid.

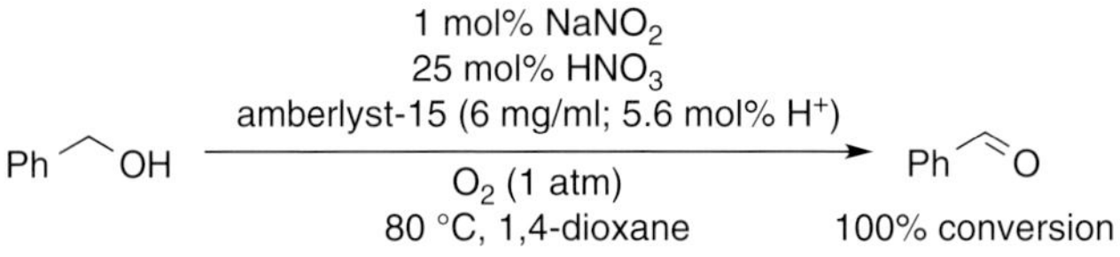

Ph–OH → [1 mol% NaNO$_2$, 25 mol% HNO$_3$, amberlyst-15 (6 mg/ml; 5.6 mol% H$^+$), O$_2$ (1 atm), 80 °C, 1,4-dioxane] → Ph–CHO
100% conversion

Scheme 15.3 Amberlyst-15- and NO$_x$-catalyzed aerobic benzylic alcohol oxidation.

alcohols in 1,4-dioxane [14]. Hermans showed that amberlyst-15 was an effective heterogeneous acid promoter for aerobic oxidation of benzyl alcohol in 1,4-dioxane (Scheme 15.3). The reaction can also be conducted in neat alcohol substrate [15]. Reaction engineering efforts led to a flow process that was demonstrated with primary benzylic alcohols, including electron-rich and electron-deficient examples, 2-octanol and hydroxyacetone [16].

Hermans provided a number of insights into the reaction mechanism (Scheme 15.4). One key observation is that benzyl nitrite, formed via the reaction between HNO$_2$ and benzyl alcohol, appears and decays over the reaction time course. Under anaerobic conditions with catalytic amberlyst-15 but without NO$_x$, benzyl nitrite fully converts to benzaldehyde and HNO, implicating its role as an intermediate (Scheme 15.4, steps I and II). In the anaerobic reaction, N$_2$O was detected as a by-product. N$_2$O can form through dimerization and then dehydration of HNO. Because N$_2$O is inert and cannot be converted back into active NO$_x$ species, it is important to inhibit its formation. The presence of NO$_2$ gas was shown to reduce N$_2$O formation, leading to the proposal that NO$_2$ is an intermediate for the oxidation of HNO back to HNO$_2$ (Scheme 15.4, step III).

15.2.2
Aerobic Alcohol Oxidation with NO$_x$ and Organic Nitroxyl Cocatalysts

Stable organic nitroxyl radicals such as TEMPO and variants thereof have been known for nearly half a century (Figure 15.1) [17]. Dimerization of these compounds is thermodynamically disfavored due to steric effects and

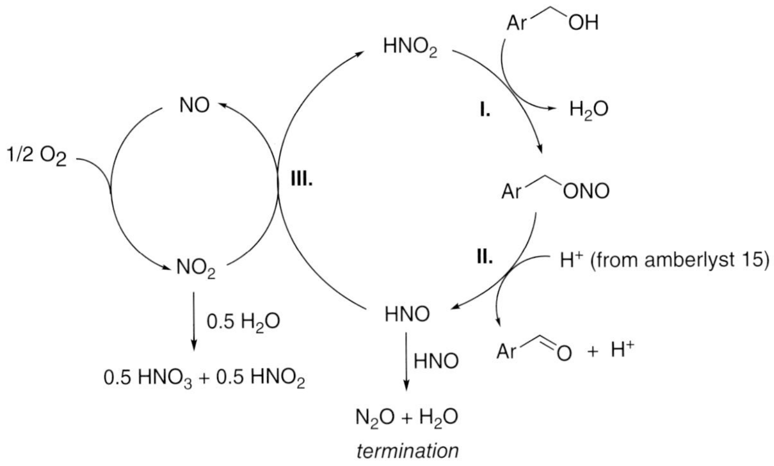

Scheme 15.4 Proposed catalytic cycle for the amberlyst-15- and NO$_x$-catalyzed aerobic oxidation of benzylic alcohols.

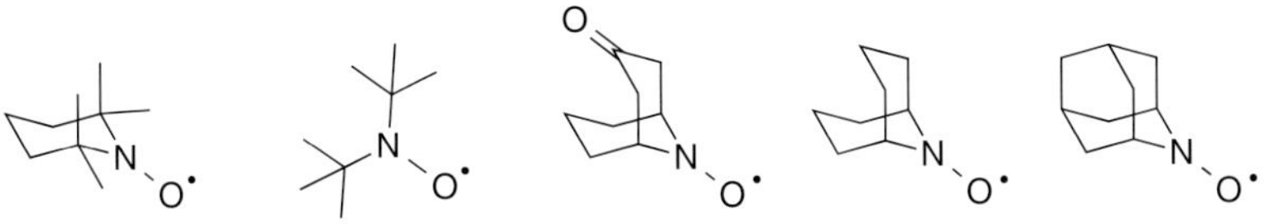

Figure 15.1 Examples of common stable nitroxyl radicals.

resonance-stabilization of the unpaired electron across the N–O bond (a two-center three-electron bond) [18]. The lack of α-hydrogen atoms in TEMPO and di-*t*-butyl nitroxyl prevents decomposition via disproportionation of two nitroxyls to generate hydroxylamine and a nitrone. This decomposition pathway is also prevented in the bicyclic nitroxyls because nitrone formation would result in the formation of a double bond involving a bridgehead carbon atom (and thus would violate Bredt's rule) [19]. These structural features underlie the bench stability of many dialkyl nitroxyl radicals.

There is extensive history on the use of stoichiometric and catalytic organic nitroxyls for alcohol oxidation, wherein the key step involves a reaction between the alcohol and an *N*-oxoammonium salt (Scheme 15.5, featuring TEMPO) [18, 20]. The *N*-oxoammonium salt can be formed *in situ* from the corresponding nitroxyl radical using various oxidants, such as NaOCl or NO$_2$ (Scheme 15.5, top left), or by acid-induced disproportionation of the nitroxyl into *N*-oxoammonium and hydroxylamine species (Scheme 15.5, bottom left) [21]. Stable *N*-oxoammonium salts have also been isolated and used directly as reagents or catalysts for alcohol oxidation [22]. The pH-dependent mechanism of the reaction of the *N*-oxoammonium salt with alcohols has been studied

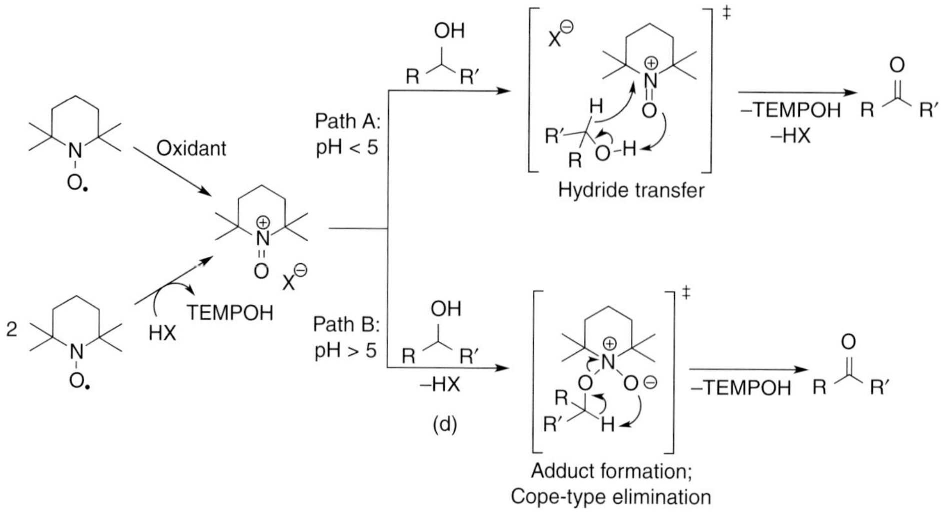

Scheme 15.5 Generation of N-oxoammonium salts from nitroxyl radicals, and the pH-dependent mechanism of oxoammonium reactivity with alcohols: hydride transfer at pH < 5 (Path A) and adduct formation/Cope-type elimination at pH > 5 (Path B).

experimentally and computationally [23]. At pH < 5, a bimolecular hydride transfer mechanism prevails. Secondary alcohols, which are better hydride donors, generally react faster than primary alcohols (Scheme 15.5, Path A). At pH > 5, a mechanism involving the formation of an adduct followed by Cope-type elimination is favored, and sterically less-hindered primary alcohols generally react faster than secondary alcohols (Scheme 15.5, Path B).

TEMPO and other organic nitroxyls have been used as catalysts in combination with numerous stoichiometric oxidants, such as sodium hypochlorite [24], PhI(OAc)$_2$ [25], and sodium chlorite [26]. A number of recent studies have shown that NO$_x$-based redox cocatalysts enable these reactions to be conducted with O$_2$ as the terminal oxidant [27]. The general catalytic cycle for these aerobic nitroxyl/NO$_x$-catalyzed alcohol oxidation reactions is depicted in Scheme 15.6a. A variation of this approach features halides as additives, in which the X$_2$/HX redox couple is believed to mediate the NO$_2$/NO and oxoammonium/hydroxylamine redox couples (Scheme 15.6b).

Minisci reported the first nitroxyl-catalyzed aerobic alcohol oxidation in the presence of NO$_x$ precursors in 2001 (Scheme 15.7a) [28]. The optimal reaction conditions included a combination of catalytic Mn(NO$_3$)$_2$ and Co(NO$_3$)$_2$ with TEMPO (10 mol%) under 1 atm of O$_2$. The authors demonstrated that Mn(NO$_3$)$_2$, Co(NO$_3$)$_2$, and Cu(NO$_3$)$_2$ were independently effective catalysts, implicating the nitrate anion as an important contributor to the reaction. Primary and secondary benzylic alcohols, primary allylic alcohols, and primary and secondary aliphatic alcohols were oxidized to the corresponding aldehydes/ketones in excellent yields.

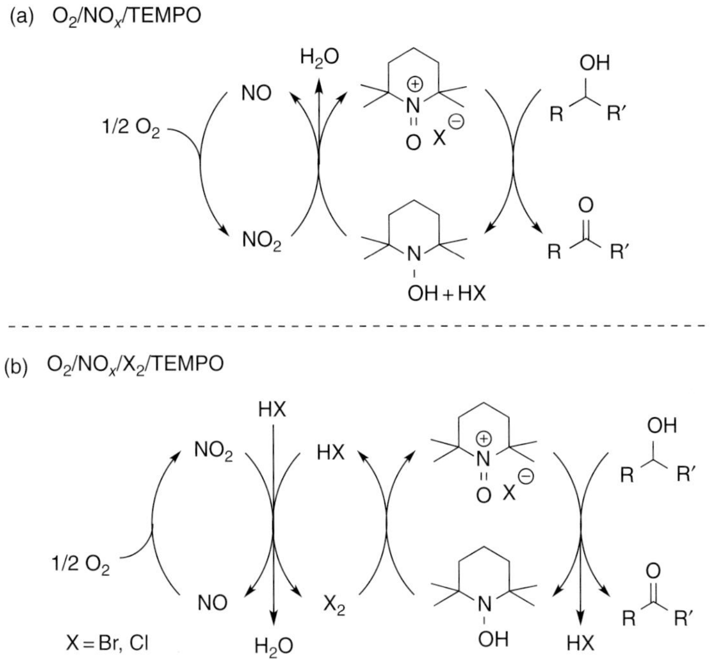

Scheme 15.6 General proposed catalytic cycles for (a) NO_x- and nitroxyl-catalyzed aerobic alcohol oxidation and (b) NO_x^-, X_2-, and nitroxyl-catalyzed aerobic alcohol oxidation.

The mechanism of the reaction was not studied, but a catalytic cycle similar to that in Scheme 15.6a is likely. Subsequently, there have been numerous other reports of nitroxyl/NO_x catalytic systems for aerobic alcohol oxidation [30], including the chemoselective oxidation of primary over secondary aliphatic alcohols [31], and application to the oxidation of lignin, in which secondary benzylic alcohols are oxidized in preference to primary aliphatic alcohols [32].

In 2004, Hu disclosed the first transition-metal-free aerobic alcohol oxidation reaction, employing catalytic $NaNO_2$, Br_2, and TEMPO (1 mol%) under 4 or 9 atm of O_2 (Scheme 15.7b) [29]. Hu proposed the mechanism depicted in Scheme 15.6b. The inspiration for Hu's conditions can be found in previously disclosed alcohol oxidation systems using catalytic TEMPO and stoichiometric X_2, in which stoichiometric weak bases were used to neutralize the by-product HX [33]. In Hu's report, HX was instead reoxidized to X_2 using NO_x and O_2.

The Hu method exhibits similar substrate scope to that reported by Minisci: primary and secondary benzylic, primary allylic, and primary and secondary aliphatic alcohols were shown to be viable substrates. The presence of HBr/Br_2 can lead to undesirable bromination side reactions, depending on the alcohol substrate. However, the authors show that in some cases, with proper choice the bromine source, these side reactions can be avoided. For example, with

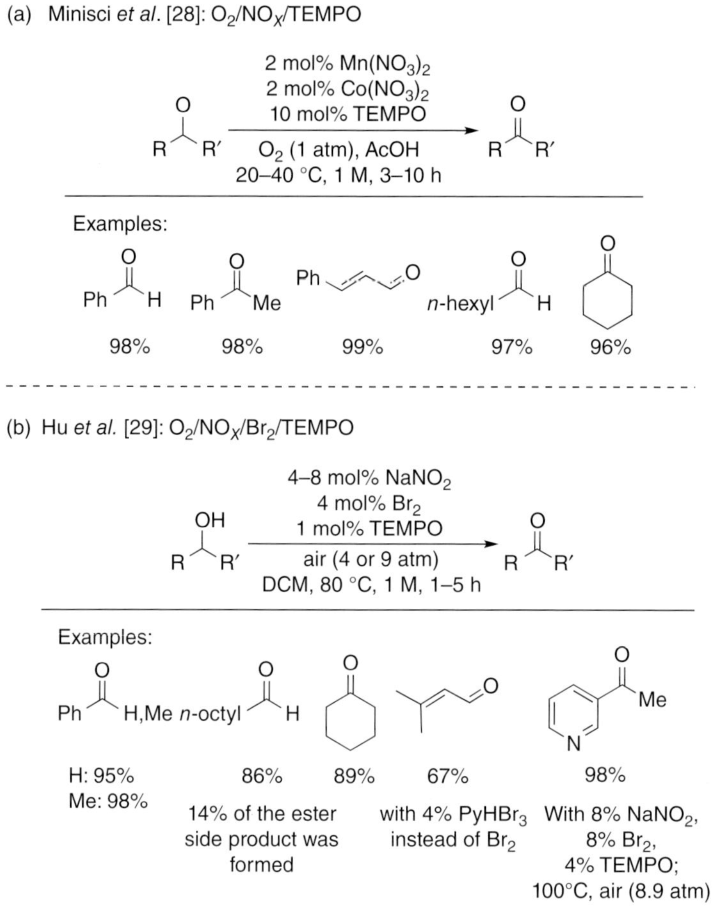

(a) Minisci et al. [28]: O_2/NO_x/TEMPO

$$R-C(=O)-R' \xrightarrow[\text{20–40 °C, 1 M, 3–10 h}]{\substack{\text{2 mol\% Mn(NO}_3)_2 \\ \text{2 mol\% Co(NO}_3)_2 \\ \text{10 mol\% TEMPO} \\ O_2 \text{ (1 atm), AcOH}}} R-C(=O)-R'$$

Examples:

Ph–CHO (98%), Ph–C(=O)–Me (98%), Ph–CH=CH–CHO (99%), n-hexyl–CHO (97%), cyclohexanone (96%)

(b) Hu et al. [29]: O_2/NO_x/Br_2/TEMPO

$$R-CH(OH)-R' \xrightarrow[\text{DCM, 80 °C, 1 M, 1–5 h}]{\substack{\text{4–8 mol\% NaNO}_2 \\ \text{4 mol\% Br}_2 \\ \text{1 mol\% TEMPO} \\ \text{air (4 or 9 atm)}}} R-C(=O)-R'$$

Examples:

Ph–C(=O)–H,Me (H: 95%, Me: 98%); n-octyl–CHO (86%, 14% of the ester side product was formed); cyclohexanone (89%); (CH$_3$)$_2$C=CH–CHO (67%, with 4% PyHBr$_3$ instead of Br$_2$); 3-acetylpyridine (98%, With 8% NaNO$_2$, 8% Br$_2$, 4% TEMPO; 100°C, air (8.9 atm))

Scheme 15.7 Minisci's Mn(NO$_3$)$_2$/Co(NO$_3$)$_2$/TEMPO-catalyzed aerobic alcohol oxidation and Hu's NaNO$_2$/Br$_2$/TEMPO-catalyzed aerobic alcohol oxidation.

the allylic alcohol 3-methyl-2-buten-1-ol, bromination of the double bond is avoided by employing PyHBr$_3$ instead of Br$_2$ (Scheme 15.7b, bottom). More recently, numerous other nitroxyl/halide/NO$_x$ catalytic systems for aerobic alcohol oxidation have been developed [34], including reactions that operate under ambient temperature and pressure [35, 36].

When the halide source was omitted from the nitroxyl/halide/NO$_x$ catalyst systems, little catalytic activity was observed, indicating that the halide-free nitroxyl/NO$_x$ catalytic cycle in Scheme 15.6a is not viable under the reported conditions. This outcome may reflect ineffective conversion of the NO$_x$ source (e.g., a nitrate or a nitrite) into NO/NO$_2$ in the absence of a halogen.

Iwabuchi demonstrated that sterically less bulky nitroxyls enable increased catalytic activity in nitroxyl/NO$_x$-catalyzed aerobic alcohol oxidation reactions. A broad array of densely functionalized and/or sterically bulky alcohol

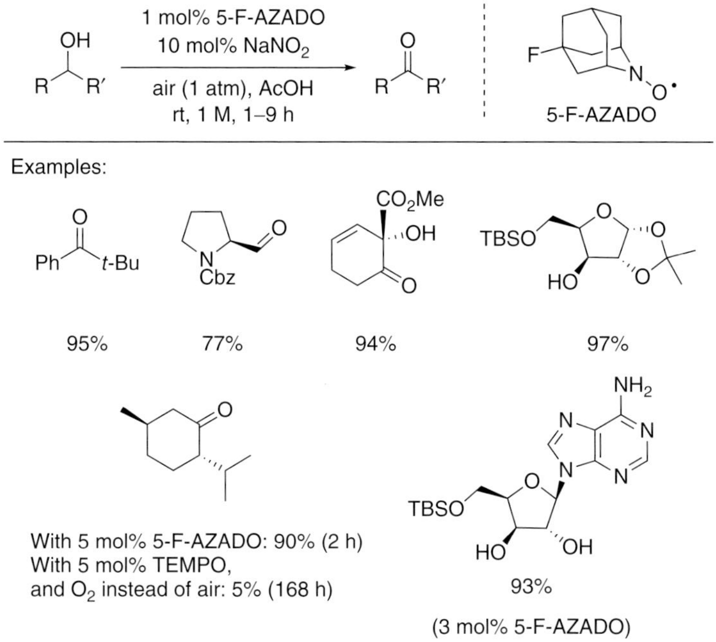

Scheme 15.8 Catalytic 5-F-AZADO and NaNO$_2$ aerobic oxidation of a broad array of alcohols.

substrates could be efficiently oxidized to aldehydes and ketones using 5-fluoro-2-azaadamantane *N*-oxyl (5-F-AZADO), ambient air pressure, and ambient temperature (Scheme 15.8) [22a]. The utility of the method is illustrated in the oxidation of menthol, which proceeds in only 5% yield with TEMPO (Scheme 15.8, bottom left), but is oxidized in 90% yield with 5-F-AZADO. Subsequently, Stahl developed a similar set of methods for aerobic alcohol oxidation with broad substrate scope for secondary alcohols, wherein 9-azabicyclo[3.3.1]nonane-*N*-oxyl (ABNO, Figure 15.1) and 9-azabicyclo[3.3.1]nonan-3-one-*N*-oxyl (keto-ABNO, Figure 15.1) were employed as catalytic nitroxyls (Scheme 15.9) [37]. An advantage of the latter methods is the commercial availability of ABNO and keto-ABNO. The Iwabuchi and Stahl reports demonstrate the broadest substrate scope for nitroxyl/NO$_x$-catalyzed aerobic alcohol oxidation to date.

15.3
Prospects for Scale-Up

The alcohol oxidation reactions described herein use inexpensive O$_2$ and NO$_x$ sources, and there is potential for their use in large-scale industrial processes. Continuous flow methods provide a particularly strategic opportunity for large-scale applications (see Chapter 23). Hermans and coworkers demonstrated a segmented-flow method for their amberlyst-15/NO$_x$-catalyzed aerobic oxidation

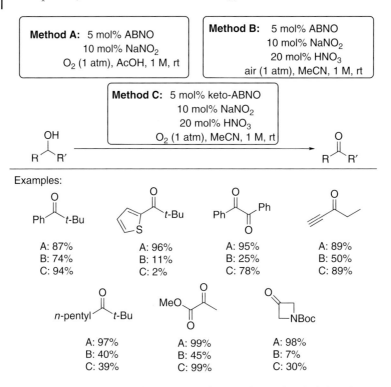

Scheme 15.9 Catalytic ABNO/keto-ABNO and NaNO$_2$ for aerobic alcohol oxidation.

of alcohols. The use of the amberlyst resin is appealing because it avoids corrosive acidic solvents that are commonly used in these reactions [12]. A small set of benzylic and aliphatic alcohols was shown to reach complete conversion in less than 30 s in the flow reactor [16b]. The extraordinary efficiency of these reactions and the low cost of the reagents make them eminently practical for scale-up, provided the method is effective for the target of interest.

For nitroxyl/NO$_x$-catalyzed aerobic alcohol oxidation processes, much attention has been directed toward developing recoverable and recyclable nitroxyl catalysts. In an early example, Minisci employed a macrocyclic catalyst with four appended 4-amino-TEMPO groups connected via 1,3,5-triazines and aliphatic linkers developed by Ciba [38, 39]. This catalyst is not genotoxic, in contrast to TEMPO, and therefore may be more amenable to use in the late stage of a synthetic sequence toward an active pharmaceutical ingredient. The amines on the macrocyclic catalyst were protonated under the reaction conditions, enabling simple separation through aqueous extraction. Various other heterogeneous nitroxyl catalysts that can be easily separated or recycled by filtration have also been developed, such as polymer-supported TEMPO catalysts (e.g., appended to poly(ethylene) glycol or FibreCat™) [40] and silica-supported adducts of TEMPO-based or bicyclic nitroxyls [41]. In another strategy, TEMPO-based ionic liquids have been used that can easily be removed from reaction mixtures via extraction [42]. Finally, progress has been made toward the application of TEMPO/NO$_x$-catalyzed aerobic alcohol oxidation in flow. For example, silica-supported TEMPO was employed toward the oxidation of benzylic and aliphatic alcohols (see Chapter 23 for details) [43].

It should be noted that nitrogen oxide species are often incompatible with certain functional groups (e.g., aliphatic amines, phosphines, electron-rich aromatics), but NO$_x$ sources have been used for nitration and nitrosation reactions, which would be undesirable in the context of alcohol oxidation [44]. Therefore, potential functional group limitations should be taken into consideration.

15.4
Conclusions

In conclusion, NO$_x$-catalyzed aerobic alcohol oxidation is a growing research area and offers the potential for metal-free aerobic oxidation reactions. Applications of nitroxyl/NO$_x$-catalyzed oxidation reactions that employ the sterically unhindered bicyclic nitroxyls (e.g., F-AZADO, ABNO, keto-ABNO) are especially effective with a broad range of substrates bearing diverse functional groups, and reaction-engineering advances that allow using these reactions in continuous processes should help to enhance reaction efficiency and promote safety.

References

1. Piera, J. and Bäckvall, J.-E. (2008) *Angew. Chem. Int. Ed.*, **47**, 3506–3523.
2. For reviews, see: (a) Fairlamb, I.J.S. (2015) *Angew. Chem. Int. Ed.*, **54**, 10415–10427; (b) Miles, K.C. and Stahl, S.S. (2015) *Aldrichim. Acta*, **48**, 8–10; (c) Wendlandt, A.E. and Stahl, S.S. (2015) *Angew. Chem. Int. Ed.*, **54**, 14638–14658.
3. For information on the properties of nitrogen oxides, with particular emphasis on NO, NO$_2$, and their complex dimers, see: Bohle, D.S. (2010) in *Stable Radicals: Fundamental and Applied Aspects of Odd-Electron Compounds* (ed R.G. Hicks), John Wiley & Sons, Ltd., Chichester, pp. 147–171.
4. Gerken, J.B. and Stahl, S.S. (2015) *ACS Cent. Sci.*, **1**, 234–243.
5. Addison, C. (1980) *Chem. Rev.*, **80**, 21–39.
6. Theimann, M., Scheibler, E., and Weigland, K.W. (2005) Nitric acid, nitrous acid, and nitrogen oxides, in *Ullmann's Encyclopedia of Industrial Chemistry*, Wiley-VCH Verlag GmbH, Weinheim.
7. Bosch, E., Rathore, R., and Kochi, J.K. (1994) *J. Org. Chem.*, **59**, 2529–2536.
8. Markovits, G.Y., Schwartz, S.E., and Newman, L. (1981) *Inorg. Chem.*, **20**, 445–450.
9. Uchiumi, S.-I., Ataka, K., and Matsuzaki, T. (1999) *J. Organomet. Chem.*, **576**, 279–289.
10. Tovrog, B., Diamond, S.E., Mares, F., and Szalkiewicz, A. (1981) *J. Am. Chem. Soc.*, **103**, 3522–3526.
11. Rodkin, M.A., Shtern, M.M., Cheprakov, A.V., Makhon'kov, D.I., Mardaleishvili, R.E., and Beletskaya, I.P. (1988) *J. Org. Chem. USSR*, **24**, 434–440.
12. For examples, see: (a) Ross, D.S., Lu, C.-L., Hum, G.P., and Malhotra, R. (1986) *Int. J. Chem. Kinet.*, **18**, 1277–1288; (b) Moodie, R. and

Richards, S. (1986) *J. Chem. Soc., Perkin Trans.* **2**, 1833–1837; (c) Dorfman, Y.A., Emel'yanova, V.S., and Shokorova, L.A. (1988) *Kinet. Catal.*, **29**, 688–692.

13. Levina, A. and Trusov, S. (1994) *J. Mol. Catal.*, **88**, L121–L123.
14. Kuang, Y., Islam, N.M., Nabae, Y., Hayakawa, T., and Kakimoto, M. (2010) *Angew. Chem. Int. Ed.*, **49**, 436–440.
15. Aellig, C., Girard, C., and Hermans, I. (2011) *Angew. Chem. Int. Ed.*, **50**, 12355–12360.
16. (a) Aellig, C., Neuenschwander, U., and Hermans, I. (2012) *ChemCatChem*, **4**, 525–529; (b) Aellig, C., Scholz, D., and Hermans, I. (2012) *ChemSusChem*, **5**, 1732–1736.
17. (a) TEMPO: Lebedev, O.L. and Kazarnovskii, S.N. (1960) *Zh. Obshch. Khim.*, **30**, 1631; Di-*tert*-butylnitroxide: (b) Hoffman, A.K. and Henderson, A.T. (1961) *J. Am. Chem. Soc.*, **83**, 4671–4672; keto-ABNO: (c) Dupeyre, R.-M. and Rassat, A. (1966) *J. Am. Chem. Soc.*, **88**, 3180–3181; ABNO: (d) Mendenhall, G.D. and Ingold, K.U. (1973) *J. Am. Chem. Soc.*, **95**, 6395–6400; AZADO: (e) Dupeyre, R.-M. and Rassat, A. (1975) *Tetrahedron Lett.*, **16**, 1839–1840.
18. Tebben, L. and Studer, A. (2011) *Angew. Chem. Int. Ed.*, **50**, 5034–5068.
19. Nilsen, A. and Braslau, R. (2006) *J. Polym. Sci., Part A: Polym. Chem.*, **44**, 697–717.
20. Other reviews: (a) de Nooy, A.E.J., Besemer, A.C., and van Bekkum, H. (1996) *Synthesis*, **1996**, 1153–1174; (b) Adam, W., Saha-Möller, C.R., and Ganeshpure, P.A. (2001) *Chem. Rev.*, **101**, 3499–3548; (c) Sheldon, R.A., Arend, I.W.C.E., ten Brink, G.-T., and Dijksman, A. (2002) *Acc. Chem. Res.*, **35**, 774–781.
21. (a) Ma, Z. and Bobbitt, J.M. (1991) *J. Org. Chem.*, **56**, 6110–6114; (b) Banwell, M.G., Bridges, V.S., Dupuche, J.R., Richards, S.L., and Walter, J.M. (1994) *J. Org. Chem.*, **59**, 6338–6343; (c) Ma, Z., Huang, Q., and Bobbitt, J.M. (1993) *J. Org. Chem.*, **58**, 4837–4843.
22. Examples: (a) Shibuya, M., Osada, Y., Sasano, Y., Tomizawa, M., and Iwabuchi, Y. (2011) *J. Am. Chem. Soc.*, **133**, 6497–6500; (b) Mercadante, M.A., Kelly, C.B., Bobbitt, J.M., Tilley, L.J., and Leadbeater, N.E. (2013) *Nat. Protoc.*, **8**, 666–676.
23. Bobbitt, J.M., Brücker, C., and Merbouh, N. (2009) *Org. React.*, **74**, 103.
24. Selected examples: (a) Anelli, P.L., Biffi, C., Montanari, F., and Quici, S. (1987) *J. Org. Chem.*, **52**, 2559–2562; (b) Anelli, P.L., Montanari, F., and Quici, S. (1990) *Org. Synth.*, **69**, 212–217.
25. Selected examples: (a) De Mico, A., Margarita, R., Parlanti, L., Vescovi, A., and Piancatelli, G. (1997) *J. Org. Chem.*, **62**, 6974–6977; (b) Epp, J.B. and Widlanski, T.S. (1999) *J. Org. Chem.*, **64**, 293–295.
26. Selected examples: (a) Zhao, M., Li, J., Mano, E., Song, Z., Tschaen, D.M., Grabowski, E.J.J., and Reider, P.J. (1999) *J. Org. Chem.*, **64**, 2564–2566; (b) Zhao, M.M., Li, J., Mano, E., Song, Z.J., and Tschaen, D.M. (2005) *Org. Synth.*, **81**, 195–203.
27. Reviews: (a) Wertz, S. and Studer, A. (2013) *Green Chem.*, **15**, 3116–3134; (b) Cao, Q., Dornan, L.M., Rogan, L., Hughes, N.L., and Muldoon, M.J. (2014) *Chem. Commun.*, **50**, 4524–4543.
28. Cecchetto, A., Fontana, F., Minisci, F., and Recupero, F. (2001) *Tetrahedron Lett.*, **42**, 6651–6653.
29. Liu, R., Liang, X., Dong, C., and Hu, X. (2004) *J. Am. Chem. Soc.*, **126**, 4112–4113.
30. For examples, see: (a) Wang, N., Liu, R., Chen, J., and Liang, X. (2005) *Chem. Commun.*, 5322; (b) He, X., Shen, Z., Mo, W., Sun, N., Hu, B., and Hu, X. (2009) *Adv. Synth. Catal.*, **351**, 89–92; (c) Yin, W., Chu, C., Lu, Q., Tao, J., Liang, X., and Liu, R. (2010) *Adv. Synth. Catal.*, **352**, 113–118; (d) Kuang, Y., Rokubuichi, H., Nabae, Y., Hayakawa, T., and Kakimoto, M.-A. (2010) *Adv. Synth. Catal.*, **252**, 2635–2642; (e) Wertz, S. and Studer, A. (2011) *Adv. Synth. Catal.*, **353**, 69–72.
31. Tanielyan, S.K., Augustine, R.L., Marin, N., Alvez, G., and Stapley, J. (2012) *Top. Catal.*, **55**, 556–564.
32. Rahimi, A., Azarpira, A., Kim, H., Ralph, J., and Stahl, S.S. (2013) *J. Am. Chem. Soc.*, **135**, 6415–6418.

33. (a) Bjørsvik, H.-R., Liguori, L., Constantino, F., and Minisci, F. (2002) *Org. Process Res. Dev.*, **6**, 197–200; (b) Miller, R.A. and Hoerrner, R.S. (2003) *Org. Lett.*, **5**, 285–287.
34. For examples, see: (a) Liu, R., Dong, C., Liang, X., Wang, X., and Hu, X. (2005) *J. Org. Chem.*, **70**, 729–731; (b) Herrerías, C.I., Zhang, T.Y., and Li, C.-J. (2006) *Tetrahedron Lett.*, **47**, 13; (c) Tanielyan, S.K., Augustine, R.L., Meyer, O., and Korell, M. (2006) Process for transition metal free catalytic aerobic oxidation of alcohols under mild conditions using stable free nitroxyl radicals. US Patent US 7.030.279 B1; (d) Xie, Y., Mo, W., Xu, D., Shen, Z., Sun, N., Hu, B., and Hu, X. (2007) *J. Org. Chem.*, **72**, 4288–4291; (e) Yang, G., Wang, W., Zhu, W., An, C., Gao, X., and Song, M. (2010) *Synlett*, **3**, 437–440.
35. Zhang, J., Jiang, Z., Zhao, D., He, G., Zhou, S., and Han, S. (2013) *Chin. J. Chem.*, **31**, 794–798.
36. Wang, X., Liu, R., Jin, Y., and Liang, X. (2008) *Chem. Eur. J.*, **14**, 2679–2685.
37. Lauber, M.B. and Stahl, S.S. (2013) *ACS Catal.*, **3**, 2612–2616.
38. Zedda, A., Sala, M., and Schneider, A. (2002) Patent WO 02/058844 A1 (Ciba Speciality Chemicals Holding, Inc.).
39. Minisci, F., Recupero, F., Rodinò, M., Sala, M., and Schneider, A. (2003) *Org. Process Res. Dev.*, **7**, 794–798.
40. (a) Gilhespy, M., Lok, M., and Baucherel, X. (2005) *Chem. Commun.*, 1085–1086; (b) Benaglia, M., Puglisi, A., Holczknecht, O., Quici, S., and Pozzi, G. (2005) *Tetrahedron*, **61**, 12058–12064; (c) Gilhespy, M., Lok, M., and Baucherel, X. (2006) *Catal. Today*, **117**, 114–119.
41. (a) Karimi, B., Biglari, A., Clark, J.H., and Budarin, V. (2007) *Angew. Chem. Int. Ed.*, **46**, 7210–7213; (b) Di, L. and Hua, Z. (2011) *Adv. Synth. Catal.*, **353**, 1253–1259; (c) Karimi, B. and Farhangi, E. (2011) *Chem. Eur. J.*, **17**, 6056–6060; (d) Karimi, B., Farhangi, E., Vali, H., and Vahdati, S. (2014) *ChemSusChem*, **7**, 2735–2741.
42. (a) Miao, C.-X., He, L.-N., Wang, J.-Q., and Wang, J.-L. (2009) *Adv. Synth. Catal.*, **351**, 2209–2216; (b) Zhu, J., Wang, P.-C., and Ming, L. (2013) *Synth. Commun.*, **43**, 1871–1881; (c) Karimi, B. and Badreh, E. (2011) *Org. Biomol. Chem.*, **9**, 4194–4198.
43. Aellig, C., Scholz, D., Conrad, S., and Hermans, I. (2013) *Green Chem.*, **15**, 1975–1980.
44. Shiri, M., Zolfigol, M.A., Kruger, H.G., and Tanbakouchian, Z. (2010) *Tetrahedron*, **66**, 9077–9106.

16
N-Hydroxyphthalimide (NHPI)-Organocatalyzed Aerobic Oxidations: Advantages, Limits, and Industrial Perspectives

Lucio Melone and Carlo Punta

16.1
Introduction

N-Hydroxyphthalimide (NHPI) is an efficient organocatalyst, capable of promoting free-radical processes via hydrogen atom transfer (HAT) reactions.

The first catalytic use of NHPI was reported in 1977 by Grochowski and co-workers for the addition of ethers to diethyl azodicarboxylate and the oxidation of 2-propanol to acetone [1]. A few years later, Masui proposed the use of NHPI as a mediator for the electrolytic oxidation of alcohols to ketones [2]. In both cases, the authors suggested the formation of the phthalimide *N*-oxyl (PINO) radical as a key intermediate, being responsible for the catalytic cycle by abstracting hydrogen atoms from activated C–H bonds. However, PINO production in the reaction medium was not experimentally proven until 1995 when, thanks to the pioneering work of Ishii *et al.*, a triplet signal, originating from PINO, was detected by electron spin resonance (ESR) technique after exposing NHPI to molecular oxygen [3]. With this work, regarding the oxidation of alkanes and alcohols, Ishii initiated the relatively recent history of oxidations catalyzed by NHPI.

Since then, and thanks to the investigation of the reaction mechanisms involving NHPI, mainly conducted by Ishii and Minisci's and Pedulli's groups, the potentiality of this organocatalyst has been significantly extended.

In the last two decades, hundreds of papers reported the use of NHPI for promoting the homogeneous selective oxidation of a wide range of organic substrates (including alcohols, ketones, ethers, amines, amides, silanes, alkynes, alkenes, alkanes, and alkyl aromatics) and the one-pot free-radical synthesis of complex molecules, involving molecular oxygen, directly or indirectly [4].

In spite of the great interest that NHPI is attracting from the scientific community and industrial companies, its concrete use for scaled productions is still

Liquid Phase Aerobic Oxidation Catalysis: Industrial Applications and Academic Perspectives,
First Edition. Edited by Shannon S. Stahl and Paul L. Alsters.
© 2016 Wiley-VCH Verlag GmbH & Co. KGaA. Published 2016 by Wiley-VCH Verlag GmbH & Co. KGaA.

limited to few examples. The reasons rely onto three partially interconnected obstacles which need to be overcome for the final launch of this organocatalyst:

1) Even if NHPI could be considered a cheap, nontoxic molecule, as it is easily produced by the reaction between phthalic anhydride and hydroxylamine, its cost would significantly affect the overall economy of the process, if the final product is not of high added value and/or the homogeneous catalyst is not completely recovered from the reaction medium and efficiently recycled. For this reason, till now, this *N*-hydroxy derivative has found industrial application only in small–medium scale productions.
2) The polar character of this molecule often requires the use of polar cosolvents to ensure its complete solubilization. This aspect not only affects the productivity and consequently, once again, the economy of the process but also raises important environmental issues. New smart-designed lipophilic derivatives analogous to NHPI need to be developed to solve this problem.
3) It has been demonstrated how PINO may undergo self-decomposition, following a first-order self-decay under classical reaction conditions (Scheme 16.1) [5].

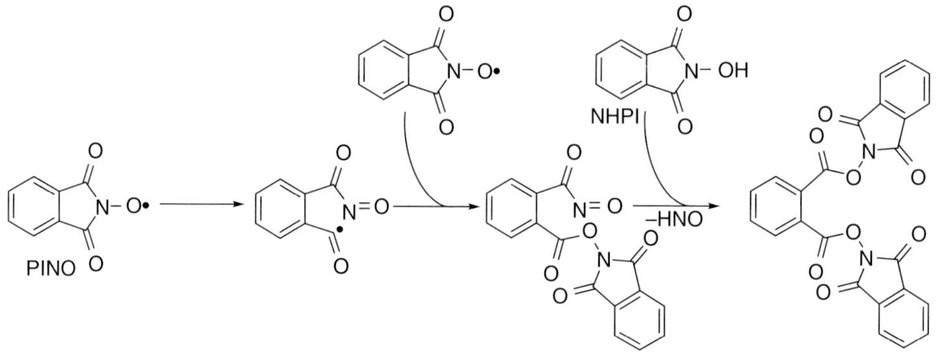

Scheme 16.1 Self-decomposition of PINO.

This phenomenon is significant when operating at temperatures higher than 80 °C, so that the activation of NHPI under mild conditions becomes a crucial aspect to be considered.

16.2
Chemistry and Catalysis

The catalytic cycle promoted by the NHPI/PINO system is reported in Scheme 16.2. Once generated *in situ* from NHPI (*Initiation*), PINO undergoes hydrogen abstraction from a generic C–H bond, forming once again NHPI and a carbon-centered radical (path i). The latter reacts with molecular oxygen, leading to the

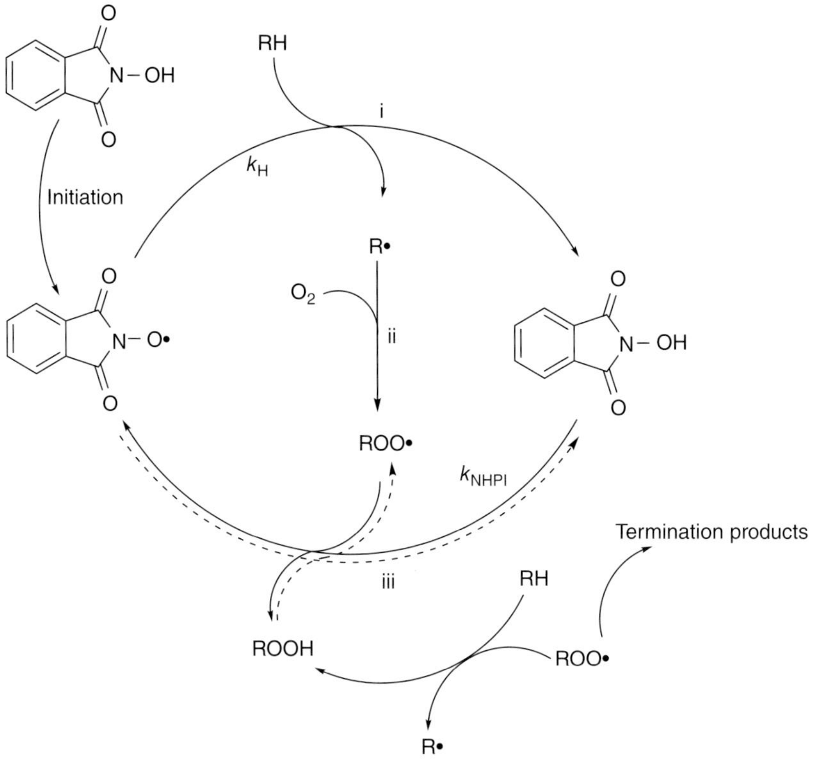

Scheme 16.2 The catalytic cycle of NHPI in the aerobic oxidation of organic substrates.

corresponding peroxyl radical (path ii), which in turn is quickly trapped by NHPI to form the hydroperoxide and a new molecule of PINO (path iii).

PINO generation being a key step of the overall process, in most cases, the use of NHPI was proposed in combination with different cocatalysts or initiators. Several examples report the beneficial effect of transition metal salts and complexes for this purpose [4]. In this context, the main role of metal salts (including Mn, Co, Cu, V, and Fe salts) is not only to accelerate classical autoxidation, by promoting the decomposition of the intermediate hydroperoxides (Scheme 16.3a), but also to bind oxygen (Scheme 16.3b), leading to the formation of PINO radical without requiring thermal treatment (Scheme 16.3c).

Eco-friendly standards, including the demand for highly selective transformations, have pushed toward the development of metal-free NHPI-mediated protocols, especially for large-scale productions, as in the case of the aerobic oxidation of hydrocarbons. So far, the activation of NHPI was obtained by means of aldehydes, quinones, nitric oxides, and enzymes and by irradiation in the presence of organic photo-mediators [6].

The high efficiency of NHPI in initiating and propagating the classical free-radical autoxidation chain finds its explanation in the concomitant manifestation

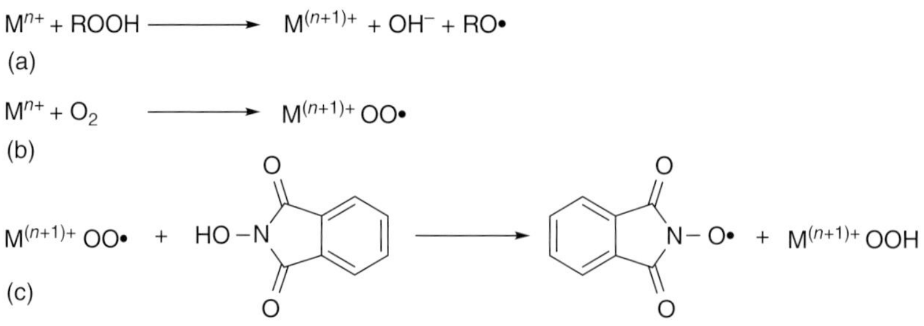

M^{n+} + ROOH ⟶ $M^{(n+1)+}$ + OH⁻ + RO•
(a)

M^{n+} + O_2 ⟶ $M^{(n+1)+}$ OO•
(b)

$M^{(n+1)+}$ OO• + HO−N(phthalimide) ⟶ (phthalimide)N−O• + $M^{(n+1)+}$ OOH
(c)

Scheme 16.3 Radical chain initiation by means of metal (M) salts.

of three favorable distinct effects. This particular and unique behavior justifies the increased attention that NHPI has attracted in the last two decades.

16.2.1
Enthalpic Effect

In 2003, Pedulli's group determined the bond dissociation enthalpy (BDE) of the O–H bond in NHPI by means of ESR radical equilibration technique [5]. The measured value of 88.1 kcal/mol (in acetonitrile, MeCN), about 18 kcal/mol higher than the corresponding O–H bond in the N-hydroxy-2,2,6,6-tetramethylpiperidine (TEMPO-H), clearly indicated that the carbonyl groups directly bonded to the nitrogen atom strongly increase the BDE values. In fact, the carbonyl group, with its electron-withdrawing character, reduces the importance of the mesomeric structure **B** of the nitroxyl radical (Scheme 16.4). As a result, the radical is less stabilized, and the corresponding O–H BDE increases.

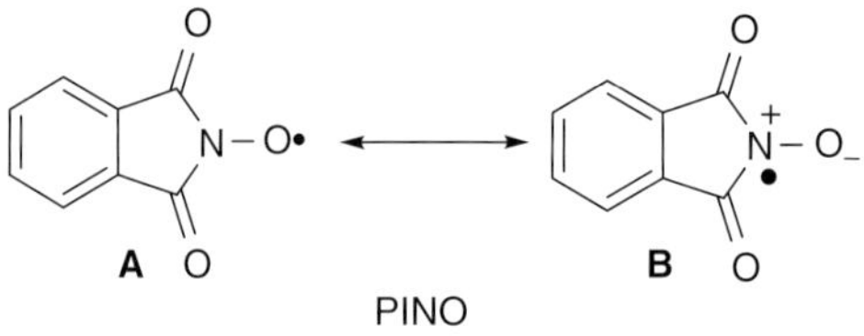

A (PINO, N−O•) ⟷ B (N⁺−O⁻ with radical)

Scheme 16.4 PINO resonance structures.

First key point: from a thermochemical point of view, path (i) in Scheme 16.2 may be in many cases exothermic or only slightly endothermic.

16.2.2
Polar Effect

In the same work [5], Pedulli and coworkers also measured the k_H value, referred to as path (i), for a wide range of substrates, demonstrating that the HAT reaction promoted by PINO is always faster than the corresponding hydrogen abstraction reaction by means of generic peroxyl radicals, occurring in classical noncatalyzed autoxidation process. This behavior cannot be ascribed to enthalpic effects, as the O–H BDE values in NHPI and generic hydroperoxides are similar, but instead to a polar effect due to a more pronounced electrophilic character of the PINO radical relative to the peroxyl one.

Second key point: PINO behaves as a good catalyst for the hydrogen atom abstraction from C–H bonds (path i).

16.2.3
Entropic Effect

To complete the rationalization of the catalytic cycle reported in Scheme 16.2, Pedulli *et al.* also determined the kinetic constant for the hydrogen atom abstraction from NHPI by peroxyl radicals (k_{NHPI} in path iii). The unexpected fairly high value obtained ($k_{NHPI} = 7.2 \times 10^3$ M^{-1} s^{-1}) [5] revealed the potential of the NHPI/PINO system. PINO shows high catalytic efficiency in the hydrogen atom abstraction (path i). NHPI behaves as a good hydrogen donor, trapping peroxyl radicals before they undergo fast termination and prolonging the propagation chain (path iii). This latter aspect justifies the need to operate in solution.

Third key point: NHPI guarantees high selectivity to the catalyzed oxidation process under homogeneous conditions.

16.3
Process Technology

16.3.1
Oxidation of Adamantane to Adamantanols

The unique structure of adamantane justifies the interest for selective functionalization of this molecule in order to develop enhanced functional materials. In particular, selective synthesis of monoalcohols or diols represents the first step for the production of photoresist materials via esterification of the hydroxyl groups with acrylic and methacrylic acids [7].

The selective oxidation of adamantane to adamantanols with molecular oxygen has found practical application by using NHPI catalysis. The reaction has been proposed and patented by Ishii in collaboration with Daicel Chemical Company [8] and consists of the aerobic oxidation of adamantane in chlorobenzene or acetic acid at temperatures ranging from 75 to 85 °C for 7 h in the presence of 10% mol of NHPI and 0.5% mol of different metal salts, including $Co(acac)_2$, $Co(OAc)_2$, $VO(acac)_2$, and V_2O_5 (Scheme 16.5).

Claimed conversion is higher than 90%, with a selectivity for alcohols depending on the cocatalyst of choice. Adamantanol production using this technology is commercialized at Daicel Arai plant in Japan.

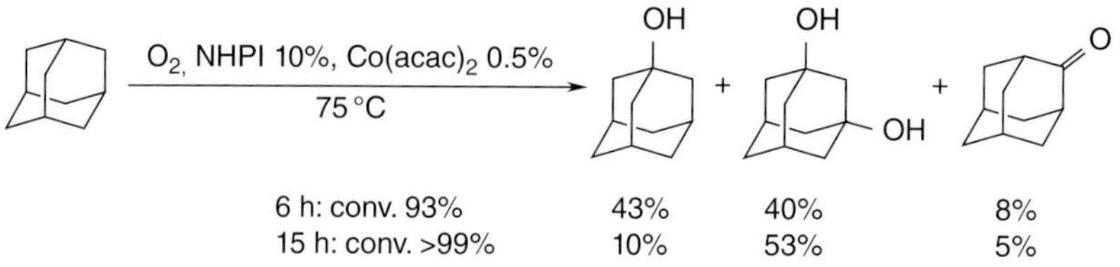

Scheme 16.5 Aerobic oxidation of adamantane catalyzed by NHPI.

16.3.2
Oxidation of Cyclohexane to Adipic Acid

Ishii and Daicel Chemical Company also patented a method for the direct aerobic oxidation of cyclohexane to adipic acid by using NHPI in combination with small amounts of Mn(acac)$_2$ and Co(OAc)$_2$ as cocatalysts (Scheme 16.6) [9].

While traces of cobalt salts seem to play a key role in reducing the induction period and accelerating the radical chain (Scheme 16.3), Mn(II) guarantees a higher selectivity for adipic acid by promoting the enolization of cyclohexanone and activating the α-carbonyl position toward oxidation [10]. The reaction occurs in acetic acid (AcOH) at 80 °C for 24 h.

As previously disclosed, the solvent is necessary to guarantee a complete solubility of the polar organocatalyst. However, in 2001 Ishii proposed the use of 4-lauryloxycarbonyl-N-hydroxyphthalimide (**1**), a lipophilic version of NHPI, in order to perform the reaction directly in neat cyclohexane [11] (Scheme 16.7).

Since 2009, the process has been under evaluation at pilot scale for further commercial application by Daicel in Aboshi (Japan).

The Ministry of Education, Culture, Sports, Science and Technology of Japan in 2003 awarded "The Third Green and Sustainable Chemistry Award" to Prof. Ishii and the Daicel Chemical Company for their relevant efforts to the development of NHPI-based industrial processes with low environmental impact.

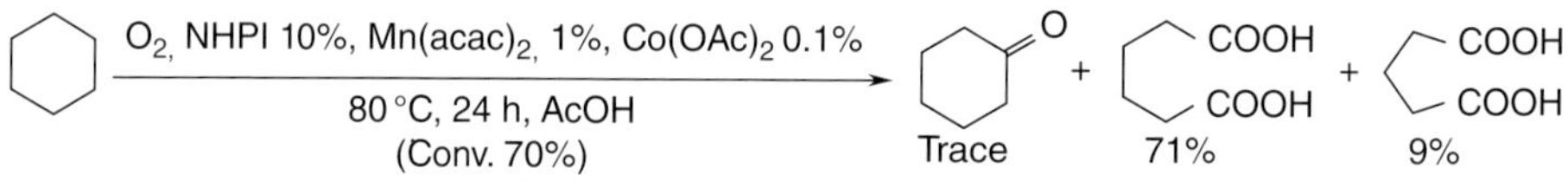

Scheme 16.6 Aerobic oxidation of cyclohexane to adipic acid catalyzed by NHPI.

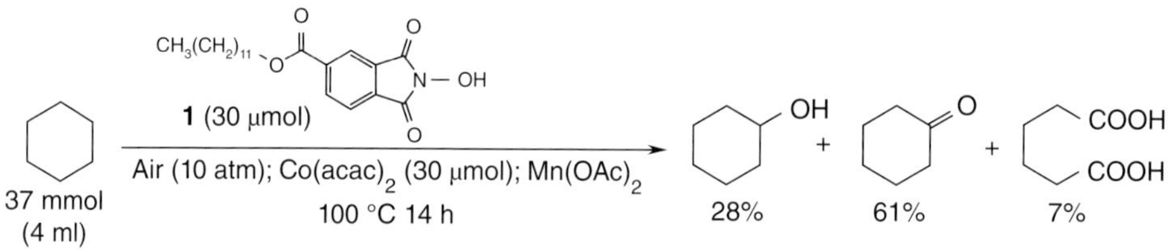

Scheme 16.7 Aerobic oxidation of cyclohexane in the presence of lipophilic NHPI.

16.3.3
Epoxidation of Olefins

In 2006, our research group proposed the selective epoxidation of α-olefins by combining, at room temperature and in MeCN, stoichiometric amounts of acetaldehyde with catalytic quantities of NHPI under oxygen atmosphere [12].

Following the catalytic mechanism reported in Scheme 16.8, we demonstrated the role of peracetic acid, generated *in situ*, in promoting the formation of PINO by undergoing molecule-induced homolysis with NHPI (path a), while the acyl peroxyl radical resulted to be the effective epoxidizing agent (path b).

The process was successfully applied to the synthesis of propylene oxide in MeCN [13].

The major drawback of this protocol is the long reaction times that are usually required in order to achieve high conversions (24–48 h). For this reason, in 2012, we started a collaboration for realizing an aerobic epoxidation catalyzed by NHPI under continuous-flow conditions by means of a new technology designed, manufactured, and developed by Prof. Biørsvik (University of Bergen) and Fluens Synthesis Company: the multijet oscillating disk (MJOD) reactor [14]. A flowchart of this reactor applied to our process is displayed in Figure 16.1.

Due to the advantageous reactor net volume versus the heating/cooling surface ratio of the MJOD reactor tube, an exceptionally good heat transfer capacity is achieved. Moreover, extremely good mixing of the components is obtained by the oscillation of the disks, resulting in an excellent mass transfer capacity. These properties usually result in a substantially increased reaction rate.

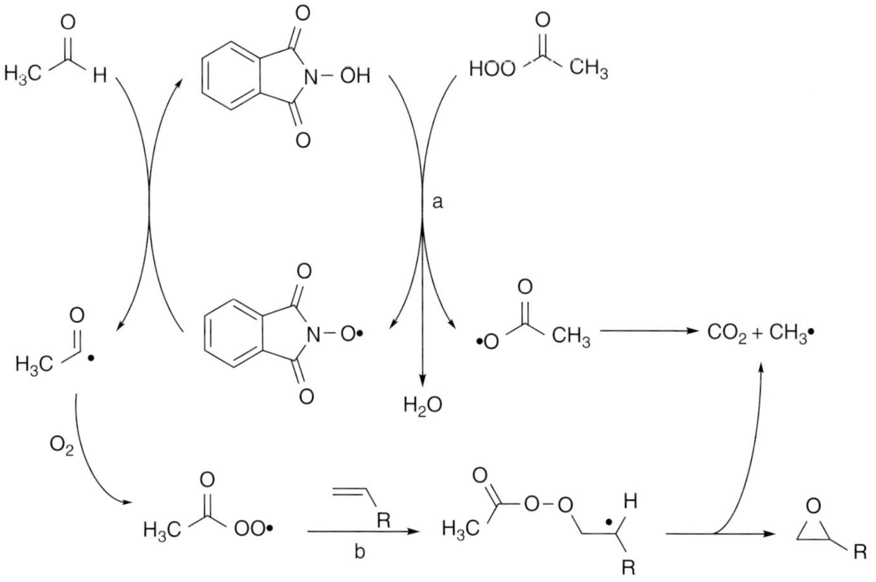

Scheme 16.8 Reaction mechanism for epoxidation of olefins.

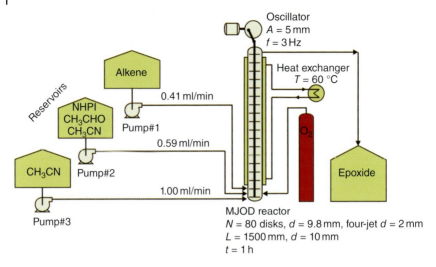

Figure 16.1 Process flowchart of NHPI-catalyzed epoxidation in MJOD millireactor system.

For all the olefins investigated under continuous-flow conditions, we obtained high conversions and yields (~80%) of the desired epoxides. Moreover, the process was substantially accelerated, shortening the residence time from 24 to 48 h (batch process) to only 1–4 h, with a standard production of about 80 g/day, which makes this protocol appealing for applications in pharmaceutical industry.

16.3.4
Oxidation of Alkylaromatics to Corresponding Hydroperoxides

Aerobic selective oxidation of alkylaromatics, including cumene (CU), ethylbenzene (EtB), and cyclohexylbenzene (CyB), to the corresponding hydroperoxides (CHPs) represents a key step for several large-scale productions, including the Hock process for the synthesis of phenol (see Chapter 2) [15] and the Shell styrene monomer/propylene oxide (SM/PO) process for the production of propylene oxide (PO) and styrene monomer (SM) [16].

In this context, the NHPI-catalyzed oxidation approach has been widely investigated as an alternative route to the classical autoxidation process due to the evident opportunity to increase conversion and selectivity in the hydroperoxide, according to the catalytic cycle reported in Scheme 16.2. In particular, autoxidation of CU usually requires high temperatures in order to favor partial homolytic decomposition of the hydroperoxide for prolonging the propagation phase of the radical chain. The use of NHPI would allow operation under milder conditions, that is, temperatures lower than 100 °C, limiting the formation of secondary products deriving from termination, such as cumyl alcohol and above all acetophenone.

The last decade has been characterized by a huge amount of patent applications in this field. Sheldon and Degussa (now Evonik) first reported in 2001 the NHPI-catalyzed oxidation of a wide range of alkyl aromatics at 100 °C in the absence of cosolvent and initiators [17]. Particular attention was devoted to the oxidation of CyB [18]. Analogous to CU, the CyB could be converted to phenol and cyclohexanone, but in this case, the dehydrogenation of the ketone could lead to the formation of a second molecule of phenol [19]. This process has attracted the interest of ExxonMobil, which developed, with a series of patent applications, a protocol for the direct synthesis of phenol and cyclohexanone starting with the direct synthesis of CyB from benzene and H_2 via hydroalkylation [20]. This approach is completely waste-free, since the stoichiometric amount of H_2 required for the hydroalkylation is afforded by dehydrogenation of cyclohexanone formed after hydroperoxide cleavage (Scheme 16.9).

In general ExxonMobil significantly contributed to the process engineering of the NHPI-catalyzed oxidation of alkylaromatics, suggesting a procedure according to which water and NHPI-deactivating organic acid impurities are stripped from a portion of the reaction medium that is continuously removed from and, after stripping, returned to the reaction zone [21].

In 2009, Fierro and Repsol Quimica S.A. reported the beneficial effect of combining NHPI with parts per million amounts of NaOH for the synthesis of hydroperoxides [22], with the dual effect of promoting the formation of PINO in the absence of transition metal salts and of neutralizing the acidic by-products of the reaction. This approach was particularly effective for increasing conversion and selectivity in the oxidation of secondary alkylaromatics [23].

In all earlier mentioned approaches, the reactions are conducted in heated solution of hydrocarbons, at temperatures higher than 100 °C in order to guarantee complete solubilization of NHPI. However, as previously disclosed, these harsh operating conditions favor the self-decomposition of the catalyst, partially limiting the applicability at industrial scale.

The results of our investigation in this field, in collaboration with Polimeri Europa (now Versalis S.p.A., Eni Group), have convinced us that the presence of variable amounts of polar cosolvents was crucial to operate under homogenous and mild conditions (<80 °C). We also proposed the possible use of tiny amounts of acetaldehyde to initiate the radical chain even at ambient temperature [24],

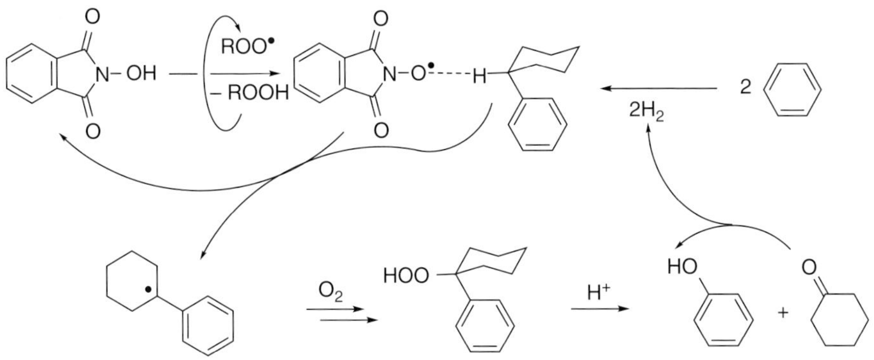

Scheme 16.9 Oxidation of cyclohexylbenzene.

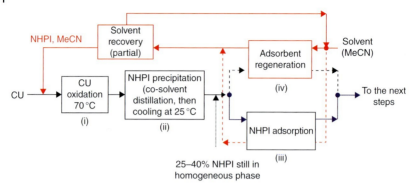

Figure 16.2 Process scheme for the NHPI-catalyzed oxidation of cumene.

according to the molecule-induced mechanism already reported in Scheme 16.8 (path a). Initiation was necessary when trying to convert the less reactive secondary alkylaromatics, such as EtB [25], but we found that no initiator was required for the NHPI-catalyzed selective oxidation of CU to the CHP. This approach was successfully applied also to the oxidation of CyB [26]. A complete design of the oxidation process should include an efficient recovery and recycle of the catalyst. A possible overall approach is reported in Figure 16.2 [27].

Hydroperoxide formation in the reaction medium progressively increases the solution polarity, with a consequent increased solubility of NHPI, so that a variable amount of NHPI still remains in solution, the quantity depending on the converted starting material. In 2009, ExxonMobil claimed the possibility to exploit the acidic characteristics of NHPI by removing the catalyst via basic aqueous extraction or treatment of the effluent with a solid sorbent having basic properties (such as metal oxides and ion exchange basic resins) [28]. However, we verified that this approach could be negatively affected by limitations in the recovery phase of the catalyst from the water solution or adsorbing bed, requiring the use of acidic solutions [27]. Moreover, NHPI can be hydrolyzed under basic conditions [29].

For this reason, we claimed a different approach, consisting of the physical adsorption of NHPI onto nonbasic solid beds, such as A26(Cl) [30, 31]. A26(Cl) guaranteed the reversibility of the adsorption process, allowing the recovery of the catalyst by washing the resin with the polar cosolvent used in the oxidation step (Figure 16.2, path iv). Figure 16.3 reports the adsorption/desorption cycles of NHPI onto A26(Cl).

16.4
New Developments

All the earlier mentioned protocols are limited by the low solubility of NHPI in apolar mediums. Ishii first opened the way for a possible solution to the problem by suggesting the introduction of lipophilic chains onto the aromatic ring of

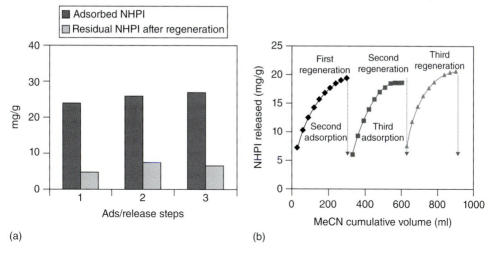

Figure 16.3 Adsorption of NHPI on A26(Cl) (milligram of catalyst adsorbed per gram of A26(Cl) – (a)) and regeneration of the adsorbing bed using MeCN. CU/CHP = 1.85/1 (mol/mol) (b). Initial NHPI concentration: 2 mg/ml.

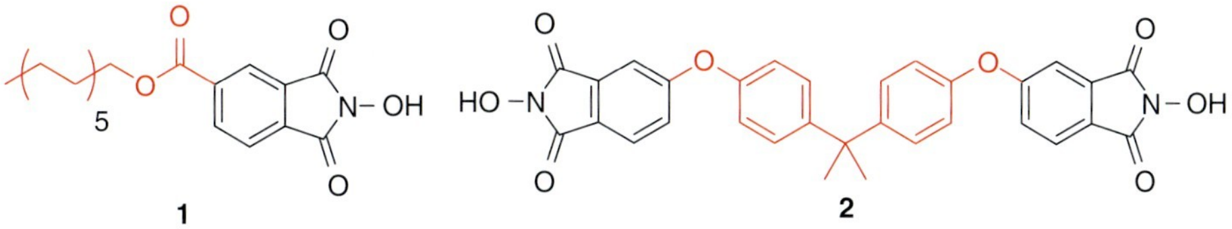

Scheme 16.10 Ishii's catalyst **1** and new organocatalyst **2**.

the N-hydroxy derivative [9]. Nevertheless, the proposed catalyst **1** suffers from a major limitation, if used in processes that require a high control of selectivity. The carboxylic group, thanks to which the alkyl tail is linked to the NHPI moiety, affects the NO–H BDE due to its electron-withdrawing character and increases the value by 0.7 kcal/mol, so that the efficiency of **1** as hydrogen donor is expected to be significantly reduced. In fact, the process requires a lower NO–H BDE value when the selective conversion to hydroperoxides is desired [32].

In this context, we have very recently proposed catalyst **2** (Scheme 16.10) as a suitable lipophilic catalyst for the aerobic oxidation of CU and other alkyl aromatics to the CHPs [33].

Our results confirmed for **2** a catalytic efficiency analogous to that observed for NHPI, and higher with respect to Ishii's catalyst **1**, for producing hydroperoxides. Even if this solution did not allow the complete removal of the polar cosolvent, it was possible to operate with reduced amounts of MeCN and to run CU oxidations under homogeneous conditions even at 45 °C, which cannot be realized in the presence of 1% NHPI catalyst. As expected, the new conditions led to an increment of CHP selectivity.

Acknowledgments

Financial support from Versalis S.p.A is gratefully acknowledged. We thank MIUR for their continual support of our research (PRIN 2010–2011, project 2010PFLRJR_005).

References

1. Grochowski, E., Boleslawska, T., and Jurczak, J. (1977) *Synthesis*, **1977**, 718–720.
2. Masui, M., Ueshima, T., and Ozaki, S. (1983) *Chem. Commun.*, 479.
3. Ishii, Y., Nakayama, K., Takeno, M., Sakaguchi, S., Iwahama, T., and Nishiyama, Y. (1995) *J. Org. Chem.*, **60**, 3934–3935.
4. (a) Recupero, F. and Punta, C. (2007) *Chem. Rev.*, **107**, 3800–3842; (b) Galli, C., Gentili, P., and Lanzalunga, O. (2008) *Angew. Chem. Int. Ed.*, **47**, 4790–4796; (c) Coseri, S. (2008) *Mini-Rev. Org. Chem.*, **5**, 222–227; (d) Coseri, S. (2009) *Catal. Rev.*, **51**, 218–292; (e) Punta, C. and Gambarotti, C. (2010) in *Ideas in Chemistry and Molecular Sciences: Advances in Synthetic Chemistry* (ed B. Pignataro), Wiley-VCH Verlag GmbH & Co. KGaA, Weinheim, pp. 3–24.
5. Amorati, R., Lucarini, M., Mugnaini, M., Pedulli, G.F., Minisci, F., Recupero, F., Fontana, F., Astolfi, P., and Greci, L. (2003) *J. Org. Chem.*, **68**, 1747–1754.
6. Melone, L. and Punta, C. (2013) *Beilstein J. Org. Chem.*, **9**, 1296–1310.
7. (a) Pasini, D., Klopp, J.M., and Frechet, J.M. (2001) *Chem. Mater.*, **13**, 4136–4146; (b) Butov, G.M., Pastukhova, N.P., Kamneva, E.A., and Saad, K.R. (2012) *Russ. J. Appl. Chem.*, **85**, 1590–1591.
8. (a) Ishii, Y., Kato, S., Iwahama, T., and Sakaguchi, S. (1996) *Tetrahedron Lett.*, **37**, 4993–4996; (b) Ishii, Y., Nakano, T., and Hirai, N. (1998) Adamantane Derivatives and Process for Producing them, Patent WO1998/40337.
9. (a) Ishii, Y., Iwahama, T., Sakaguchi, S., Nakayama, K., and Nishiyama, Y. (1996) *J. Org. Chem.*, **61**, 4520–4526; (b) Iwahama, T., Shoujo, K., Sakaguchi, S., and Ishii, Y. (1998) *Org. Process Res. Dev.*, **2**, 255–260.
10. Minisci, F., Punta, C., and Recupero, F. (2006) *J. Mol. Catal. A: Chem.*, **251**, 129–149.
11. Sawatari, N., Yokota, T., Sakaguchi, S., and Ishii, Y. (2001) *J. Org. Chem.*, **66**, 7889–7891.
12. Minisci, F., Gambarotti, C., Pierini, M., Porta, O., Punta, C., Recupero, F., Lucarini, M., and Mugnaini, V. (2006) *Tetrahedron Lett.*, **47**, 1421–1424.
13. Punta, C., Moscatelli, D., Porta, O., Minisci, F., Gambarotti, C., and Lucarini, M. (2008) in *Mechanisms in Homogeneous and Heterogeneous Epoxidation Catalysis* (ed S.T. Oyama), Elsevier, pp. 217–229.
14. Spaccini, R., Liguori, L., Punta, C., and Bjørsvik, H.-R. (2012) *ChemSusChem*, **5**, 261–265.
15. (a) Hock, H. and Lang, S. (1944) *Ber. Dtsch. Chem. Ges.*, **B77**, 257–264; (b) Jordan, W., van Barneveld, H., Gerlich, O., Boymann, M.K., and Ullrich, J. (1985) *Ullmann's Encyclopedia of Industrial Organic Chemicals*, vol. **A9**, Wiley-VCH Verlag GmbH, Weinheim, pp. 299–312.
16. (a) Buijink, J.K.F., Lange, J.-P., Bos, A.N.R., Horton, A.D., and Niele, F.G.M. (2008) in *Mechanisms in Homogeneous and Heterogeneous Epoxidation Catalysis* (ed S.T. Oyama), Elsevier, Amsterdam, pp. 355–371.
17. Kühnle, A., Duda, M., Sheldon, R.A., Sasidharan, M., Arends, I., Schiffer, T., Fries, G., and Kirchhoff, J. (2001) Method for oxidizing hydrocarbons. Patent WO2001/74742A1.
18. Kühnle, A., Duda, M., Tanger, U., Sheldon, R.A., Arends, I.W.C.E., and Manickam, S. (2001) Method for

18. producing aromatic alcohols, especially phenol, Patent WO2001/74767A1.
19. Arends, I.W.C.E., Sasidharan, M., Kühnle, A., Duda, M., Jost, C., and Sheldon, R.A. (2002) *Tetrahedron*, **58**, 9055–9061.
20. (a) Dakka, J.M., Buchanan, J.S., Cheng, J.C., Chen, T.-J., Decaul, L.C., Helton, T.E., Stanat, J.E., and Benitez, F.N. (2009) Process for producing phenol and/or cyclohexanone, Patent WO2009/131769A1; (b) Dakka, J.M., Buchanan, J.S., Lattner, J.R., and Mizrahi, S. (2009) Process for producing cyclohexanone, Patent WO2009/134514A1; (c) Cheng, J.C., Chen, T.-J., Benitez, F.N., Helton, T.E., and Stanat, J.E. (2009) Process for producing cyclohexylbenzene, Patent WO2009/134516A1; (d) Buchanan, J.S., Stanat, J.E., Cheng, J.C., Dakka, J.M., and Lattner, J.R. (2010) Process for producing phenol, Patent WO2010/024975A1.
21. Lattner, J.R., Hagemeister, M.P., Stanat, J.E., Buchanan, J.S., Dakka, J.M., and Zushma, S. (2010) Oxidation of hydrocarbons, Patent WO2010/042273A1.
22. de Frutos, M.P., Toribio, P.P., Martos Calvente, R., Campos-Martin, J.M., and Fierro, J.L.G. (2009) Process for preparation of hydroperoxides, Patent EP1,520,853B1.
23. Toribio, P.P., Gimeno-Gargallo, A., Capel-Sanchez, M.C., de Frutos, M.P., Campos-Martin, J.M., and Fierro, J.L.G. (2009) *Appl. Catal., A*, **363**, 32–39.
24. (a) Minisci, F., Porta, O., Recupero, F., Punta, C., Gambarotti, C., and Pierini, M. (2008) Process for the preparation of phenol by means of new catalytic systems, Patent WO2008/037435A1; (b) Melone, L., Gambarotti, C., Prosperini, S., Pastori, N., Recupero, F., and Punta, C. (2011) *Adv. Synth. Catal.*, **353**, 147–154.
25. (a) Minisci, F., Porta, O., Recupero, F., Punta, C., Gambarotti, C., and Spaccini, R. (2009) Catalytic process for the preparation of hydroperoxides of alkylbenzenes by aerobic oxidation under mild conditions, Patent WO2009/115275A1; (b) Melone, L., Prosperini, S., Gambarotti, C., Pastori, N., Recupero, F., and Punta, C. (2012) *J. Mol. Catal. A: Chem.*, **355**, 155–160.
26. Bencini, E., Melone, L., Pastori, N., Prosperini, S., Recupero, F., Punta, C., and Gambarotti, C. (2011) Process for the preparation of phenol and cyclohexanone, Patent WO2011/001244A1.
27. Melone, L., Prosperini, S., Ercole, G., Pastori, N., and Punta, C. (2014) *J. Chem. Technol. Biotechnol.*, **89**, 1370–1378.
28. Dakka, J.M., Vartuli, J.C., and Zushma, S. (2009) Oxidation of hydrocarbons, Patent WO2009/058527A1.
29. Hamdah, W., Ahmad, W., Sim, Y.L., and Khan, M.N. (2013) *Monatsh. Chem.*, **144**, 1299–1305.
30. Minisci, F., Porta, O., Recupero, F., Punta, C., Gambarotti, C., and Spaccini, R. (2009) Process for the production of alkylbenzene hydroperoxides under mild conditions and in the presence of new catalytic systems, Patent WO2009/115276A1.
31. Recupero, F., Punta, C., Melone, L., Prosperini, S., and Pastori, N. (2011) Process for the oxidation of alkylaromatic hydrocarbons catalyzed by n-hydroxy derivatives, Patent WO2011/161523A1.
32. Annunziatini, C., Gerini, M.F., Lanzalunga, O., and Lucarini, M. (2004) *J. Org. Chem.*, **69**, 3431–3438.
33. Petroselli, M., Franchi, P., Lucarini, M., Punta, C., and Melone, L. (2014) *ChemSusChem*, **7**, 2695–2703.

17
Carbon Materials as Nonmetal Catalysts for Aerobic Oxidations: The Industrial Glyphosate Process and New Developments

17.1
Introduction
Mark Kuil and Annemarie E. W. Beers

Activated carbon is being applied as a nonmetal oxidation catalyst in the last step of the production of glyphosate (i.e., N-(phosphonomethyl)glycine (PMG)). Glyphosate is a very active herbicide, in particular when being applied on plant leaves. Moreover, glyphosate has negligible harmful effects on the environment [1]. A well-known glyphosate formulation is Monsanto's Roundup. In industry, three main routes exist toward the production of glyphosate (Figure 17.1) [2]. The first route is the glycine route in which glycine is transformed in several steps into glyphosate. The second route includes the HCN–IDA process. In this process, HCN is first being converted into iminodiacetic acid (IDA), and then IDA is transformed into N-(phosphonomethyl)iminodiacetic acid (PMIDA) and finally to glyphosate. The third route is the DEA–IDA process, in which ethylene oxide with ammonia produces diethanolamine (DEA). DEA is then converted to IDA, to PMDA, and then to glyphosate. The last steps are analogous to the second route. In these last two production routes, activated carbon is applied as an oxidation catalyst (see also Figure 17.1).

The major differences between these three production routes have been depicted in Table 17.1. The process costs are relatively high for the DEA–IDA route. In addition, the environmental impact is high for the glycine process (formation of waste NH_4Cl salt) and for the HCN–IDA process (HCN is a toxic intermediate) [2]. The quality (or purity) of the glyphosate end product is higher for the production routes using IDA as an intermediate. The demand for high-quality glyphosate is expected to increase in the coming years, and thus the demand for catalytically active activated carbon will also increase.

Activated carbon is thus being applied in the last step of the glyphosate production: the oxidative decarboxylation of PMIDA toward glyphosate uses oxygen gas as the oxidant, forming carbon dioxide and formaldehyde as side products (Scheme 17.1).

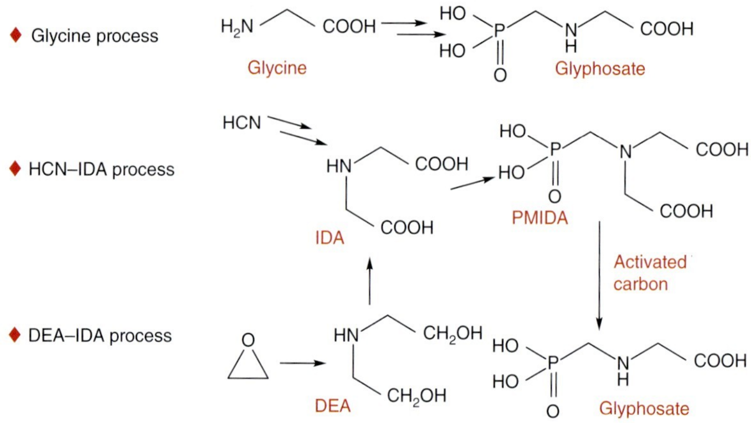

Figure 17.1 Three routes toward the production of glyphosate.

Table 17.1 Major differences between the three production routes toward glyphosate [2].

	Glycine	HCN–IDA	DEA–IDA
Process costs	Moderate	Moderate	High
Environmental impact	High	High	Moderate
Quality glyphosate	Low	High	High
Activated carbon applied	No	Yes	Yes

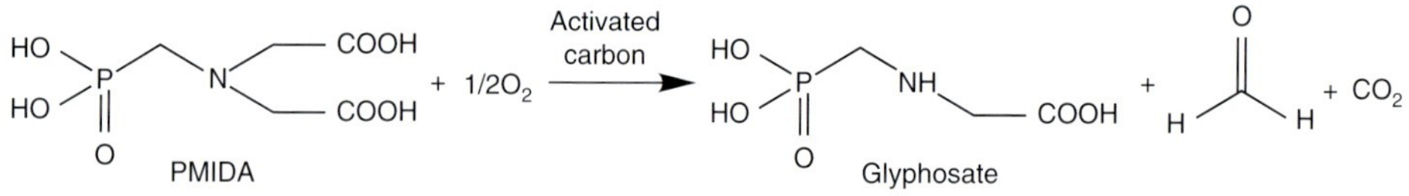

Scheme 17.1 The oxidative decarboxylation of N-(phosphonomethyl)iminodiacetic acid (PMIDA) toward glyphosate catalyzed by activated carbon.

17.2
Chemistry and Catalysis
Mark Kuil and Annemarie E. W. Beers

Besides the desired main reaction, several side reactions can occur during the glyphosate production (Scheme 17.2). For instance, the side product formaldehyde may react with glyphosate affording N-formyl-N-(phosphonomethyl)glycine (NFG) and N-methyl-N-(phosphonomethyl)glycine (NMG). In order to prevent formaldehyde side reactivity and the buildup of CO_2 gas in the reactor vessel (which leads to an undesired decrease in partial O_2 pressure), a constant oxygen flow is applied over the reaction mixture to remove these gaseous side products formaldehyde and carbon dioxide [3]. Another noncatalyzed side reaction

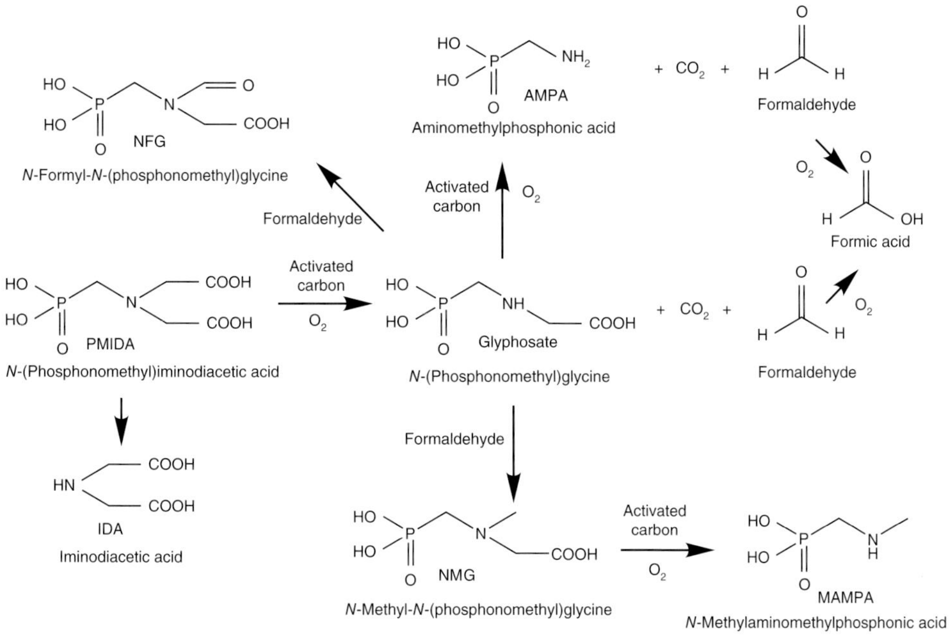

Scheme 17.2 Possible side reactions during the glyphosate production.

comprises the dephosphorylation of the starting compound toward IDA under the reaction conditions. However, the most common side reaction is overoxidation: a second catalytic oxidative decarboxylation takes place, transforming glyphosate into aminomethylphosphonic acid (AMPA). In particular, this side reaction occurs at the end of the main reaction when almost all PMIDA have been converted to glyphosate and the glyphosate concentration is high (i.e., the reaction rate for this second catalytic oxidative decarboxylation reaction is much lower compared with the main catalytic oxidative decarboxylation reaction) [4, 5].

An early patent that describes the use of activated carbon as a catalyst for the production of glyphosate originates from Monsanto Company [6] and uses different commercial activated carbons in this process. Subsequent patents from the Monsanto Company elaborated on this application using other tertiary amines and improved activated carbon catalysts wherein surface oxides have been removed [7–9]. The replacement of an oxygen-containing gas as the oxidant by hydrogen peroxide has been reported by the Hampshire Chemical Company [10]. The combination of hydrogen peroxide and an activated carbon catalyst afforded high glyphosate yields and the subsequent oxidation of the formaldehyde by-product to formate and CO_2. Also, the Sankyo Company reported the use of hydrogen peroxide, affording a process that would be suitable for industrial upscaling [11]. The use of an activated carbon catalyst having a specific particle size distribution to facilitate the recycling process of the carbon catalyst has been described [12]. Calgon Carbon Corporation reported on an improved process for glyphosate production in the presence of oxygen using a catalytically active activated carbon capable of rapidly decomposing hydrogen peroxide in an aqueous solution [13]. A continuous process for glyphosate production with an activated carbon catalyst afforded a high-yield and cost-effective glyphosate end product [5]. As it is important to stop the glyphosate reaction in time to prevent overoxidation of the formed glyphosate product, methods for the control of the PMIDA conversion as reported by Monsanto are key to obtain optimal glyphosate yields [14]. Recently, Cabot Norit Nederland B.V. described the production of catalytically active activated carbons from charcoal as sustainable carbonaceous feedstock suitable for use as an oxidation catalyst in gas/liquid applications, such as the glyphosate production [15].

An early study on the influence of various properties of different activated carbon catalysts for the glyphosate production has been reported by Pinel and coworkers [16, 17]. Activated carbons from different precursors and modified by thermal treatments have been tested in the glyphosate reaction. The catalytic activity of an activated carbon catalyst was increased by the presence of nitrogen-containing functional groups either introduced from the carbon precursors or by thermal treatments under NH_3 of the activated carbons. A well-known example comprises the use of urea. In this case, urea is incorporated in the carbon structure [18]. Under activation conditions, the urea decomposes, and the nitrogen functionality is built into the carbon structure, affording a catalytically active carbon material. If the activation process is not stopped on time, the ongoing pore formation will ultimately lead to a decrease in the presence of the nitrogen functionality and thus to a decrease in catalytic activity. Unfortunately, the exact nature of the basic sites as well as the mechanism of the glyphosate reaction were not identified during this study [19].

A macrokinetic model of the glyphosate reaction has been developed by Hu, *et al.* based on the primary reaction of PMIDA toward glyphosate and on the further oxidation of glyphosate to the undesired side product AMPA [4]. They concluded that an increase in the reaction temperature had an almost equal effect on the rate constant for the main reaction and the side reaction. The oxygen concentration also influenced both the main as well as the side reaction. However, a higher oxygen concentration accelerated the main reaction more compared with the side reaction under the reaction conditions.

17.3
Process Technology
Mark Kuil and Annemarie E. W. Beers

Extensive research, performed at Cabot Norit Activated Carbon, has been done to determine the most optimal conditions for the glyphosate process as well as to study the influence of various properties of the activated carbon catalyst on the glyphosate production. The impact of these quantitative and qualitative parameters on the glyphosate reaction is discussed later. These studies were carried out in a 400 ml aqueous 2–3 wt% PMIDA solution containing 1–5 wt% activated carbon at 60–95 °C applying a constant oxygen flow of 0–500 ml/min.

17.3.1
Oxygen Pressure

Using Norit SXRO as the activated carbon catalyst, the increase in oxygen pressure from 1.0 to 5.0 bar reduced the time to achieve 98% PMIDA conversion by a factor of about 1.7. Unfortunately, this increase in rate was accompanied by a glyphosate selectivity decrease as measured by the formation of the side product AMPA. The formation of other side products such as NMG, NFG, MAMPA, and IDA was much smaller than 1% under the employed reaction conditions. The selectivity dropped from 95% at 3.0 bar O_2 to only 84% at 5.0 bar O_2. Apparently, the rate of the second oxidative decarboxylation of glyphosate into AMPA is accelerated more by increasing the oxygen pressure than the first glyphosate-yielding oxidative decarboxylation. An oxygen pressure of 3.0 bar provided the best compromise between rate increase and selectivity optimization.

17.3.2
Oxygen Flow

As expected, an increase in oxygen flow from 0 to 500 ml/min afforded a corresponding increase in the glyphosate reaction rate. This reaction rate increase leveled off at about 300 ml O_2/min on the tested 400 ml scale.

17.3.3
Activated Carbon Pore Size Distribution

Four Norit SXRO activated carbon catalysts that only differ in their porosity have been tested in the glyphosate process. Other carbon properties, such as the H_2O_2 activity times, were more or less the same. The H_2O_2 activity time specifies the catalytic activity of an activated carbon sample in the decomposition of a certain amount of H_2O_2 to H_2O and O_2. Accordingly, a lower H_2O_2 time implies a higher catalytic activity of the activated carbon sample. The reaction rate increased while optimizing the pore size distribution of the activated carbon catalyst but leveled off above a pore size distribution of 4 a.u. (Figure 17.2a). Figure 17.2b shows that the glyphosate selectivity also leveled off when the pore size distribution exceeded 4 a.u. It is noteworthy that the carbon with a low value of 2 a.u. already showed a significant decrease in the selectivity after about 20% PMIDA conversion and afforded only 80% selectivity at 98% PMIDA conversion. Thus, an optimal pore size distribution is required both for high activity and selectivity.

17.3.4
Activated Carbon H_2O_2 Time

Four other Norit SXRO activated carbon catalysts that essentially differ only in their H_2O_2 activity time have been tested in the glyphosate process. As expected, the reaction rate increased with a decreasing H_2O_2 time (Figure 17.3a).

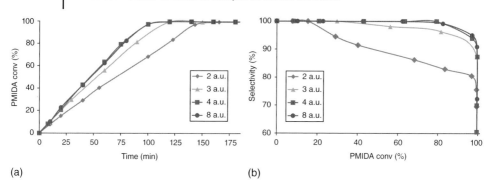

Figure 17.2 Activity (a) and selectivity (b) profiles for glyphosate production by four Norit SXRO activated carbon samples with different pore size distributions ranging from 2 to 8 a.u.

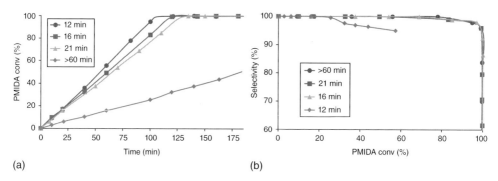

Figure 17.3 Activity (a) and selectivity (b) profiles for glyphosate production by four Norit SXRO activated carbon samples with different H_2O_2 times ranging from 12 to 60 min (pore size distributions kept constant at 3 a.u.).

Glyphosate selectivity, on the other hand, hardly depended on the H_2O_2 time of the catalyst, except for the carbon sample with a high H_2O_2 time (>60 min) which showed a significant decrease in selectivity after only 25% PMIDA conversion (Figure 17.3b).

17.3.5
Activated Carbon Nitrogen Content

Pinel et al. [16] showed that the catalytic activity of an activated carbon catalyst can be increased by nitrogen-containing functional groups, either introduced from the carbon precursors or by thermal treatments under NH_3 of the activated carbons. Therefore, the relation between the addition of various N-donors to the carbon precursor and the performance of eight Norit SXRO activated carbon materials have been studied. These carbons varied in their H_2O_2 times and degree of optimal pore size distribution due to the use of different nitrogen donors. Figure 17.4a shows that the glyphosate reaction rate is lowest for activated

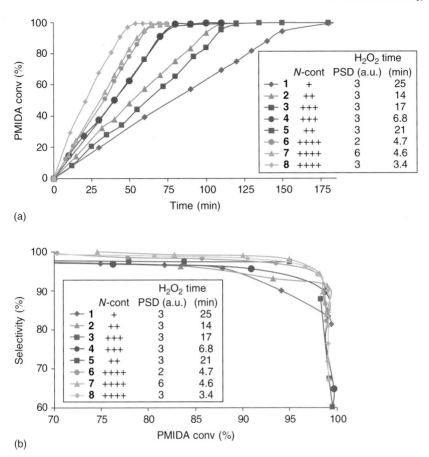

Figure 17.4 Activity (a) and selectivity (b) profiles for glyphosate production by eight Norit SXRO activated carbon samples 1–8 prepared with different N-donors (PSD is degree of optimal pore size distribution).

carbon **1** with a low N-content. Activated carbons **2** and **5** with a moderate N-content show a higher catalytic activity, which increases further for carbons **3** and **4** with a high N-content. Surprisingly, carbons **3** and **4** have significantly different H_2O_2 times (17 and 6.8 min, respectively). Apparently, the H_2O_2 time is not a good indicator for catalytic activity for carbons prepared in a different way. Carbons **6** and **7** with a high N-content show a much higher catalytic activity, with sample **7** being slightly more active due to a more optimal pore size distribution. Carbon **8** containing a high N-content and prepared under optimized activation conditions affording an H_2O_2 time of only 3.4 min has the highest activity in the glyphosate reaction. Concerning the glyphosate selectivity (Figure 17.4b), carbon **1** with the lowest N-content shows the largest decrease after a PMIDA conversion of 90%. For the other carbons **2–8**, the glyphosate selectivity is essentially the same. Overall, the reaction rate in the glyphosate process can thus be significantly enhanced by proper optimization of the N-content of the carbon in combination with the activation conditions. However, these optimization studies have only limited influence on the selectivity in the process. Overall, it can be concluded that to achieve high catalytic activities in the glyphosate process, it is required to have an optimal pore size distribution and to have a low H_2O_2 activity time present in the activated carbon sample. In addition, the H_2O_2 time is a good indicator for the catalytic activity. Furthermore, it is required to have an optimal pore size distribution and to have a low H_2O_2 activity time present in the activated carbon sample to achieve high selectivities in the glyphosate process. Hereby, the H_2O_2 time is not an indicator for the glyphosate selectivity. From this research study, it is clear that tunable activated carbon catalysts can be prepared by the application of nitrogen-containing modifiers. Future work will be aimed at the development of a further improved catalytically active activated carbon catalyst for the glyphosate production.

17.4
New Developments
Paul L. Alsters

This research clearly demonstrates the attractiveness of catalytically active activated carbons for application in industrially relevant aerobic transformations. Catalysis by carbon materials (carbocatalysis), including aerobic oxidation catalysis, is a rapidly developing area. The following presents an overview of the wide range of aerobic transformations of organic synthetic interest catalyzed by these materials. Space limitations prevent us from detailed discussions on the synthesis and morphology control of these materials, their structural and mechanistic aspects, and other transformations outside the organic synthetic domain, such as water splitting, the oxygen reduction reaction, and wastewater treatment. For further background information, the reader is referred to recent reviews [20–40]. The first part deals with applications of carbon materials such as active carbon (AC), graphite oxide (GiO) or its exfoliated counterpart graphene oxide (GeO), carbon nanotubes (CNTs), and nanoshell carbon (NSC). This part includes catalysis by boron- or nitrogen-doped versions of these materials. The second part focuses on graphitic carbon nitride (g-CN), which is treated separately because of its well-defined C_3N_4 structure built up of tri-*s*-triazine units, cross-linked by trigonal nitrogen atoms. This makes it distinct from N-doped carbons, as also evidenced by the fact that g-CN, in contrast to the aforementioned carbon materials, most commonly acts as a heterogeneous photocatalyst. As a medium bandgap (2.7 eV) semiconductor with its conduction band (CB) at -1.3 V and its valence band (VB) at 1.4 V versus the NHE, g-CN has the proper redox characteristics for visual light-driven aerobic oxidations, without suffering from reduced selectivities by formation of indiscriminate hydroxyl radicals by oxidation of water or hydroxide [41, 42]. In both parts, synthetic applications have been subdivided into oxygenations (including oxidative cleavages) and dehydrogenations (including dehydrogenative couplings). Charts 17.1 and 17.2 provide graphical overviews of the structural diversity accessible via oxygenative or dehydrogenative conversions catalyzed by carbon materials or g-CN, respectively. For GiO and GeO, true aerobic oxidation catalysis is not always unambiguously demonstrated when these materials are used at very high loadings relative to the substrate. Such examples are nevertheless included, because of their synthetic relevance and because catalytic behavior may be targeted by further material structure and reaction parameter control. Figure 17.5 presents the structure of a hypothetical hybrid graphene-type material containing various regions ranging from g-CN to graphene, including B- and N-doped graphene and graphene oxide.

17.4.1
Aerobic Carbon Material Catalysis

17.4.1.1 Oxygenations and Oxidative Cleavage Reactions

The methylene group of activated alkylarenes such as fluorenes, xanthenes, and anthrone could be oxygenated up to high conversion with high loadings (100 wt%) of commercially available ACs [43]. Of synthetic interest is the stability of the sulfur atom in thioxanthene and the aldehyde group in fluorene-2-carbaldehyde under these oxidizing conditions. Moreover, the AC catalyst is reusable after the reaction. Using only 1 wt% (relative to the substrate) of an N-doped layered carbon catalyst prepared from GeO, benzylic ketonization of ethylbenzene proceeded to high conversions under elevated O_2 pressures (40–50 bar) with substoichiometric TBHP as initiator [44]. Benzylic oxygenations under neat conditions at low catalyst loadings (1 wt%) were also catalyzed by other N- or B-doped graphenes. In addition, these were active for oxygenation of cyclooctane and of styrene, the latter affording a mixture of benzaldehyde and styrene oxide [45].

Cis-stilbene could be oxidized to benzil with large amounts of GiO (200–400 wt%). The yield hardly changed on replacing air by a nitrogen atmosphere [46]. Using similar high loading conditions, benzylic oxidation of diarylmethanes to ketones and of methylbenzenes to aldehydes took place in good and low yields, respectively. In addition, GiO has been used for sulfoxidation of various aliphatic or aromatic sulfides [47].

With DMF as solvent, a large variety of fluorenes was oxidized in very high yields to the corresponding fluorenones under the influence of a graphene-supported KOH composite catalyst [48]. The latter is prepared from stoichiometric KOH and a small amount of commercially available graphite via exfoliation under sonication. The method, including catalyst recycling, has been successfully scaled up to multikilogram scale.

Synthetic ACs catalyzed the oxidative cleavage of cyclohexanone to a mixture of mostly C_4, C_5, and C_6 dicarboxylic acids in aqueous solution [49, 50]. Among the metal-free carbons, the highest adipic acid selectivity (34%) was obtained with an air-activated carbon containing phosphorus [49]. Catalytic activity is

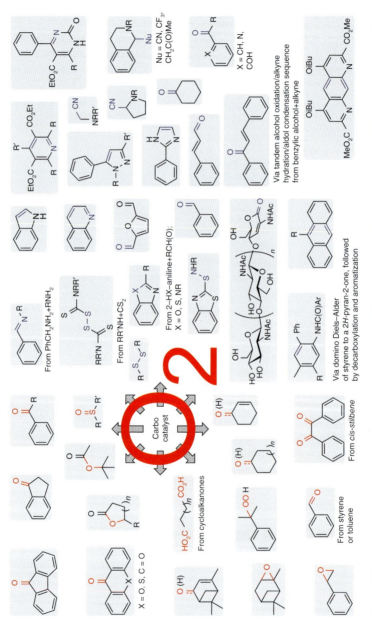

Chart 17.1 Structures accessible via oxygenative and dehydrogenative conversions catalyzed by carbon materials (excluding g-CN). Red and blue bonds refer to oxygenative and dehydrogenative bond formation, respectively. Phenyl rings may contain additional substituents.

17.4 New Developments | 277

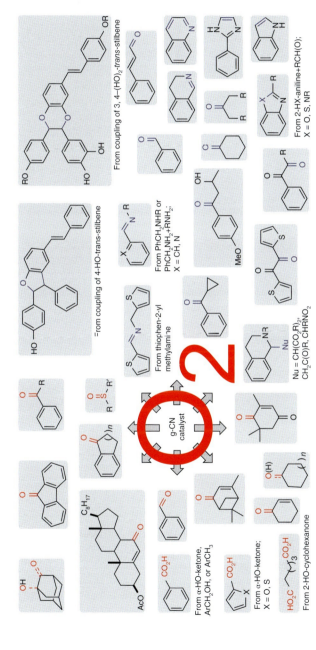

Chart 17.2 Structures accessible via oxygenative and dehydrogenative conversions catalyzed by g-CN. Red and blue bonds refer to oxygenative and dehydrogenative bond formation, respectively. Phenyl rings may contain additional substituents.

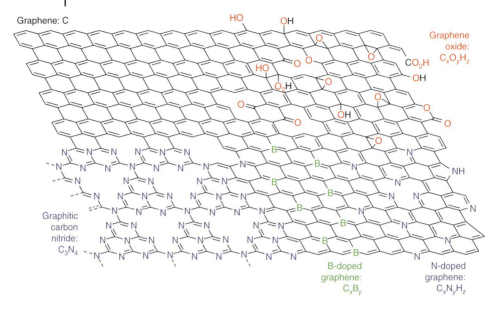

Figure 17.5 Hypothetical hybrid graphene-type material composed of regions ranging from g-CN to graphene, including B- and N-doped graphene and GeO. The GeO region is based on the Lerf–Klinowski model, g-CN is depicted as its tri-s-triazine-based allotrope. Defect holes are included in the structures of GeO and N-doped graphene.

associated with the presence of oxygenated groups. Cyclododecanone and 3,3,5-trimethylcyclohexanone were cleaved similarly [17, 51]. In the presence of a Ketjen Black carbon catalyst and sacrificial benzaldehyde, cyclic, and linear ketones underwent aerobic Baeyer–Villiger oxidation to the corresponding esters [52].

CNTs have gained significant interest recently as catalysts for a variety of aerobic oxygenations, including radical chain oxidation of cyclohexane to cyclohexanol/one [53], ethylbenzene to acetophenone [54], and cumene to cumene hydroperoxide [55]. The CNTs revealed good recyclability. Introduction of oxygenated functional groups on the CNTs negatively influenced these conversions. On the other hand, nitrogen and phosphorous doping of CNTs had a beneficial effect on oxygenation of cyclohexane to a mixture of primarily cyclohexanol, cyclohexanone, and adipic acid [56, 57]. N-doping led to the highest mass-normalized activity (761 mmol·g^{-1}·h^{-1}), whereas the highest surface-normalized activity (28 mmol·m^{-2}·h^{-1}) was obtained by P-doping [56]. N-doping also promoted allylic oxidation of cyclohexene [58] and α-pinene [59], though conversion of the latter generated an equimolar mixture of epoxide and allylic oxidation products. Plasma-chemical bromination of CNTs also boosted catalytic activity for aerobic hydrocarbon oxidation, including oxidation of a naphthenic oil fraction to petroleum acids [60]. For cyclohexane oxidation catalyzed by materials composed of sp^2 or sp^3 carbons, it was shown that the highest activity was obtained with sp^2 carbons, in particular mesoporous graphene [61].

17.4.1.2 Dehydrogenations and Dehydrogenative Coupling Reactions

Using a commercially available AC (~50 wt%), various secondary benzylic alcohols were dehydrogenated aerobically to the corresponding ketones [62, 63]. The preparative relevance of this method is illustrated by the synthesis of ketones from substrates containing functional groups prone to undergo oxidation with other methods, such as pyridine and phenol groups. It also allowed the preparation of complex 4-O-β-N-acetylchitooligosyl lactones from the corresponding chitooligosaccharides [64].

N-doping may be used to boost the activity for aerobic alcohol oxidation of an otherwise poorly active carbon material [65, 66], although the synthetic scope was still limited to activated substrates such as benzyl alcohol, cinnamyl alcohol, and 5-hydroxymethyl-2-furaldehyde (HMF). These substrates were selectively oxidized to the corresponding aldehydes. Aerobic oxidation of primary and secondary benzyl alcohols to aldehydes and ketones, respectively, was also achieved with HNO_3 as a promotor under CNT [67] or NSC [68] catalysis. Catalytic activity of CNTs for this transformation was further enhanced by N-doping [69]. By using DMF as solvent, efficient conversion was also achieved without promotor [70]. Though overoxidation to benzoic acid occurs with increasing conversion, an advantageous aspect of these CNT catalysts is their outstanding recyclability.

Various benzylic alcohols, as well as cinnamyl alcohol and cyclohexanol, were dehydrogenated using large amounts of GiO under sonication [71] or GeO [72]. Despite the high GeO loadings (200 wt%), the material acted as an aerobic oxidation catalyst, since hardly any reaction occurred under nitrogen and the material could be recycled many times. Interestingly, GiO acted as a multifunctional tandem alcohol oxidation–alkyne hydration–aldol condensation catalyst for chalcone formation directly from phenylacetylenes and benzyl alcohols [73].

Using a mesoporous carbon catalyst prepared by pyrolysis of phthalocyanine, (heterocyclic) benzylamines were aerobically dehydrogenated to the corresponding N-benzylidene products. Catalyst recyclability on preparative, solvent-free 0.1 mol scale was excellent [74]. The same transformation was also catalyzed by nanocomposites of CNTs and negatively charged N,B-codoped graphene [75], by N,B-codoped holey graphene monoliths [76], or under solvent-free conditions by GiO [77] or GeO [78]. In the latter case, high catalytic activity was imparted on GeO by base–acid treatment, allowing catalyst loadings as low as 5 wt%. With GiO as carbocatalyst, oxidative cross condensations of various anilines and aliphatic amines with benzyl amine to the corresponding N-benzylidene derivatives have also been achieved by suppressing homocondensation of benzyl amine by using an excess of the cross-coupling partner [77]. Reduced GeO served as a reusable aerobic dehydrogenation catalyst for the preparation of azo compounds from their hydrazo precursors [79].

A diverse range of carbocycles and heterocycles has been aromatized by aerobic dehydrogenation under AC catalysis [80]. Substrates include 9,10-dihydroanthracenes [81], a 1,8-diaza-9,10-dihydroanthracene [82], Hantzsch 1,4-dihydropyridines [83], pyrazolines [83], 2-arylimidazolines [84, 85], indolines [86], 3,4-dihydropyrimidin-2(1H)-ones [87, 88], and various

1,2,3,4-tetrahydroaromatic compounds such as 1,2,3,4-tetrahydroquinoline [89]. In some cases, the AC-catalyzed aerobic approach clearly outperformed other dehydrogenation methods [82, 89]. Similar aromatizations have been achieved by using large loadings of GiO [46], GiO in the presence of molecular sieves [90], or multiwalled CNTs [91], the latter being under an oxygen atmosphere (1 bar). A simple one-step synthesis of benzoxazoles [92, 93], benzimidazoles [94], and benzothiazoles [95] is based on the condensation of anilines carrying *ortho*-X–H groups (X = O, NH, S) with aldehydes, followed by AC-catalyzed aromatization of the *in situ* generated heterocycles [80, 96, 97]. AC-catalyzed aromatization has also been applied as one of the steps in domino sequences involving a Diels–Alder cycloaddition, with the dehydrogenation either generating the final substituted benzene obtained from a 2*H*-pyran-2-one and a styrene via decarboxylation [98] or generating an anthracene as acceptor for a maleimide [99]. It has been shown that not only molecular oxygen but also the alkene dienophiles serve as hydrogen acceptors [98–100].

Cross-dehydrogenative coupling of aliphatic, aromatic, or heteroaromatic thiols to the corresponding disulfides has been carried out aerobically under AC catalysis [101]. Aerobic AC-catalyzed cross-dehydrogenative coupling of mercaptobenzothiazole with amines has been developed as a one-step, salt-free alternative for the classical two-step approach based on chlorine [102]. The resulting sulfonamides are used as antiscorching compounds for rubbers. Typically, good yields of bis(aminothiocarbonyl)disulfides were obtained in one step by reduced GeO-catalyzed aerobic cross-dehydrogenative coupling of secondary amines in the presence of CS_2 [103]. Finally, GeO in combination with rose bengal [104, 105] or as a hybrid material with polythiophene [105, 106] acted as a visible light-driven photooxidation catalyst for aerobic generation of iminium intermediates from α-C–H bond containing tertiary amines. This triggered subsequent coupling to various carbon nucleophiles, such as cyanide or trifluoromethyl from TMSX (X = CN, CF_3) [104, 105], or the enol derived from acetone, and the latter via tandem organocatalysis with proline [105, 106].

17.4.2
Aerobic Graphitic Carbon Nitride Catalysis

17.4.2.1 Oxygenations and Oxidative Cleavage Reactions

Bare g-CN is not an efficient catalyst for the oxyfunctionalization of inert hydrocarbons [34]. Fluorene, a substrate with weak benzylic C–H bonds, was successfully converted into fluorenone under thermal g-CN catalysis. Boron doping results in lowering of the HOMO (VB) that defines the energy level of the positive holes generated by excitation and thus increases the oxidation power. Such a B-doped catalyst allowed oxygenation of activated alkylarenes such as fluorene, diphenylmethane, indane, and tetraline [107]. Less activated alkylarenes such as toluene or ethylbenzene were inert under these conditions. Interestingly, (*para*-substituted) toluene(s) could be converted with very high selectivity to benzaldehyde(s) by using mpg-CN instead of conventional bulk g-CN [108]. This represents an example of morphology-dependent control of activity and selectivity. Binding of superoxide to radical cations in the VB as a spin couple is suggested to assure unusual selectivity by preventing unselective oxidations via nonsurface-bound superoxide. The role of mpg-CN in this reaction is better described as a moderator rather than a catalyst, since autoxidation of toluene under similar conditions (160 °C/10 bar O_2) proceeded with higher conversion. Thermally driven oxygenation of both activated and unactivated C–H bonds was enabled by synergism between graphene sheets and g-CN in nanocomposite material, resulting in enhanced oxidation power of g-CN by HOMO lowering [109]. Thermally induced aerobic oxidation of benzylic and allylic C–H bonds has also been accomplished by synergistic cooperation of g-CN and NHPI (N-hydroxyphthalimide) as organocatalysts [110]. The C–H abstracting N-oxyl PINO radical probably resulted from the reaction of superoxide with NHPI.

The same combination of organocatalysts also allowed photocatalytic C–H bond oxyfunctionalization of various substrates but under much milder conditions (35–60 °C/1 bar O_2) compared with the thermally induced examples (130 °C/12 bar O_2) [111]. Noteworthy is the absence of alcohols and acids as by-products of the oxidations of cyclohexane and toluene. The role of g-CN in the catalytic cycle is thought to go beyond the facilitation of PINO radical generation via the redox properties of g-CN. Its Lewis base properties probably promote decomposition of organic hydroperoxide intermediates. The photocatalytic aerobic oxidation of β-isophorone to ketoisophorone under mpg-CN catalysis was found to benefit from covalent attachment of various organic groups on the surface via Prato's reaction. These branching groups are believed to break up the sheet-like structure of the surface, thereby enhancing the interaction of the substrates with catalytically active sites [112]. Conventional mpg-CN allowed photocatalytic aerobic oxidative cleavage of a variety of α-hydroxy ketones to the corresponding carboxylic acids [113]. Cooxidation with isobutyraldehyde greatly increased the efficiency for sulfoxide formation under mpg-CN photocatalysis, very likely via intermediate acylperoxy species as oxygen donors [114].

17.4.2.2 Dehydrogenations and Dehydrogenative Coupling Reactions

Primary benzylic or aliphatic and secondary aliphatic alcohols could be dehydrogenated to the corresponding carbonyl compounds under (mesoporous) g-CN photocatalysis. Initially, it was believed that the dehydrogenation benefitted from the catalyst's basic character that may assist in alkoxide formation as a primary step [115]. Subsequent work demonstrated a beneficial effect of acidity, either by running the reaction in water at low pH [116] or by pretreating g-CN with H_2SO_4 and running the reaction in an organic solvent [117]. Acidic surface sites may participate in proton-coupled electron transfer to molecular oxygen from the CB [116] and promote β-hydride elimination from the alcohol [117]. Acid pretreatment also induces favorable morphology changes of the catalyst [117]. In line with a superoxide-driven radical mechanism, the rate of the reaction was enhanced by both electron-withdrawing and electron-donating substituents on benzyl alcohol [115], though under acidic conditions electron-withdrawing substituents contributed only marginally, suggesting the involvement of electron-deficient intermediates [116]. Under aqueous conditions at moderate temperatures, H_2O_2 is initially formed during alcohol dehydrogenation [116, 118]. H_2O_2 is quickly decomposed by g-CN at higher temperatures by participation in two-electron cycles [115].

Integration of thiophene motifs into mpg-CN allowed dehydrogenation of benzyl alcohols to the benzaldehydes even at longer wavelengths. Besides superoxide, also singlet oxygen ($^1O_2$) is generated from molecular oxygen by this catalyst, and alcohol oxidation is thought to involve their combined action [119]. (Benzylic) α-hydroxy ketones have been converted into 1,2-diketones via mpg-CN controlled photocatalysis [120]. Thermally induced oxidation of benzylic alcohols under mpg-CN catalysis was promoted by CO_2, which not only increased the conversion but also switched selectivity toward benzoic acids instead of benzaldehydes in case of primary alcohol substrates [121]. Formation of CO is indicative of the oxygenating role of CO_2, which may be activated by the catalyst via formation of carbamate groups [121].

(Heterocyclic) benzylamines have also been aerobically dehydrogenated by mpg-CN as a photocatalyst, yielding the corresponding N-benzylidene products [122]. Coordination of the imine to a hole in the VB is thought to facilitate intermediate aminal formation by nucleophilic attack on the C=N double bond. Dehydrogenative aromatizations of several N-heterocycles have also been carried out with mpg-CN [122].

Aerobic dehydrogenative coupling reactions hold promise for industrial applications given their high atom efficiency, with only water being formed as by-product. The preparative power of this approach was demonstrated by easy access to dihydrobenzofurans with relevant bioactivities via intermolecular photocatalytic coupling of a variety of 4-hydroxy-*trans*-stilbenes catalyzed by mpg-CN [123]. Benzodioxane dimers were obtained with 3,4-dihydroxy-*trans*-stilbenes. High yields require the presence of a pyridine base, which allows generation of phenolate anions that undergo one-electron transfer to the VB, thus forming highly delocalized radicals that dimerize to the product. These radicals may also be generated by direct phenolic hydrogen abstraction by superoxide formed upon irradiation [123].

Initial imine or iminium generation from amines can be used to allow C–X (X = heteroatom) or C–C bond formation. Thus, anilines substituted in the *ortho*-position with X–H functionalities (X = O, NMe, S) have been dehydrogenatively condensed with benzylamines under mpg-CN photocatalysis to yield the corresponding benzofused heterocycles [122]. Via iminium intermediates, N-aryl-1,2,3,4-tetrahydroisoquinolines could be dehydrogenatively coupled photocatalytically by mpg-CN to various carbon nucleophiles derived from nitroalkanes, malonates, and 2-alkanones, and the latter via tandem organocatalysis with proline [124].

17.5
Concluding Remarks

Liquid phase oxidations catalyzed by (activated) carbon materials have already found industrial applications, and more processes based on this approach are expected to be implemented given the strong progress that has been achieved recently. Carbon materials (including g-CN) catalyze an amazingly diverse range of oxygenative and dehydrogenative transformations. The scope and efficiency of these continuously improve, driven by progress in engineering, the properties of carbocatalysts. In many cases, these catalysts have proven to be robust and recyclable. Several useful transformations, in particular aromatizations, are efficiently catalyzed by active carbons that are already commercially available. Especially for dehydrogenations and dehydrogenative couplings, carbocatalysts have demonstrated applicability for preparing products of high molecular complexity [64, 123], and aerobic carbocatalysis sometimes has proven superiority over other methods [62, 63, 82]. In addition, the combination of both redox and acid/base properties in one material enables unique tandem reaction sequences [73]. Newer catalyst generations are expected to be available at low cost, since they are built up of earth-abundant elements and are usually accessible from readily available precursors. Kilogram-scale applications of such new materials have already been demonstrated [48]. Although many literature protocols employ high carbocatalyst loadings, lowering the latter down to levels as low as 5 wt% has been achieved by a simple chemical treatment [78]. In principle, carbocatalysts are true heterogeneous organocatalysts devoid of (toxic and/or precious) metals. This makes them particularly suitable for being applied late in a synthetic sequence close to, for example, an active pharmaceutical ingredient. As has been emphasized frequently in the literature, however, a word of caution is in order here. Although the AC-catalyzed glyphosate process is a clear industrial example of metal-free carbocatalysis, in other cases it is not always convincingly demonstrated that the observed catalytic behavior originated from true carbocatalysis instead of catalysis by metal or other impurities. This being said, we are convinced that catalysis by carbon materials (including g-CN) holds great promise for future industrial applications.

References

1. Duke, S.O. and Powles, S.B. (2008) *Pest Manage. Sci.*, **64**, 319–325 and references cited therein.
2. Tian, J., Shi, H., Li, X., Yin, Y., and Chen, L. (2012) *Green Chem.*, **14**, 1990–2000.
3. See for example examples 6 and 8 in patent Monsanto Company (1974) US 3.969.398.
4. Bao, Y., Wu, J., and Hu, A. (2012) *Adv. Mater. Res.*, **455–456**, 872–879.
5. Dow Agrosciences LLC (2002) Patent WO 03/099831.
6. Monsanto Company (1974) Patent US 3.969.398.
7. Monsanto Company (1976) Patent US 4.072.706.

8. Monsanto Company (1976) Patent US 4.264.776.
9. Monsanto Company (1976) Patent US 4.696.772.
10. Hampshire Chemical Company (1995) Patent WO 9.638.455.
11. Sankyo Company, Limited (1997) Patent US 5.948.938.
12. SKW Trostberg AG (1999) Patent DE 19938622.
13. Calgon Carbon Corporation (1998) Patent WO 9.958.537.
14. Monsanto Company (2010) Patent US 2011/0097810.
15. Cabot Norit Nederland B.V. (2012) Patent US 2014/0037536.
16. Pinel, C., Landrivon, E., Lini, H., and Gallezot, P. (1999) *J. Catal.*, **182**, 515–519.
17. Besson, M., Gallezot, P., Perrard, A., and Pinel, C. (2005) *Catal. Today*, **102–103**, 160–165 and references cited therein.
18. Adib, F., Bagreev, A., and Bandosz, T.J. (2000) *Langmuir*, **16**, 1980–1986.
19. See for a nice overview on the catalytic properties of nitrogen-containing carbons: Boehm, H.-P. (2009) in *Carbon Materials for Catalysis* (eds P. Serp and J.L. Figueiredo), John Wiley & Sons Inc., Hoboken, NJ, pp. 219–265.
20. Cao, S., Low, J., Yu, J., and Jaroniec, M. (2015) *Adv. Mater.*, **27**, 2150–2176.
21. Gong, Y., Li, M., Li, H., and Wang, Y. (2015) *Green Chem.*, **17**, 715–736.
22. Hu, H., Xin, J.H., Hu, H., Wang, X., and Kong, Y. (2015) *Appl. Catal., A*, **492**, 1–9.
23. Chua, C.K. and Pumera, M. (2015) *Chem. Eur. J.*, **21**, 12550–12562.
24. Zhu, J., Xiao, P., Li, H., and Carabineiro, S.A.C. (2014) *ACS Appl. Mater. Interfaces*, **6**, 16449–16465.
25. Lang, X., Chen, X., and Zhao, J. (2014) *Chem. Soc. Rev.*, **43**, 473–486.
26. García-Bordejé, E., Pereira, M.F.R., Rönning, M., and Chen, D. (2014) *Catalysis*, **26**, 72–108.
27. Kong, X.-K., Chen, C.-L., and Chen, Q.-W. (2014) *Chem. Soc. Rev.*, **43**, 2841–2857.
28. Dreyer, D.R., Todd, A.D., and Bielawski, C.W. (2014) *Chem. Soc. Rev.*, **43**, 5288–5301.
29. Centi, G., Perathoner, S., and Su, D.S. (2014) *Catal. Surv. Asia*, **18**, 149–163.
30. Navalon, S., Dhakshinamoorthy, A., Alvaro, M., and Garcia, H. (2014) *Chem. Rev.*, **114**, 6179–6212.
31. Haag, D.R. and Kung, H.H. (2014) *Top. Catal.*, **57**, 762–773.
32. Su, D.S., Perathoner, S., and Centi, G. (2013) *Chem. Rev.*, **113**, 5782–5816.
33. Sun, X., Wang, R., and Su, D. (2013) *Chin. J. Catal.*, **34**, 508–523.
34. Wang, Y., Wang, X., and Antonietti, M. (2012) *Angew. Chem. Int. Ed.*, **51**, 68–89.
35. Wang, X., Blechert, S., and Antonietti, M. (2012) *ACS Catal.*, **2**, 1596–1606.
36. Schaetz, A., Zeltner, M., and Stark, W.J. (2012) *ACS Catal.*, **2**, 1267–1284.
37. Dreyer, D.R. and Bielawski, C.W. (2011) *Chem. Sci.*, **2**, 1233–1240.
38. Su, D.S., Zhang, J., Frank, B., Thomas, A., Wang, X., Paraknowitsch, J., and Schlögl, R. (2010) *ChemSusChem*, **3**, 169–180.
39. Kawashita, Y. and Hayashi, M. (2009) *Molecules*, **14**, 3073–3093.
40. Hayashi, M. (2008) *Chem. Rec.*, **8**, 252–267.
41. Cui, Y., Ding, Z., Liu, P., Antonietti, M., Fu, X., and Wang, X. (2012) *Phys. Chem. Chem. Phys.*, **16**, 1455–1462.
42. Khan, M.A., Teixeira, I.F., Li, M.M.J., Koito, Y., and Tsang, S.C.E. (2016) *Chem. Commun.*, **52**, 2772–2775.
43. Kawabata, H. and Hayashi, M. (2004) *Tetrahedron Lett.*, **45**, 5457–5459.
44. Gao, Y., Hu, G., Zhong, J., Shi, Z., Zhu, Y., Su, D.S., Wang, J., Bao, X., and Ma, D. (2013) *Angew. Chem. Int. Ed.*, **52**, 2109–2113.
45. Dhakshinamoorthy, A., Primo, A., Concepcion, P., Alvaro, M., and Garcia, H. (2013) *Chem. Eur. J.*, **19**, 7547–7554.
46. Jia, H.-P., Dreyer, D.R., and Bielawski, C.W. (2011) *Tetrahedron*, **67**, 4431–4434.
47. Dreyer, D.R., Jia, H.-P., Todd, A.D., Geng, J., and Bielawski, C.W. (2011) *Org. Biomol. Chem.*, **9**, 7292–7295.
48. Zhang, X., Ji, X., Su, R., Weeks, B.L., Zhang, Z., and Deng, S. (2013) *ChemPlusChem*, **78**, 703–711.

49. Besson, M., Gauthard, F., Horvath, B., and Gallezot, P. (2005) *J. Phys. Chem. B*, **109**, 2461–2467.
50. Besson, M., Blackburn, A., Gallezot, P., Kozynchenko, O., Pigamo, A., and Tennison, S. (2000) *Top. Catal.*, **13**, 253–257.
51. Gauthard, F., Horvath, B., Gallezot, P., and Besson, M. (2005) *Appl. Catal., A*, **279**, 187–193.
52. Nabae, Y., Rokubuichi, H., Mikuni, M., Kuang, Y., Hayakawa, T., and Kakimoto, M. (2013) *ACS Catal.*, **3**, 230–236.
53. Yang, X., Wang, H., Li, J., Zheng, W., Xiang, R., Tang, Z., Yu, H., and Peng, F. (2013) *Chem. Eur. J.*, **19**, 9818–9824.
54. Luo, J., Peng, F., Yu, H., Wang, H., and Zheng, W. (2013) *ChemCatChem*, **5**, 1578–1586.
55. Liao, S., Peng, F., Yu, H., and Wang, H. (2014) *Appl. Catal., A*, **478**, 1–8.
56. Cao, Y., Yu, H., Tan, J., Peng, F., Wang, H., Li, J., Zheng, W., and Wong, N.-B. (2013) *Carbon*, **57**, 433–442.
57. Yu, H., Peng, F., Tan, J., Hu, X., Wang, H., Yang, J., and Zheng, W. (2011) *Angew. Chem. Int. Ed.*, **50**, 3978–3982.
58. Cao, Y., Yu, H., Peng, F., and Wang, H. (2014) *ACS Catal.*, **4**, 1617–1625.
59. Cao, Y., Li, Y., Yu, H., Peng, F., and Wang, H. (2015) *Catal. Sci. Technol.*, **5**, 3935–3944.
60. Zeynalov, E., Friedrich, J., Meyer-Plath, A., Hidde, G., Nuriyev, L., Aliyeva, A., and Cherepnova, Y. (2013) *Appl. Catal., A*, **454**, 115–118.
61. Cao, Y., Luo, X., Yu, H., Peng, F., Wang, H., and Ning, G. (2013) *Catal. Sci. Technol.*, **3**, 2654–2660.
62. Sano, Y., Tanaka, T., and Hayashi, M. (2007) *Chem. Lett.*, **36**, 1414–1415.
63. Tanaka, T. and Hayashi, M. (2008) *Synthesis*, **21**, 3361–3376.
64. Ogata, M., Takeuchi, R., Suzuki, A., Hirai, H., and Usui, T. (2012) *Biosci. Biotechnol., Biochem.*, **76**, 1362–1366.
65. Watanabe, H., Asano, S., Fujita, S.-I., Yoshida, H., and Arai, M. (2015) *ACS Catal.*, **5**, 2886–2894.
66. Long, J., Xie, X., Xu, J., Gu, Q., Chen, L., and Wang, X. (2012) *ACS Catal.*, **2**, 622–631.
67. Luo, J., Peng, F., Yu, H., and Wang, H. (2012) *Chem. Eng. J.*, **204–206**, 98–106.
68. Kuang, Y.B., Islam, N.M., Nabae, Y., Hayakawa, T., and Kakimoto, M. (2010) *Angew. Chem. Int. Ed.*, **49**, 436–440.
69. Luo, J., Peng, F., Wang, H., and Yu, H. (2013) *Catal. Commun.*, **39**, 44–49.
70. Luo, J., Yu, H., Wang, H., Wang, H., and Peng, F. (2014) *Chem. Eng. J.*, **240**, 434–442.
71. Mirza-Aghayan, M., Kashef-Azar, E., and Boukherroub, R. (2012) *Tetrahedron Lett.*, **53**, 4962–4965.
72. Dreyer, D.R., Jia, H.-P., and Bielawski, C.W. (2010) *Angew. Chem. Int. Ed.*, **49**, 6813–6816.
73. Jia, H.-P., Dreyer, D.R., and Bielawski, C.W. (2011) *Adv. Synth. Catal.*, **353**, 528–532.
74. Chen, B., Wang, L., Dai, W., Shang, S., Lu, Y., and Gao, S. (2015) *ACS Catal.*, **5**, 2788–2794.
75. Wang, H., Zheng, X., Chen, H., Yan, K., Zhu, Z., and Yang, S. (2014) *Chem. Commun.*, **50**, 7517–7520.
76. Li, X.-H. and Antonietti, M. (2013) *Angew. Chem. Int. Ed.*, **52**, 4572–4576.
77. Huang, H., Huang, J., Liu, Y.-M., He, H.-Y., Cao, Y., and Fan, K.-N. (2012) *Green Chem.*, **14**, 930–934.
78. Su, C., Acik, M., Takai, K., Lu, J., Hao, S.-J., Zheng, Y., Wu, P., Bao, Q., Enoki, T., Chabal, Y.J., and Loh, K.P. (2012) *Nat. Commun.*, **3**, 1298–1306.
79. Bai, L.-S., Gao, X.-M., Zhang, X., Sun, F.-F., and Ma, N. (2014) *Tetrahedron Lett.*, **55**, 4545–4548.
80. Hayashi, M. and Kawashita, Y. (2006) *Lett. Org. Chem.*, **3**, 571–578.
81. Nakamichi, N., Kawabata, H., and Hayashi, M. (2003) *J. Org. Chem.*, **68**, 8272–8273.
82. Singleton, M.L., Castellucci, N., Massip, S., Kauffmann, B., Ferrand, Y., and Huc, I. (2014) *J. Org. Chem.*, **79**, 2115–2122.
83. Nakamichi, N., Kawashita, Y., and Hayashi, M. (2004) *Synthesis*, **7**, 1015–1020.
84. Haneda, S., Okui, A., Ueba, C., and Hayashi, M. (2007) *Tetrahedron*, **63**, 2414–2417.

85. Gan, Z., Kawamura, K., Eda, K., and Hayashi, M. (2010) *J. Organomet. Chem.*, **695**, 2022–2029.
86. Nomura, Y., Kawashita, Y., and Hayashi, M. (2007) *Heterocycles*, **74**, 629–635.
87. Okunaga, K.-I., Nomura, Y., Kawamura, K., Nakamichi, N., Eda, K., and Hayashi, M. (2008) *Heterocycles*, **76**, 715–726.
88. Michigami, K., Uchida, S., Adachi, M., and Hayashi, M. (2013) *Tetrahedron*, **69**, 595–599.
89. Tanaka, T., Okunaga, K.-I., and Hayashi, M. (2010) *Tetrahedron Lett.*, **51**, 4633–4635.
90. Zhang, X., Xu, L., Wang, X., Ma, N., and Sun, F. (2012) *Chin. J. Chem.*, **30**, 1525–1530.
91. Bégin, D., Ulrich, G., Amadou, J., Dang Sheng Su, D.S., Pham-Huu, C., and Ziessel, R. (2009) *J. Mol. Catal. A.*, **302**, 119–123.
92. Kawashita, Y., Nakamichi, N., Kawabata, H., and Hayashi, M. (2003) *Org. Lett.*, **5**, 3713–3715.
93. Adachi, K., Michigami, K., and Hayashi, M. (2010) *Heterocycles*, **82**, 857–865.
94. Haneda, S., Adachi, Y., and Hayashi, M. (2009) *Tetrahedron*, **65**, 10459–10462.
95. Dhayalan, V. and Hayashi, M. (2012) *Synthesis*, **44**, 2209–2216.
96. Haneda, S., Gan, Z., Eda, K., and Hayashi, M. (2007) *Organometallics*, **26**, 6551–6555.
97. Kawashita, Y., Yanagi, J., Fujii, T., and Hayashi, M. (2009) *Bull. Chem. Soc. Jpn.*, **82**, 482–488.
98. Juranovic, A., Kranjc, K., Polanc, S., and Kocevar, M. (2014) *Synthesis*, **46**, 909–916.
99. Krivec, M., Kranjc, K., Polanc, S., and Kocevar, M. (2014) *Curr. Org. Chem.*, **18**, 1520–1527.
100. Krivec, M., Gazvoda, M., Kranjc, K., Polanc, S., and Kocevar, M. (2012) *J. Org. Chem.*, **77**, 2857–2864.
101. Hayashi, M., Okunaga, K.-I., Nishida, S., Kawamura, K., and Eda, K. (2010) *Tetrahedron Lett.*, **51**, 6734–6736.
102. Riley, D., Stern, M., and Ebner, J. (1993) in *The Activation of Dioxygen and Homogeneous Catalytic Oxidation* (ed D.H.R. Barton), Plenum Press, New York, pp. 31–44.
103. Wang, M., Song, X., and Ma, N. (2014) *Catal. Lett.*, **144**, 1233–1239.
104. Pan, Y., Wang, S., Kee, C.W., Dubuisson, E., Yang, Y., Loh, K.P., and Tan, C.H. (2011) *Green Chem.*, **13**, 3341–3344.
105. Su, C. and Loh, K.P. (2013) *Acc. Chem. Res.*, **46**, 2275–2285.
106. Wang, S., Nai, C.T., Jiang, X.-F., Pan, Y., Tan, C.-H., Nesladek, M., Xu, Q.-H., and Loh, K.P. (2012) *J. Phys. Chem. Lett.*, **3**, 2332–2336.
107. Wang, Y., Li, H., Yao, J., Wang, X., and Antonietti, M. (2011) *Chem. Sci.*, **2**, 446–450.
108. Li, X.-H., Wang, X., and Antonietti, M. (2012) *ACS Catal.*, **2**, 2082–2086.
109. Li, X.-H., Chen, J.-S., Wang, X., Sun, J., and Antonietti, M. (2011) *J. Am. Chem. Soc.*, **133**, 8074–8077.
110. Liu, G., Tang, R., and Wang, Z. (2014) *Catal. Lett.*, **144**, 717–722.
111. Zhang, P., Wang, Y., Yao, J., Wang, C., Yan, C., Antonietti, M., and Li, H. (2011) *Adv. Synth. Catal.*, **353**, 1447–1451.
112. Zhang, P., Li, H., and Wang, Y. (2014) *Chem. Commun.*, **50**, 6312–6315.
113. Zhan, H., Liu, W., Fu, M., Cen, J., Lin, J., and Cao, H. (2013) *Appl. Catal., A*, **184**, 184–189.
114. Zhang, P., Wang, Y., Li, H., and Antonietti, M. (2012) *Green Chem.*, **14**, 1904–1908.
115. Su, F., Mathew, S.C., Lipner, G., Fu, X., Antonietti, M., Blechert, S., and Wang, X. (2010) *J. Am. Chem. Soc.*, **132**, 16299–16301.
116. Long, B., Ding, Z., and Wang, X. (2013) *ChemSusChem*, **6**, 2074–2078.
117. Zhang, L., Liu, D., Guan, J., Chen, X., Guo, X., Zhao, F., Hou, T., and Mu, X. (2014) *Mater. Res. Bull.*, **59**, 84–92.
118. Shiraishi, Y., Kanazawa, S., Sugano, Y., Tsukamoto, D., Sakamoto, H., Ichikawa, S., and Hirai, T. (2014) *ACS Catal.*, **4**, 774–780.
119. Chen, Y., Zhang, J., Zhang, M., and Wang, X. (2013) *Chem. Sci.*, **4**, 3244–3248.
120. Zheng, Z. and Zhou, X. (2012) *Chin. J. Chem.*, **30**, 1683–1686.

121. Ansari, M.B., Jin, H., and Park, S.-E. (2013) *Catal. Sci. Technol.*, **3**, 1261–1266.
122. Su, F.Z., Mathew, S.C., Möhlmann, L., Antonietti, M., Wang, X.C., and Blechert, S. (2011) *Angew. Chem. Int. Ed.*, **50**, 657–660.
123. Song, T., Zhou, B., Peng, G.-W., Zhang, Q.-B., Wu, L.-Z., Liu, Q., and Wang, Y. (2014) *Chem. Eur. J.*, **20**, 678–682.
124. Möhlmann, L., Baar, M., Rieß, J., Antonietti, M., Wang, X., and Blechert, S. (2012) *Adv. Synth. Catal.*, **354**, 1909–1913.

Part V
Biocatalytic Aerobic Oxidation

18
Enzyme Catalysis: Exploiting Biocatalysis and Aerobic Oxidations for High-Volume and High-Value Pharmaceutical Syntheses

Robert L. Osborne and Erika M. Milczek

18.1
Introduction

Reactions yielding oxygenated or hydroxylated organic compounds are of great value for organic synthesis. However, selective oxidations of organic substrates can pose significant challenges for the organic chemist in pharmaceutical and fine chemical industries. The tools most often employed by the organic synthetic chemist when tasked with achieving stereoselective transformations include chiral transition metal catalysts [1–7] and organocatalysts [8–13]. Furthermore, the implementation of many metallo- or organocatalysts into the large-scale synthetic route of complex substrates can be accompanied by expensive reagents, the production of undesired side products, and reaction conditions that are environmentally taxing. Enzymatic catalysis is increasingly becoming a viable and often superior tool for accomplishing the same task [14–18]. Enzymes offer an attractive alternative for many chemical oxidative transformations that are critical for industrial applications. Specifically, enzyme-catalyzed reactions can provide chemo-, regio-, and stereoselectivity advantages that often cannot be matched chemosynthetically. However, the implementation of enzyme-catalyzed transformations in industrial processes has been limited due to assumed disadvantages including narrow substrate scope and limited rate and stability under process-relevant reaction conditions. Directed evolution has emerged as a protein engineering method that often eliminates the limitations assumed to be prohibitive for the application of enzymes [19–26]. Directed evolution incorporates iterative cycles of mutagenic diversification, screening/selection, and diversity ranking [27, 28]. The best variant (hit) is identified and is used as the template (backbone) for the next iteration of mutagenesis, expression, and screening. In this chapter, we review successful examples of Codexis' directed evolution technology (CodeEvolver™)

applied to engineer oxidative enzymes resulting in their implementation at industrial scale in collaboration with pharmaceutical partners such as Merck & Co., Inc. The evolution, strategy, and scale-up of a Baeyer–Villiger monooxygenase (BVMO) for the production of esomeprazole and a monoamine oxidase (MAO) for the production of boceprevir are summarized in detail and compared with the chemosynthetic routes. The evolution and scale-up of MAO and the emerging technologies discussed herein are the result of a continuing collaboration between Codexis, Inc. and Merck & Co., Inc. In addition, advancements and challenges remaining for the application of alternate monooxygenases (i.e., cytochrome P450s, and styrene monooxygenases for industrial processes are discussed.

A subset of flavin monooxygenases that catalyze the Baeyer–Villiger reaction was first isolated nearly 50 years ago [29]. These soluble, yellow flavoproteins were designated BVMOs because they incorporate one atom of molecular oxygen into a ketone substrate with concomitant reduction of the other oxygen atom to water (Scheme 18.1). BVMOs require a reduced flavin cofactor (either FMN or FAD) and NADPH as a source of electrons for catalysis [30]. BVMOs function in microbial catabolic and biosynthetic pathways [31–34]. Nature has evolved BVMOs for specific metabolic steps, yet these enzymes retain the ability to accommodate numerous substrates [35]. Cyclohexanone monooxygenase (CHMO) from *Acinetobacter* sp. is the most extensively studied BVMO (Scheme 18.2), and the mechanism has been comprehensively investigated [36–46]. BVMOs require reduced flavin to activate molecular oxygen *en route* to the formation of a stable C4a-(hydro)peroxyflavin adduct. The rate-determining step is nucleophilic attack by the flavin peroxide on the carbonyl group of ketones and formation of the Criegee intermediate (Scheme 18.3) [47]. Rearrangement of the Criegee intermediate yields the ester or lactone, an equivalent of water, and the oxidized flavin. The Release of the product, NADP+, and reduction of the oxidized flavin completes turnover. The Criegee intermediate is electrophilic, and thus BVMOs catalyze the oxidation of soft nucleophiles such as sulfur, selenium, phosphorous, and in some cases nitrogen [40, 48, 49].

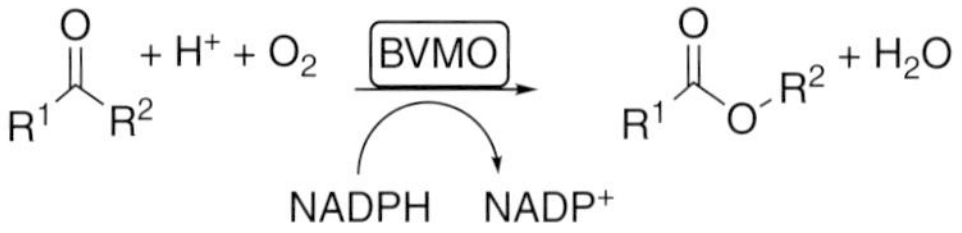

Scheme 18.1 Baeyer–Villiger oxidation of carbonylic substrates catalyzed by BVMOs.

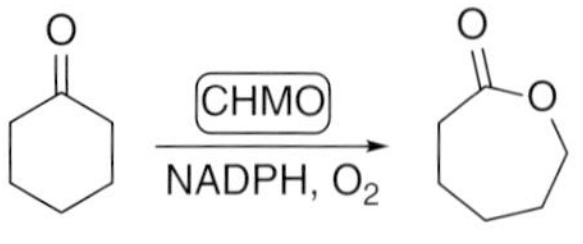

Scheme 18.2 Conversion of cyclohexanone to hexane-6-lactone catalyzed by cyclohexanone monooxygenase (CHMO).

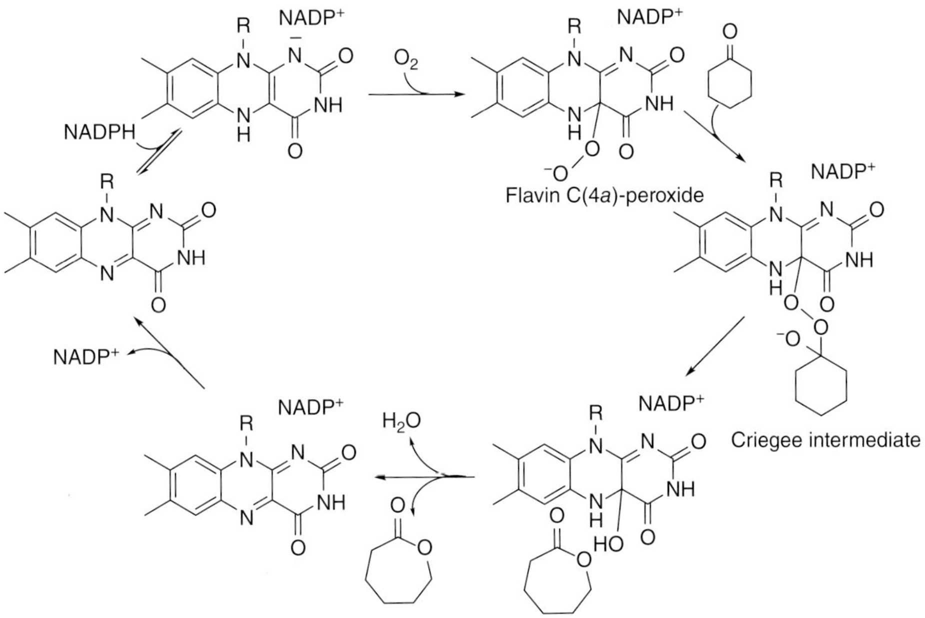

Scheme 18.3 Mechanism of Baeyer–Villiger monooxygenases [36–46].

18.2
Chemistry and Catalysis

Numerous asymmetric sulfoxidation reactions are catalyzed by chiral organometallic complexes or by organocatalysts though biological oxidations have been reported. Herein, the development of chemosynthetic methods related to those used for the synthesis of esomeprazole is reviewed. We recognize that these methods account for a significantly small representation of methods developed and contributed for synthesizing enantiomerically enriched sulfoxides. A comprehensive review providing detail on the development of chemosynthetic asymmetric sulfide oxidation catalysts is available [50].

The sulfoxide moiety can be found in many classes of drug targets [50]. The chirality of the sulfoxide and the emphasis in determining the pharmacological activity of each component in a drug necessitates the separation of enantiomers. Traditional approaches to resolve enantiomers include resolution methods, derivatization from a chiral auxiliary, asymmetric catalysis, and more recently asymmetric biocatalysis. The best-known example of a drug target for which enantiomeric resolution of the sulfoxide was achieved is that of the proton pump inhibitor omeprazole (Figure 18.1). Omeprazole was introduced in 1990 by AstraZeneca under the brand name Prilosec and quickly became the most widely used and best-selling drug for the treatment of gastroesophageal reflux disease, peptic ulcers, and erosive esophagitis [51]. Omeprazole was marketed as a racemate, and the final step in its industrial synthesis is a racemic oxidation

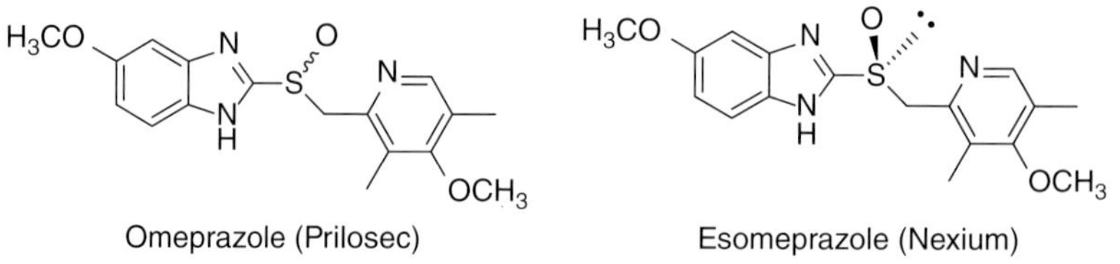

Figure 18.1 Structures of the proton pump inhibitors omeprazole (racemate) and esomeprazole (S-enantiomer).

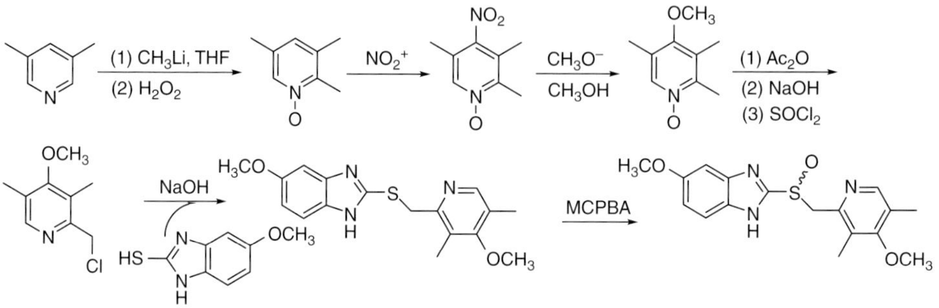

Scheme 18.4 Synthetic route for the production of omeprazole [52].

by *m*-chloroperbenzoic acid (MCPBA) (Scheme 18.4) [52, 53]. The impact of synthesizing the racemate was thought to be minor due to the conversion of the prodrug to the active achiral sulfenamide in the acidic environment of the stomach. The combination of patent lifetime and pharmaceutical focus on the safety and efficacy of enantiomerically pure drugs led to the exploration of esomeprazole, the *S*-enantiomer of omeprazole (Figure 18.1). Synthesis of esomeprazole was first achieved by a resolution of racemic omeprazole [54]. The Kagan catalytic oxidation method was employed for the synthesis of esomeprazole in order to produce an industrially scalable process [55]. The Kagan method is derived from Sharpless' titanium-catalyzed asymmetric epoxidation reaction [56–59] (Scheme 18.5).

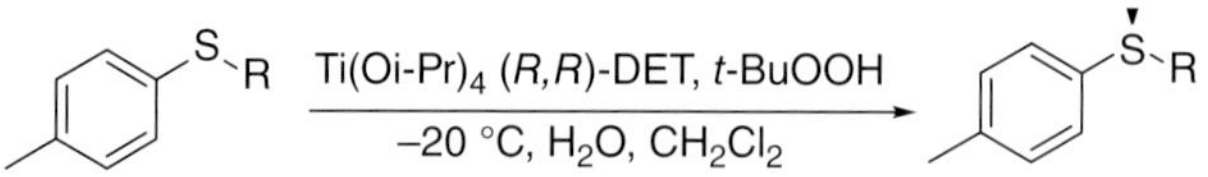

Scheme 18.5 Oxidation method developed by Kagan [56, 57] for catalyzing asymmetric sulfoxidation reactions.

Application of the Kagan method as reported in the literature effectively led to the production of racemic omeprazole. Formation of the racemate was rationalized by the similar steric demand of the two substituents of the sulfide precursor (pyrmetazole). Preparation of the catalyst complex at 30 °C, in the presence of water and the sulfide, and the use of *N,N*-diisopropylethylamine (DIEA) as an additive improved enantioselectivity. These slight modifications to the Kagan method afforded esomeprazole in yields exceeding 90% and ee of 94% (Scheme 18.6). This process was scaled-up for the industrial production of esomeprazole, marketed as Nexium™.

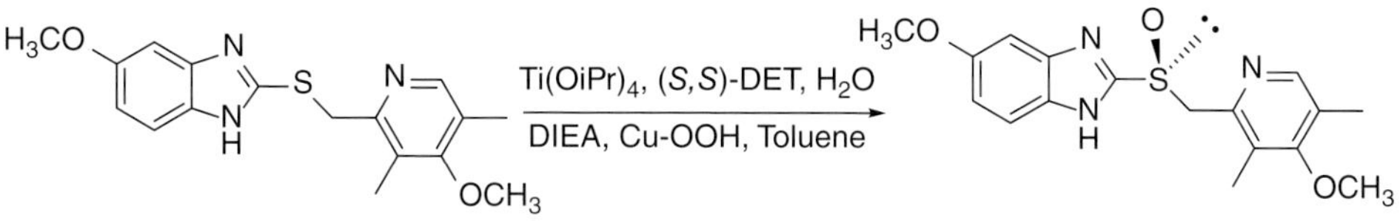

Scheme 18.6 Modified Kagan catalytic oxidation for the industrial scale-up of esomeprazole production.

18.2.1
Directed Evolution of BVMOs for the Manufacturing of Esomeprazole

BVMOs have long been identified as a promising target for applications in pharmaceutical chemistry due to their ability to catalyze epoxidations and enantioselective sulfoxidations in addition to catalyzing the regio- and enantioselective Baeyer–Villiger oxidations of a wide range of ketones [60, 61]. Reetz and coworkers published the first account on the directed evolution of a BVMO using CHMO as the target enzyme. CHMO catalyzes the Baeyer–Villiger oxidation of several 4-substituted cyclohexanones (i.e., 4-methylcyclohexanone, 4-ethylcyclohexanone, and 4-chlorocyclohexanone) with S-selectivity exceeding 95% ee. However, specific substituents at the 4-position (4-hydroxycyclohexanone) resulted in reversed enantioselectivity with an ee of 9%. Based on these results, 4-hydroxycyclohexanone was chosen to provide the selection pressure to enhance both R- and S-selectivity (Scheme 18.7) [62]. Roughly 12 000 mutants were generated by epPCR, and variants with enhanced R-selectivity up to 90% ee were identified. In addition, reversal of enantioselectivity was achieved with the identification of S-selective variants (up to 79% ee). In this study, reversal of enantioselectivity was attributed to a single point mutation (Phe432Ser). This mutation was also identified to enhance the enantioselectivity in the sulfoxidation of thioethers catalyzed by CHMO [63].

Considering the magnitude of production for the industrial scale-up of esomeprazole, the ability to improve the chiral purity and suppress overoxidation

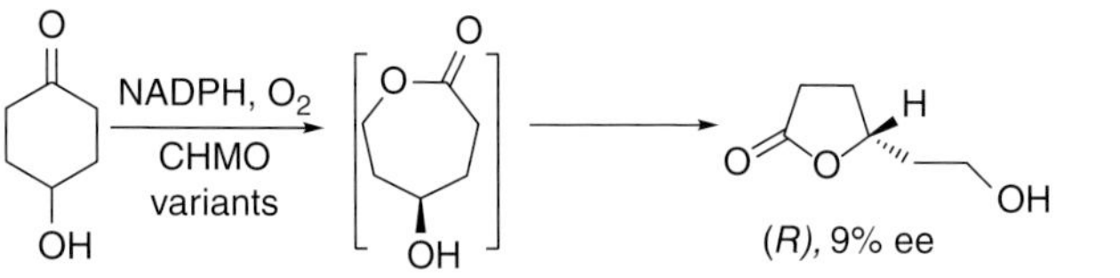

Scheme 18.7 Screening reaction used in the first directed evolution study of a BVMO; (R)-5-(2-hydroxyethyl)dihydrofuran-2(3H)-one is formed with 9% ee [62].

of the sulfoxide (to yield the undesired sulfone) presented an opportunity for a biocatalytic/enzymatic production route. The starting BVMO selected was wild-type (WT) CHMO from *Acinetobacter* sp. In order to provide a production route that was cost competitive, WT CHMO had to be evolved for an optimized process capable of producing a greater than 98% yield with a greater than 99.7% ee and less than 0.3% of the sulfone by-product. Initial experiments showed that approximately 0.3% of the sulfoxide product could be produced in 24 h at a substrate loading of 2 g/l pyrmetazole and an enzyme loading of 20 g/l WT CHMO. Furthermore, the undesired *R*-enantiomer was the major product generated.

A multidimensional evolution strategy was employed to identify variants that could overcome the challenges encountered while developing an enzymatic process for the preparation of esomeprazole (Tables 18.1 and 18.2). During CHMO evolution, a recycling system was used to regenerate the cofactor, NADPH. The recycling system was composed of a ketone reductase that converted isopropanol to acetone with the concomitant production of NADPH. The recycling system is advantageous for an enzymatic process due to the cost of the nicotinamide cofactor. However, the recycling system was incorporated only after demonstrating that CHMO remained the limiting enzyme when combined with an enzyme-based recycling system. Substrate/product solubility, uncoupling of electron transfer/O_2 reduction/substrate oxidation, substrate degradation, and overoxidation of the product were some of the process challenges (Table 18.2) that could jeopardize the success of an enzymatic process and could be addressed by evolution (Scheme 18.8).

Specifically, 20 rounds of evolution were required and multiple selection pressures applied to produce a process-capable enzyme (Table 18.1). The volumetric productivity of WT CHMO was so poor that the initial objective was to identify and incorporate diversity that significantly improved the rate of the enzyme for

Table 18.1 Evolution challenges and strategies for evolving a CHMO enzyme suitable for the industrial production of esomeprazole.

Challenges	Major strategies
Initial activity	Identify and evaluate several enzyme candidates; rational design large combinatorial library
Activity and stability (thermo; solvent; product)	Screen in the presence of saturating substrate (6 g/l pyrmetazole)
	Exert additional selection pressures by screening at increasing temperature, at increasing % cosolvent, and in the presence of saturating product
Enantioselectivity	Screening at low substrate loading (1 g/l pyrmetazole) and chiral separation
Overoxidation	Screening in the presence of product (1 g/l esomeprazole)
Uncoupling (H_2O_2 formation)	Screening in the absence of substrate

Table 18.2 Process challenges and strategies for developing an industrial process implementing an evolved CHMO enzyme for the production of esomeprazole.

Challenges	Major strategies
Cofactor recycling (two-enzyme system)	Ensure CHMO is the limiting enzyme and cofactor regeneration system is compatible with CHMO
Mass transfer limitation	Good stirring with proper choice of stirrer blade Utilizing an oxygen atmosphere for O_2 source
Low solubility of substrate	Incorporation of a cosolvent
Uncoupling (H_2O_2 formation)	Use of catalase
Degradation and overoxidation	Alkaline pH (8.5–9) and room temperature (25–30 °C)

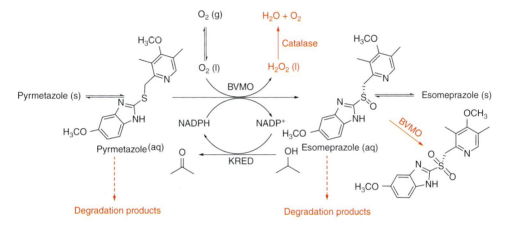

Scheme 18.8 Schematic representation of BVMO-mediated enzymatic process for esomeprazole production. During process development, pyrmetazole degradation, esomeprazole degradation and overoxidation, and the production of hydrogen peroxide were identified as side reactions/processes that needed to be minimized or eliminated for successful implementation.

the conversion of the sulfide precursor to the sulfoxide product without applying an enantiomeric enhancement selection pressure. From round 1 to round 2, eight mutations were incorporated and the volumetric productivity improved by 3 orders of magnitude (Table 18.3). During the next several rounds of evolution, multiple screening strategies and selection pressures were incorporated. For example, screening in the presence of increasing percentages of cosolvent was incorporated to identify solvent-tolerant variants due to the low solubility of the substrate. Furthermore, screening at temperatures and alkaline pH mimicking the manufacturing process was incorporated and improved variants were identified. These selection pressures were chosen and required to minimize the degradation of pyrmetazole and esomeprazole and the overoxidation of esomeprazole. After completion of seven rounds of evolution, the evolved CHMO backbone was

Table 18.3 Evolution summary, selection pressures, and results.

CHMO	Productivity (g/g$_{CHMO}$·h)	% ee	% sulfone	[NADP$^+$] (g/l)	Mutations (#) from previous round(s)
WT	0.000013	R-selective	n.d.	(GDH recycling)	—
Rd 2	0.01	−95	n.d.	2	8
Rd 7	0.08	96.5	0.5	0.5	18
Rd 13	1.07	97.8	0.5	0.5	9
RD 20 (CDX-003)	1.79	99.8	0.1	0.1	16

capable of producing the desired S-enantiomer with a 96.5% ee at a rate nearly 4-orders of magnitude greater than WT CHMO while producing only 0.5% of the undesired sulfone by-product. Furthermore, NADP$^+$ loading was reduced fourfold, which is significant due to the cost of the cofactor. An additional 13 rounds of evolution were completed to generate an enzyme that could achieve the volumetric productivity and ee to be a cost-competitive, process-ready enzyme. Furthermore, the generation of the sulfone by-product was further reduced to 0.1%, and the amount of NADP$^+$ loading was further decreased to 0.1 g/l (Table 18.3) [64]. The final evolved CHMO variant (Table 18.3; CDX-003) has been used for the manufacture of esomeprazole in 100 kg batches.

18.2.2
Directed Evolution and Incorporation of a Monoamine Oxidase for the Manufacturing of Boceprevir

MAOs are flavin-dependent enzymes that are responsible for the oxidative deamination of biogenic amines by oxidation to the imine followed by nonenzymatic hydrolysis to afford the corresponding ketone or aldehyde (Scheme 18.9). Upon oxidation of the amine substrate, the reduced flavin cofactor (FADH$_2$)

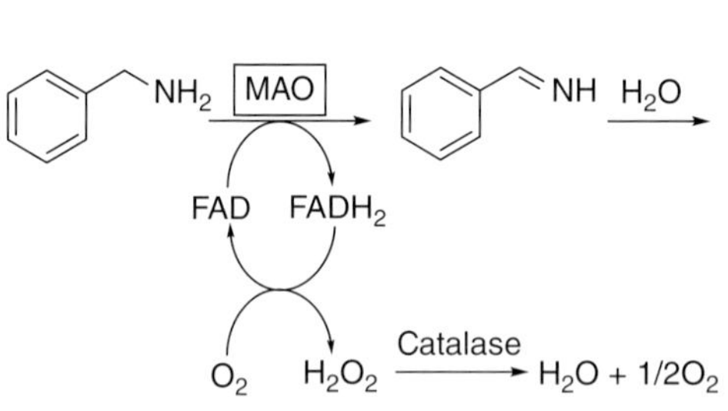

Scheme 18.9 Monoamine oxidase-catalyzed oxidation of amines coupled with catalase disproportionation of hydrogen peroxide to form O$_2$ and water.

is reoxidized by molecular oxygen to afford oxidized flavin (FAD) with the concomitant production of hydrogen peroxide. Catalase is often employed to prevent buildup of hydrogen peroxide in these reactions and prevent subsequent deleterious side reactions. While there is much interest in the human isoforms (MAO A and MAO B) for therapeutic purposes [65], there is little interest in these human isoforms from a biotechnology perspective. This is largely due to the fact that MAO A and MAO B are membrane-bound proteins that demonstrate poor stability when expressed in their soluble forms (C-termini truncated).

The more recently characterized MAO N is a flavin-dependent MAO that is naturally expressed in both *Aspergillus niger* and *Aspergillus oryzae*. This enzyme has approximately 24% sequence homology to the human MAOs and is believed to follow a similar reaction mechanism [66]. This soluble, 55.6 kDa enzyme expresses well in *Escherichia coli* (*E. coli*) [67] providing an excellent starting point for biotechnology applications. The WT enzyme has a somewhat limited substrate scope oxidizing only simple low molecular weight primary amines such as butylamine, amylamine, and benzylamine. Given the facile expression of MAO N in *E. coli* in combination with the recently published crystal structure [68], MAO N is an excellent candidate for enzyme engineering through directed evolution. The amine oxidase family is particularly an interesting candidate for biocatalyst development as there are relatively few enzyme classes that can be employed for the asymmetric generation of new C–C bonds. Therefore, MAO presents an exciting prospect for novel synthetic tools.

Intrigued by the promise of MAO oxidations, researchers at Codexis, in collaboration with Schering-Plough (now Merck & Co.), sought to incorporate an MAO-catalyzed oxidative desymmetrization in the synthesis of a key pyrroline intermediate in the manufacturing route for boceprevir (Victrelis®) [69]. Boceprevir is a peptidomimetic protease inhibitor marketed for the treatment of chronic hepatitis C. This first-in-class treatment prevents viral replication by binding to the NS3 protease. An efficient, convergent synthesis can be envisaged by disconnecting the molecule at the three peptide bonds (Scheme 18.10), resulting in the (*S*)-*tert*-leucine derivative, bicyclic [3.1.0]proline moeity, and racemic diketoamine. However, the asymmetric synthesis of the dimethylcyclopropylproline methyl ester proved challenging. The first-generation route featured an early-stage desymmetrization of caronic anhydride to set the two stereocenters followed by

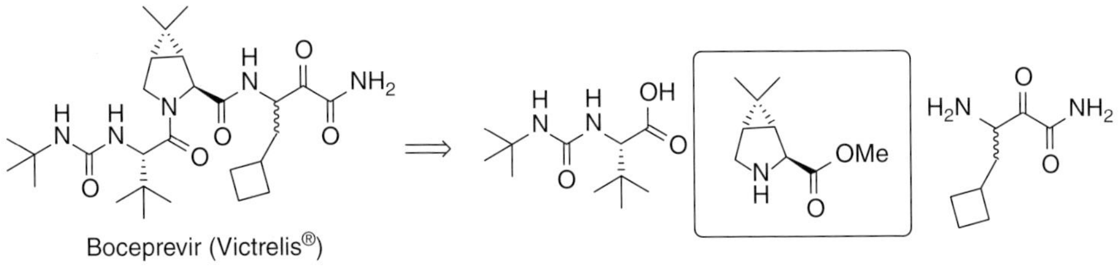

Scheme 18.10 Key disconnections of the peptide bonds of Boceprevir arriving at the bicyclic [3.1.0]proline core.

eight linear steps, which included a diastereospecific cyanation to set the third chiral center [70]. Again starting from caronic anhydride, the second-generation process streamlined the synthesis by maintaining the molecule's symmetry until the final step thereby reducing the overall step count (Scheme 18.11) [71–73]. To set the cyclopropyl stereocenters and access the desired methyl ester diastereomer, a classic resolution was employed, which resulted in approximately 50% loss in material late in the synthesis of the bicyclic [3.1.0]proline core. Process chemists instead pursued a direct desymmetrization using phase-transfer catalysis (PTC) of the fused cyclopropylpyrrolidine to recover this synthetic route. These efforts were not met with success as the catalysts evaluated provided only 20% ee product [73]. With these results in hand, a biocatalytic option was pursued for the oxidative desymmetrization step.

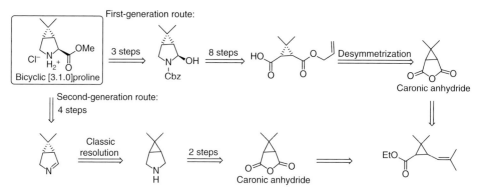

Scheme 18.11 Comparison of the first- and second-generation retrosynthetic analysis of the bicyclic [3.1.0]proline core.

Utilizing MAO for the oxidative desymmetrization is highly attractive in terms of both step count and theoretical yield (Scheme 18.12). Researchers at Codexis, Inc. first demonstrated the ability to produce asymmetric proline derivatives from pyrrolidines using an evolved MAO [74]. Reports from the Turner lab subsequently showed that desymmetrization of meso-pyrrolidines could provide access to chiral proline analogs [75]. However, low enzymatic activity at process-relevant substrate concentrations prevented this from being a viable process for a manufacturing scale. To improve volumetric productivities, enzyme stability, and turnover numbers (TONs), a collaboration between Schering Plough, now part of Merck & Co., and Codexis, Inc. was initiated for the directed evolution of MAO

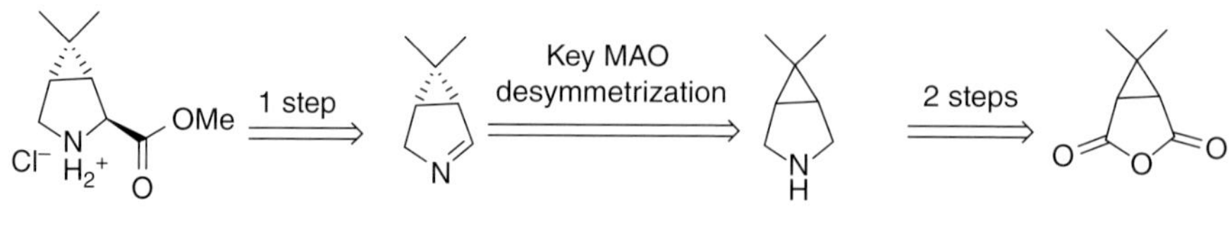

Scheme 18.12 Third-generation process for the manufacture of the bicyclic [3.1.0]proline featuring an MAO-catalyzed desymmetrization.

N. To test the feasibility of an MAO-catalyzed desymmetrization of the fused cyclopropyl pyrrolidine, MAO N isolated from *A. niger* and *A. oryzae* was tested for activity and absolute stereochemistry. While both WT variants demonstrated activity and greater than 99% ee for the desired enantiomer, *A. niger* was selected as the parent sequence for evolution on the basis of higher stability in comparison to the *A. oryzae* variant. When developing this chemistry, a number of limitations were identified: poor product solubility/volatility, O_2 dependence, pH decreases over time, and H_2O_2 production. Because the reaction is O_2 dependent, the use of organic cosolvents to manage product solubility was not desirable as this presented a potential fire hazard on scale. The issue of pH was easily addressed by using free base starting material to maintain pH. The disproportionation of H_2O_2 was carried out on scale using catalase, a highly efficient enzyme commercially available in bulk quantities. Moreover, the generation of hydrogen peroxide was exploited as a tool for high-throughput (HTP) identification of activity variants. Peroxide served as a reporter for reaction progress by coupling the pyrrolidine oxidation to a horseradish peroxidase–Amplex Red spectrophotometric assay (Scheme 18.13) [76].

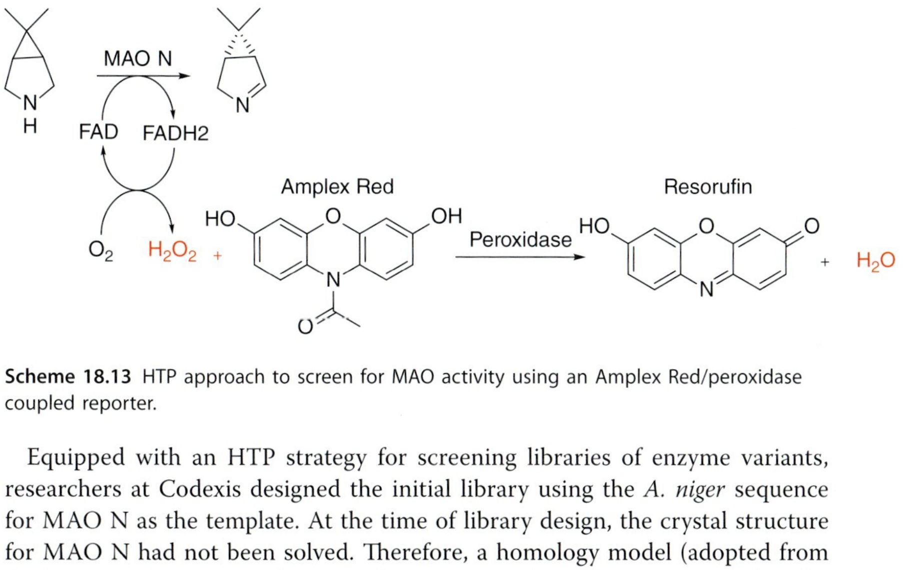

Scheme 18.13 HTP approach to screen for MAO activity using an Amplex Red/peroxidase coupled reporter.

Equipped with an HTP strategy for screening libraries of enzyme variants, researchers at Codexis designed the initial library using the *A. niger* sequence for MAO N as the template. At the time of library design, the crystal structure for MAO N had not been solved. Therefore, a homology model (adopted from the human MAO structure) was used to target the desired active site residues for activity improvements, and mutagenesis or saturation mutagenesis was employed to generate diversity. This strategy yielded a variant with a 2.4-fold improvement based solely on active site mutations. Subsequent libraries were generated using two approaches: (i) a family shuffling approach using diversity from the *A. oryzae* variant and (ii) recombination of diversity identified in the round 1 library. In this round of evolution, variants with three- and six-fold improvements in activity were observed over the backbone selected from the round one library. In addition, these variants demonstrated thermal stability up to 40 °C by virtue of retaining 80–100% of their original activity after overnight incubation.

Two additional rounds of evolution were carried out specifically targeting substrate tolerance, product inhibition, and thermal stability. The round three library was generated using a family shuffling approach to combine the diversity from the two top variants identified during round two screening. This approach generated three variants with improved stability in comparison to the round two parents at 50 °C. This diversity was carried into a fourth round of evolution to generate the final variant that met the needs of the process chemists at Merck (then Schering-Plough). This final variant demonstrated stability when incubated at 40 °C for 3 days retaining approximately 85% of its initial activity. Moreover, this final MAO N variant (CDX-616) was 2.8-fold more active than the round three backbone used for the final round of evolution. In summary, CDX-616 was shown to have an 8.4-fold increase in specific activity over WT MAO N.

18.3
Process Technology

Using an O_2-dependent enzyme comes with additional challenges for process optimization. At ambient temperature and pressure, the solubility of O_2 in water is 240 µM. As the salt concentration increases, the solubility of O_2 decreases resulting in mass transfer issues of the second substrate, oxygen. CDX-616 has a K_{oxygen} of 210 µM, which suggests that at air saturation the enzyme is only functioning at half its productivity. With this in mind, the reaction was fed O_2 gas to increase the dissolved oxygen concentration in the reaction (O_2 saturation of water is 960 µM). Poor solubility of oxygen was not the only solubility issue experienced by process chemists. The oxidized imine tended to vaporize due to its poor solubility in water, which introduced additional problems when using a gas dispersion tube to introduce O_2. These problems were mitigated by using a condenser held at -10 °C to trap the imine product. However, the volatility of the product in combination with the use of O_2 presented a fire hazard that needed to be addressed on scale.

To circumvent these issues, the imine product was converted to a water soluble intermediate *in situ* by forming a bisulfite adduct. The amino sulfonate was generated during the course of the reaction by employing sodium bisulfite to trap the imine intermediate (Scheme 18.14). The addition of sodium bisulfite also assuaged the product inhibition observed in the absence of a chemical trapping agent. Furthermore, 94% product mass balance suggested that little of the iminium product

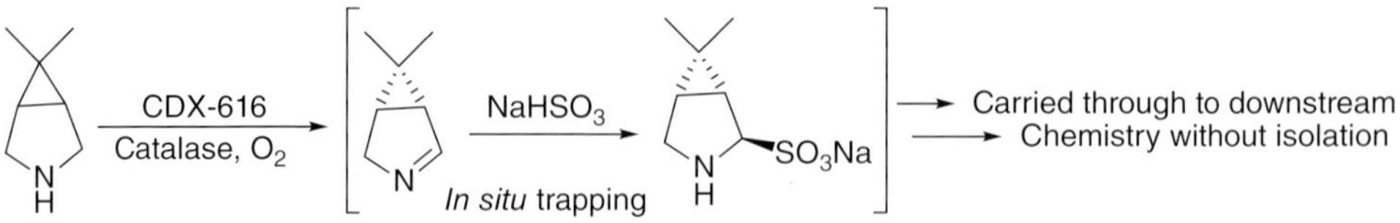

Scheme 18.14 Trapping the volatile imine with sodium bisulfite to make the stable sulfone intermediate.

was lost due to volatility. While the use of sodium bisulfite mitigated many of the safety concerns, background oxidation of bisulfite was a competing pathway that depleted the available bisulfite in solution (Scheme 18.15). It was determined that the equilibrium could be shifted in favor of sulfonate formation by adding a solution of starting pyrrolidine and sodium bisulfite to the reaction mixture. The kinetic parameters of MAON401 were determined and used to optimize the rate of substrate–bisulfite addition over time to favor the productive sulfonation and limit buildup of the product imine which irreversibly inhibits the MAON401 (Scheme 18.15).

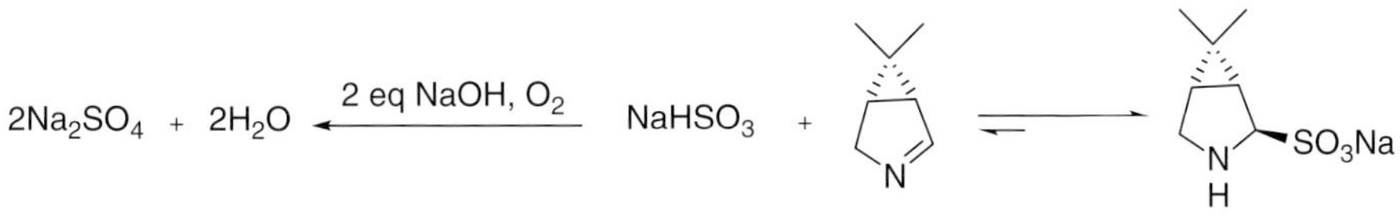

Scheme 18.15 Competition between sulfonation and oxidation of bisulfite.

For the manufacture of the sulfonate, the reaction was run in a closed, pressurized vessel under 4 psi of oxygen with the agitation set to maintain a predetermined $K_L a$ to maximize the enzyme rate. The activity of MAON401 increased threefold when oxygen sparging was replaced with pressurized O_2. Slow addition of the substrate–bisulfite solution to a solution of 6 wt% MAON401 and catalase allowed for a final substrate concentration of 65 g/l. Using cyclopentylmethyl ether (CPME) as an extraction solvent, the sulfonate was immediately converted to the nitrile without purification of the reaction stream, which provided an overall 90% yield starting from the pyrrolidine. Methanolysis of the nitrile under Pinner conditions followed by recrystallization provided the desired bicyclic [3.1.0]proline core in 56% overall yield and greater than 99% ee (Scheme 18.16).

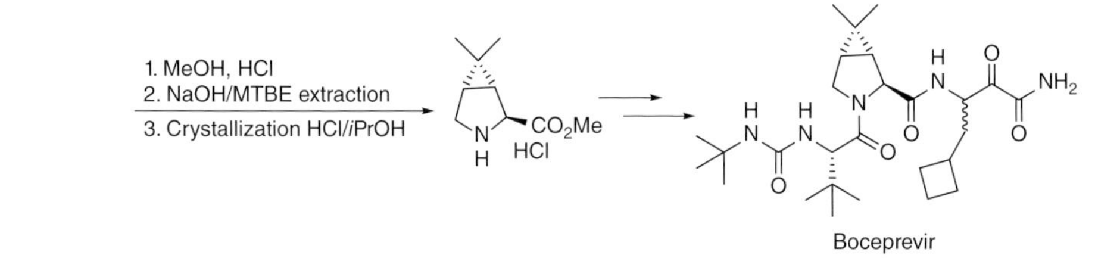

Scheme 18.16 Synthetic sequence to prepare boceprevir from caronic anhydride.

18.4 New Developments

The examples summarized herein represent the successful scale-up and implementation of a BVMO for the production of an active pharmaceutical ingredient (API) and an MAO for an intermediate *en route* to an API, respectively. While these examples merit celebration, heteroatom oxidation and amine oxidation are milder oxidations, which benefit from having chemocatalytic solutions. However, enzymes catalyze significantly more challenging oxidative transformations, and these reactions have biocatalytic potential as the chemocatalytic counterpart can be much more difficult to undertake. For example, selective oxidative functionalization of hydrocarbon moieties and asymmetric epoxidations (without allylic hydroxylation) are challenging using purely synthetic means due to modest selectivity and/or poor reactivity. The excellent regio- and stereoselectivity of enzymes offers an opportunity for the ongoing development of oxidative biocatalysis through the use of technology such as directed evolution. The potential utility of cytochrome P450s (CYPs) and styrene monoxygenases (SMOs) as biocatalysts for industrial-scale syntheses has long been recognized due to their ability to catalyze regio- and stereoselective hydroxylations of nonactivated C–H bonds (Scheme 18.17) and epoxidations of olefins (Scheme 18.18) using molecular oxygen as the oxidant. In fact, hydrocortisone is produced at a scale of approximately 100 t/year with a CYP-mediated process [77]. In addition, the successful utilization of a CYP expressed from an engineered *E. coli* strain for the biotransformation of compactin to pravastatin has been reported (Scheme 18.19) [78]. A whole-cell process using recombinant *E. coli* expressing SMO for the production of (*S*)-styrene oxide has been reported at pilot scale (Scheme 18.18) [79]. These reports highlight examples of the successful scale-up and utilization of monooxygenases for the production of pharmaceuticals and enantiopure chemical building blocks.

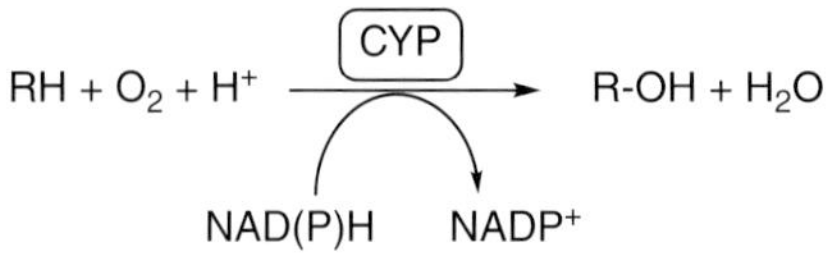

$$RH + O_2 + H^+ \xrightarrow[\text{NAD(P)H} \quad \text{NADP}^+]{\text{CYP}} R\text{-OH} + H_2O$$

Scheme 18.17 Hydroxylation reaction catalyzed by cytochrome P450s.

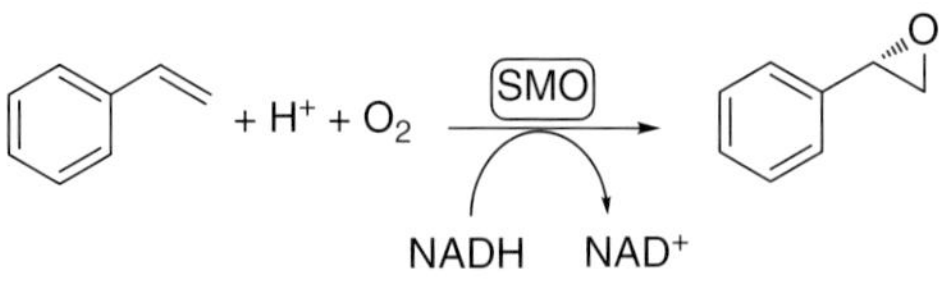

$$\text{PhCH=CH}_2 + H^+ + O_2 \xrightarrow[\text{NADH} \quad \text{NAD}^+]{\text{SMO}} \text{PhCH(O)CH}_2$$

Scheme 18.18 Asymmetric epoxidation reaction catalyzed by styrene monooxygenase.

The application of CYPs and SMOs for industrial applications will continue to be broadened as we improve our understanding of these enzymes and thus overcome the technical limitations on scale. Reviews summarizing many of the

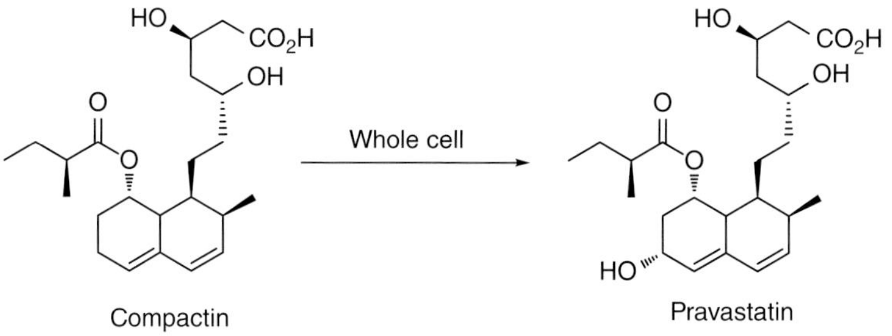

Scheme 18.19 Biosynthesis of pravastatin using an engineered *E. coli* strain expressing a cytochrome P450 gene.

technical challenges that have been encountered using CYPs as biocatalysts are available [80]. These limitations apply to SMOs as well. For example, the cost of cofactors, substrate/product solubility, low activity, and the use of redox partners are often described as limitations for CYP enzymes, but these challenges have been or can be overcome using directed evolution. Cofactor recycling systems have been used at the industrial scale for MAO (in addition to other enzyme classes) as well as at HTP scale for screening enzyme variants including P450s. Implementation of a cosolvent for substrate solubility and evolving enzymes for solvent tolerance is well-established. Self-sufficient CYP and SMO enzymes, in which the full complement of hydroxylation/epoxidation and redox domains are expressed in a single polypeptide, provide a significant advantage in activity and ease of experimental manipulation [81, 82]. CYP102A1 (P450-BM3) was the first self-sufficient CYP reported [81], and novel self-sufficient CYPs have been discovered since [83].

For CYPs to achieve industrially relevant TONs, the coupling of electron transfer to product hydroxylation must be optimized. The efficiency of a CYP-catalyzed process demonstrating the importance of coupling and its relationship to TON is elucidated for the production of propanol from propane by an evolved CYP102A1 variant [84]. To what extent coupling is a substrate-specific property has not been comprehensively investigated or established. Perhaps the greater challenge facing CYP enzymes is overcoming the limitation of volumetric productivity. Most reports of CYP-catalyzed biotransformations are carried out at substrate concentrations that are at least an order of magnitude less than would be needed for a cost-competitive biocatalytic route. Fundamental factors limiting the ability of CYP enzymes to tolerate high concentrations of a broad range of substrates is not well-established and warrants further investigation. The range of olefins that SMOs can react with needs to be expanded. Substrate choice should be guided based on known chemosynthetic routes using Sharpless, Jacobson, or Katsuki catalysts not able to achieve full enantioselectivity in combination with identifying industrial epoxidations for which the application of an engineered enzyme is cost competitive. Engineered SMOs with enhanced activity and (*R*)-selectivity have been reported [85], and continued evolution of SMOs will enhance the breadth of potential SMO catalyzed applications.

Oxidative transformations comprise a small percentage of the reactions used on the preparative scale of APIs. Safety and the production of undesired waste streams as a result of using reagents that significantly enhance selectivity are limitations to incorporating oxidation chemistry into synthetic routes at manufacturing scale. The ability of enzymes to utilize molecular oxygen to achieve exquisite stereo- and regioselectivities under benign reaction conditions could result in the paradigm shift needed to include oxidative chemistry in process route design.

References

1. Jacobsen, E.N. (2000) *Acc. Chem. Res.*, **33** (6), 421–431.
2. Noyori, R. (2002) *Angew. Chem. Int. Ed.*, **41** (12), 2008–2022.
3. Sharpless, K.B. (2002) *Angew. Chem. Int. Ed.*, **41** (12), 2024–2032.
4. Walsh, P.J. and Kozlowski, M.C. (eds) (2009) *Fundamentals of Asymmetric Catalysis*, University Science Books, Sausalito, CA.
5. Corey, E.J. and Kürti, L. (eds) (2011) *Enantioselective Chemical Synthesis: Methods, Logic and Practice*, Direct Book Publishing, LLC, Dallas, TX.
6. Zhou, Q.-L. (2011) *Privileged Chiral Ligands and Catalysts*, Wiley-VCH Verlag GmbH, Weinheim.
7. Blaser, H.-U. and Federsel, H.-J. (eds) (2011) *Asymmetric Catalysis on Industrial Scale: Challenges, Approaches and Solutions*, Wiley-VCH Verlag GmbH, Weinheim.
8. Briére, J.-F., Oudeyer, S., Dalla, V., and Levacher, V. (2012) *Chem. Soc. Rev.*, **41** (5), 1696–1707.
9. Müller, T.E., Hultzsch, K.C., Yus, M., Foubelo, F., and Tada, M. (2008) *Chem. Rev.*, **108** (9), 3795–3892.
10. Berkessel, A. and Gröger, H. (eds) (2006) *Asymmetric Organocatalysis*, Wiley-VCH Verlag GmbH, Weinheim.
11. MacMillan, D.W.C. (2008) *Nature*, **455** (7211), 304–308.
12. List, B. (2010) *Angew. Chem. Int. Ed.*, **49** (10), 1730–1734.
13. Allen, A.E. and MacMillan, D.W.C. (2012) *Chem. Sci.*, **3** (3), 633–658.
14. Drauz, K., Gröger, H., and May, O. (eds) (2012) *Enzyme Catalysis in Organic Synthesis*, 3rd edn, vol. **1–3**, Wiley-VCH Verlag GmbH, Weinheim.
15. Faber, K. (2011) *Biotransformations in Organic Chemistry*, 6th edn, Springer-Verlag, Berlin.
16. Liese, A., Seelbach, K., and Wandrey, C. (eds) (2006) *Industrial Biotransformations*, 2nd edn, Wiley-VCH Verlag GmbH, Weinheim.
17. Tao, J., Lin, G.-Q., and Liese, A. (eds) (2009) *Biocatalysis for the Pharmaceutical Industry*, Wiley-VCH Verlag GmbH, Weinheim.
18. Gotor, V., Alfonso, I., and García-Urdiales, E. (eds) (2008) *Asymmetric Organic Synthesis with Enzymes*, Wiley-VCH Verlag GmbH, Weinheim.
19. Lutz, S. and Bornscheuer, U.T. (eds) (2012) *Protein Engineering Handbook*, vol. **1–2**, Wiley-VCH Verlag GmbH, Weinheim.
20. Turner, N.J. (2009) *Nat. Chem. Biol.*, **5** (8), 567–573.
21. Jäckel, C., Kast, P., and Hilvert, D. (2008) *Annu. Rev. Biophys.*, **37**, 153–173.
22. Bershtein, S. and Tawfik, D.S. (2008) *Curr. Opin. Chem. Biol.*, **12** (2), 151–158.
23. Romero, P.A. and Arnold, F.H. (2009) *Nat. Rev. Mol. Cell Biol.*, **10** (12), 866–876.
24. Reetz, M.T. (2008) in *Asymmetric Organic Synthesis with Enzymes* (eds V. Gotor, I. Alfonso, and E. García-Urdiales), Wiley-VCH Verlag GmbH, Weinheim, pp. 21–63.
25. Otten, L.G., Hollmann, F., and Arends, I.W.C.E. (2010) *Trends Biotechnol.*, **28** (1), 46–54.

26. Khersonasky, O. and Tawfik, D.S. (2010) *Annu. Rev. Biochem.*, **79**, 471–505.
27. Arnold, F.H. and Volkov, A.A. (1999) *Curr. Opin. Chem. Biol.*, **3** (1), 54–59.
28. Fox, R.J., Davis, S.C., Mundorff, E.C., Newman, L.M., Gavrilovic, V., Ma, S.K., Chung, L.M., Ching, C., Tam, S., Muley, S., Grate, J., Gruber, J., Whitman, J.C., Sheldon, R.A., and Huisman, G.W. (2007) *Nat. Biotechnol.*, **25** (3), 338–344.
29. Conrad, H.E., DuBus, R., Namtvedt, M.J., and Gunsalus, I.C. (1965) *J. Biol. Chem.*, **240**, 495–503.
30. van Berkel, W.J.H., Kamerbeek, N.M., and Fraaije, M.W. (2006) *J. Biotechnol.*, **124** (4), 670–689.
31. Cheng, Q., Thomas, S.M., and Rouviére, P. (2002) *Appl. Microbiol. Biotechnol.*, **58** (6), 704–711.
32. Wright, J.L.C., Hu, T., MacLachlan, J.L., Needham, J., and Walter, J.A. (1996) *J. Am. Chem. Soc.*, **118** (36), 8757–8758.
33. Damtoft, S., Franzyk, H., and Jensen, S.R. (1995) *Phytochemistry*, **40** (3), 773–784.
34. Townsend, C.A., Christensen, S.B., and Davis, S.G. (1982) *J. Am. Chem. Soc.*, **104** (22), 6154–6155.
35. Trudgill, P.W. (1994) in *Biochemistry of Microbial Degradation* (ed C. Ratledge), Kluwer Academic, Dordrecht, pp. 33–61.
36. Turfitt, G.E. (1948) *Biochem. J.*, **42** (3), 376–383.
37. Donoghue, N.A., Norris, D.B., and Trudgill, P.W. (1976) *Eur. J. Biochem.*, **63** (1), 175–192.
38. Schwab, J.M., Li, W.B., and Thomas, L.P. (1983) *J. Am. Chem. Soc.*, **105** (14), 4800–4808.
39. Ryerson, C.C., Ballou, D.P., and Walsh, C. (1982) *Biochemistry*, **21** (11), 2644–2655.
40. Walsh, C.T. and Chen, Y.-C. (1988) *Angew. Chem. Int. Ed. Engl.*, **27** (3), 333–343.
41. Sheng, D., Ballou, D.P., and Massey, V. (2001) *Biochemistry*, **40** (37), 11156–11167.
42. Brzostowicz, P.C., Walters, D.M., Thomas, S.M., Nagarajan, V., and Rouviére, P.E. (2003) *Appl. Environ. Microbiol.*, **69** (1), 334–342.
43. Fraaije, M.W. (2006) *J. Biotechnol.*, **124** (4), 670–689.
44. Kayser, M.M. (2009) *Tetrahedron*, **65** (5), 947–974.
45. Torres Pazmiño, D.E., Winkler, M., Glieder, A., and Fraaije, M.W. (2010) *J. Biotechnol.*, **6** (1–2), 9–24.
46. Orru, R., Dudek, H.M., Martinoli, C., Torres Pazmiño, D.E., Royant, A., Weik, M., and Fraaije, M.W. (2011) *J. Biol. Chem.*, **286** (33), 29284–29291.
47. Criegee, R. (1948) *Justus Liebigs Ann. Chem.*, **560** (1), 127–135.
48. Branchaud, B.P. and Walsh, C.T. (1985) *J. Am. Chem. Soc.*, **107** (7), 2153–2161.
49. Latham, J.A., Branchaud, B.P., Chen, Y.-C., and Walsh, C. (1986) *J. Chem. Soc., Chem. Commun.*, (7), 528–530.
50. O'Mahony, G.E., Kelly, P., Lawrence, S.E., and Maguire, A.R. (2011) *Arkivoc*, **1**, 1–110.
51. Agranat, I. and Caner, H. (1999) *Drug Discovery Today*, **4** (7), 313–321.
52. Lindberg, P., Brändström, A., Wallmark, B., Mattsson, H., Rikner, L., and Hoffman, K.-J. (1990) *Med. Res. Rev.*, **10** (1), 1–54.
53. Carlsson, E.I., Junggren, U.K., Larsson, H.S., and von Wittken Sundell, G.W. (1981) Pharmaceutical compositions containing [(pyridyl-methyl0thio]benzimidazole derivatives. European Patent EP 0074341, filed Aug. 13, 1981 and issued Mar. 16, 1983.
54. Erlandsson, P., Isaksson, R., Lorentzon, P., and Lindberg, P. (1990) *J. Chromatogr.*, **532** (2), 305–319.
55. Cotton, H., Elebring, T., Larsson, M., Li, L., Sörenson, H., and von Unge, S. (2000) *Tetrahedron: Asymmetry*, **11** (18), 3819–3825.
56. Pitchen, P. and Kagan, H.B. (1984) *Tetrahedron Lett.*, **25** (10), 1049–1052.
57. Pitchen, P., Dunach, E., Deshmukh, M.N., and Kagan, H.B. (1984) *J. Am. Chem. Soc.*, **106** (26), 8188–8193.
58. Katsuki, T. and Sharpless, K.B. (1980) *J. Am. Chem. Soc.*, **102** (18), 5974–5976.

59. Sharpless, K.B. (1986) *Chem. Br.*, **22** (1), 38–40.
60. Mihovilovic, M., Müller, B., and Stanetty, P. (2002) *Eur. J. Org. Chem.*, **2002** (22), 3711–3730.
61. de Gonzalo, G., Mihovilovic, M.D., and Fraaije, M.W. (2010) *ChemBioChem*, **11**, 2208–2231.
62. Reetz, M.T., Brunner, B., Schneider, T., Schulz, F., Clouthier, C.M., and Kayser, M.M. (2004) *Angew. Chem. Int. Ed.*, **43** (31), 4075–4078.
63. Reetz, M.T., Daligault, F., Brunner, B., Hinrichs, H., and Deege, A. (2004) *Angew. Chem. Int. Ed.*, **43** (31), 4078–4081.
64. Bong, Y.K., Clay, M.D., Collier, S.J., Mijts, B., Vogel, M., Zhang, X., Zhu, J., Nazor, J., Smith, D., and Song, S. (2011) Synthesis of prazole compounds. International Publication Number WO2011/071982 A2, filed Dec. 8, 2010 and issued June 16, 2011.
65. Youdim, M.B.H., Edmondson, D., and Tipton, K.F. (2006) *Nat. Rev. Neurosci.*, **7** (4), 295–309.
66. Sablin, S.O., Yankovskaya, V., Bernard, S., Cronin, C.N., and Singer, T.P. (1998) *Eur. J. Biochem.*, **253** (1), 270–279.
67. Schilling, B. and Lerch, K. (1995) *Mol. Gen. Genet.*, **247** (4), 430–438.
68. Atkin, K.E., Reiss, R., Koehler, V., Bailey, K.R., Hart, S., Turkenburg, J.P., Turner, N.J., Brzozowski, A.M., and Grogan, G. (2008) *J. Mol. Biol.*, **384** (5), 1218–1231.
69. Li, T., Liang, J., Ambrogelly, A., Brennan, T., Gloor, G., Huisman, G., Lalonde, J., Lekhal, A., Mijts, B., Muley, S., Newman, L., Tobin, M., Wong, G., Zaks, A., and Zhang, X. (2012) *J. Am. Chem. Soc.*, **134** (14), 6467–6472.
70. Park, J., Sudhakar, A., Wong, G.S., Chen, M., Weber, J., Yang, X., Kwok, D.-I., Jeon, I., Raghavan, R.R., Tamarez, M., Tong, W., and Vater, E.J. (2004) Process and intermediates for the preparation of (1R,2S,5S)-6,6-dimethyl-3-azabicyclo[3,1,0]hexane-2-carboxylates or salts thereof via asymmetric esterification of caronic anhydride. Patent 2004-US19135, 2004113295, 20040615, filed June 15, 2004 and issued Dec. 29, 2004.
71. Wu, G., Chen, F.X., Rashatasakhon, P., Eckert, J.M., Wong, G.S., Lee, H.-C., Erickson, N.C., Vance, J.A., Nirchio, P.C., Weber, J., Tsai, D.J.-S., and Nanfei, Z. (2007) Process for the preparation of 6,6-dimethyl-3-azabicyclo[3.1.0]hexane compounds and enantiomeric salts thereof. Patent 2006-US48613, 2007075790, 20061220, filed Dec. 12, 2006 and issued June 25, 2010.
72. Berranger, T. and Demonchaux, P. (2008) Process for preparation of optically pure 6,6-dimethyl-3-azabicyclo[3.1.0]hexane derivatives. 2007-US25809, 2008082508, 20071218, filed Dec. 18, 2007 and issued July 10, 2008.
73. Kwok, D.-L., Lee, H.-C., and Zavialov, I.A. (2009) Dehydrohalogenation process for preparation of intermediates useful in providing 6,6-dimethyl-3-azabicyclo[3.1.0]hexane compounds. Patent 2008-US84174, 2009073380, 20081120, filed Nov. 20, 2008 and issued June 11, 2009.
74. Mijts, B., Muley, S., Liang, J., Newman, L.M., Zhang, X., Lalonde, J., Clay, M.D., Zhu, J., Gruber, J.M., Colbeck, J., Munger, J.D. Jr.,, Mavinahalli, J., and Sheldon, R. (2008) Biocatalytic process for the preparation of substantially steromerically pure fused bicyclic proline compounds. Patent WO/2010/008828, filed June 24, 2008 and issued Jan. 21, 2010.
75. Köhler, V., Bailey, K.R., Znabet, A., Raftery, J., Helliwell, M., and Turner, N.J. (2010) *Angew. Chem. Int. Ed.*, **49** (12), 2182–2184.
76. Zhou, M. and Panchuk-Voloshina, N. (1997) *Anal. Biochem.*, **253** (2), 169–174.
77. Sonomoto, K., Hoq, M.M., Tanaka, A., and Fukui, S. (1983) *Appl. Environ. Microbiol.*, **45** (2), 436–443.
78. Fujii, T., Fujii, Y., Machida, K., Ochiai, A., and Ito, M. (2009) *Biosci. Biotechnol., Biochem.*, **73** (4), 805–810.
79. Panke, S., Held, M., Wubbolts, M.G., Witholt, B., and Schmid, A. (2002) *Biotechnol. Bioeng.*, **80** (1), 33–41.

80. Bernhardt, R. and Urlacher, V.B. (2014) *Appl. Microbiol. Biotechnol.*, **98** (14), 6185–6203.
81. Fulco, A.J. (1991) *Annu. Rev. Pharmacol.*, **31**, 177–203.
82. Tischler, D., Eulberg, D., Lakner, S., Kaschabek, S.R., van Berkel, W.J.H., and Schlömann, M. (2009) *J. Bacteriol.*, **191** (15), 4996–5009.
83. Choi, K.-Y., Jung, E., Jung, D.-H., Pandey, B.P., Yun, H., Park, H.-Y., Kazlauskas, R.J., and Kim, B.-G. (2012) *FEBS J.*, **279** (9), 1650–1652.
84. Fasan, R., Meharenna, Y.T., Snow, C.D., Poulos, T.L., and Arnold, F.H. (2008) *J. Mol. Biol.*, **383** (5), 1069–1080.
85. Lin, H., Tang, D.-F., Ahmed, A.A.Q., Liu, Y., and Wu, Z.-L. (2012) *J. Biotechnol.*, **161** (3), 235–241.

Part VI
Oxidative Conversion of Renewable Feedstocks

19
From Terephthalic Acid to 2,5-Furandicarboxylic Acid: An Industrial Perspective

Jan C. van der Waal, Etienne Mazoyer, Hendrikus J. Baars, and Gert-Jan M. Gruter

19.1
Introduction

With the foreseeable rise of bio-based platform chemicals from starch and sucrose initially and lignocellulosic or waste feedstock streams somewhat later, the question can be raised whether new methods for defunctionalization or further functionalization will be required or can we rely on the reuse of existing catalytic conversions as a strategy? To start answering this question, we need to realize that bio-based molecules can, but most likely will not, be the same as the current petrochemical-based platform chemicals. And even if the same molecules are targeted ("drop-in"), the production routes will not be the same. Carbohydrates are the most abundant and most frequently used bio-based feedstock. They are rich in oxygen atoms, whereas the petro-based feedstocks are hydrocarbons, exclusively made up of C and H. In many cases, bio-based feedstock is overoxidized and thus a clear need for reduction, hydrogenation, dehydration, decarboxylation, and other chemistries will be needed. Perhaps an even more challenging task for these molecules is to selectively oxidize them further, as these bio-based molecules are inherently more activated due to the already present oxygen groups.

This chapter provides an industrial perspective on several oxidation routes to new bio-based molecules. In particular, it focuses on the use of Co/Mn/Br catalyst systems in air oxidations, based on the Amoco Mid-Century catalyst system used for *para*-xylene oxidation (also see Chapter 4), as an efficient methodology for the conversion of 5-(hydroxymethyl)furfural (HMF) and 5-(methoxymethyl)furfural (MMF) to 2,5-furandicarboxylic acid (FDCA) in Avantium's YXY process. In addition, other less-studied conversions, such as methyl levulinate (ML) to succinic acid (SA), lignin to a variety of aromatic and phenolic carboxylic acids, are discussed as well.

19.1.1
The Avantium YXY Technology to Produce PEF, a Novel Renewable Polymer

The YXY technology is currently developed by Avantium Chemicals in the Netherlands with the aim of efficiently converting sugars to FDCA and its application as a monomer in new and previously described condensation polymers [1]. Especially, the polyester obtained from FDCA and ethylene glycol, resulting in polyethylene furandicarboxylate or PEF, has attracted widespread attention as it has high potential to replace polyethylene terephthalate (PET) in bottle, film, and fiber applications. In addition, the unique barrier properties of PEF will also allow this new material to be used in applications not reachable for PET without additional technology such as multilayer or scavenger technologies. Recently, The Coca-Cola Company, Danone, and ALPLA announced their collaborations with Avantium on the development and application of PEF for use in carbonated soft drink (CSD) and water bottles [2, 3].

The starting point in the YXY technology is the dehydration of hexoses to give rise to a group of chemicals called *furanics* (molecules that contain a furan ring). Dehydration of sugars is long known and especially HMF and its decomposition product levulinic acid (LA) are the most commonly observed and researched compounds [4]. Both these compounds have been identified by Werpy and Petersen in the beginning of the twenty-first century as important building blocks of a future renewable-based chemical industry [5]. Of these two, HMF has received most of the attention as it can be oxidized to FDCA, an interesting monomer that can be used for the production of PEF.

The general approach of Avantium [6] to the production of FDCA starts from fructose syrup. Today, fructose syrup is predominantly made from dextrose (glucose), which is obtained from the hydrolysis of starch (from corn or wheat). Although several companies have recently started the production and application of lignocellulosic ("nonfood") carbohydrates (for making ethanol), to date no processes have been commercialized that can produce dextrose with a quality and with economics comparable to its conventional first-generation reference. As Avantium does believe that in the future high-quality, cost-efficient nonfood carbohydrates should be available for fuels and materials production, it has initiated its own program to produce high-quality dextrose from cellulose hydrolysis. Fructose is dehydrated in methanol to MMF, the methyl ether of HMF, while suppressing the formation of LA and its esters [7]. In a next separate step, after purification, the MMF is oxidized to FDCA (see Figure 19.1). In a final process step, FDCA can be polymerized to a large number of target polyesters or polyamides. With ethylene glycol PEF is obtained.

19.2
Chemistry and Catalysis

During the last 150 years, several routes to FDCA have been described. After the initial synthesis via dehydration of mucic acid (galactaric acid) with fuming HBr

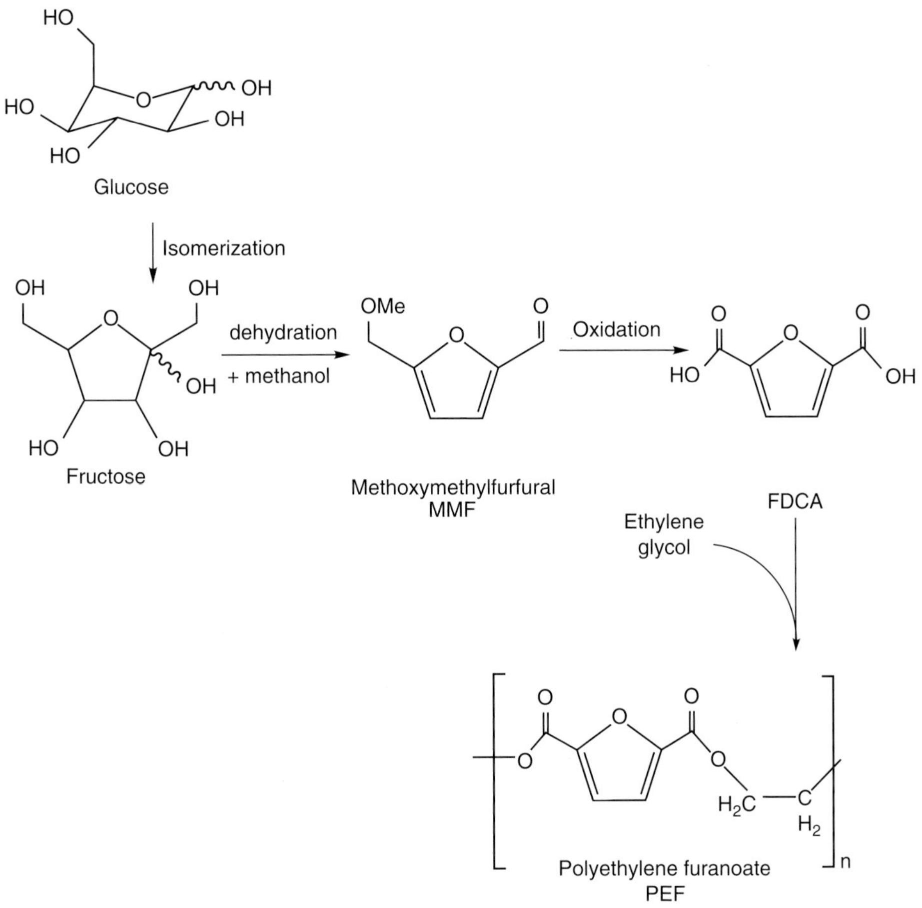

Figure 19.1 The five steps that are required to produce PEF starting from glucose.

in 1876 [8] and of its isomer saccharic acid in 1888 [9], the oxidation of HMF and derivatives was first reported in 1899 [10] followed by another approach based on the condensation of glyoxal with dimethyl diglycolate in the presence of sodium methoxide in 1936 [11].

In the last decades, renewed industrial interest in FDCA occurred which resulted in the filing of dozens of patent applications on routes to FDCA. Some relevant examples are worth mentioning in this overview.

The oxidation of HMF and MMF can be done by adaptation of several industrially applied oxidation methods. The oxidation of HMF in particular has several features that need to be taken into consideration. First, HMF is a rather unstable molecule. Prolonged heating at even moderate temperatures and/or the presence of acids or bases should be avoided, thereby limiting the operation and process window for the oxidation. Second, similar to PET monomer, that is, terephthalic acid (TA), the solubility of the produced FDCA in almost all solvents is very low. Though solubility increases with temperature, HMF instability limits the process window with respect to maximum allowable temperature. The addition of bases has often been used to overcome the low solubility, but is undesired due to the high amounts of coproduced waste salts, and the additional cost for the required stoichiometric amounts of base, and acids used for neutralization. Lastly, the oxidation needs to be performed with very high selectivity, as even small amounts of intermediates or of decarboxylated products are detrimental for the polymerization chemistry to follow.

Here, we further focus only on those oxidations that use molecular oxygen as the oxidant. The reason for this is that these processes show the best economics and in order to apply FDCA in applications with bulk potential, its price should be in line with that of PET. Ierapetritou [12] showed in a study on the production of FDCA from HMF using aqueous acetic acid as solvent and Pt/ZrO_2 as catalyst that only by using cheap air or pure oxygen a price target for FDCA close to that of TA can be reached. Here, we discuss heterogeneous and homogeneous catalyst systems.

19.2.1
Production of 2,5-Furandicarboxylic Acid Using Heterogeneous Catalysts

Many industrial processes use heterogeneous catalysts as they have several advantages over homogeneous systems. It is thus not surprising that many research efforts focus on using heterogeneous catalysts. Heterogeneously catalyzed oxidations to FDCA have only been reported with HMF as the starting material. HMF has both an alcohol and an aldehyde group in contrast to MMF, which has a relatively unreactive ether group and an aldehyde group. The required conversion of HMF involves the selective oxidation of both the alcohol and the aldehyde to the corresponding carboxylic acid. Both reactions are well known in industrial practice using heterogeneous catalysts. Several heterogeneously catalyzed oxidation methods have been described in the literature so far. They are summarized in Table 19.1.

In 1988, Hoechst filed a patent on the oxidation of HMF in water with oxygen using a platinum on carbon catalyst [15]. In this and most other heterogeneously catalyzed HMF oxidations, at least 2 equiv. of base were needed to keep the FDCA (and intermediates) in solution and prevent precipitation on the catalyst [13]. Although the Hoechst method is relatively easy on a laboratory scale, it is not scalable due to high catalyst cost (fast catalyst deactivation) and stoichiometric salt formation. Recently, examples were reported using supported gold catalysts for HMF oxidation. Gorbanev et al. [18] demonstrated that Au/TiO_2 could oxidize HMF into FDCA in 71% yield at near room temperature. Gupta et al. [19] reported the base-free oxidation over gold catalysts supported on hydrotalcites. However, the basic hydroxyl groups of the hydrotalcite probably interact with the acidic FDCA product. Casanova et al. [20] used Au/CeO_2 as an active and selective catalyst for the base-free oxidative esterification of 1 wt% HMF in methanol to the more soluble dimethyl ester of FDCA, thus circumventing solubility problems of FDCA.

Table 19.1 Oxidation of furanics to FDCA over heterogeneous catalysts.

Reaction conditions	Temperature (°C)	Conversion HMF	Yield FDCA	References
HMF as substrate				
Pt/Al$_2$O$_3$ (pH 9), air	60	n.a.	100	[13]
Pb-Pt/C (NaOH), air	60	100	81	[14]
Pt/C (NaOH), air	60	100	95	[15]
Pt-ZrO$_2$; Pt/Al$_2$O$_3$ (Na$_2$CO$_3$), air	100	100	98	[16, 17]
Au/TiO$_2$, 20 equiv. base, O$_2$	30	100	71	[18]
Au/hydrotalcite, air	95	100	100	[19]
Au/CeO$_2$, diluted in MeOH	130	100	100[a]	[20]
Pt/C, Pd/C, Au/C, Au/TiO$_2$, air	22	100	79	[21]
NiO$_2$/OH anode (basic conditions, electrochemical oxidation)	22		71	[22]
Other substrates (fructose)				
Co(acetylacetonate)-SiO$_2$ (from fructose), air	160	72	99	[14]
PtBi/C (from fructose, solid acid, H$_2$O/MIBK), air	80	50	25	[23]
(5-Methyl furfural)		(5-Methyl furoic acid)		
Ag$_2$O+CuO on Al$_2$O$_3$, or Cr$_2$O$_3$, air	room temperature	—	84	[24–26]

a) As FDCA dimethyl ester.

The requirement for (more than) stoichiometric amounts of base in most heterogeneously catalyzed HMF oxidations is clearly very disadvantageous from a bulk industrial application point of view. The required base and neutralizing acid (typically NaOH and H$_2$SO$_4$, respectively) not only increase raw material costs, but they also have a negative environmental impact because of the large amount of salt that is coproduced with FDCA. In addition, heat management for this strongly exothermic conversion is negatively impacted by the low reaction temperatures usually employed because of instability of HMF under the basic reaction conditions. Also often high oxygen partial pressures (10–20 bar of air) are reported to be required, which may pose a safety risk at industrial scale.

The oxidation of 5-methyl furfural over Ag$_2$O shows another disadvantage generally observed for heterogeneous oxidation catalysts, namely, the inability to selectively activate unreactive groups such as methyl or methoxymethyl substituents [24, 25]. At Avantium, we were not successful in oxidizing MMF with the Pt/C catalyst system, which worked well for HMF. This limits the use of more stable furanics, such as dimethylfuran or the ether stabilized methoxymethyl group in MMF, as feedstock in oxidation reactions with these catalyst systems. To our knowledge, no active and selective heterogeneous catalysts for methylfurfural or for MMF have been reported.

An interesting approach is to combine both the formation of HMF from fructose with the oxidation to FDCA in one pot. The aim is to convert the unstable HMF before it can react further to levulinates or degrade into humins. An FDCA yield of 25% was reported using a PtBi/C oxidation catalyst in combination with a solid acid in water/methyl *iso*-butyl ketone biphasic reaction medium [23]. Recently, Ribeiro and Schuchardt [14] reported a 72% fructose conversion and excellent greater than 99% selectivity to FDCA when using Co(acac)–SiO$_2$ as a bifunctional acid–redox catalyst at 160 °C and 20 bar air pressure.

19.2.2
Production of 2,5-Furandicarboxylic Acid Using Homogeneous Catalysts

In analogy with the oxidation of *para*-xylene to TA in the so-called Amoco Mid-Century process, the Co/Mn/Br homogeneous catalyst system has also been used for producing FDCA. In March 2000, Dupont filed a provisional patent application, which was granted in 2013 on the stepwise air oxidation of HMF in acetic acid using a Co/Mn/Br catalyst system [27]. In this stepwise oxidation, 2,5-diformylfuran (DFF), formylfuroic acid, and FDCA were obtained respectively, depending on the conditions. In May 2009, ADM filed a provisional patent application, which was granted in 2013, on the air oxidation of 5-(butoxymethyl)furfural (BMF) with a Co/Mn/Br catalyst in acetic acid to predominantly the monobutyl ester of FDCA [28].

Avantium showed in their patent application filed in October 2009 and granted in 2013 that the air oxidation of MMF using a Co/Mn/Br catalyst in acetic acid predominantly leads to FDCA in a single step [29]. In 2013, a subsequent patent was filed that described the continuous oxidation of MMF to FDCA [30]. These processes have been scaled to multiton scale in Avantium's pilot plant in Geleen, the Netherlands. At this moment, Avantium seems to be the only company that can produce purified FDCA on multiton scale.

In Table 19.2, an overview of the existing literature for the oxidation of substituted furan compounds with Co/Mn/Br catalyst systems is provided. The commonly reported FDCA yields are usually in the mid-50s, and 90% at best. This contrasts with the *para*-xylene to TA process, for which greater than 98% yield has been reported (also see Chapter 4) [34, 35]. The temperatures applied are generally around 100–130 °C, which is considerably lower than the 180–220 °C typical for the TA process [36]. The only exception here is the process claimed by Sanborn for ADM, which gives high yields at 180 °C by apparently lowering the amount of Mn and Br present in the process [28]. The need for especially lower Mn amounts is in line with results reported by Janka *et al.* [32] for 5-(acetoxymethyl)furfural (AcMF) and 5-(ethoxymethyl)furfural (EMF), though in this case the reaction temperature is not as high as claimed by Sanborn [28].

In some reports, the mechanism of the oxidations has been discussed. Grushin *et al.* [27] reported both DFF and 5-formyl furancarboxylic acid (FFCA) as isolable intermediates in the oxidation of HMF to FDCA using the Co/Mn/Br system.

Table 19.2 FDCA obtained with Co/Mn/Br catalyst systems.

Substrate	Catalyst	Co:Mn:Br (mol:mol:mol)	Oxidant	Temperature (°C)	Pressure	Time (h)	Conditions	Substrate concentration (%)	Yield (%)	Selectivity	References
HMF	CoAc/MnAc/HBr/ZrAc (Zr: 20 ppmw)	1:1:2	Air	105	70 bar	12	Batch	—	59	—	[27]
AcMF	CoAc/MnAc/NaBr	1:1:2	O_2	100	35–55 bar	2	Batch	10	54	—	[28, 31]
HMF + HMF ester	CoAc/MnAc/NaBr	—	—	100	—	—	—	—	—	—	[31]
HMF	CoAc/MnAc/NaBr	1:1:2	O_2	100	55 bar	5	Batch	7	49	—	[28]
HMF	CoAc/MnAc/NaBr	1:1:2	O_2	100	55 bar	4	Batch	20	48	—	[28]
HMF	CoAc/MnAc/HBr	1:0.03:1	O_2:N_2 (1:1)	160	60 bar	0.3 + 0.2	Fed batch	4.5	66	—	[28]
HMF	CoAc/MnAc/HBr	1:0.015:0.5	O_2:CO_2 (1:1)	180	30 bar	0.3 + 0.2	Fed batch	4.5	90	—	[28]
BMF	CoAc/MnAc/NaBr	1:1:2	O_2	100	70 bar	5	Batch	7	—	—	[28]
BMF	CoAc/MnAc/NaBr	1:1:2	O_2	100	40 bar	5	Batch	19.50	16 (BFDCA)	—	[28]
AcMF	CoAc/MnAc/HBr	1:0.05:1.1	Air (flow)	130	9 bar	1 + 1	Fed batch	—	90	—	[32]
EMF	CoAc/MnAc/HBr	1:0.05:1.1	Air (flow)	130	9 bar	1 + 1	Fed batch	—	88	—	[32]
5-Methylfurfural	CoAc/MnAc/HBr	1:0.05:1.1	Air (flow)	130	9 bar	1 + 1	Fed batch	—	61	—	[32]
5-Methylfurfural	CoAc/MnAc/NH_3Br	1:0.133:0.05	Air (flow)	1) 118 2) 130	1) 20 atm 2) 30 atm	1) 4.5 2) 1.5	Batch	6.60	36	—	[33]
5-Methylfurfural	CoAc/MnAc/NH_3Br	1:0.125:0.05	Air (flow)	116	10 atm	4	Batch	6.60	29	—	[33]

Slavinskaya et al. [24] also reported the presence of minor amounts of the intermediate DFF in the product when oxidizing 5-methyl furfural. Given that in general no significant amounts of by-products have been detected, the yields of at most 90% indicate that overoxidation to CO_2 is the major side reaction. Of mechanistic interest is the oxidation of functionalized HMF derivatives (see Figure 19.2). The acetyl derivative (AcMF) can be converted with this catalyst system, though under the acidic conditions employed, hydrolysis of this ester function may be fast enough to liberate HMF *in situ*.

Both ADM [31] and Avantium [29, 30] have reported on the oxidation of MMF, the more stable methyl ether of HMF. Avantium is currently scaling up its process based on an optimized process in its pilot plant, yielding greater than 96 mol% FDCA + FDCAMe from MMF. Though this compound could not be oxidized over heterogeneous catalyst systems (discussed earlier), facile oxidation using the Co/Mn/Br catalyst system has been reported (see Table 19.3). Since ethers are hydrolytically very stable, it clearly indicates that the activation of MMF must proceed via a direct oxidation of the methylene carbon of the methoxymethyl moiety.

5-Formyl furancarboxylic acid (FFCA) and its monomethyl ester (FFCAMe) are key intermediates of the reaction (see Figure 19.2) [37]. Similar to 4-carboxybenzaldehyde (4-CBA), the analogous intermediate in the *para*-xylene to TA process [38], FFCA is problematic due to its structural similarity with FDCA. It combines with FDCA in the solid state, making a small part inaccessible for further oxidation to FDCA. Great care has to be taken to maximize its conversion and to remove any residual FFCA during the FDCA purification as FFCA causes color and FFCA is monofunctional and thus a chain termination agent.

19.3
Process Technology

19.3.1
Process Economics and Engineering Challenges

Currently no industrial process for the oxidation of HMF or other furanics to FDCA exists although today bio-based FDCA can be obtained from technology developers and toll manufacturers (mainly in China). Though no details are available for these kilogram-scale processes, it is not expected that these are based on air oxidations with the Co/Mn/Br catalyst system. It is likely that instead this FDCA is produced using heterogeneous catalysts and stoichiometric amounts of bases or using nitric acid or other stoichiometric oxidants. Clearly, this will not be the way forward for an industrial-scale process at hundreds of kilotons of scale. Avantium has a continuous 20 t/year MMF pilot plant and a continuous 40 t/year FDCA oxidation pilot plant in operation in Geleen, the Netherlands. These pilot plants are used for scale-up process development and for producing FDCA

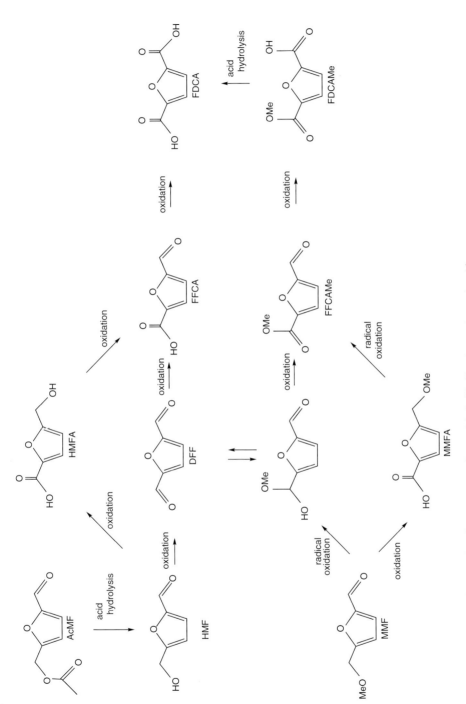

Figure 19.2 Pathways for three key furanic intermediates with Co/Mn/Br mixtures in acetic acid.

Table 19.3 Oxidation of 5-(methoxymethyl)furfural (MMF).

Substrate	Temperature (°C)	Catalyst	Co:Mn:Br (mol:mol:mol)	Oxidant	Pressure (bar)	Time (h)	Conditions	Substrate concentration (%)	Yield	Selectivity	References
MMF	180	CoAc/MnAc/NaBr	1:3:20	Air	20	1	Batch	1.23	72% FDCA	11% FDCAMe	[31]
MMF	145	CoAc/MnAc/HBr	1:1:1.4	Lean air (flow)	14	2	Fed batch	20	84% FDCA	14% FDCAMe	[30]

for application development. Between 5 and 10 t of purified FDCA is isolated currently on a yearly basis. Though the choice for the Co/Mn/Br catalyst system is directly related to the use of MMF as the furanic feedstock, this is not the only reason. Both economical drivers to get FDCA that is cost competitive with TA and technical drivers for safe operation requirements play an important role in the decisions made for the process layout. The economic drivers simply define the need for high feedstock efficiency and a process that operates with as high feedstock concentration as possible. In addition, it is clear that the use of oxygen is essential and that processes using a heterogeneous or enzymatic oxidation catalyst using sacrificial bases (for solubility) and acids (needed for subsequent neutralization) or using alternative stoichiometric oxidants should be avoided.

19.3.1.1 Gas Composition Control

For safe operation of an air oxidation process, it is important to consider the entire process and evaluate possible hazardous conditions. The oxygen that is needed for the oxidation reaction is fed to the reactor as (diluted) air. Due to the high temperature in the reactor, the gas flow is quickly saturated with acetic acid and thereby too rich in fuel to ignite. During the reaction oxygen is consumed and the main safety concern lies on the vent gas composition. Downstream of the reactor the acetic acid is refluxed in a condenser, and since the condensation of liquid acetic acid lowers the concentration of acetic acid in the gaseous phase, the effluent gas stream becomes poor in fuel and could pass through the flammable region. Since a complete gradient of acetic acid composition can be found in this condenser, safe operation must be maintained by staying below the limiting oxygen concentration (LOC). This is carried out by making sure that the oxidation reaction consumes enough O_2 from the air fed to keep the vent gas composition below a safe operating limit of 10% oxygen (LOC reported to be 14.9% for AcOH/water/air mixtures at 11 bar and 155 °C) [39]. A safety feedback loop will stop the air flow in case the oxygen conversion is not sufficient and the LOC is reached.

19.3.1.2 Temperature Control

Similar to any oxidation reaction, the MMF oxidation is an exothermic reaction. Thus, the first technical challenge in the design of an oxidation reactor deals with the methodology to remove the generated heat. Although the heat in the production of FDCA from HMF and MMF (~850 kJ/mol)[1] is considerably less than that of TA from *para*-xylene (~1300 kJ/mol) [40], still the amount of heat is so high that very efficient cooling is required to avoid thermal runaways.

This task is made complex by the nature of the product. In order to operate an oxidation line in the most profitable way, it should run with the highest concentration of substrate. Aromatic dicarboxylic acids such TA or FDCA generally exhibit very low solubility, and at high feedstock concentrations, the product will precipitate during the reaction. Their poor solubility prevents the use of a direct cooling system, such as a cold jacket or a cooling coil, since the product would precipitate on the cold surfaces and progressively reduce the efficiency of the cooling.

The removal of heat in the TA production has been solved by an elegant reactor design (see also Chapter 4). As (diluted) air is continuously introduced to the reactor, the nonconsumed oxygen and the remaining nitrogen as well as the CO and CO_2 formed need to be vented. This considerable gas flow is stripping acetic acid from the reactor, and by doing so, the equivalent heat of vaporization is removed from the reaction mixture. The saturated gas flow is then fed to a condenser where most of the solvent is condensed to the liquid phase. The cooled solvent is returned to the reactor, either directly or indirectly after treatment (e.g., to remove water), further cooling down the reactor. For the TA process, it has been shown that overall two-thirds of the heat of reaction is taken off the reactor by the evaporation and one-third by the return of the cold solvent.

Since the heat of evaporation of acetic acid of +23.7 kJ/mol is low compared with the overall heat formed in the oxidation (~1300 kJ/mol for TA), it is clear that considerable amounts of acetic acid need to be evaporated in order to maintain thermal stability. The amount of acetic acid that can be evaporated is directly related to the vapor pressure in the effluent gas flow. Since the vapor pressure is only a function of temperature, the actual operation temperature in the oxidation reactor is determined by the temperature required to obtain the required vapor pressure for safe heat removal in the reactor. Though counterintuitive, it means that higher operation temperatures are typically safer to operate for this catalyst system. For the TA process, this leads to a minimum operation temperature of around 180 °C. Higher temperatures are even better from a heat integration perspective but yield more unwanted by-products and overoxidation to CO_2.

The safe industrial oxidation of furanics using the Co/Mn/Br catalyst system needs a similar heat removal system as FDCA, like TA, is a very insoluble diacid, preventing the use of jacketed cooling. In contrast to *para*-xylene, HMF, MMF, and AcMF are already partially oxidized on the benzylic positions, and consequently less heat is formed in their oxidation to FDCA, and reactor temperature may be

---

1) Calculated based on heat of formation simulated by Aspen Plus and heat of combustion measured experimentally.

lower than for the TA process. But even in the FDCA case sufficiently high reaction temperature will be required to ensure stable operation by efficient heat removal via acetic acid vaporization. Considering the reported catalytic results for various furanics, HMF in particular (Tables 19.2 and 19.3), the typical reaction temperature is 100–140 °C. This is very close to the boiling point of acetic acid, and one could wonder whether sufficient heat can be removed and safe operation guaranteed. Since in contrast to MMF, HMF is a very unstable molecule and quickly degrades to unwanted products such as LA and humins, its oxidation cannot be carried out at higher temperatures without loss in selectivity.

19.3.1.3 Oxygen Mass Transfer Limitations

It is clear from the previous section that the gas flow, temperature, and pressure are bound both by the chemical reaction and by safety considerations. As was already noted, the optimized reaction temperature for selective oxidation of furanic compounds is considerably lower than that for *para*-xylene (see Tables 19.2 and 19.3). Operating at lower temperature will lower the vapor pressure of the solvent (acetic acid) and thus also lower overall reactor pressure. Consequently, the partial oxygen pressure is also reduced, which can be detrimental to the performance of the oxidation process as the oxygen partial pressure is the driving force to transfer the oxygen from the gas to the liquid phase (N_A) following Eq. (19.1):

$$N_A = K_L \alpha (P_{O_2}/H_c - [O_2]_{liq}) \tag{19.1}$$

where

N_A	=	volumetric O_2 transfer rate
K_L	=	mass transfer coefficient at phase boundary
α	=	specific gas–liquid interfacial area
P_{O_2}	=	partial pressure oxygen
H_c	=	Henry's law coefficient

To compensate for a reduced gas- to liquid phase oxygen transfer driving force that results from lower temperatures, technical solutions to improve the oxygen mass transfer from gas to liquid will have to be used (e.g., efficient stirrer designs).

19.3.1.4 Overall Safety Operation

The foregoing illustrates that reactor pressure, temperature, partial oxygen pressure, and gas flow are all bound within tight limits and intimately linked. For instance, using a fixed gas flow, decreasing the pressure will automatically decrease the temperature, as more solvent will be stripped from the reactor. Or in the case of a runaway scenario, the heat generated would significantly increase but at the same time the increasing reactor temperature would exponentially increase solvent evaporation, thereby increasing the cooling capacity of the reactor and mitigating the risk of a runaway scenario. Furthermore, the reaction rate will be limited by the amount of air fed, thus mitigating risks of too much heat generation that could result from accidental addition of too much substrate. Overall the air oxidation with the Co/Mn/Br catalyst system, despite being a

19.4
New Developments

19.4.1
Outlook for Co/Mn/Br in the Air Oxidation of Biomass-Derived Molecules

After the established industrial use of the Co/Mn/Br catalysis system for *para*-xylene oxidation and its anticipated industrial use for oxidation of HMF and derivatives, we can ask the question whether this catalyst and process can also be applied to other bio-based target molecules. Surprisingly, few Co/Mn/Br oxidations to other bio-based targets have been described so far. Only LA oxidation to SA [41] and oxidative treatment of lignin have been reported to date [42].

The oxidation of LA to SA using Co/Mn/Br has first been reported by Venkitasubramaniam [41] in the coproduction with FDCA from HMF. Since LA is a degradation product of HMF, coproducing SA and FDCA could offer significant economic gains as it may simplify the expensive purification of HMF. Based on the examples provided, greater than 99% of the LA was converted to products including SA. But a very low SA yield of only 12.0% was reported as the best result. Apparently, the selectivity of this radical reaction is not exclusively selective for the methyl group at C5 but can also attack methylene groups at C2 and C3 in LA resulting in malonates and oxalates (see Figure 19.3). In addition, further oxidation of succinic and malonic acid to acetic acid and CO_2 leads to a complex mixture of products. The foregoing points out that when considering the use of a Co/Mn/Br-catalyzed oxidation, all C–H bonds prone to radical hydrogen abstraction should be taken into account when judging the selectivity. Recent work has shown that selectivities up to 61.3 mol% at 95.6% conversion for oxidation of LA to SA can be achieved by using only $Mn(OAc)_3$ as catalyst without Co or Br [43]. A hydrogen abstraction mechanism was proposed, and the high selectivity can be explained by the higher abundance of protons in the CH_3 group at C5 compared with the CH_2 at C3, in combination with steric constraints on the less accessible CH_2 positions.

Another aspect to consider is that the selectivity in the TA and FDCA systems in part may also be due to the insolubility of the diacids formed. Thus, the product is removed before it can be oxidized further.

Partenheimer has suggested the Co/Mn/Br-catalyzed oxidation of lignin as a method to break down this recalcitrant biopolymer in valuable products [42]. The major products formed are vanillin, vanillic acid, syringaldehyde, and syringic acid. None of these are present in very high yields. The amounts vary with catalyst composition and source of lignin, but the total yield of products did not exceed 10.9 wt% with typical ratio between syringic/vanillic compounds of around 2. Considering that many products were not quantified, the authors consider this

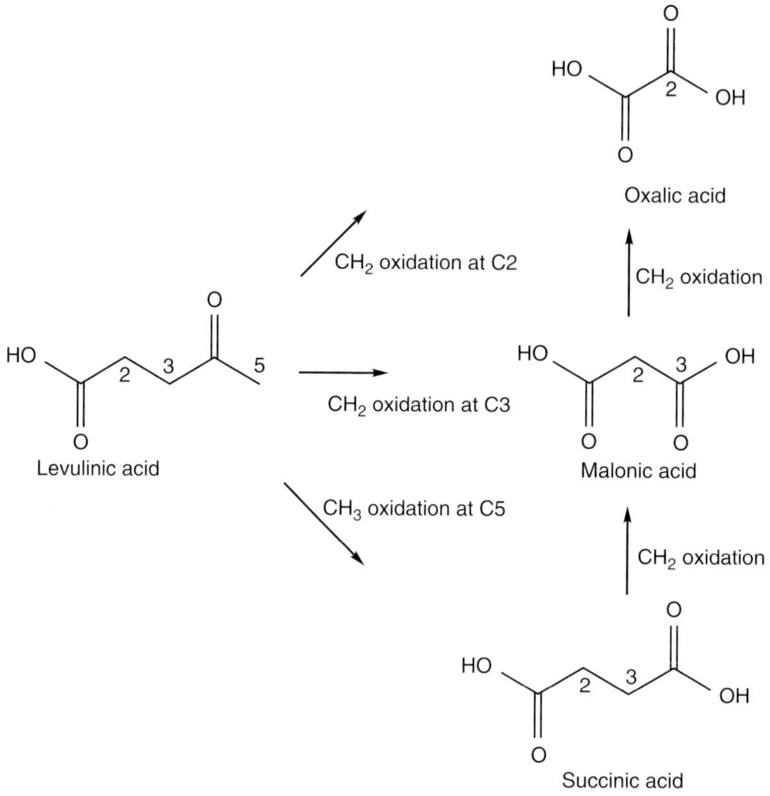

Figure 19.3 Primary radical oxidation pathways for levulinic acid.

a very promising route, although no new articles have since been published on this topic.

When considering the Co/Mn/Br system as oxidation catalyst for bio-based target molecules, one should take the fundamental characteristics of the oxidizing radical species into account. Only for substrates where the C–H bonds to be oxidized are much more prone to undergo radical chain C–H abstraction because of lowest C–H bond strength, ideally combined with high abundancy and little steric hindrance, the Co/Mn/Br may offer a way forward. Considering the often cited DoE Top-12 of biomass-derived platform molecules of the future [5], no molecule on this list besides FDCA is a good fit for a process using this catalyst system. FDCA fits well because it resembles *para*-xylene in that its C–H bonds on the aromatic furan ring, like those on the benzene ring for *para*-xylene, are much less prone to C–H abstraction by bromine radicals than those on the side groups. However, the suggestion by Partenheimer to investigate the selective oxidation of lignin, another bio-based substrate for which preferential benzylic side chain oxidation over aromatic ring functionalization can be expected, may extend application of Co/Mn/Br catalysis for bio-feedstock conversion beyond FDCA.

19.5 Conclusion

The Co/Mn/Br system that was originally developed and further optimized by Amoco, Mid-Century, BP, Eastman Chemicals, and others for the oxidation of *para*-xylene to TA is a very powerful oxidation tool, but its implementation is not straightforward and restricted by the operational window, whereas its application is limited to specific feedstock that allow sufficient selectivity in the radical chain mechanism. There is an opportunity for the highly selective HMF and derivatives oxidation to FDCA and there may be a potential use in lignin oxidation.

List of Abbreviations

4-CBA	4-carboxybenzaldehyde
AcMF	5-(acetoxymethyl)furfural
BMF	5-(butoxymethyl)furfural
CoAc	cobalt acetate
CSD	carbonated soft drink
DFF	2,5-diformylfuran
EMF	5-(ethoxymethyl)furfural
FDCA	2,5-furandicarboxylic acid
FFCA	5-formyl furancarboxylic acid
H_c	Henry's law coefficient
HMF	5-(hydroxymethyl)furfural
K_L	mass transfer coefficient at phase boundary
LA	levulinic acid
ML	methyl levulinate
MMF	5-(methoxymethyl)furfural
MnAc	manganese acetate
LOC	lower oxygen concentration for combustion
N_A	volumetric O_2 transfer rate
PEF	polyethylene furandicarboxylate
PET	polyethylene terephthalate
P_{O_2}	partial pressure oxygen
SA	succinic acid
TA	terephthalic acid
ZrAc	zirconium acetate
α	specific gas–liquid interfacial area

References

1. van der Waal, J.C. and de Jong, E. (2013) in *Producing Fuels and Fine Chemicals from Biomass Using Nanomaterials*, Chapter 9 (eds A.M. Balu and R. Luque), CRC Press, p. 223vv.

2. Neuman, W. (2011) Race to greener bottles could be long, printed on December 16, 2011, on page b1 of the New York edition. Press release, December 15, http://www.thecoca-colacompany

.com/dynamic/press_center/2011/12/
plantbottle-partnerships.html (accessed
27 January 2016).
3. Danone se lance dans les bouteilles en
plastique bio, http://www.lemonde
.fr/economie/article/2012/03/22/
danone-se-lance-dans-les-bouteilles-
en-plastique-bio_1674241_3234.html,
http://avantium.com/news/Avantium-
and-Danone-sign-development-
partnership-for-Next-Generation-bio-
based-plastic-PEF.html (accessed July
2012), https://www.avantium.com/press-
releases/avantium-raises-e36m-
investment-swire-pacific-coca-cola-
company-danone-alpla/.
4. van Putten, R.-J., van der Waal,
J.C., de Jong, E., Rasrendra,
C.B., Heeres, H.J., and de Vries,
J.G. (2013) *Chem. Rev.*, **113**,
1499–1597.
5. Werpy, T. and Petersen, G. (2004) *Top
Value Added Chemicals From Biomass*,
Pacific Northwest National Laboratory
(PNNL) and the National Renewable
Energy Laboratory (NREL).
6. Gruter, G.-J. and de Jong, E. (2009)
Biofuels Technol., **1**, 11–17.
7. Dias, A.S., Gruter, G.J.M., and van
Putten, R.-J. (2012) Process for the con-
version of a carbohydrate-containing
feedstock. Patent WO2012091570, to
Furanix Technologies B.V.
8. Fittig, R. and Heinzelman, H. (1876) *Ber.
Dtsch. Chem. Ges.*, **9**, 1198.
9. Sohst, O. and Tollens, B. (1888) *Justus
Liebig's Ann. Chem.*, **245**, 1.
10. Fenton, H.J. and Gostling, M. (1899)
J. Chem. Soc., **95**, 423.
11. Hoehn, W.H. (1936) *Iowa State Coll. J.
Sci.*, **11**, 66.
12. Triebl, C., Nikolakis, V., and Ierapetritou,
M. (2013) *Comput. Chem. Eng.*, **52**,
26–34.
13. Vinke, P., Poel, W.V., and van Bekkum,
H. (1991) *Stud. Surf. Sci. Catal.*, **59**, 385.
14. Ribeiro, M.L. and Schuchardt, U. (2003)
Catal. Commun., **4**, 83.
15. Leupold, E.I., Wiesner, M., Schlingmann,
M., and Kapp, K. (1988) Process for the
oxidation of 5-hydroxymethylfurfural.
Patent EP0356703, to Hoechst AG.
16. Lew, B.W. (1967) Method of producing dehydromucic acid. US Patent
3,326,944, to Atlas Chemical Industries,
Inc.
17. (a) Lilga, M.A., Hallen, R.T., White, J.F.,
and Frye, G.J. (2008) Hydroxymethyl
furfural oxidation methods. US Patent
20080103318, to Battelle Memorial Institute.; (b) Lilga, M.A., Hallen, R.T., and
Gray, M. (2010) *Top. Catal.*, **53**, 1264.
18. Gorbanev, Y.Y., Klitgaard, S.K., Woodley,
J.M., Christensen, C.H., and Riisager, A.
(2009) *ChemSusChem*, **2**, 672.
19. Gupta, N.K., Nishimura, S., Takagaki, A.,
and Ebitani, K. (2011) *Green Chem.*, **13**,
824.
20. Casanova, O., Iborra, S., and Corma, A.
(2009) *J. Catal.*, **265**, 109.
21. Davis, S.E., Houk, L.R., Tamargo, E.C.,
Datye, A.K., and Davis, R.J. (2011) *Catal.
Today*, **160**, 55.
22. Grabowski, G., Lewkowski, J., and
Skowroński, R. (1991) *Electrochim.
Acta*, **36**, 1995.
23. Kröger, M., Prüsse, U., and Vorlog, K.-E.
(2000) *Top. Catal.*, **13**, 237.
24. Slavinskaya, V.A., Kreile, D.R., Dziluma,
E., Eglite, D., Milmanis, I., Korchogova,
E.Kh., Avots, A., and Pinka, U. (1976)
Incomplete catalytic oxidation and
ammoxidation of furan and other
compounds. Tezisy Dokladov Respub-
likanskoi Konferentsii Dokladov
Okislitel'nomu Geterogennomu Katalizu
3rd, p. 199.
25. Slavinskaya, V.A., Kreile, D., Sile, D.,
Eglite, D., and Kruminya, L.Y. (1979)
React. Kinet. Catal. Lett., **11** (3),
215–220.
26. Masamune, T., Ono, M., and Matsue,
H. (1975) *Bull. Chem. Soc. Jpn.*, **48**,
491–496.
27. (a) Grushin, V., Partenheimer, W.,
and Manzer, L.E. (2000) Processes for
preparing diacids, dialdehydes and polymers. US Patent 8524923 (B2); (b) also
published as: Grushin, V., Partenheimer,
W., and Manzer, L.E. (2000) Oxidation of 5-(hydroxymethyl) furfural to
2,5-diformylfuran and subsequent decarbonylation to unsubstituted furan. Patent
WO0172732 (A3), to Du Pont.
28. Sanborn, A. (2009) Oxidation of furfural compounds. US Patent Application
8558018B2, WO2010132740, to Archer
Daniels Midland Co.

29. Munoz De Diego, C., Schammel, W.P., Dam, M.A., and Gruter, G.J.M. (2012) Method for the preparation of 2,5-furandicarboxylic acid and esters thereof. US Patent 8519167B2, to Furanix Technologies BV.
30. Mazoyer, E., De Sousa Dias, A.S.V., McKay, B., Baars, H.J., Vreeken, V.P.C., Gruter, G.J.M., and Sikkenga, D.L. (2014) Process for the preparation of 2,5-furandicarboxylic acid. Patent WO2014163500, to Furanix Technologies BV.
31. Sanborn, A. and Howard, S. (2009) Conversion of carbohydrates to hydroxymethylfurfural (hmf) and derivatives. Patent WO2009076627, To Archer Daniels Midland Co.
32. Janka, M.E., Lange, D.M., Morrow, M.C., Bowers, B.R., Parker, K.R., Shaikh, A., Partin, L.R., Jenkins, J.C., Moody Paula, P., Shanks, T.E., and Sumner, C.E. (2012) An oxidation process to produce a crude and/or purified carboxylic acid product. Patent WO/2012/161967, to Eastman Chem Co.
33. Slavinskaya, V.A., Kreile, D., Hillers, S., Krumina, L., Eglite, D., and Strautina, A. (1974) 2,5-furandicarboxylic acid. SU448177.
34. Belmonte, F.G., Sikkenga, D.L., Ogundiran, O.S., Abrams, K.J., Leung Linus, L.K., Meller, C.G., Figgins, D.A., and Mossman, A.B. (2005) Staged countercurrent oxidation. Patent WO2005051881 (a1), to BP Corporation North America.
35. An overview of alternative TA oxidation strategies is given in: Loydd, L. (2011) *Handbook of Industrial Catalysis*, Springer-Verlag294 v.v.
36. Tomás, R.A.F., Bordado, J.C.M., and Gomes, J.F.P. (2013) *Chem. Rev.*, **113**, 7421–7469.
37. Partenheimer, W. and Grushin, V.V. (2001) *Adv. Synth. Catal.*, **343**, 102.
38. Sheehan, R.J. (2011) Terephthalic acid, dimethyl terephthalate, and isophthalic acid, in *Ullmann's Encyclopedia of Industrial Chemistry*, Wiley-VCH.
39. Hoyle, M. and Astbury, G.R. (1997) *Investigation of the Flammability Limits of Acetic Acid at Elevated Temperature and Pressure*, Symposium Series, IChemE, vol. **141**, pp. 293–304.
40. Li, M. (2013) A spray reactor concept for catalytic oxidation of *p*-xylene to produce high-purity terephthalic acid. PhD Thesis. University of Kansas, p. 14.
41. Subramaniam, B., Zuo, X., Busch, D.H., and Venkitasubramaniam, P. (2013) Process for producing both biobased succinic acid and 2,5-furandicarboxylic acid. Patent WO 2013033081 A2, to University of Kansas, to Archer Daniels Midland Co.
42. Partenheimer, W. (2009) *Adv. Synth. Catal.*, **351** (3), 456–466.
43. Liu, J., Du, Z., Lu, T., and Xu, J. (2013) *ChemSusChem*, **6**, 2255–2258.

20
Azelaic Acid from Vegetable Feedstock via Oxidative Cleavage with Ozone or Oxygen

Angela Köckritz

20.1
Introduction

Azelaic acid (AA), 1,9-nonanedioic acid or heptane-1,7-dicarboxylic acid, is used in manifold application areas. AA is offered in the pharmaceutical field for the treatment of acne and rosacea due to its antibacterial effect [1, 2]. However, AA is much more widely applied as a monomer for the production of polymers such as alkyd resins, polyamides, and impact-proof polyesters or for the synthesis of plasticizers, lubricants, lithium complex greases, corrosion inhibitors, dielectric fluids, heat-transfer fluids, metal to glass fluxes, emulsion breakers, waxes, tobacco sheet plasticizers, hot-melt coatings and adhesives, water-soluble coating resins, hydraulic fluids, fungicides, insecticides, and so on [3–7].

The lower melting point of azelaic acid (107 °C) compared with sebacid acid and adipic acid caused by the linear odd-carbon structure and the resulting lower crystallizing ability is of interest for special polyamides and other polymers [8]. Thus, 6,9 nylons based on AA possess excellent abrasion resistance, low water absorption, and high impact resistance [6]. Various esters of AA are used as components in urethanes and as a source of soft segments in copolyester fibers in order to achieve good low-temperature performance and hydrolytic resistance. Poly(methylene azelamide) produced from azelanitrile and formaldehyde show properties similar to those of poly(hexamethylene adipamide) [4]. Isodecyl esters of different dicarboxylic acids such as AA are useful plasticizers for thermoplastics. They impart such polymers an excellent resistance to aging even at higher temperatures and a good flexibility at low temperatures to a certain extent [3].

Main renewable feedstock for the production of AA is oleic acid (OA), which is cleaved to AA and pelargonic acid (PA) by ozonolysis [7]. Although various catalysts and oxidants [5, 9] for the cleavage of C18 fatty acids and their derivatives to AA have been described in the scientific literature, these have not been used

---

Liquid Phase Aerobic Oxidation Catalysis: Industrial Applications and Academic Perspectives, First Edition. Edited by Shannon S. Stahl and Paul L. Alsters.
© 2016 Wiley-VCH Verlag GmbH & Co. KGaA. Published 2016 by Wiley-VCH Verlag GmbH & Co. KGaA.

industrially owing to insufficient activities and the formation of problematic and inefficient waste products. Other methods of synthesis of AA such as carbonylation of 1,5-cyclooctadiene [10], oxidation of 1,9-nonanedial with O_2 [11, 12], and oxidative cleavage of 2-cyanoethylcyclohexanone [13] and cyclononanone [14] were also not established, possibly due to the poor accessibility of the starting materials or insufficient selectivities.

New interesting developments regarding technical processes for the preparation of AA such as biocatalytic oxidation of PA and the two-stage cleavage of oleic acid using H_2O_2 and O_2 as oxidants are reported later in the text. However, ozonolysis coupled with an aerobic oxidative work-up of intermediates is the most important industrial process at the present time [5, 7].

20.1.1
Current Technical Process: Ozonolysis

The synthesis of AA by ozonolysis of OA or ricinoleic acid is known since the beginning of the twentieth century [15–17]. The key intermediates and products of the method are in accordance with the still-accepted mechanism developed by Criegee [18] (Scheme 20.1).

Using modern analytical methods, a number of transient intermediates and by-products could be verified [19, 20]. The first step in the mechanism of ozonolysis is the 1,3-dipolar cycloaddition of the dipole ozone to the double bond of OA. A 1,2,3-trioxolane is formed, the unstable primary ozonide or molozonide. The primary ozonide collapses in a 1,3 dipolar cycloreversion to a carbonyl compound and a carbonyl oxide, the so-called Criegee zwitterion. Since OA is substituted with two diverse groups at the double bond, two different opportunities exist for the formation of carbonyl compound and carbonyl oxide. Again, a 1,3-dipolar cycloaddition of these intermediates leads to three different pairs of 1,2,4-trioxolane derivatives (*cis/trans*), the secondary ozonides, which are more stable than the primary ones. Their oxidative cleavage results in AA and PA.

A first patent for the technical implementation of ozonolysis of OA was claimed by Rieche [21]. According to a BASF patent filed in 1941, AA (97%) and PA (95%) are accessible by treatment of the ozonide with alkaline lye and Ag_2O [22]. However, the technical manufacturing of AA succeeded first by Emery Industries in the year 1953 [23]. Their successor Emery Oleochemicals is still the world's largest producer of AA by ozonolysis of OA with a production volume of about 10 000 t/year [9, 24]. The achievable yield of AA is 70–80% [24], based on the amount of applied OA. AA yields of more than 90% were obtained in a pilot plant [25]. The reaction is carried out in PA/water (70/30 v/v) as solvent. The admixture of water reduces the formation of side products by removal of the heat of reaction via evaporative heat loss, and in addition, undesired reactive radicals are quenched by reaction with water [25, 26].

Ozone is applied in a concentration of about 1–3 vol% in air [25]. Another process uses the ozone synthesis from oxygen obtained via air separation in a closed circular flow. Ozone is then adsorbed on silica and desorbed subsequently

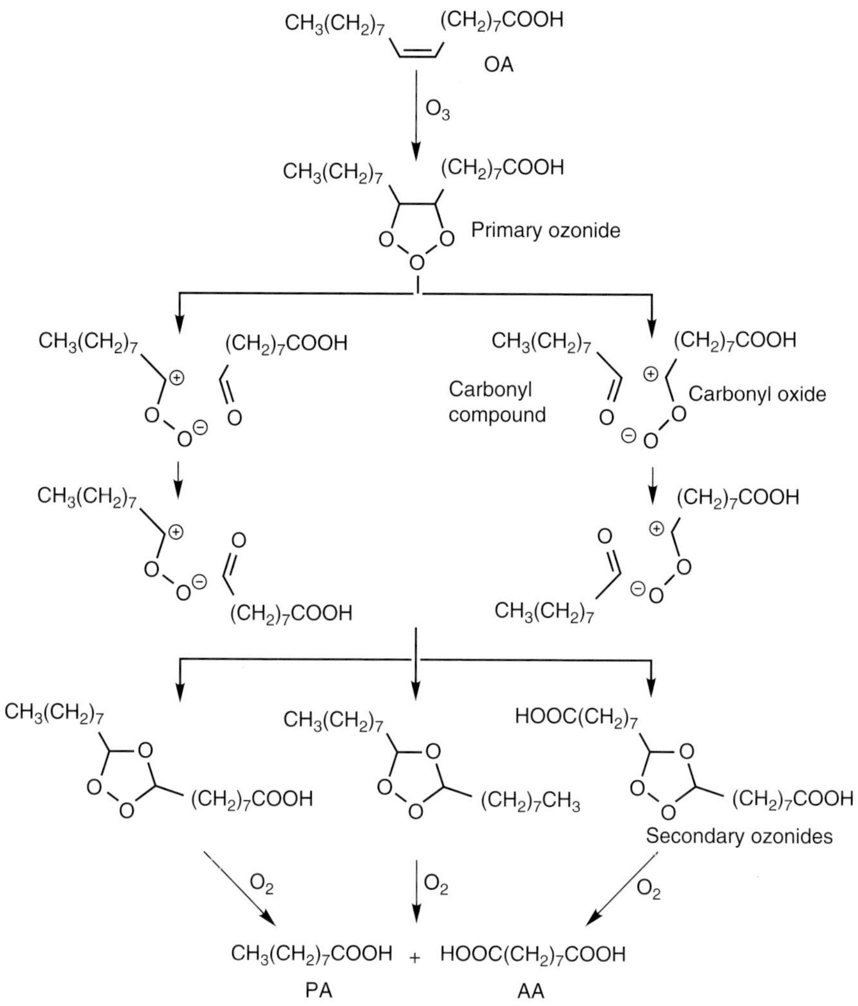

Scheme 20.1 Mechanism of ozonolysis.

by nitrogen from the air separation. The ozone/nitrogen mixture is fed into the reactor. The necessary oxygen for the oxidative cleavage of the ozonide is added so that the reactive gas consists of an ozone/air mixture [25]. The residence time of the fatty acid or its ester is about 10 min. PA and AA are separated in a first purification step from higher boiling species by distillation. AA is then extracted from the distillate to remove PA and lower monocarboxylic acids. A final distillation of the crude AA solution results in pure AA [7].

The products of different quality categories of AA are available in the market under the brand EMEROX® azelaic acids [4]. Products containing 79% (19% other dicarboxylic acid, 2% monocarboxylic acids) [6] and 85% AA are used for the preparation of low-temperature and polymeric type vinyl plasticizers, ester-based synthetic lubricants (e.g., for aircraft engines), and lithium complex grease. The product containing approximately 89% AA and 11% of other dibasic acids was developed as starting material for polymers [4, 6].

The benefits of ozonolysis are (i) its high selectivity to give a sufficiently pure AA for polymer production, (ii) the low reaction temperature of the first step (20–45 °C), and (iii) moderate temperature in the oxidation step (70–120 °C) [7, 23, 27]. The coupling product PA is a valuable material that can be merchandized itself (e.g., as lubricant [4]), or as described later, it may be used in the bioenzymatic synthesis of AA in the future.

However, drawbacks are the required high energy use as well as the toxicity and explosivity of ozone and ozonides. In past times, and also today, proposals for the improvement of safety and process control of ozonolysis have been claimed [28–31]. The carbonyl oxide is converted with water into a hydroperoxide, which reacts to an aldehyde and hydrogen peroxide [32]. The latter is used for the subsequent oxidation of the aldehyde group to the carboxylic acid group. It is claimed that an avoided formation of secondary ozonides makes the process safer and diminishes the amount of side products.

Regardless of intensive studies on the industrial synthesis of AA, the improvement of the process is still of interest. However, only contributions of the last 15 years were considered apart from which the scope of this review would be exceeded. References have also been taken into account where methyl oleate (MO) was used as starting material. The product monomethyl azelate (MMA) obtained besides PA can either be hydrolyzed to AA or can be used directly for certain applications.

With respect to polymer synthesis and to the development of biologically active compounds in recent years, the ozonolysis/reductive cleavage of OA or MO or the fragmentation of the intermediary carbonyl oxide to 9-oxononanoic acid and its respective methyl ester gained attention again [33–39]. 9-Oxononanoic acid can be easily converted to 9-amino- or 9-hydroxynonanoic acid as sources for polyamides and polyesters [40]. A French patent claimed the synthesis of oxononanoic acid via homometathesis of OA, followed by reductive ozonolysis of the unsaturated C18 diacid obtained [41].

An improved processing and optimization of different reaction steps in the ozonolysis was described in several publications and patents [42–46]. Sun *et al.* reported an optimization using the replacement of oxygen by hydrogen peroxide for oxidative splitting of the secondary ozonide [47]. The application of solid catalysts such as sulfated $ZrO_2-Al_2O_3$ and TiO_2 or phosphotungstic acid-loaded macroporous molecular sieves in combination with a phase-transfer catalyst and H_2O_2 was described in [48]. *In situ* formed cetyltrimethylammonium peroxotungstophosphate from cetyltrimethylammonium halide, tungstophosphoric acid, and hydrogen peroxide was found to be identified as the active catalyst in the ozonolysis/oxidation of OA [49–52]. In a variation of this method, tetraalkylammonium peroxotungstate species were formed as active catalysts [53]. A

V_2O_5–polyaniline (PANI) nanocomposite was utilized as catalyst in the oxidative work-up of the ozonide to AA and PA. V_2O_5/PANI was twice as selective to AA as bulk V_2O_5 and even as nanostructured V_2O_5 (53% compared with 28% and 32%, respectively) [54]. The application of a non-nucleophilic solvent was claimed in [55]. An improved purification and recrystallization of crude AA obtained from ozonolysis resulted in suitable starting material for polyamide synthesis [56]. Cai *et al.* reported the application of MoO_3 as a catalyst for the oxidative splitting of the ozonide and a yield of AA of about 72% [57]. Sun *et al.* showed that V_2O_5 was the most active but least selective catalyst in the cleavage step; they investigated a series of metal oxide catalysts. However, the yield of AA amounted to 75% using MoO_3 or PbO_2 as a catalyst [58]. The oxidative cleavage was carried out using various iron, manganese, tungsten, or cobalt catalysts in [59, 60]. Sawut *et al.* used a mixture of acetic acid and hexane in the ozonolysis of MO and received 78% of MMA with a purity greater than 96% [61]. The ozonolysis of MO in liquid CO_2 and CO_2-expanded solvents and the support of ultrasonics effectuated a better solubility of ozone and a 100% conversion of MO. Besides PA and MMA, also 9-oxononanoic acid and nonanal had been isolated as products [62]. In contrast, no conversion was observed if MO was reacted with ozone in supercritical CO_2 [63], but OA could be converted with a yield of 64% to AA and 35% to PA [64].

Pure ozone for ozonolysis of OA was generated from pure carbon dioxide by electric corona-discharge treatment, thus offering an industrially affordable process [65].

Some authors demonstrated alternatives of process technology. The ozonolysis of OA to AA at 150 °C in a Bach bubbling reactor with fine bubbles in the absence of any catalyst or any solvent was carried out by Kadhum *et al.* with a yield of 20% after 2 h [66]. The decomposition of the ozonide was accelerated using microwave radiation, and yields of AA between 69.7% and 80% were received [67, 68]. The products of ozonolysis of MO were separated by nanofiltration, whereas the ionic liquid used as reaction medium was retained. The highest rejection, up to 96%, was achieved by polyimide-based DuraMem membranes [69]. A yield of 52.7% of AA could be obtained in the ozonolysis of a technical OA in a 64-channel microstructured falling film reactor at 20 °C (ozonolysis) and 95 °C and an oxygen pressure of 1.3 bar (oxidation) [70]. The application of such microstructured reactors can reduce the hazard potential of chemical processes significantly due to their very small volume compared with conventional reactors. Furthermore, the evolved heat of reaction can be dissipated fast. This technology enables an efficient use and transfer of energy, and often an improved product selectivity is possible by adjusting optimal residence times and concentration gradients. An integrated continuous processing with coupling of ozonolysis and oxidation or reduction is claimed in [71].

All these variations or improvements might be useful in regard to a patentable solution or to a safer and probably less energy-demanding process, if any, but the yield of received AA did not exceed that of the current technical process.

20.1.1.1 Analytical Investigations of the Mechanism of Ozonolysis

Confirmation of the Criegee mechanism and new insights into ozonolysis reaction were gained using modern analytical methods. Often, the reason for these studies was the reaction of oleic acid aerosols with naturally occurring ozone in the stratosphere and not the investigation of industrial ozonolysis. Using photoelectron resonance capture ionization (PERCI) mass spectrometry, 1-nonanal, PA, 9-oxononanoic acid, and AA were identified as main products [72]. These four products were also detected simultaneously with the aerosol chemical ionization mass spectrometry (Aerosol CIMS) in different yields, suggesting that secondary reactions of the Criegee intermediates occur. 9-Oxooctadecanoic acid was identified as a further product [73]. An extractive electrospray ionization (EESI) source for the online mass spectrometry analysis using the ozonolysis of oleic acid aerosols as a model reaction was investigated by Gallimore and Kalberer [74]. Ozonolysis products of MO were identified by coupling of normal and nonaqueous reversed-phase (NARP) liquid chromatography (LC) coupled to electrospray ionization (ESI) mass spectrometry (MS) and tandem mass spectrometry (MS/MS) using a hybrid quadrupole-time of flight instrument[75]. Known ozonolysis reaction pathways could be confirmed. Sage *et al.* investigated the ozonolysis of OA aerosols applying MS techniques [76]. They observed that in addition to the Criegee mechanism previously unrecognized secondary chemical reactions occur that possibly involve the carbon backbone of the fatty acid. The ozonolysis of OA and MO in different solvents was examined by Ledea *et al.* via ^{1}H NMR spectroscopy [77]. They observed hydroperoxides among others as side products depending from starting material and solvent.

20.2
Chemistry and Catalysis

20.2.1
Direct Aerobic Cleavage of the Double Bond of Oleic Acid or Methyl Oleate

In general, the direct cleavage of OA to AA using molecular oxygen as "green" and atom economic oxidant would be the "dream" reaction in this field. However, the diradical nature of oxygen often causes a low selectivity of aerobic oxidations. In the case of fatty acids and their derivatives, the competing allylic oxidation and degradation of the fatty acid chain may be the main reason for disappointing yields and selectivities of target products (Scheme 20.2).

Dapurkar *et al.* [78] reported the aerobic cleavage of OA in supercritical carbon dioxide. The authors used doped mesoporous and microporous molecular sieves such as CrMCM-41, MnMCM-41, CoMCM-41, CrAlPO-5, CoMFI, and MnMFI as catalysts, which are stable materials in oxidations, an advantage of the method. The highest selectivity in regard to AA was found in the reaction using Cr-MCM-41 (conversion >99%, yield 32%), but a metal leaching of about 10 wt% was observed. A decrease in conversion from 99% to 95% was noticed in a second

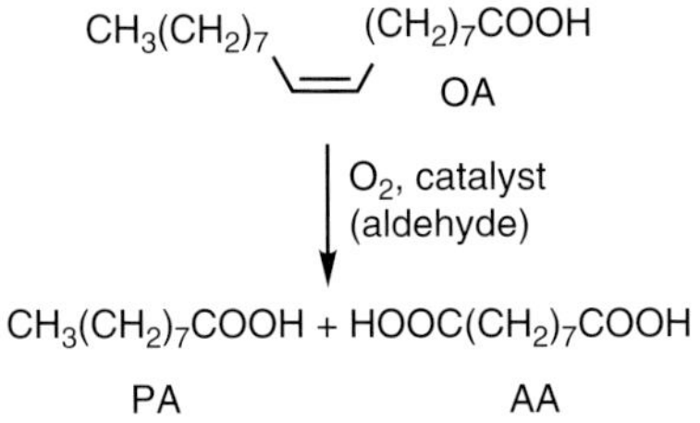

$$CH_3(CH_2)_7\text{—}\!\!=\!\!\text{—}(CH_2)_7COOH$$
$$OA$$

↓ O_2, catalyst (aldehyde)

$$CH_3(CH_2)_7COOH + HOOC(CH_2)_7COOH$$
$$\quad\quad PA \quad\quad\quad\quad\quad\quad AA$$

Scheme 20.2

run of the recycled catalyst, whereas the yield of AA and PA was not reduced. A Japanese patent described a similar catalyst, Co-AlPO-5, and an AA yield of 34% was mentioned [79].

A Chinese patent claims that a mixture of H_2O_2 and bubbling O_2 oxidizes OA to AA in the presence of the doped mesoporous molecular sieve W-MCM-41 up to a yield of 94% [80]. The same reaction was published in [81]. The usage of t-BuOH as solvent suggests the idea that *in situ* formed t-BuOOH might be the active oxidant. Alternatively, the formation of 9,10-dihydroxystearic acid (DSA) using H_2O_2 followed by an aerobic cleavage may occur in a one-pot reaction. The latter reaction pathway may also be assumed for a reaction described in a French patent [82]. Tungstic acid was used in a system of H_2O_2/bubbling O_2 for the cleavage of technical mixtures of fatty acids. AA and PA besides 9-oxononanoic acid were obtained as reaction products in the case of OA-rich substrates.

Other metal species, which are known to be active in aerobic radical reactions such as vanadium compounds [83], Co naphthenate [84], Co(acetate)$_2$ [85], Co oleate [86], and metal porphyrins [87], were also reported to give low yields of AA with molecular oxygen. When the aerobic scission of olive oil fatty acids in a ball mill in the presence of CeO_2 stopped at the aldehyde stage, 9-oxononanoic acid and nonanal besides a range of other aldehydes were received [88].

The aerobic cleavage of OA or MO with aldehydes as scarifying agents was performed more selectively with AA or MMA yields up to 95%. Suitable catalysts for these oxidations were OsO_4 or osmate, RuO_2, or $Ru/(NH_4)_6H_3PV_6Mo_6O_{40}$ [89–91]. However, the stoichiometric formation of acids as coupling products formed from the aldehyde is a substantial drawback of such processing.

20.2.2
Aerobic Oxidation Step within a Two-Stage Conversion of Oleic Acid or Methyl Oleate

Only the second step, the catalytic cleavage of DSA or methyl dihydroxystearate (MDS), was carried out using oxygen when two-stage processes for the synthesis of AA or MMA from fatty acids or esters were developed (Scheme 20.3). The first step, a direct dihydroxylation or an epoxidation/hydrolytic ring scission of an epoxide formed as intermediate, may be performed using H_2O_2 or performic acid as oxidant, for example.

At first, DSA was prepared without any catalyst with performic acid *in situ* generated from H_2O_2 and formic acid. Then, the cleavage to AA and PA was executed subsequently with a heterogeneous $Ru(OH)_x/Al_2O_3$ catalyst and air as the oxidant

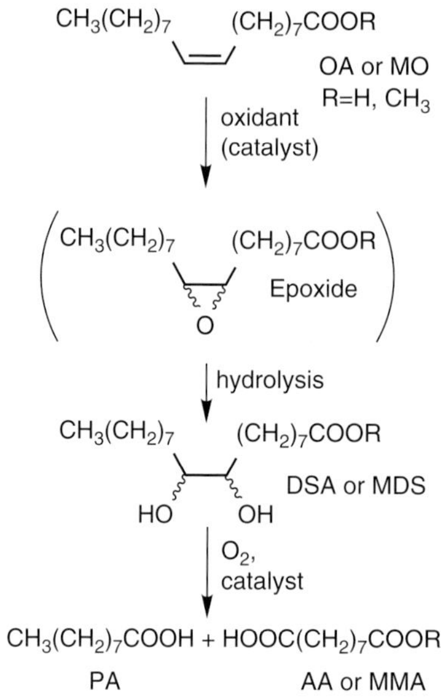

CH₃(CH₂)₇ —CH=CH— (CH₂)₇COOR   OA or MO, R=H, CH₃

↓ oxidant (catalyst)

(CH₃(CH₂)₇ —CH—CH— (CH₂)₇COOR) Epoxide
                  \\O/

↓ hydrolysis

CH₃(CH₂)₇ —CH(OH)—CH(OH)— (CH₂)₇COOR   DSA or MDS

↓ O₂, catalyst

CH₃(CH₂)₇COOH + HOOC(CH₂)₇COOR
      PA   AA or MMA

Scheme 20.3

[92]. The yield of AA was 44–54%. The method has the advantage that the use of *in situ*-generated peracid is an established method in the oleochemical industry. Fujitani *et al.* prepared pure DSA as reported in [93] and cleaved it using Co(OAc)$_2$ to 95% yield of dicarboxylic acids (AA and suberic acid). The method failed if technical DSA was used. However, this substrate was converted to AA and suberic acid in the presence of Co(OAc)$_2$–Mn(OAc)$_2$–HBr with a yield of 84%. A similar method was claimed in a Soviet patent [94].

The groups of Woodward and coworkers [95] and Santacesaria and coworkers [96, 97] published the application of H$_2$WO$_4$ as catalyst precursors in combination with H$_2$O$_2$ for the preparation of DSA. The catalytically active species may be a peroxoisopolytungstate formed from the precursor. Subsequently, the scission of DSA was carried out using Co(acac)$_3$/N-hydroxyphthalimide (NHPI) or Co(OAc)$_2$ and molecular oxygen as oxidant. Only 15% yield of AA was obtained if OA was used as single substrate. The yield could be increased up to 56% if a 1 : 1 mixture of OA and erucic acid was used as starting material [94]. Santacesaria *et al.* [95, 96] synthesized 9-hydroxynonanoic acid besides AA and PA. The authors discussed a lacunary polyoxometalate in which cobalt is sequestered by the tungstate anion groups. It needs a certain induction period to be formed as active species for the second step of the reaction. The occurrence of a vicinal hydroxyketone as an intermediate of the mechanism was proved. A similar reaction procedure was claimed in a Chinese patent [98].

This two-stage catalytic cleavage using environmentally benign oxidants such as hydrogen peroxide and oxygen is also the principle of the new Matrìca process for the production of AA described in Section 20.3.

The gold-catalyzed splitting of DSA as substrate applying heterogeneous gold catalysts in basic aqueous solution was investigated in our own group; the highest yields of AA and PA obtained were 86% and 99%, respectively [99, 100]. As a drawback, catalyst aging was observed, which is expected to prevent technical application. An agglomeration of gold particles took place even at storage in the dark at low temperatures and under argon. The catalytic cleavage of AA was used as a model reaction; a sample of the stored catalyst was subjected to the same reaction conditions every 3 months. The yield of AA decreased from 61% to 28–49% in this experimental series after 1 year of storage time depending on storage conditions [101].

Behr and Tenhumberg [9] considered security requirements for explosion protection in such a liquid oxidation system. The authors investigated the example of the aerobic cleavage of MDS to PA and MMA with acetic acid and PA as a solvent in the loop reactor. They applied Co(acac)$_3$/NHPI as a catalyst and air as oxidant and received up to 67% MMA.

The aerobic cleavage of DSA up to 72% of AA was carried out using a mixture of Co-, Mn-, Ce-, and Ni-acetates, NaBr, and HBr [102].

20.2.3
Aerobic Oxidation Step within a Three-Stage Conversion of Oleic Acid or Methyl Oleate

The Ru-catalyzed ethenolysis of a Δ9 fatty acid forming 9-decenoic acid (9-DA) was utilized as a first step in a multistage synthesis of AA that included at least one step applying molecular oxygen or air (Scheme 20.4).

Thus, 9-DA was splitted by ozonolysis into AA and formic acid; the reaction can be also transferred to fatty esters [103].

9-DA was also reported by Warwel *et al.* [104] as the substrate for Wacker oxidation (O$_2$, PdSO$_4$/heteropolyacid). The formed α-ketone was then cleaved using Mn-stearate and O$_2$ in acetic acid (formed peroxy radical or peracid might be the active oxidants) to a mixture of AA and suberic acid. Using 10-undecenoic acid, which can be easily made by pyrolysis of castor oil, a mixture of AA and sebacic acid could be synthesized.

20.2.4
Biocatalysis

Biotechnological syntheses of α,ω-dicarboxylic acids succeeded by the oxidation of the terminal alkyl groups of alkanes or monocarboxylic acids with the same number of carbon atoms as the target diacid. Petrochemical feedstocks as well as vegetable oils were exploited. Monooxygenases of the CYP52 type catalyze the hydroxylation of the terminal methyl group, followed by the reaction of an alcohol oxidase and aldehyde dehydrogenase with the intermediates under consumption of oxygen to the diacid in a sequence of single steps [105]. Both isolated enzymes and whole cells containing several required enzymes were investigated. The latter allow the implementation of multistage processes. A first patent that described

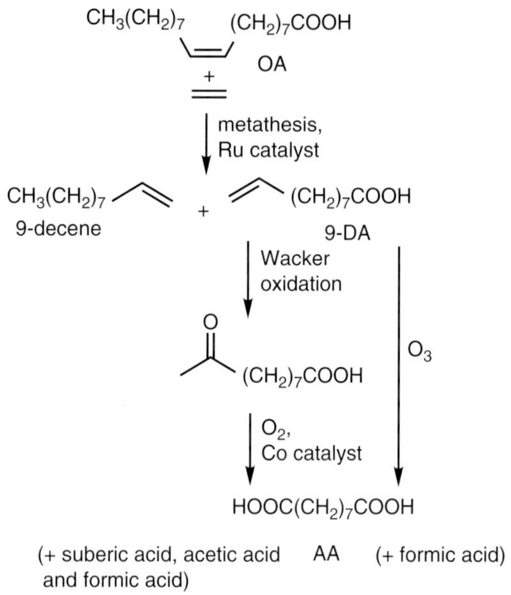

$$CH_3(CH_2)_7\text{—}\text{=}\text{—}(CH_2)_7COOH \quad \text{OA}$$

$$+$$

$$\text{=}$$

↓ metathesis, Ru catalyst

$$CH_3(CH_2)_7\text{—}\text{=} \quad + \quad \text{=}\text{—}(CH_2)_7COOH$$

9-decene 9-DA

↓ Wacker oxidation

$$CH_3\text{—CO—}(CH_2)_7COOH$$

→ O_3

↓ O_2, Co catalyst

$$HOOC(CH_2)_7COOH$$

(+ suberic acid, acetic acid and formic acid)   AA   (+ formic acid)

Scheme 20.4

the synthesis of AA from PA using animal visceral organ tissue homogenates was filed in 1960 [106]. The "normal" β-oxidation pathway for oxidative degradation of such microorganisms or enzymes has to be blocked. Mostly, they were genetically engineered in order to achieve higher productivities as well as product selectivities and yields [107, 108].

The yield of diacid strongly depends on the chain length of the substrate used, the catalyst, and reaction conditions. Fatty acids with chain lengths from C11 to C18 and their derivatives appear to be favored. Such diacids are produced via fermentation using genetically optimized *Candida tropicalis* strains offered by Cathay Industrial Biotech Co., Ltd. (C11–C16) [109] and Shandong Hilead Biotechnology Co., Ltd. (C11–C14, C16, C18) [110] at an industrial scale. The total diacid production of these companies amounts to about 25 000 t/year, and it is planned to be expanded [108]. AA is to the best of my knowledge not yet available commercially on a biotechnological basis.

Henkel/Cognis (today BASF Personal Care and Nutrition) developed novel fermentation processes of vegetable oil fatty acids using *C. tropicalis* strains several years ago [111]. Especially, processes for the syntheses of C18 dicarboxylic acids were engineered [112]. Octadec-9-ene-1,18-dioic acid produced by fermentation of OA or triglycerides with *C. tropicalis* enzymes was cleaved at the double bond via ozonolysis or catalytic oxidation to AA using strong oxidants. Yields of AA of about 45–67% were observed and the reaction time needed amounted to 110–180 h. Advantageously, this reaction pathway avoids the formation of PA as a coupling product. PA can be also converted directly by this enzymatic ω-oxidation method to AA [111]. A process that utilized both alkenes and fatty acids as substrates and that produced AA among others is described in [113, 114]. Nonane and other paraffins were oxidized by *Torulopsis candida* strains resulting in the formation of AA and other dicarboxylic acids [115, 116].

The oxidation of the terminal methyl group of PA produced an AA yield of 6% if *Debaryomyces pfaffii* was used as biocatalyst [117]. Liu and Yi [118] investigated the diterminal oxidation of 1-decene and 1-dodecene by the mutant SD6 from *C. tropicalis*. Besides other dicarboxylic acids, AA was detected in both product mixtures. Accordingly, the authors proposed the microbial metabolic pathway.

Another more complex route for the preparation of dicarboxylic acids such as AA from long-chain fatty acid derivatives has been investigated by several authors. Song *et al.* [119, 120] reported the synthesis of ω-hydroxycarboxylic acids. Such compounds can be obtained by oxidative scission of fatty acids via enzymatic reaction cascades involving a fatty acid double bond hydratase, an alcohol dehydrogenase (ADH), a Baeyer–Villiger monooxygenase, and an esterase. The intermediary products are oxidized in a consecutive step to α,ω-dicarboxylic acids by ADH and AlkJ from *Pseudomonas putida* GPo1. AA was obtained in a yield of 48% from OA via 10-hydroxystearic acid. Otte *et al.* demonstrated the whole-cell biotransformation of linoleic acid in a multistage cascade [121]. In the introductory step, the lipoxygenase St-LOX1 catalyzed the hydroperoxidation of linoleic acid. Subsequently, the hydroperoxy intermediate was cleaved by the action of a hydroperoxide lyase Cs-9/13HPL to yield 9-oxononanoic acid. Finally, 9-oxononanoic acid was oxidized by an endogenous *Escherichia coli* oxidoreductase to AA. The best results were achieved in a two-phase system with cyclohexane as organic phase. At 1 mM substrate concentration, 34% of linoleic acid was converted and an AA yield of 16% (product titer of 29 mg/l) was obtained after 8 h.

A *Malassezia* strain SA20 (KFCC 11252) was also used for the synthesis of AA from unsaturated fatty derivatives [122]. *Sarcina lutea* ICR2010 was observed to oxidize isobutyl oleate to AA (0.09 mg/ml) besides 10-oxostearic acid and 4-oxolauric acid [123].

A Chinese patent claimed the lipase-catalyzed dihydroxylation of oleic acid to DSA followed by the conventional cleavage using peracid. The yield of AA obtained was about 60–70% [124].

20.3
Prospects for Scale-Up

Looking at the different approaches to the production of AA described earlier, the question arises, why ozonolysis is still the established manufacturing process? But attempts were done in recent years to bring a concept to industrial scale, which were already discussed in Section 20.2.2 [94–96]. Matrìca, a 50 : 50 joint venture between ENI Versalis and Novamont, started the manufacture of bio-based products in Porto Torres (Italy) in mid-2014 [125]. Information about production capacity is not provided. The company offers AA and PA from the cleavage of OA/MO in at least 98% or 99% purity (after purification). Obviously, the production process is based upon various patents of Novamont. The two-stage reaction of MO or OA to MDS or DSA with 60 wt% H_2O_2 and H_2WO_4 or other suitable tungsten compounds as catalyst precursors in the first step and the following cleavage of MDS/DSA to MMA/AA and PA with cobalt catalysts (e.g., $Co(OAc)_2$) and O_2 was already claimed since 1992 [126]. The crude yield for AA of approximately 75–80% can be increased by subsequent purification steps to high-purity products. Triglycerides with higher oleic acid radical concentrations are also suitable as raw material [127]. The method is simple, cost-efficient, and environmentally benign, since it is carried out without organic solvent using inexpensive, green oxidants (H_2O_2; O_2), while also employing inexpensive tungsten and cobalt catalysts without requiring onium-type phase-transfer catalysts [128]. The catalysts can be recycled, and classical phase-transfer catalysts can be omitted in the first step by adding a small amount of the MDS or DSA product of the first step at the start of the reaction [129]. In addition, continuous processing is described [130]. The technical process may be conducted in two continuous reactors connected in series (e.g., continuously stirred tank reactor (CSTR), step 1, and jet loop reactor, step 2, respectively) at mild reaction temperatures (60–62 °C), followed by a consecutive distillative separation of monocarboxylic acids (mainly PA) from AA or MMA. Triglycerides may also be used as feed [131].

This method could become an economical and ecological alternative to ozonolysis. The development of heterogeneous catalysts for the continuous process would be desirable but seems to be difficult on the basis of known active structures.

However, the biocatalytic production of AA at an industrial scale appears to be possible rather at a medium- to long-term time scale. Definitely, space-time yields and downstream processing have to be improved. In addition, *C. tropicalis* is classified as a pathogenic microorganism in Europe. This fact raises the investment for a production site possessing a high security standard for an industrial process [110]. Such problems may occur repeatedly in regard to other microorganisms in restrictive legislation.

20.4
Concluding Remarks and Perspectives

20.4.1
New Promising Developments

The properties of polyoxometalates (POMs) can be adjusted on the basis of their composition in a wide range. As inorganic compounds they are significantly more stable to oxidation than transition metal complexes with organic ligands. Therefore, POM catalysts should show a better long-term operating time and should be better suited for a scale-up of processes. Their ability of oxygen activation makes them similar to biomimetic systems.

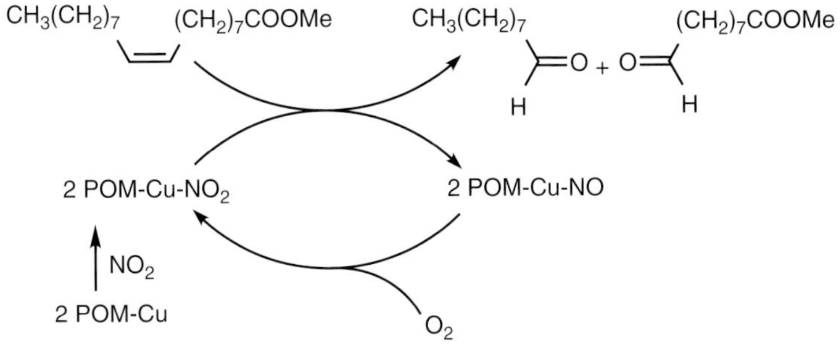

Scheme 20.5

An interesting approach for the direct oxidative scission of MO was published by Neumann and coworkers (Scheme 20.5) [132].

They used a NO_2-binding copper-substituted sandwich-type POM as an active catalyst for the conversion of MO to nonanal and methyl 9-oxononanoate. The catalyst could be generated either (i) *in situ* from $Q_{12}\{ZnWCu_2(NO)_2(ZnW_9O_{34})_2\}$ (Q = methyltrioctylammonium) and oxygen (2 bar), (ii) via decomposition of a nitroalkane under reaction conditions, or (iii) using a minimal amount of NO_2. The obtained yields of aldehydes were 93% (2 h, 125 °C, solvent propionic acid) and 99% (6 h, 85 °C, solvent nitroethane). However, the achieved turnover numbers might be improved. The possibility of heterogenization of POM in the liquid phase appears to be limited. Therefore, processes in CSTRs may be alternatives for technical applications with retention of POM in the reactor via suitable membranes.

The activation of oxygen by $Pd(OAc)_2$ in an acidic aqueous medium for the cleavage of different alkenes [133] is another interesting approach for environmentally benign processes as alternatives to ozonolysis. Although most compounds tested possessed activated double bonds, 2-nonene could be cleaved to form heptanal in 68% yield. The method should be tested using OA.

Other alternative reaction pathways to AA or MAA via hydroformylation of 7-octenoic acid or the respective ester and consecutive oxidation of the formed aldehydic group may fail due to the poor accessibility of the C8 compound needed as starting material. This is unfortunate because recent development in the selective hydroformylation to linear aldehydes using sophisticated phosphite ligands [134, 135] may enable such reaction sequence.

20.4.2
Summary

Probably, the use of AA in particular for the production of new high-performance polyamides and polyesters for high-quality applications and thus its production volume might be larger (production volumes of several 10 000 t/year), if the ozonolysis could be replaced by an alternative manufacturing process. However, opportunities for optimization were developed even for ozonolysis, which make the process safer, for example, processing in microstructured reactors. The new

Table 20.1 Comparison of different approaches for the synthesis of AA.

References	Substrate	Oxidant	Catalyst	Yield$_{AA}$ (%)
[24]	OA	O_3/O_2	Only for oxidation using O_2, if any	70–80
[131]	High oleic sunflower oil	1) H_2O_2 2) O_2	H_2WO_4 Co(acetate)$_2$	80
[119, 120]	10-Hydroxy-stearic acid (from OA)	O_2	Multistep biocatalysis	48 (22)[a]
[132]	MO	O_2	POM-Cu-NO$_2$	99[b]

a) Isolated yield.
b) 9-Oxononanoic acid.

Matrìca process appears promising; however, the future will show whether those expectations will come true. The direct aerobic cleavage of OA or MO using metal-doped ordered porous catalysts in liquid or supercritical phase is guessed to be less suitable for a scale-up. Besides insufficient long-term stability due to the unavoidable leaching of catalytically active, partly toxic heavy metals, deactivation of micro- and mesoporous catalysts by deposition of adsorbates was often observed. Probably, the required long-term stability in a continuous process also might not be achieved. Executing biocatalytic procedures, space-time yield, and downstream processing have to be improved for applications at a large scale. A summary of the best results in the synthesis of AA can be found in Table 20.1.

The aim of further research should be the development of robust catalysts, which activate oxygen in a nonradical manner, probably via biomimetic cascades, and which are also able to provide an oxidative cleavage of unactivated internal double bonds in unsaturated fatty acids.

References

1. Spellman, M.C. and Pincus, S.H. (1998) *Clin. Ther.*, **20**, 711–721.
2. Jones, D.A. (2009) *J. Clin. Aesthetic Derm.*, **2**, 26–30.
3. Thieme Römpp (2014) Römpp Online, Georg Thieme Verlag, Stuttgart, http://www.roempp.com/prod/ (accessed 23 September 2014).
4. Emery (2014) Information Brochure, www.emeryoleo.com/biz_biolubricant.php (accessed 24 September 2014).
5. Köckritz, A. and Martin, A. (2011) *Eur. J. Lipid Sci. Technol.*, **113**, 83–91.
6. Janakiefski, N.G. (1997) *NLGI Spokesman*, **61**, 14–24.
7. Cornils, B. and Lappe, P. Aliphatic Dicarboxylic Acids. (2010) *Ullmann's Encyclopedia of Industrial Chemistry, Electronic Release*, 7th edn, Wiley-VCH Verlag GmbH, Weinheim.
8. Heidbreder, A., Hofer, R., Grützmacher, R., Westfechtel, A., and Blewett, C.W. (1999) *Fett/Lipid*, **101**, 418–424.
9. Behr, A. and Tenhumberg, N. (2012) *Chem. Ing. Tech.*, **84**, 1559–1567.

10. Zakharin, L.I. and Guseva, V.V. (1983) Patent SU 1092150, Institute of Heteroorganic Compounds, Academy of Sciences.
11. Matsumoto, M., Yoshimura, N., and Tamura, M. (1982) Patent JP S58140038, Kuraray.
12. Saito, H., Shinpo, C., and Tokito, Y. (1995) Patent JP 07133250, Kuraray.
13. Minisci, F., Maggioni, P., and Citterio, A. (1980) Patent DE 3027111, Brichima.
14. Osowska-Pacewicka, K. and Alper, H. (1988) *J. Org. Chem.*, **53**, 808–10.
15. Harries, C. and Tank, L. (1907) *Ber. Dtsch. Chem. Ges.*, **40**, 4555–4559.
16. Haller, A. and Brochet, A. (1910) *Compt. Rend.*, **150**, 496–503.
17. Hill, J.W. and McEwen, W.L. (1933) *Org. Synth.*, **13**, 4.
18. Criegee, R. (1975) *Angew. Chem.*, **21**, 765–771.
19. Ziemann, P.J. (2005) *Faraday Discuss.*, **130**, 469–490.
20. Hearn, J.D., Lovett, A.J., and Smith, G.D. (2005) *Phys. Chem. Chem. Phys.*, **7**, 501–511.
21. Rieche, A. (1931) Patent DE 565158.
22. BASF (1941) Patent DE 868148.
23. Brown, A.C., Goebel, C.G., Oehlschlaeger, H.F., and Rolfes R.P. (1953) Patent US 2813113, Emery Industries.
24. Tenhumberg, N. (2013) PhD thesis. Katalytische oxidative Spaltung des Ölsäuremethylesters. TU Dortmund.
25. Witthaus, M. (1986) Schriftenreihe des Fonds der Chemischen Industrie, Frankfurt am Main, issue 26, pp. 19–25.
26. Naudet, M. and Pelloquin, A. (1973) *Rev. Franc. Corps Gras*, **20**, 89–94.
27. Weissermel, K. and Arpe, H.-J. (1998) *Industrielle Organische Chemie*, Wiley-VCH Verlag GmbH, Weinheim, p. 229.
28. Hann, V.A. (1956) Patent US 2874164, Welsbach.
29. Walker, T.C. (2012) Patent WO 2012103317, Emery Oleochemicals.
30. Walker, T.C. (2012) Patent WO 2012103310, Emery Oleochemicals.
31. Ullrich, M., Hannen, P., and Roos, M. (2012) Patent EP 2573062, Evonik Degussa.
32. Hannen, P., Häger, H., and Roos, M. (2012) Patent EP 2502899, Evonik Degussa.
33. Pryde, E.H., Anders, E., Teeter, H.I., and Cowan, J.C. (1960) *J. Org. Chem.*, **25**, 618–621.
34. Pryde, E.H., Anders, E., Teeter, H.I., and Cowan, J.C. (1962) *J. Org. Chem.*, **27**, 3055–3059.
35. Tamura, S., Kaneko, M., Shiomi, A., Yang, G.-M., Yamaura, T., and Murakami, N. (2010) *Bioorg. Med. Chem. Lett.*, **20**, 1837–1839.
36. Schwartz, C., Raible, J., Mott, K., and Dussault, P.H. (2006) *Org. Lett.*, **8**, 3199–3201.
37. Willand-Charnley, R., Fisher, T.J., Johnson, B.M., and Dussault, P.H. (2012) *Org. Lett.*, **14**, 2242–2245.
38. Wang, J., Chen, G., Wu, J., Weng, W., Ye, J., Shi, Z., Jiang, H., and Wang, L. (2012) Patent CN 201110349814, Hangzhou Youbang Flavors and Fragrances.
39. Wang, J., Chen, G., Wu, J., Weng, W., Ye, J., Shi, Z., Jiang, H., and Wang, L. (2011) Patent CN 102351697, Hangzhou Youbang Flavor and Fragrances.
40. Louis, K., Vivier, L., Clacens, J.-M., Brandhorst, M., Dubois, J.-L., De Oliveira Vigier, K., and Pouilloux, Y. (2014) *Green Chem.*, **16**, 96–101.
41. Dubois, J.-L. (2010) Patent FR 2933696, Arkema France.
42. Wang, Y.Z. (2007) Patent CN 101062891, Hangzhou Youbang Flavor and Fragrances.
43. Li, Y. and Song, Z. (2003) *Jingxi Shiyou Huagong*, 37–39.
44. Zhu, J., Du, J., Zhou, Y., and Xie, X. (2008) Patent CN 101250101, Sichuan Sipo Chemical Industry.
45. Yu, Z., Wang, J., and Shi, Z. (2007) Patent CN 101062891, Hangzhou Youbang Flavor and Fragrances.
46. Macho, V., Mravec, D., Vojtko, J., Kralik, M., Gattnar, O., Varga, I., Snuparek, V., and Skoda, A. (2001) Patent SK 281462, Slovakofarma.
47. Sun, Y., Yan, L., and Zhang, M. (2012) *Guangzhou Huagong*, **40**, 47–49.

48. Chen, Y., Shi, C., and Chen, X. (2007) Patent CN 101077856, University of Shanghai.
49. Chen, H., Shi, C., and Chen, X. (2006) Patent CN 1927806, University of Shanghai.
50. Shi, C., Yan, Y., and Yin, W. (2009) *Huagong Keji*, **17**, 16–19.
51. Shi, C., Wang, Y., and Chen, Y. (2008) *Liaoning Shiyou Huagong Daxue Xuebao*, **28**, 1–3.
52. Shi, C., Chen, Y., and Jia, R. (2007) *Huaxue Shijie*, **48**, 612–614, 611.
53. Shi, C. and Chen, Y. (2006) *Xiandai Huagong*, **26**, 285–287.
54. Khadijeh, B.G., Nilofar, A., and Yarmo, M.A. (2012) *Adv. Mater. Res.*, **364**, 217–221.
55. Zhu, J., Du, J., Zhou, Y., and Xie, X. (2008) Patent CN 101244998, Sichuan Sipo Chemical Industry.
56. Volkheimer, J., Frische, R., and Hegwein, K. (2001) Patent EP 1074540, Dr. Frische (Alzenau).
57. Cai, Z., Le, Q., and Wu, M. (2010) *Huaxue Shijie*, **51**, 362–366.
58. Sun, Z.-C., Wumanjiang, E., Xu, T.-Y., and Zhang, Y.-G. (2006) *Yingyong Huaxue*, **23**, 161–164.
59. Sun, Z., Zhang, Y., Fan, L., Hu, S., Gao, J., and Dong, X. (2005) Patent CN 1680254, Xinjiang Physical and Chemical.
60. Sun, Z., Zhang, Y., Wumanjiang, E., Hu, S., and Gao, J. (2006) *Zhongguo Youzhi*, **31**, 40–42.
61. Sawut, A., Eli, W., and Nurulla, I. (2009) *Jingxi Shiyou Huagong*, **26**, 43–46.
62. Subramaniam, B., Busch, D., Danby, A.M., and Binder, T.P. (2008) Patent US 20090118498.
63. Sparks, D.L., Estevez, L.A., and Hernandez, R. (2009) *Green Chem.*, **11**, 986–993.
64. Yoshida, M. (2010) Patent WO 2010095669, University Utsonomiya.
65. Moran, E.F. and Piasecki, E.V. (2002) Patent WO 2002064498, Du Pont.
66. Kadhum, A., Amir, H., Wasmi, B.A., Mohamad, A.B., Al-Amiery, A.A., and Takriff, M.S. (2012) *Res. Chem. Intermed.*, **38**, 659–668.
67. Chen, Y. and Cao, S. (2008) *Fujian Shifan Daxue Xuebao, Ziran Kexueban*, **24**, 58–61.
68. Wumanjiang, A., Hu, S., Zhang, Y., Dong, X., Gao, J., and Fan, L. (2004) Patent CN 1616393, Xinjiang Physical and Chemical Technology Institute, Chinese Academy of Sciences.
69. Van Doorslaer, C., Glas, D., Peeters, A., Cano Odena, A., Vankelecom, I., Binnemans, K., Mertens, P., and De Vos, D. (2010) *Green Chem.*, **12**, 1726–1733.
70. Kloeker, M., Franzen, S., and Gutsche, B. (2006) Patent DE 102006021438, Cognis.
71. Foley, P. and Yang, Y. (2013) Patent WO 2014015290, P2 Science.
72. LaFranchi, B.W., Zahardis, J., and Petrucci, G.A. (2004) *Rapid Commun. Mass Spectrom.*, **18**, 2517–2521.
73. Hearn, J.D. and Smith, G.D. (2004) *J. Phys. Chem. A*, **108**, 10019–10029.
74. Gallimore, P.J. and Kalberer, M. (2013) *Environ. Sci. Technol.*, **47**, 7324–7331.
75. Sun, C., Zhao, Y.-Y., and Curtis, J.M. (2012) *Rapid Commun. Mass Spectrom.*, **26**, 921–930.
76. Sage, A.M., Weitkamp, E.A., Robinson, A.L., and Donahue, N.M. (2009) *Phys. Chem. Chem. Phys.*, **11**, 7951–7962.
77. Ledea, O., Diaz, M., Molerio, J., Jardines, D., Rosado, A., and Correa, T. (2003) *Revista CENIC, Cienc. Quim.*, **34**, 3–8.
78. Dapurkar, S.E., Kawanami, H., Yokoyama, T., and Ikushima, Y. (2009) *Top. Catal.*, **52**, 707–713.
79. Ikeda, T., Kiyozumi, Y., Komura, K., Mizukami, F., Nagase, T., and Yokoyama, T. (2004) Patent JP 2005255652, National Institute of Advanced Industry & Technology.
80. Zhao, P., Peng, Z., Ni, P., and Suo, J. (2002) Patent CN 1415593, Lanzhou Institute of Chemical Physics.
81. Peng, Z., Zhao, P., and Suo, J. (2003) *Huaxue Tongbao*, **66**, w089/1–w089/2.
82. Brandhorst, M. and Dubois, J.-L. (2012) Patent WO 2013079849, Arkema.
83. Svenska Oljeslageraktiebolage (1946) Patent GB 652355.
84. Celanese (1955) Patent GB 809451.

85. Ooi, T.L., Mizukami, F., and Niwa, S. (1999) *J. Oil Palm Res.*, **11**, 53–61.
86. Swern, D., Knight, H.B., Scanlan, J.T., and Ault, W.C. (1945) *J. Am. Chem. Soc.*, **67**, 1132–1135.
87. Guo, C. and Li, Y. (2004) Patent CN 1629120, Hunan University.
88. Blair, R.G. (2010) Patent WO 2011046883, University of Central Florida Research Foundation.
89. Köckritz, A., Blumenstein, M., and Martin, A. (2009) *Eur. J. Lipid Sci. Technol.*, **112**, 58–63.
90. Kawamoto, K. and Yoshioka, T. (1980) Patent JP 21118, Mitsui Petrochemical Industry.
91. Yonehara, K. and Sumita, Y. (2004) Patent JP 2005325035, Nippon Shokubai.
92. Lemaire, M., Metay, E., Sutter, M., Debray, J., Raoul, Y., and Duguet, N. (2012) Patent FR 2994178, Societe Interoleagineuse d'Assistance et de Developpement, Université Claude Bernard Lyon, Centre National de la Recherche Scientifique, Novance.
93. Fujitani, K., Manami, H., Nakazawa, M., Oida, T., and Kawase, T. (2009) *J. Oleo Sci.*, **58**, 629–637.
94. Sapunov, V.N., Yu Litvintsev, I., Egorenkov, A.A., Vardanyan, V.D., Oganyan, N., Denisenkov, V., and Rostomyan, L. (1992) Patent SU 1766905, Mendeleyev Institute of Chemical Technology Moscow, Erevanskii Zavod Khimisheskich Reaktivov.
95. Oakley, M.A., Woodward, S., Coupland, K., Parker, D., and Temple-Heald, C. (1999) *J. Mol. Catal. A: Chem.*, **150**, 105–111.
96. Santacesaria, E., Ambrosio, M., Sorrentino, A., Tesser, R., and Di Serio, M. (2003) *Catal. Today*, **79–80**, 59–65.
97. Santacesaria, E., Sorrentino, A., Rainone, F., Di Serio, M., and Speranza, F. (2000) *Ind. Eng. Chem. Res.*, **39**, 2766–2771.
98. Chen, J., Song, H., and Tong, J. (2004) Patent CN 1746145, Lanzhou Chemical Physics Institute, Chinese Academy of Sciences.
99. Kulik, A., Janz, A., Pohl, M.-M., Martin, A., and Köckritz, A. (2012) *Eur. J. Lipid Sci. Technol.*, **114**, 1327–1332.
100. Janz, A., Köckritz, A., and Martin, A. (2011) Patent DE 102010002603, Leibniz Institute for Catalysis.
101. Kulik, A., Martin, A., Pohl, M.-M., Fischer, C., and Köckritz, A. (2014) *Green Chem.*, **16**, 1799–1806.
102. Nakazawa, M., Fujitani, T., and Sanami, H. (1984) Patent EP 0128484, New Japan Chemical.
103. Abraham, W., Kaido, H., Lee, C.W., Pederson, R.L., Schrodi, Y., Tupy, M.J., and Uptain, K.D. (2007) Patent WO 2008060383, Cargill, Materia.
104. Warwel, S. and Rüsch gen. Klaas, M. (1997) *Lipid Technol.*, **9**, 10–14.
105. Schörken, U. and Kempers, P. (2009) *Eur. J. Lipid Sci. Technol.*, **111**, 627–645.
106. Robbins, K.C. (1960) Patent US 3080296, Armor Pharmaceutical Company.
107. Weiss, A. (2007) in *Modern Biooxidation. Enzymes, Reactions and Applications* (eds R.D. Schmid and V.B. Urlacher), Wiley-VCH Verlag GmbH, Weinheim, pp. 193–209.
108. Huf, S., Krügener, S., Hirth, T., Rupp, S., and Zibek, S. (2011) *Eur. J. Lipid Sci. Technol.*, **113**, 548–561.
109. Cathay Industrial Biotech Ltd Dibasic Acids (Dicarboxylic Acids), http://www.cathaybiotech.com/en/products/dibasic (accessed 09 October 2014).
110. HILEAD Europe GmbH http://www.hilead-europe.com/index.php?id=5 (accessed 09 October 2014).
111. Anderson, K.W., Wenzel, J.D., Fayter, R.G., and McVay, K.R. (1998) Patent US 5962285, Henkel.
112. Kroha, K. (2004) *Inform*, **15**, 568–571.
113. Qiu, Y., Li, N., Hu, B., and Zhong, J. (2004) Patent CN 1570124, Shanghai Kaisai Biotechnology.
114. Akabori, S., Shiio, I., and Uchio, R. (1971) Patent DE 2140133, Ajinomoto.
115. Kaneyuki, H. and Ogata, K. (1973) Patent US 3912586, Mitsui Petrochemical Ind.
116. Ogata, K., Kaneyuki, H., Kato, N., Tani, Y., and Yamada, H. (1973) *Hakko Kogaku Zasshi*, **51**, 227–235.
117. Taoka, A., Uchida, S., and Urawa, S. (1978) Patent DE 2853847, Bio Research Center Tokio.

118. Liu, Z. and Yi, Z. (1992) *Weishengwu Xuebao*, **32**, 340–345.
119. Song, J.-W., Lee, J.-H., Bornscheuer, U.T., and Park, J.-B. (2014) *Adv. Synth. Catal.*, **356**, 1782–1788.
120. Song, J.-W., Jeon, E.-J., Song, D.-H., Jang, H.-Y., Bornscheuer, U.T., Oh, D.-K., and Park, J.-B. (2013) *Angew. Chem.*, **125**, 2594–2597.
121. Otte, K.B., Kittelberger, J., Kirtz, M., Nestl, B.M., and Hauer, B. (2014) *ChemCatChem*, **6**, 1003–1009.
122. Baek, H.D. and Kim, G.T. (2001) Patent KR 2001025411, Proco Biotech.
123. Blank, W., Takayanagi, H., Kido, T., Meussdoerffer, F., Esaki, N., and Soda, K. (1991) *Agric. Biol. Chem.*, **55**, 2651–2652.
124. Wumaijiang, E., Ayixiamui, N., Xiamuxikaramarl, M., and Halidan, M. (2007) Patent CN 101200735, Xinjiang Physical and Chemical Technology Institute, Chinese Academy of Sciences.
125. Matrica www.matrica.it (accessed 15 October 2014).
126. Foa, M., Gardano, A., and Sabarino, G.) (1992) Patent WO 9312064, Novamont.
127. Bastioli, C., Borsotti, G., Merlin, A., and Milizia, T. (2008) Patent WO 2008138892, Novamont.
128. Bastioli, C., Milizia, T., and Borsotti, G. (2006) Patent WO 2007039481, Novamont.
129. Gardano, A., Strologo, S., and Foa, M. (1994) Patent IT 1276314, Novamont.
130. Bieser, A., Borsotti, G., Digioia, F., Ferrari, A., and Pirocco, A. (2010) Patent WO 2011080297, Novamont.
131. Bieser, A., Borsotti, G., Digioia, F., Ferrari, A., and Pirocco, A. (2010) Patent WO 2011080296, Novamont.
132. Rubinstein, A., Jimenez-Lozanao, P., Carbo, J.J., Poblet, J.M., and Neumann, R. (2014) *J. Am. Chem. Soc.*, **136**, 10941–10948.
133. Wang, A. and Jiang, H. (2010) *J. Org. Chem.*, **75**, 2321–2326.
134. Dydio, P., Detz, R.J., and Reek, J.N.H. (2013) *J. Am. Chem. Soc.*, **135**, 10817–10828.
135. Dydio, P., Dzik, W.I., Lutz, M., de Bruin, B., and Reek, J.N.H. (2011) *Angew. Chem. Int. Ed.*, **50**, 396–400.

21
Oxidative Conversion of Renewable Feedstock: Carbohydrate Oxidation

Cristina Della Pina, Ermelinda Falletta, and Michele Rossi

21.1
Introduction

Carbohydrates are the most abundant resource for the conversion of renewable feedstocks in useful chemicals and energy. Approximately 180 billion tons of biomass is produced from photosynthesis each year, including about 180 million tons of edible sugars [1] and more than 1 billion tons of starch from grains [2].

Carbohydrates can be transformed into fuels such as ethanol and ethyl *tert*-butylether, but triglycerides represent a better energy source owing to the high enthalpy of their conversion and thus challenge petrol in diesel engines. The low enthalpy of carbohydrates does not suggest particular advantages in their direct utilization as a fuel. In recent applications, however, glucose has been successfully employed as a renewable feed for fuel cells [3].

On the other hand, the six-carbon unit of glucose is one of the most prominent components in biomasses, is easy to produce, and represents one of the most attractive building blocks for the preparation of a variety of chemical intermediates. Thus, glucose will play a central role in feeding the so-called biorefinery, a future plant where chemical and biochemical processes are advantageously employed for the synthesis of intermediates and fine chemicals as an alternative to fossil-derived chemicals (Figure 21.1).

Crystalline glucose and, preferably, its concentrated solution (syrup) are the starting materials for the production of several chemical intermediates, which represent the platform for other important chemicals derived from renewable biological sources according to standard chemical and biochemical processes. Figure 21.2 shows the important molecules whose synthesis requires at least one oxidative step. A variety of natural and genetically modified microorganisms allow the aerobic fermentation of glucose toward different products, whereas

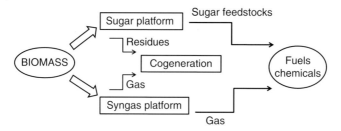

Figure 21.1 Biorefinery concept.

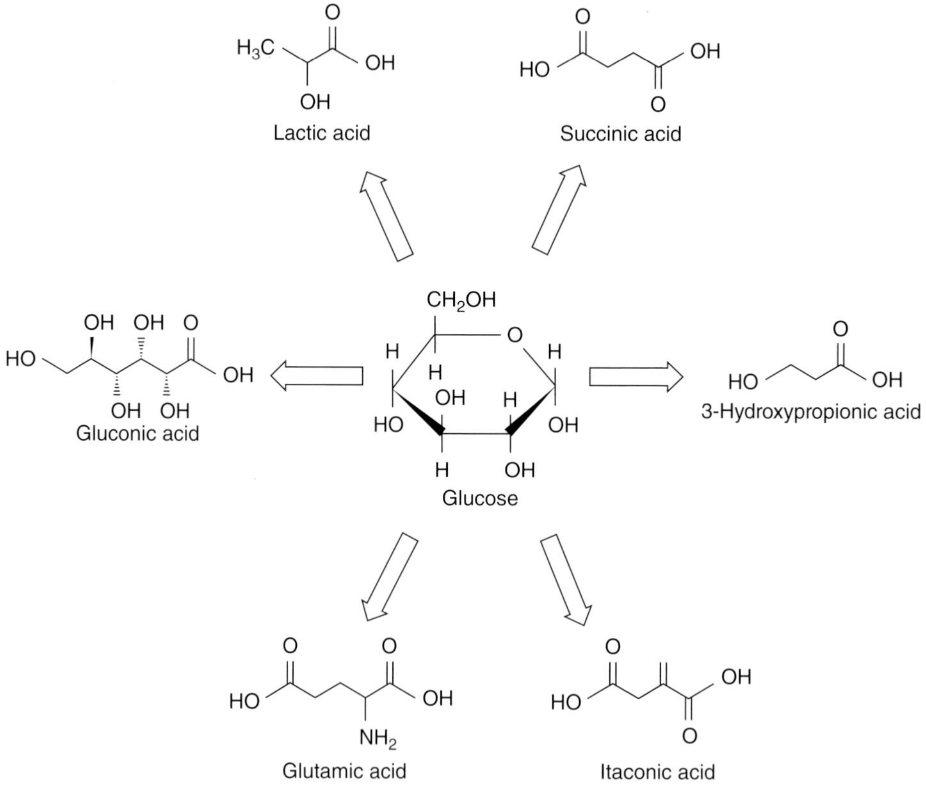

Figure 21.2 Products derived by aerobic oxidation of glucose.

only a limited number of synthetic catalysts have been proven to be active in glucose-selective oxidation [4, 5].

As mentioned earlier, glucose is a strategic material for technological applications including chemical processes and alternative energy resources. Therefore, industrial and academic research is deeply involved in optimizing current technologies as well as developing novel processes. A focus of studies is the oxidation of glucose to gluconic acid and gluconates, currently accomplished by the enzymatic process based on *Aspergillus niger* mold. Despite the positive aspects of this latter process (e.g., mild and environmentally acceptable "quasi-green" conditions), some drawbacks arise mainly from the limited glucose concentration tolerated by the enzyme and the toxic effects of *A. niger*, thus necessitating large plants and the deep purification from the enzyme [6].

This chapter focuses on the catalytic conversion of carbohydrates, in particular aerobic oxidation, with the aim of highlighting the potential of new emerging technologies.

21.2 Chemistry and Catalysis

All carbohydrates are polyhydroxyaldehydes, polyhydroxyketones, or molecules that yield polyhydroxy aldehydes or ketones on hydrolysis. Monosaccharides are the smallest carbohydrate molecules and include five- and six-carbon sugars, namely, pentoses and hexoses. Polysaccharides, such as starch and cellulose, lead to many monosaccharides upon hydrolysis.

Considering the complexity of even the simplest monosaccharides, the conversion of carbohydrates into valuable chemicals involves crucial aspects of chemo-, regio-, and stereoselectivity. Owing to environmental constraints, the use of stoichiometric reagents such as nitric acid, periodic acid, and chlorinating agents is limited to a few processes, as in the case of the production of the sweetener sucralose (Figure 21.3) [7].

Considering oxidations employing air or molecular oxygen as the oxidizing reagent, the reactivity of different chemical groups in carbohydrates can be predicted on the basis of thermodynamic and kinetic models. Under ambient conditions, monomeric molecules can be more easily oxidized than polymeric materials but, in any case, the kinetics are very slow, and most of the processes require catalysis. On the basis of reaction models underpinned by experimental data, the reactivity of carbohydrates is more predictable in the case of chemical catalysis than in the case of enzymatic catalysis.

Thus, several experiments have shown that the aldehyde reacts faster than the alcohols, and primary alcohols are more reactive than the secondary ones with gold catalysts [4]. Moreover, it has emerged that mild conditions, allowed

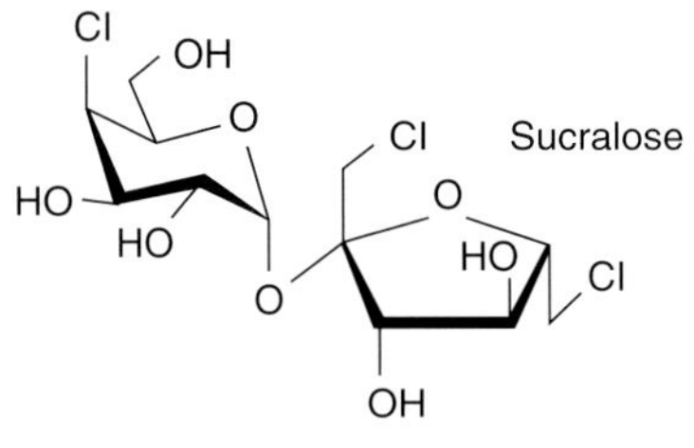

Figure 21.3 Sucralose.

by several inorganic catalysts, ensure the preservation of the chiral centers not involved in the chemical attack.

Molecular mechanisms, supported by experimental evidence, can be helpful for improving catalytic performance and designing enhanced catalysts.

For instance, the determination of the precise stoichiometry of glucose oxidation by gold catalysis has highlighted the two-electron reduction of oxygen to hydrogen peroxide, according to Eq. (21.1) [8]:

$$C_6H_{12}O_6 + O_2 + H_2O \rightarrow C_6H_{12}O_7 + H_2O_2 \qquad (21.1)$$

Considering the ability of basic conditions in favoring aldehyde hydration (Eq. (21.2)) [9],

$$RCHO + H_2O \rightarrow RCH(OH)_2 \qquad (21.2)$$

and the strong promoting effect of bases, the following mechanism has been proposed for glucose oxidation (Figure 21.4) [8]. According to this mechanism, oxygen added to glucose is derived from water instead of molecular oxygen.

In the case of palladium- and platinum-catalyzed oxidation of glucose, bismuth plays a crucial role as a cocatalyst, as discussed in the next section, which has been interpreted by the redox cycle depicted in Figure 21.5 [10].

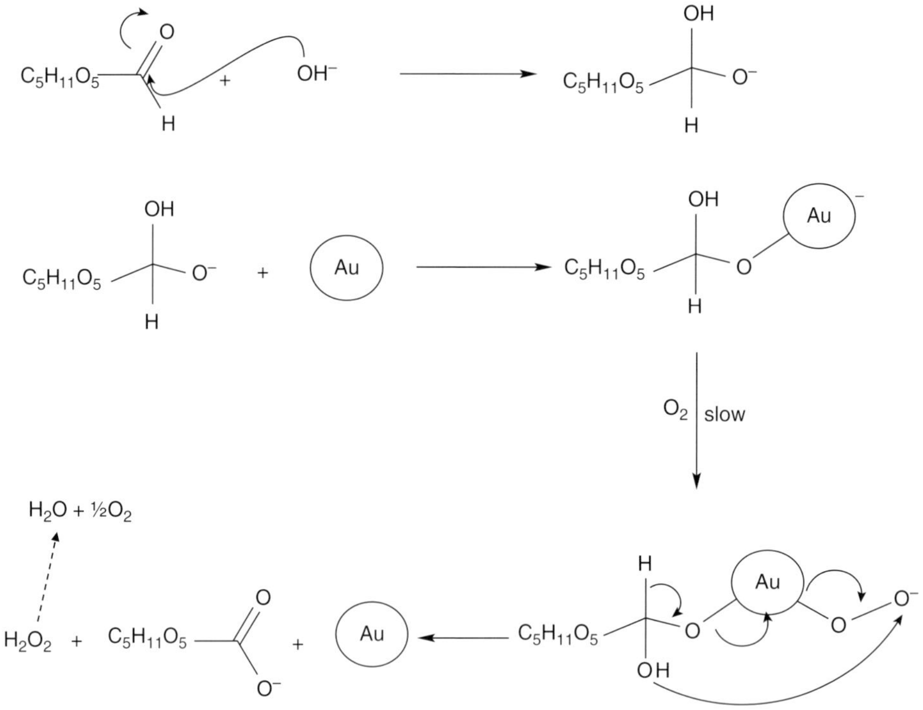

Figure 21.4 Molecular mechanism of glucose oxidation on gold catalyst in the presence of alkali.

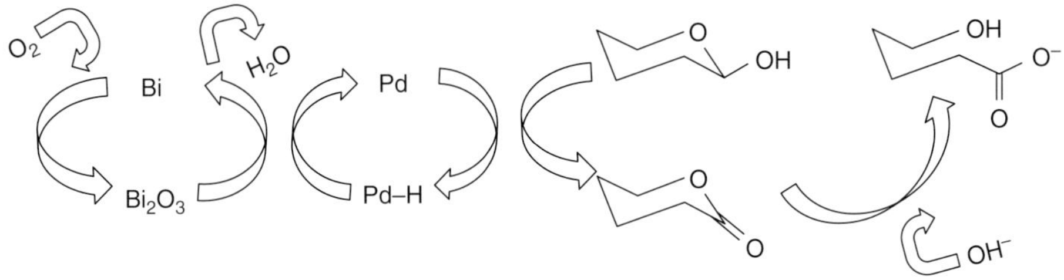

Figure 21.5 Mechanism of glucose oxidation on Bi–Pd catalyst.

Experimental kinetic investigations are useful tools for scaling up and optimizing industrial processes. In the case of carbohydrates, following the exciting application of gold catalysis, kinetic studies have been performed on glucose oxidation with different supported and unsupported gold catalysts. In the investigation by Önal et al. on the liquid phase oxidation of glucose using Au/C [11], a negligible effect of glucose concentration on the initial reaction rate has been found, and a Langmuir–Hinshelwood model has been proposed in which both glucose and oxygen are adsorbed on the catalyst.

The kinetic investigation of Beltrame et al. [12] is related to the aerobic oxidation of glucose using an unsupported gold catalyst, namely, colloidal gold, in order to avoid shielding effects of the support. This catalytic system, operating under quasi-homogeneous conditions, allowed a closer comparison with the homogeneous enzymatic catalysis investigated by the same academic group [13]. In the case of colloidal gold catalysis, an Eley–Rideal mechanism has been proposed, which is characterized by the adsorption of the hydrated form of glucose on gold. The rate equation has been interpreted using a mechanism where the rate-determining step is the oxidation of glucose by oxygen dissolved in the liquid phase. This justifies the observed first-order dependence on oxygen concentration. The derived apparent activation energy was 47 kJ/mol.

The kinetic investigation of Haruta's group in the presence of Au/Al_2O_3 catalyst revealed a strong effect of the initial glucose concentration on the reaction rate for low initial glucose concentrations. At higher concentrations, this effect was minimized, whereas with a limiting oxygen feed, a first order in oxygen was determined [14].

More recently, Prüße et al. employed a similar Au/Al_2O_3 catalyst [15] for carrying out an accurate analysis of the oxygen reaction order. They found a competitive adsorption of oxygen and glucose on the catalyst. Experimental data suggested a Langmuir–Hinshelwood model, and the derived activation energy was 54 kJ/mol.

The analytical data show that gold catalysis and enzymatic catalysis allow fast and selective aerobic oxidation of glucose according to the same stoichiometry characterized by the formation of hydrogen peroxide as the by-product (Eq. (21.1)) [8]. However, it is not surprising that completely different catalytic systems adopt different reaction mechanisms as shown by the kinetic studies on commercial enzymatic preparations containing *glucose oxidase* and *catalase* [13]. The results of the research support a Michaelis–Menten type mechanism where the kinetic parameters are limited by the reaction between glucose and oxidized enzyme. Although the reaction mechanisms are different, gold and enzymatic catalysis are characterized by close values of the activation energy (47 and 49.6 kJ/mol).

21.2.1
Oxidation of Monosaccharides

Along with other carbohydrates, the oxidation of monosaccharides by supported metal catalysts has been reviewed [16]. Glucose is the most investigated substrate for the synthesis of a variety of products and intermediates (Figure 21.6).

Both enzymatic catalysis and chemical catalysis can be applied under proper experimental conditions in order to prepare the desired product, and literature data may assist in making the best choice. According to the pioneering work of Heyns and Paulsen, the chemical oxidation of carbohydrates with finely divided platinum is first directed toward the anomeric carbon.

When only secondary hydroxyl groups are present in the cyclic system, the axial hydroxyl groups are oxidized in preference to the less reactive equatorial hydroxyl groups [17].

Laboratory-scale liquid phase oxidation with heterogeneous catalysis is generally performed under mild conditions by bubbling air or oxygen at 20–50 °C and atmospheric pressure in the presence of Pd, Pt, Au, or derived bimetallic catalysts [4].

Slightly alkaline conditions (pH 8–9.5) promote the oxidation reaction, favoring the formation of the gem-diol intermediate (Eq. (21.2)) and desorption of carboxylate from the catalyst (Figure 21.4). With Pt group metals, glucose can be completely converted with moderate selectivity. On recycling, however, the catalyst undergoes deactivation owing to self-poisoning and overoxidation. According to the invention at Degussa [18], modified mono- and bimetallic

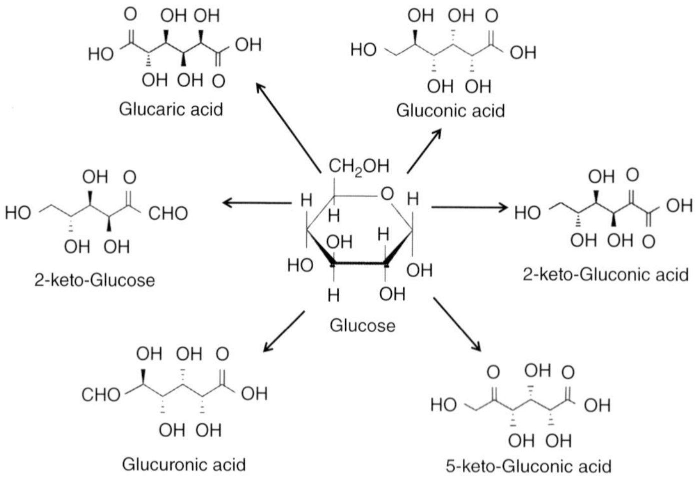

Figure 21.6 Main chemicals achieved via glucose oxidation.

catalysts have been disclosed to overcome this problem. A detailed procedure for preparing Bi-doped Pd/C catalyst containing 1 to 2 nm Pd particles via a surface redox reaction has been later described by Gallezot et al. [10]. According to comparative tests, a Bi-promoted catalyst allowed 99.3% yield with more than 99% selectivity, while an unpromoted Pd/C catalyst led to gluconate in 82.6% yield with 94.6% selectivity. In the latter experiment, overoxidation products were detected, such as 2-ketogluconate, 5-ketogluconate, and glucarate besides the isomerization product fructose. It is noteworthy that the catalyst can be recycled without loss of activity and selectivity.

Although gold has been traditionally considered catalytically inert, the "yellow metal" has recently been found to be unusually active in the liquid phase aerobic oxidation of organic compounds including glucose and other carbohydrates, under mild conditions [19–21]. The intrinsic activity of nanometric gold particles in glucose oxidation, expressed as turnover frequency (TOF = mole of oxidized substrate per mole of Au per hour), is superior to that of Pt and Pd, as derived by evaluation of the respective colloidal nanoparticles having similar size (Figure 21.7) [22].

Gold deposited on supports such as C, Al_2O_3, TiO_2, or CeO_2 was found to be very active and selective for the oxidation of many monosaccharides and disaccharides [20–22].

Unlike from palladium and platinum, gold does not require any promoter. However, with gold catalysts the activity is very sensitive to the size of the metal particles, with activity in the narrow range of 3–7 nm [22].

Due to thermodynamic and kinetic reasons, the aerobic conversion of glucose is generally carried out in alkaline solution, thus producing metal gluconates, which find application in the food and pharmaceutical industries sharing an over 60 000 t/year market. Figure 21.8 displays the principal products and their uses. Important manufacturers of gluconates are Benckiser, Bristol-Myers Squibb,

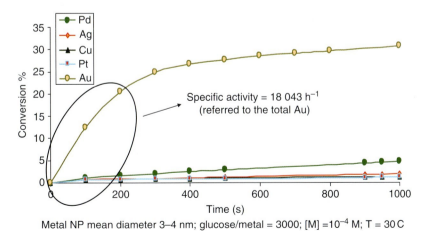

Figure 21.7 Comparison among Au, Pd, and Pt colloidal nanoparticles in glucose oxidation.

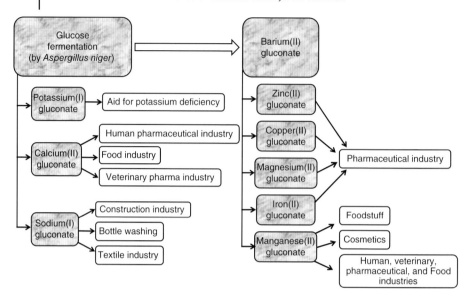

Figure 21.8 Commercial gluconates and uses.

Pfizer Inc., Premier Malt Products Inc., Roquette Frères, Fujisawa, and Kyowa Hakko.

No direct methods are known for the large-scale production of free gluconic acid, which is today manufactured from calcium gluconate and sulfuric acid. A major drawback of this process is the production of a large amount of the by-product calcium sulfate. A more sustainable synthesis of gluconic acid, which avoids the acidification step, could be of commercial interest owing to an attractive market based on its use in the alimentary industry, calcium scales, stain and rust removal, metal-cleaning solutions, tanning processes, and as a precursor for other specific gluconates.

During the attempts to overcome kinetic limitations in acidic solution, mono- and bimetallic catalysts (Au, Pt, Pd, and Rh) in the form of supported particles were tested in glucose oxidation in the absence of alkali. Whereas the activity of single metals was weak in the case of Au and Pt (TOF = 51 and 60 h^{-1}) and very low in the case of Rh and Pd (TOF < 2 h^{-1}), the bimetallic formulations were found to be enhanced by combining Au with Pd or Pt. In the latter case, high TOFs (924 h^{-1}) led to the production of free gluconic acid in high yield [23].

Figure 21.9 displays the synergistic effect produced by alloying Au and Pt metals.

In order to compare the kinetic results using morphologically similar catalytic systems, metal particles should be sized in a narrow range of diameters. This was realized by tailoring the particles' dimensions in the form of colloidal dispersions and depositing the resulting particles onto a selected activated carbon (see Section 21.3.3.1). As shown in Table 21.1, different supported catalysts contain very small metallic particles having similar size (2–5 nm). The progressive shift of

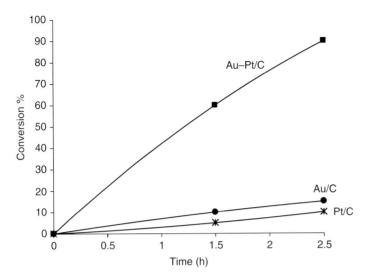

Figure 21.9 Comparison between supported mono- and bimetallic gold platinum catalysts.

Table 21.1 Mono- and bimetallic catalysts tested in glucose oxidation under acidic conditions.

Metals	$2\theta°$ angle	d (nm)
Rh	43.2	2.0
Pd	40.2	2.9
Pt	39.7	5.0
Au	38.1	3.4
Au/Pt = 0.25	39.1	3.0
Au/Pt = 0.5	39.0	3.2
Au/Pt = 1	38.9	3.3
Au/Pt = 2	38.6	3.8
Au/Pt = 4	38.4	4.8

$2\theta°$ angle in Au–Pt catalysts having different compositions indicates the presence of alloyed metals.

Besides the industrially relevant selective oxidation of glucose to gluconate, other products deriving from glucose oxidation could be of economic interest. According to Abbadi and Van Bekkum, gluconic acid and other aldonic acids can be further oxidized to the corresponding 2-keto acids over lead- or bismuth-modified platinum catalysts. This reaction is influenced by pH: fairly good selectivity to 2-ketogluconic acid could be obtained only below pH 6 [24].

On the other hand, Besson *et al.* reported that when a concentrated solution of glucose and a well-dispersed Pt catalyst was used, the oxidation can be forced to produce the dicarboxylic acid (glucaric acid). The yield of this process is, however, low (55%) [25].

21.2.2
Oxidation of Disaccharides

Glucose, fructose, and galactose are the building blocks for the most common disaccharides, namely, sucrose (glucose + fructose), lactose (glucose + galactose), and maltose (glucose + glucose) (Figure 21.10).

Among these three disaccharides, beet- and cane-derived sucrose displays the highest world production, reaching the amount of over 175 million metric tons per year, with Brazil (2014: 38 t/year) and India (2014: 27 t/year) being the leading producers. Compared with sucrose, lactose (1.2 million tons per year) and maltose (0.6 million tons per year) are produced in much smaller amounts.

Because of the wide availability of sucrose, oxidation of this disaccharide seems to be industrially attractive. However, aside from serving as a substrate for fermentation, very little sucrose is used as an industrial raw material. Nevertheless a wide range of products may be expected, as shown in Figure 21.11.

Studies on the aerobic oxidation of sucrose with noble metal catalysts revealed low selectivity and, surprisingly, the involvement of all of the three primary alcoholic functions at C_6, C_1', and C_6' while preserving the disaccharide backbone (Figure 21.12). This is not a problem, because mono- and polycarboxylated sugars are both economically attractive as biodegradable calcium sequestering agents;

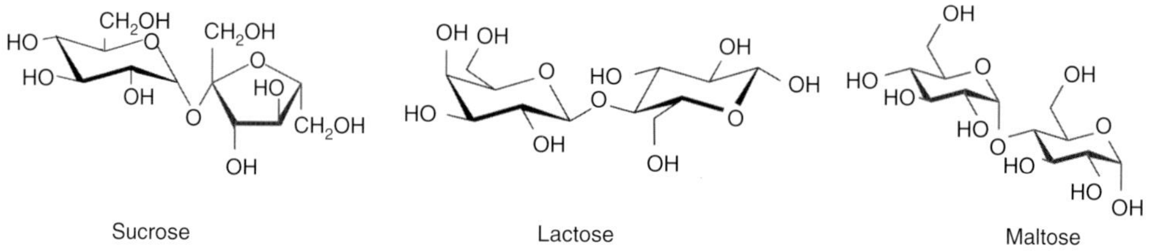

Figure 21.10 Structure of disaccharides (sucrose, lactose, and maltose).

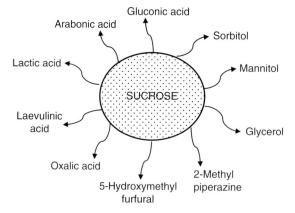

Figure 21.11 Products deriving from sucrose.

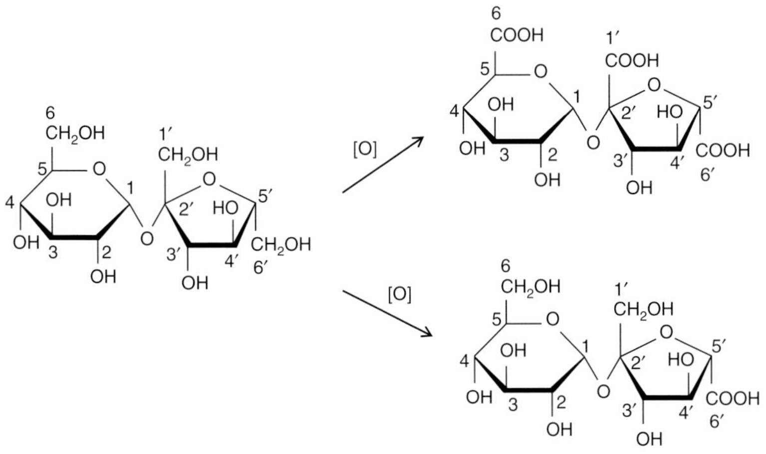

Figure 21.12 Tricarboxylated sucrose and one of the three possible monocarboxylated isomers.

hence, industrial research, mainly in Germany, has been involved in this application [26–28].

The attention at Hoechst [26, 27] was mainly focused on the preparation of tricarboxylated sucrose (Figure 21.12) using Pt/C catalysts. In order to push the reaction to oxidize multiple alcohols, the strategy was to use a programmed temperature ramp from room temperature to 70–80 °C during several hours, followed by adding alkali to pH 9. This procedure allowed the formation of a product rich in tricarboxylated sucrose and showing excellent properties such as low fabric encrustation during the use as detergent builder in laundering.

Differently from Hoechst, Zuckerindustrie Verein [28] was mainly interested in the preparation of monocarboxylated sucrose, thus adopting a different strategy, consisting of operating at low temperature with a limited oxygen concentration, while removing the product by electrodialysis. Accordingly, the oxidation of sucrose with 5% Pt/C catalyst at 35 °C using a 4 : 1 N_2/O_2 dilute gas mixture while buffering the pH at 6.5 by addition of $NaHCO_3$ produces a solution rich in monocarboxylic and bicarboxylic acids. After purification in electrodialysis cells, sucrose monocarboxylate could be obtained in high purity (98%).

Gold catalysis has been recently applied to the selective oxidation of cellobiose to gluconic acid over carbon nanotube-supported gold (Figure 21.13). According to Tan *et al.* [29], the aerobic oxidation afforded 68% yield at 81% cellobiose conversion, producing free gluconic acid as the main product. The key point of this process is the bifunctional catalyst containing acidic functional groups on carbon nanotubes, which catalyzes hydrolysis of the substrate, while gold nanoparticles ensure the oxidation of the monomeric sugar. Although this interesting example refers only to the oxidation of cellobiose, a similar catalytic behavior could be expected with other di and polysaccharides, making this discovery very attractive.

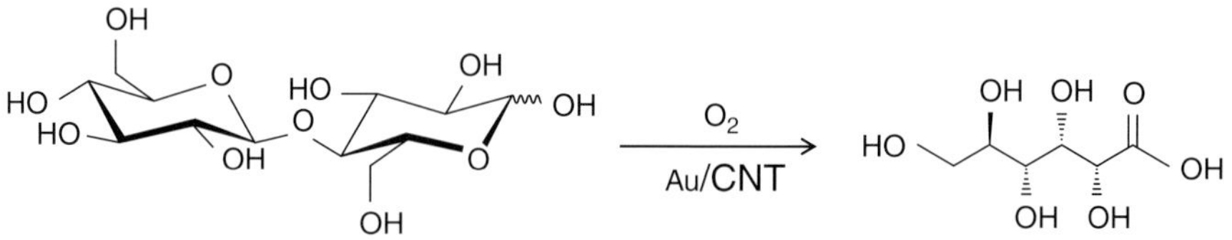

Figure 21.13 Gold-catalyzed oxidation of cellobiose.

Many other disaccharides can be successfully oxidized in the presence of Pt, Pd, and Au catalysts according to the exhaustive study of Mirescu and Prüße [21], describing a general procedure for preparing aldonic acids. The comparative oxidation of monosaccharides (arabinose, ribose, xylose, lyxose, mannose, rhamnose, glucose, galactose, N-acetyl-glucosamine) and disaccharides (lactose, maltose, cellobiose, melibiose) on Au, Pd, and Pt catalysts showed that the gold catalyst demonstrates complete selectivity toward aldonic acids and displays high catalytic activity for all the investigated sugars (Table 21.2).

A new elegant chemoenzymatic approach to sugar oxidation has been recently developed [30]. *Laccase*, a blue copper-protein oxidase from *Trametes pubescens*, and the chemical mediator TEMPO have been combined to catalyze the selective oxidation of the primary hydroxyl groups of complex sugars and their derivatives. A schematic picture of the reaction mechanism is shown in Figure 21.14. This system shows the ability to catalyze the selective monocarboxylation of disaccharides while preserving the sugar backbone. However, the poor regioselectivity in sucrose oxidation (possessing three nonequivalent primary hydroxyls) and raffinose oxidation (with four nonequivalent primary hydroxyls) results in the formation of a mixture of many monocarboxylated compounds which are difficult to separate.

Table 21.2 Selective oxidation of disaccharides on Au, Pd, and Pt catalysts.

Saccharide	0.45% Au/TiO$_2$		4.6% Pd/Al$_2$O$_3$		5% Pt/Al$_2$O$_3$	
	S.A. (m^2/g)	Selectivity (%)	S.A. (m^2/g)	Selectivity (%)	S.A. (m^2/g)	Selectivity (%)
Maltose	54	99.5	6	96	5	91
Cellobiose	50	99.5	1	99	5	87
Lactose	18	99.5	2	98	2	90
Melibiose	9	99.5	1	81	2	94

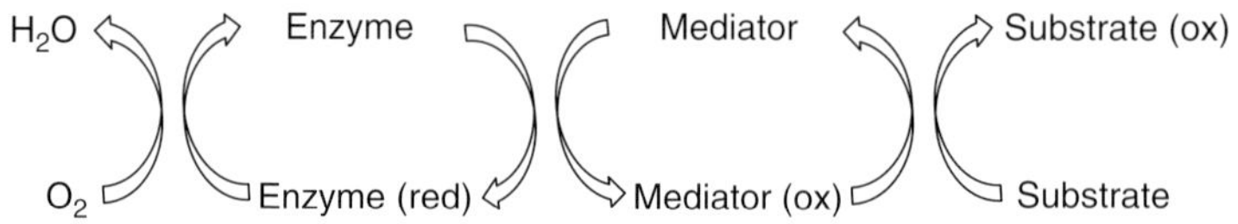

Figure 21.14 Chemoenzymatic oxidation of sugars.

More interestingly, the chemoenzymatic oxidation can be efficiently applied to complex molecules such as water-soluble cellulose, thus justifying further scientific interest and developments.

21.2.3
Polysaccharide Oxidation

Selective processes for useful transformations of polysaccharides are of strategic interest for food and chemical industries, considering the economic importance of such renewable resources [31–33].

From a structural point of view, polysaccharides may be oxidized according to a predictable pathway [4] (Figure 21.15).

In fact, in the presence of a free primary alcohol (e.g., at C_6 in cellulose or amylose), this chemical group undergoes a preferential oxidation to aldehyde or further toward carboxylic acid. The oxidation to the carboxylic acid level produces uronic acid units resembling the structure of natural polyuronic acids such as pectin and alginate.

The aerobic oxidation employing traditional supported catalysts, such as Pt/C, is quite slow in the case of insoluble substrates, owing to adsorption–desorption impediments, as reported in the research of Van Bekkum *et al.* [34]. While the monomeric unit of inulin in Me α-D-fructo-furanoside is promptly oxidized at the expected C_6 position, the rate of oxidation of inulin oligosaccharides decreases upon increase of the chain length of various substrates, thus leading to unsatisfactory conversions.

In principle, homogeneous catalysis can overcome this latter problem. This has been demonstrated by further research reporting that the homogeneous catalyst TEMPO could be used for the chemoselective oxidation of the primary alcohols in polysaccharides to give the corresponding polyuronic acids. As expected, this method is selective for primary alcohols [35].

The oxidation of different cellulosic materials by TEMPO catalysis has highlighted the importance of the substrate morphology [36]. Crystalline native celluloses were unaffected even after long reaction times, whereas amorphous celluloses were promptly oxidized at C_6.

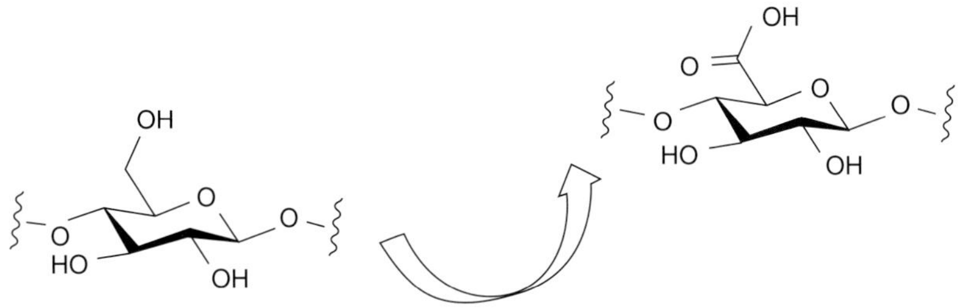

Figure 21.15 Oxidation pattern of polysaccharides.

The enzymatic oxidation of polysaccharides has been investigated for a long time mainly using the powerful *galactose-6-oxidase*. This enzyme is able to catalyze the C_6 oxidation of a galactose unit to an aldehyde in a variety of mono- and polysaccharides by using oxygen as the oxidant, generating hydrogen peroxide as the reduced by-product, as expected [32]. Unfortunately, this specific enzyme is inert toward other different sugar units.

It has been shown that the performance of *galactose-6-oxidase* can be enhanced when used in combination with two other enzymes, *catalase* and *peroxidase* [37].

21.3
Prospects for Scale-Up

Basic and applied research efforts show that the oxidative transformation of renewable feedstocks represents a promising approach for the production of high-value chemicals under environmentally acceptable conditions. Many laboratory-scale processes are carried out under mild conditions, and the employment of molecular oxygen ensures ecological and economic advantages. In these processes, a crucial role is played by the catalytic step that requires a specific selectivity toward the desired product. In principle, either enzymatic catalysis or chemical catalysis can be effectively employed in a given process and, in some cases, the performance of the different processes is quite similar as presented in the following sections.

21.3.1
Enzymatic Process *versus* Chemical Process: Glucose Oxidation as a Model Reaction

As reported earlier, the aerobic oxidation of glucose has been widely investigated either for commercial applications or as a model reaction. Notable progress has been achieved by using transition metal catalysts, and critical evaluation of these studies reveals that Pd–Pt–Bi-based catalysts and Au-based catalysts afford similar results in terms of activity and selectivity. However, considering its excellent catalytic properties and durability, gold seems to be today the metal of choice [20–22]. Thus, a gold-based chemical catalytic process can be proposed that is competitive with the biological oxidation, which is presently performed using *A. niger* mold-derived enzymes. In this context, the commercially available enzymatic extract of *A. niger* and a supported gold catalyst have been compared at Milan University [38]. For this scope, hyderase (from Amano Enzyme Co, UK), an enzymatic preparation containing glucose *oxidase* and *catalase* as active components and FAD (1.3×10^{-6} mol/g) as the rate-controlling factor, and the homemade catalyst 0.5% Au on carbon ($1200 \, m^2/g$) were used under the optimized conditions reported in Table 21.3.

Although the specific activity of the enzyme, with respect to the content of FAD, was shown to be higher compared to that of gold, the productivity obtained using a similar weight of catalyst was about fourfold higher in the case

Table 21.3 Comparison between enzymatic and chemical oxidation of glucose.

Catalyst	Glucose (mol/l)	Cat/glucose (g/kg)	T (°C)	pH	Stirring (rpm)	TOF (h^{-1})	Productivity (kg/m^3/h)
Hyderase	1	6	30	6	900	5.5×10^5	122
Au	3	5	50	9.5	39 000	1.5×10^5	514

of gold catalysis because of the low FAD concentration and the threefold higher concentration allowed by gold. A key parameter in these processes is the stirring speed, which controls the gas phase–liquid phase transfer of oxygen: the very high speed (39 000 rpm) tolerated by the gold catalyst allows a faster reaction rate, while the enzymatic catalyst suffers from degradation above 900 rpm.

21.3.2
Enzymatic Oxidation: Industrial Process and Prospects

Sodium gluconate is the most important commercial product deriving from the enzymatic oxidation of glucose. It is obtained when solutions of glucose are appropriately supplemented with other nutrients and subjected to fermentation with *A. niger*, a mold possessing many desirable characteristics for this purpose. With this organism, it is necessary to neutralize the fermenting medium basic compounds having sodium counterions. While information about detailed operations in industrial plants are difficult to collect, pilot plant procedures are available from the scientific literature. In particular, a fundamental paper of Blom *et al.* describes the optimization of the experimental parameters using a 1 m^3 stainless steel fermentor provided with a propeller-type blade, air sparger, and automatic control of temperature and pH [39]. The spore suspension to be used *as inoculum* in the pilot plant fermentation is separately prepared using spores of *A. niger* incubated at 30 °C for 7 days in the appropriate germination medium containing the conventional nutrients.

Pilot plant fermentation is carried out with 22% glucose concentration in the presence of several nutrients such as $MgSO_4$, KH_2PO_4, NH_2CONH_2, and $(NH_4)_2HPO_4$. When inocula of a good fermentation are used, no lag period is observed, and sugar oxidation occurs immediately at a high rate.

Agitation, fermentation pressure, and aeration are important factors governing the oxidation rate. An increase in any of these operating variables is accompanied by an increase in the rate of sugar utilization. However, foam formation, a typical problem of enzymatic catalysis, increases with agitation and aeration rate. Thus, more antifoam agent is required in the case of high reaction rate.

As a compromise, the following conditions have been adopted: aeration, 1.5 volumes per minute per volume of medium; agitation, 220 rpm; and fermentation pressure, 2 atm.

Sodium gluconate can be recovered by either crystallization or drying. For most applications, the less expensive drum drying process is adopted. In each case, the mycelium must be removed by accurate filtration. The sterilization of plant and reacting mixture, performed by thermal treatments at up to 135 °C, is a relevant problem of enzymatic catalysis encountered throughout the entire process.

As shown in the aforementioned description, fermentation is a very complex process, and it is often difficult to obtain full control of the process. Progress in traditional enzymatic oxidation is expected by basic studies concerning kinetic models for the traditional *A. Niger* oxidation [40, 41], as well as by process innovation. Progress in the biotechnological production of gluconic acid has been reviewed by Singh and Kumar [42], reporting on the present patent literature, which is proof of the strong interest in industrial applications. Short-time developments are expected in many different stages of glucose oxidation, as strain improvements in industrial microorganisms by mutagenesis and genetic engineering, modifications in microbial culture conditions, optimization in submerged and solid-state surface fermentation, immobilized cell fermentation, cell-free process, and product recovery.

21.3.3
Chemical Oxidation: Industrial Process and Prospects

According to several studies in the last 30 years, the chemical oxidation of glucose and other simple sugars is a challenging process, yet chemical methods can overtake the enzymatic process in terms of global efficiency. While the enzymatic process is today based on a single active species, *A. niger*, Pt, Pd, and Au nanometric particles in optimized environments share comparable performances that, together with other important aspects, could determine the choice of the catalyst for large-scale application.

21.3.3.1 Metal Catalysts: Concepts Guiding Choice and Design

Starting from the first application of supported Pd and Pt catalysts in glucose oxidation suffering from selectivity and leaching problems, new sophisticated Bi-promoted Pd and Bi-promoted Pt derivatives have been developed mainly at Gallezot's and van Bekkum's laboratories [10, 16, 34] and finally commercialized by Johnson Matthey. Liquid phase oxidation with air can match or surpass enzymatic processes, having comparatively high productivities (8 mol/h/g_{Pd}) for glucose oxidation on Pd–Bi catalysts [34]. These processes present the important advantage of the high simplicity of their operation ("one-pot" reaction). Although leaching of noble metals and low selectivity have been largely overcome by doping the catalyst with Bi, the leaching of Bi itself seems to be a serious problem.

For industrial applications, metal catalysts should be repeatedly recycled or used in continuous mode for a long time. The introduction of gold catalysis in the aerobic oxidation of glucose has opened exciting perspectives: Au is a biocompatible, nontoxic metal, which allows even superior productivities with respect to enzymatic catalysis [38], and no leaching problems have been observed using nanometric particles dispersed on different supports [43]. Compared with chemical oxidations, enzymatic catalysis suffers from more plant complexities including mold growth and sterilization sections contrasting the cheapness of the catalytic system.

Several methods are available for preparing supported metal catalysts. Commonly used methods include impregnation to incipient wetness, deposition–precipitation (DP), coprecipitation, and deposition/immobilization of colloidal metals. A useful description of such techniques is reported for gold catalysts [43].

In the impregnation method, a controlled amount of the metal solution is added up to fill the pore volume, and the resulting product is calcined in order to promote the thermal reduction of the metal. Owing to thermal stress, this method does not allow the formation of small particles (<10 nm). Recently, a modified method has been developed in which the resulting impregnated material is dried under mild conditions and then reduced with H_2 at a lower temperature. The resulting catalyst displayed improved activity, selectivity, and long-term stability in the liquid phase oxidation of glucose [44, 45].

In the DP procedure, the active metal is deposited on the surface of the support in the form of a hydroxide (or a hydrated oxide) by gradual pH increase of the mother solution. This procedure has been developed particularly for gold deposition starting from hydrogen tetrachloroaurate, which undergoes hydrolysis to yield various species ($[AuCl_4]^-$, $[AuCl_3(H_2O)]$, $[AuCl_3OH]^-$, $[AuCl_2(OH)_2]^-$, $[AuCl(OH)_3]^-$, $[Au(OH)_4]^-$), their composition depending on pH value and chloride concentration. Typical precipitating agents are either sodium hydroxide (DP NaOH) or urea (DP urea). Baatz et al. [44] showed that the DP method leads to quite active and long-term stable catalysts for glucose oxidation.

Coprecipitation is applied to support the noble metal to the in situ-prepared oxidic material, such as Fe_2O_3, Co_3O_4, NiO, Al_2O_3, and others. In the case of gold, coprecipitation with Fe_2O_3 can be performed by addition of sodium carbonate to a solution containing $HAuCl_4$ and $Fe(NO_3)_3$, followed by noble metal reduction [46].

Particular attention has been devoted to the immobilization of preformed gold colloidal dispersions (sol) on activated carbons and various oxides. This method allows a good nanoparticle size control and high metal dispersion. The colloid protecting agent (mainly polyvinylalcohol (PVA) and polyvinylpyrrolidone (PVP) but also polyhydroxylated compounds such as glucose itself), the reducing agent ($NaBH_4$), pH value, gold sol concentration, and total gold loading are important factors controlling the final particle size and catalytic performance [19, 20]. With this technique, the preparation of bimetallic systems (e.g., Au–Pd, Au–Pt, Au–Cu) can be easily performed, thus extending the availability of a wide series of catalysts [23].

The nature of the supporting material can determine the final catalytic performance. Interestingly, gold catalysts supported on various carbons displayed differences even in the same reaction, which implies that the catalytic performance was affected not only by the nature of the carbon but also by the preparation method [47]. One of the most relevant parameters for the choice of the support seems to be the type and distribution of surface groups, as these can favor or inhibit the grafting of the nanoparticles. Hence, acidic or basic treatments of supporting materials are often required in order to "activate" the surface before depositing the metal, to optimize catalytic performances in terms of activity, selectivity, and durability. In conclusion, the choice of a catalytic system candidate for the chemical aerobic oxidation of carbohydrates, supported by several laboratory tests and demonstration plants, seems to converge toward gold catalysis. However, besides scientific considerations, commercial constraints, such as patent rights, could force the use of other metals, such as palladium and platinum.

21.4
Concluding Remarks and Perspectives

Today, the oxidation of carbohydrates represents an industrially undervalued technology with respect to the potential as demonstrated by academic and industrial achievements. In fact, the global market of organic acids produced by fermentation amounts to 4 billion in 2013 and represents the third largest category, after antibiotics and amino acids, in the global market of fermentation [48]. Citric acid dominates the market of organic acids due to its application in various fields, while the market of gluconic acid is comparatively much smaller. However, besides the annual production of 60 000 t, the present technology could expand the success of organic acids to a variety of other aldonic acids derived from inexpensive natural sources.

Most processes for oxidation reactions require a catalytic stage, which represents often the critical step in which either biological or chemical oxidation can be applied. Owing to the simple, one-step synthesis commonly allowed by chemical catalysis, this latter process seems to be preferable in the case of similar space-time yields and economic evaluation.

However, as the discussed aerobic oxidation of glucose teaches, a deep investigation on different alternatives must be performed, and the choice should be made after a critical comparison: until few decades ago, noble metal catalysis was scarcely considered due to the cost of the catalyst and many prejudices.

A point of strategic importance is represented by the transfer of information between basic research and applied technology. In fact, progress in selectivity and activity of a catalytic system could suddenly make possible a commercial application. While innovation proceeds rapidly in academia, rapid application of these innovations is also rare because of the poor dissemination of information. In some cases, however, a synergistic collaboration occurs, as witnessed by Prüße's group (German Institute of Agriculture Technology) and Südzucker, whose project on the previously discussed sugar oxidation by gold catalysis has afforded a demonstration plant operating since 2010 [49].

Other interesting targets for applications of catalytic oxidation of sugars may include the synthesis of free gluconic acid [23] and antiscaling polycarboxylated sugar [26–28].

Long-term perspectives for extending the market of sugars include fuel cell energy production [3]. However, in the absence of electrochemical processes efficiently oxidizing glucose to CO_2 and H_2O, the mentioned processes based on glucose/gluconate appear today to be limited to a mere fundamental interest.

References

1. United States Department of Agriculture (November 2011) Sugar: World Markets and Trade.
2. White, J.W. and McGrew, W. (2012) in *Renewable Resources and Renewable Energy: A Global Challenge*, 2nd edn (eds P. Fornasiero and M. Graziani), CRC Press, Boca Raton, FL, p. 3.
3. Pasta, M., La Mantia, F., Ruffo, R., Peri, F., Della Pina, C., and Mari, C.M. (2011) *J. Power Sources*, **196**, 1273–1278.
4. Della Pina, C., Falletta, E., Prati, L., and Rossi, M. (2008) *Chem. Soc. Rev.*, **37**, 2077–2095.
5. Corma, A., Iborra, S., and Velty, A. (2007) *Chem. Rev.*, **107**, 2411–2502.
6. Blumenthal, C.Z. (2004) *Regul. Toxicol. Pharm.*, **39**, 214–228.
7. http://cheminfo2011.wikispaces.com/Yiwei+Wang+Final (accessed 27 January 2016).
8. Comotti, M., Della Pina, C., Falletta, E., and Rossi, M. (2006) *Adv. Synth. Catal.*, **348**, 313–316.
9. Allinger, N.L., Cava, M.P., De Jongh, D.C., Johnson, C.R., Lebel, N.A., and Stevens, C.L. (1971) *Organic Chemistry*, Worth Publishers, Inc., New York, p. 475.
10. Besson, M., Lahmer, F., Gallezot, P., Fuertes, P.G., and Fleche, G. (1995) *J. Catal.*, **152**, 116–121.
11. Önal, Y., Schimpf, S., and Claus, P. (2004) *J. Catal.*, **223**, 122–133.
12. Beltrame, P., Comotti, M., Della Pina, C., and Rossi, M. (2006) *Appl. Catal., A*, **297**, 1–7.
13. Beltrame, P., Comotti, M., Della Pina, C., and Rossi, M. (2004) *J. Catal.*, **228**, 282–287.
14. Okatsu, H., Kinoshita, N., Akita, T., Ishida, T., and Haruta, M. (2009) *Appl. Catal., A*, **369**, 8–14.
15. Prüße, U., Herrmann, M., Baatz, C., and Decker, N. (2011) *Appl. Catal., A*, **406**, 89–93.
16. Besson, M. and Gallezot, P. (2000) *Catal. Today*, **57**, 127–141.
17. Heyns, K. and Paulsen, H. (1962) *Adv. Carbohydr. Chem.*, **17**, 169–221.
18. Deller, K., Krause, H., Peldszus, E., and Despeyroux, B. (1989) Catalytic oxidation of glucose to gluconic acid. Patent DE 3823301.
19. Prati, L. and Rossi, M. (1998) *J. Catal.*, **176**, 552–560.
20. Biella, S., Prati, L., and Rossi, M. (2002) *J. Catal.*, **206**, 242–247.
21. Mirescu, A. and Prusse, U. (2007) *Appl. Catal., B*, **70**, 644–652.
22. Comotti, M., Della Pina, C., Matarrese, R., and Rossi, M. (2004) *Angew. Chem. Int. Ed.*, **43**, 5812–5815.
23. Comotti, M., Della Pina, C., and Rossi, M. (2006) *J. Mol. Catal. A: Chem.*, **251**, 89–92.
24. Abbadi, A. and Van Bekkum, H. (1995) *Appl. Catal., A*, **124**, 409–417.
25. Besson, M., Fleche, G., Fuertes, P., Gallezot, P., and Lahmer, F. (1996) *Recl. Trav. Chim. Pays-Bas*, **115**, 217–221.
26. Fritsche-Lang, W., Leupold, E.I., and Schlingmann, M. (1987) Preparation of sucrose tricarboxylic acid. Patent DE 3535720.
27. Leupold, E.I., Schoenwaelder, K.H., Fritsche-Lang, W., Schlingmann Linkies, M.A., Heinz, G., Werner, D., and Franz, J. (1990) Preparation of sucrose oxidation product containing tricarboxy derivative for use in detergents. Patent DE 3900677.
28. Kunz, M., Puke, H., Recker, C., Scheiwe, L., and Kowalczyk, J. (1994) Process and apparatus for preparation of of monoxidized products from carbohydrates, carbohydrates derivatives and primari alcohols. Patent DE 4307388.
29. Tan, X., Deng, W., Liu, M., Zhang, Q., and Wang, Y. (2009) *Chem. Commun.*, 7179–7181.
30. Marzorati, M., Danieli, B., Haltrich, D., and Riva, S. (2005) *Green Chem.*, **7**, 310–315.
31. Yalpani, M. (1985) *Tetrahedron*, **41** (15), 2957–3020.

32. Van de Vyver, S., Geboers, J., and Jacobs, P.A. (2011) *ChemCatChem*, **3**, 82–94.
33. Cumpstey, I. (2013) *ISRN Org. Chem.*, **2013**, 417672–417700.
34. Verraest, D.L., Peters, J.A., and van Bekkum, H. (1998) *Carbohydr. Res.*, **306**, 197–203.
35. De Nooy, A.E.J., Besemer, A.C., and van Bekkum, H. (1994) *Recl. Trav. Chim. Pays-Bas*, **113**, 165–166.
36. Isogai, A. and Kato, Y. (1998) *Cellulose*, **5**, 153–164.
37. Parikka, K. and Tenkanen, M. (2009) *Carbohydr. Res.*, **344**, 14–20.
38. Comotti, M., Della Pina, C., Falletta, E., and Rossi, M. (2006) *J. Catal.*, **244**, 122–125.
39. Blom, R.H., Pfeifer, V.F., Moyer, A.J., Traufler, D.H., Conway, H.F., Crocker, C.K., Farison, R.E., and Hannibal, D.V. (1952) *Ind. Eng. Chem.*, **44**, 435–440.
40. Takamatsu, T., Shioya, S., and Furuya, T. (1981) *J. Chem. Technol. Biotechnol.*, **31**, 697–704.
41. Jian-Zhong, L., Li-Ping, W., Qian-Ling, Z., Hong, X., and Liang-Nian, J. (2003) *Biochem. Eng. J.*, **14**, 137–141.
42. Singh, O.V. and Kumar, R. (2007) *Appl. Microbiol. Biotechnol.*, **75**, 713–722.
43. Bond, G.C., Louis, C., and Thompson, D.T. (2006) in *Catalysis by Gold*, Catalytic Science Series, vol. **6** (ed G.J. Hutchings), Imperial College Press, p. 72.
44. Baatz, C., Decker, N., and Prüße, U. (2008) *J. Catal.*, **258**, 165–169.
45. Baatz, C. and Prüße, U. (2007) *J. Catal.*, **249**, 34–40.
46. Haruta, M., Kageyama, H., Kamijio, N., Kobayashi, T., and Delannay, F. (1988) *Stud. Surf. Sci. Catal.*, **44**, 33–42.
47. Prati, L. and Porta, F. (2005) *Appl. Catal., A*, **291**, 199–203.
48. BCC Research LLC http://www.bccresearch.com/market-research/food-and-beverage/fermentation-ingredients-fod020c.html (accessed 27 January 2016).
49. Thünen http://www.ti.bund.de/en/startseite/institutes/agricultural-technology/research-areas/renewables-conversion-technology/closed-projects/oxidation-of-carbohydrates.html (accessed 27 January 2016).

Part VII
Aerobic Oxidation with Singlet Oxygen

22
Industrial Prospects for the Chemical and Photochemical Singlet Oxygenation of Organic Compounds

Véronique Nardello-Rataj, Paul L. Alsters, and Jean-Marie Aubry

22.1
Introduction

Oxygen in the singlet excited state $^1\Delta_g$, subsequently noted $^1O_2$, was detected for the first time in the upper atmosphere. However, it has been regarded as an important reactive oxygen species (ROS) by chemists only after its fortuitous detection in 1963 by Kasha and Khan [1] of its formation through the oxidation of hydrogen peroxide by sodium hypochlorite (Eq. (22.1)):

$$H_2O_2 + ClO^- \longrightarrow O_2 + H_2O + Cl^- \qquad (22.1)$$

Using this reaction, Foote demonstrated the following year that $^1O_2$ was the reactive species involved in the photooxidation of terpenes and polycyclic aromatic hydrocarbons studied by Schenck [2] and Dufraisse [3], respectively, many years ago. Since then, photochemically generated $^1O_2$ has been widely used in biology and organic synthesis on a laboratory scale to selectively oxidize a wide variety of electron-rich organic substrates including olefins [4–6], aromatic hydrocarbons [7, 8], phenols [9], amines [10], sulfides [11], and various heterocycles [12] such as furans [13, 14] or indoles [15]. Although the number of scientific works published on singlet oxygen grows exponentially, the number of patents is much smaller and increases only linearly.

Actually, despite its selectivity and its versatility, photooxidation is problematic on an industrial scale because it requires a specifically designed photoreactor. H_2O_2/ClO^- is not suitable as a nonphotochemical source of $^1O_2$ for organic synthesis since most of the short-lived $^1O_2$ is released as bubbles, and ClO^- itself induces side reactions with organic substrates. However, this chemical source of gaseous $^1O_2$ has found a practical application in feeding chemically pumped iodine lasers developed in the "Star Wars" program in the 1980s [16].

Liquid Phase Aerobic Oxidation Catalysis: Industrial Applications and Academic Perspectives,
First Edition. Edited by Shannon S. Stahl and Paul L. Alsters.
© 2016 Wiley-VCH Verlag GmbH & Co. KGaA. Published 2016 by Wiley-VCH Verlag GmbH & Co. KGaA.

Since then, there has been increasing interest in the role of $^1O_2$ in biology since it was discovered that several enzymatic and photochemical processes generate this ROS *in vivo*. In particular, it was demonstrated that $^1O_2$ is formed during the so-called photodynamic effect, that is, the simultaneous effect of oxygen, dye, and visible light on a biological target. This finding allows cancerous tumors to be destroyed through the design of photosensitizers capable of selective adsorption on cancer cells and irradiation with a red laser beam [17]. Meanwhile, many other chemical sources of $^1O_2$ were discovered [18–20]. In particular, a series of water-soluble naphthalene endoperoxides were designed to mimic the photodynamic effect (Eq. (22.2)) [7]:

$$\text{naphthalene endoperoxide} \underset{5\,°C}{\overset{37\,°C}{\rightleftarrows}} \text{naphthalene} + {}^1O_2 \qquad (22.2)$$

The incubation at 37 °C of such endoperoxides in a biological system triggers the release of precise amounts of pure $^1O_2$ free from any other ROS, which are inevitably formed besides $^1O_2$ by the current photochemical method. Although very useful to elucidate the exact role of $^1O_2$ with respect to biological targets [21], these $^1O_2$ carriers have no practical interest in organic synthesis since they only release a single molecule of $^1O_2$ per molecule of endoperoxide and require a "primary" source of $^1O_2$ to peroxidize the naphthalenic carriers (Eq. (22.2)). In 1982, we discovered a series of new chemical $^1O_2$ sources based on the decomposition of H_2O_2 catalyzed by oxoanions or metal hydroxides [18] such as MoO_4^{2-} [22], WO_4^{2-} [23], $Ca(OH)_2$ [24, 25], or $La(OH)_3$ [26–28]. In particular, we proved that MoO_4^{2-} is able to induce H_2O_2 disproportionation giving 100% of $^1O_2$ and two molecules of water (Eq. (22.3)) [29]. This reaction allows the formation of huge amounts of $^1O_2$ in the absence of light source. Therefore, it offers a valuable alternative to the conventional photochemical method:

$$2H_2O_2 \xrightarrow[\text{water}]{MoO_4^{2-}} {}^1O_2 + 2H_2O\,(100\%) \qquad (22.3)$$

In this chapter, we first describe the peculiar electronic structure of $^1O_2$ and its impact on its chemical reactivity that is opposite from that of ordinary oxygen $^3O_2$. Then, we compare the respective advantages and limitations of photochemical and chemical methods to generate $^1O_2$ in a context of industrial development. In particular, we detail the criteria for choosing a reaction medium compatible with both the organic substrate and water-soluble chemical sources of $^1O_2$. Finally, the main reactions of $^1O_2$ in organic chemistry are listed and illustrated with two industrially relevant examples recently developed in the fields of perfumery (synthesis of rose oxide) and pharmacy (synthesis of artemisinin).

22.2
Chemistry and Catalysis

22.2.1
Comparison of Singlet and Triplet Oxygen

Molecular dioxygen in its ground state exhibits an atypical electronic structure that explains its main features. For most molecules, the HOMO consists of a single orbital filled with two electrons of opposite spins. Electronic excitation of such molecules can be achieved by promoting an electron from (usually) the HOMO to an unoccupied level of higher energy. In general, this electron transfer requires significant energy attainable either photochemically with UV radiation or chemically through the decomposition of very energetic molecules (e.g., dioxetanes). Ground-state N_2 exhibits such a typical electronic structure with a σ_{2p} HOMO (Figure 22.1a).

In contrast, the HOMOs of ground-state molecular oxygen consist of two degenerate orbitals filled with only two electrons. Each of them lies in one orbital with parallel spins in accordance with Hund's rule leading to a triplet state, $^3O_2$, $^3\Sigma_g^-$ (Figure 22.1b). To obtain the lowest excited states, it is no longer necessary to promote an electron up to a higher empty orbital. It suffices to change the distribution of electrons within the degenerate Π^* orbitals. The first excited state is a singlet state ($^1O_2, ^1\Delta_g$), since it has two electrons with opposite spins in the same Π^* orbitals (Figure 22.1c). There is another excited state of higher energy, which is also in a singlet state ($^1O_2, ^1\Sigma_g^+$) since it has one electron in each Π^* orbital with opposite spins (Figure 22.1d). However, this species decays rapidly (10 ps) giving $^1O_2, ^1\Delta_g$ and does not play a significant role in the oxidation processes operating in condensed media. Thereafter, the symbols $^1O_2$ and $^3O_2$ will always refer to the lowest excited state $^1\Delta_g$ and to the ground-state $^3\Sigma_g^-$ of oxygen, respectively.

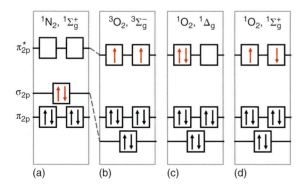

Figure 22.1 Higher occupied molecular orbitals of (a) ground-state nitrogen ($^1N_2, ^1\Sigma_g^+$), (b) ground-state oxygen ($^3O_2, ^3\Sigma_g^-$), and the (c) and (d) two lowest excited singlet states of oxygen ($^1O_2, ^1\Delta_g$) and ($^1O_2, ^1\Sigma_g^+$).

This rough description of the electronic structure is sufficient to rationalize the main physicochemical properties of $^3O_2$ and $^1O_2$. Ground-state $^3O_2$ is a biradical that reacts very rapidly with free radicals and triplet excited molecules, whereas it reacts slowly with usually singlet organic molecules despite its high redox potential ($E° = 1.23$ V/NHE). This behavior relies on the conservation rule of the spin quantum number (Wigner's rule), which states that for both radiative and non-radiative transitions, the transitions between species of the same multiplicity are spin allowed, while transitions between terms of different multiplicity are spin-forbidden. Therefore, the direct addition of $^3O_2$ to an organic molecule R in a singlet ground state giving a peroxide RO_2 in a singlet state violates the Wigner's spin conservation rule (Eq. (22.4)). In contrast, $^3O_2$ readily reacts with a doublet state molecule, as illustrated by the formation of peroxyl radical $RO_2^\bullet$ from $R^\bullet$ (Eq. (22.5)), or with the triplet excited state of a photosensitizer ^{3}Sens, giving $^1O_2$ and the sensitizer in the singlet ground state (Eq. (22.6)). Unlike $^3O_2$, $^1O_2$ is very reactive with many electron-rich substrates in a singlet ground state according to a spin-allowed process (Eq. (22.7)):

$$^3O_2(\uparrow\uparrow) + {}^1R(\uparrow\downarrow) \not\rightarrow {}^1RO_2(\uparrow\downarrow) \tag{22.4}$$

$$^3O_2(\uparrow\uparrow) + {}^2R^\bullet(\downarrow) \rightarrow {}^2RO_2^\bullet(\uparrow) \tag{22.5}$$

$$^3O_2(\uparrow\uparrow) + {}^3\text{Sens}(\downarrow\downarrow) \rightarrow {}^1O_2(\uparrow\downarrow)^1 + {}^1\text{Sens}(\uparrow\downarrow) \tag{22.6}$$

$$^1O_2(\uparrow\downarrow) + {}^1R(\uparrow\downarrow) \rightarrow {}^1RO_2(\uparrow\downarrow) \tag{22.7}$$

$$^3O_2(\uparrow\uparrow) + h\nu \not\rightarrow {}^1O_2(\uparrow\downarrow) \text{ spin forbidden} \tag{22.8}$$

For similar reasons, the direct electronic transition between $^3O_2$ and $^1O_2$ is spin forbidden (low probability) in both directions (Eq. (22.8)). This feature has two main consequences:

- Direct excitation of $^3O_2$ into $^1O_2$ is very ineffective since the extinction coefficient corresponding to the maximum IR absorption of $^3O_2$ is 7–8 orders lower than that of a photosensitizer. Therefore, an effective photochemical production of $^1O_2$ requires a photosensitizer able to give high yield of the triplet excited state ^{3}Sens to transfer its energy to $^3O_2$ according to Eq. (22.6).
- The reverse process, that is, the radiative decay of $^1O_2$, is also highly forbidden. The lifetime of $^1O_2$ is, therefore, exceptionally long since it varies from 45 min in vacuum to a few micro- or milliseconds in organic solvents.

Finally, the simple rearrangement of the distribution of electron spins accompanying the excitation of $^3O_2$ into $^1O_2$ requires only 22.5 kcal (94 kJ). Such a low energy is readily available either photochemically with visible or even IR radiations or chemically with moderately exergonic reactions. Actually, the spin conservation rule makes the formation of $^1O_2$ by chemical excitation even more likely. When a ground-state molecule AO_2 suffers a thermal cleavage into O_2 and A, Wigner's rule foresees the conservation of the total spin and thus oxygen should be released in a singlet state (Eqs. (22.9) and (22.10)). Nevertheless, this rule may be circumvented through the formation of an intermediate able to undergo an intersystem crossing (ISC) [30]:

$$^1AO_2(\uparrow\downarrow) \longrightarrow {}^1O_2(\uparrow\downarrow) + {}^1A(\uparrow\downarrow) \text{ spin allowed} \qquad (22.9)$$

$$^1AO_2(\uparrow\downarrow) \longrightarrow {}^3O_2(\uparrow\uparrow) + {}^1A(\uparrow\downarrow) \text{ spin forbidden} \qquad (22.10)$$

22.2.2
Photochemical Generation of $^1O_2$

The direct excitation of $^3O_2$ with 1270 nm IR radiation provides $^1O_2$ with a very low efficiency (Eq. (22.8)). Though not relevant for organic synthesis, it can be extremely useful for mechanistic studies because it specifically generates $^1O_2$ and requires no additional compound (photosensitizer or chemical precursors of $^1O_2$) that may induce side reactions. Recently, powerful lasers emitting at precisely this IR wavelength have been developed to generate known amounts of $^1O_2$ in biological media [31].

The most common method for generating $^1O_2$ is photosensitization of $^3O_2$ in the presence of a suitable dye ^{1}Sens. It must be able to undergo efficient ISC from the first excited singlet state ^{1}Sens* to the first excited triplet state ^{3}Sens*. The latter then transfers its energy $^3O_2$ according to a spin-allowed process giving $^1O_2$ (Eq. (22.6) and Figure 22.2). This process, named type II photooxidation, is widely used at the lab scale to oxidize selectively a wide range of organic substrates. Alternatively, ^{3}Sens* may follow a radical pathway called *type I photooxidation*. In that

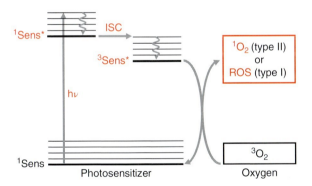

Figure 22.2 Jablonski diagram showing the excitation of a photosensitizer ^{1}Sens into the excited singlet state ^{1}Sens*, which suffers intersystem crossing (ISC) giving excited triplet state ^{3}Sens*. The latter may transfer its energy to ground-state oxygen $^3O_2$ providing $^1O_2$ (type II photooxidation). Alternatively, it may abstract one electron or a H atom from the substrate releasing organic free radicals as well as various reactive oxygen species (ROS) such as $O_2^{\bullet-}$, $OH^\bullet$, or H_2O_2 (type I photooxidation).

case, it abstracts an electron or a hydrogen atom from the organic substrate generating organic free radicals as well as various ROS such as superoxide anion $O_2^{\bullet-}$, hydrogen peroxide H_2O_2, and hydroxyl radical $OH^{\bullet}$.

The photosensitized generation of $^1O_2$ is very versatile because it can be conducted in a wide range of polar or nonpolar solvents and at low temperature if unstable oxidized products have to be prepared. However, it requires dedicated photochemical reactors, which are not always available in research laboratories or industrial plants. The choice of a suitable sensitizer Sens is very important to ensure effective photosensitization. It should absorb most of the incident radiation and must undergo ISC in high yield to populate effectively the triplet excited state ^{3}Sens*. The lifetime of ^{3}Sens* must be sufficiently long to allow its interaction with dissolved oxygen. The absorption band of Sens must be in a spectral region where the substrate does not absorb to avoid unwanted side reactions. Finally, it should not generate other ROS than $^1O_2$ and must be inert to this species. A wide variety of compounds have been used as sensitizers for the photooxidation, but the most commonly used are organic dyes such as rose bengal, methylene blue, or tetraphenylporphyrin (Scheme 22.1).

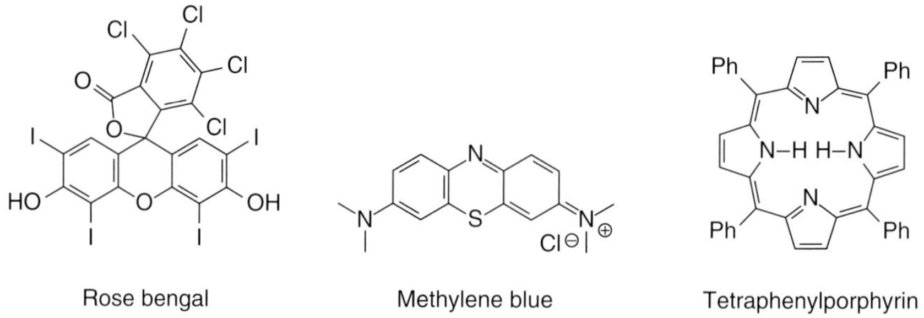

Scheme 22.1 Main photosensitizers used to generate photochemically $^1O_2$.

22.2.3
Chemicals Sources of $^1O_2$ Based on the Catalytic Disproportionation of H_2O_2

Because of the particular electronic structure of $^1O_2$ and its low excitation energy, many chemical reactions were found to generate this species in high yield [18–20]. Among the most efficient (yields > 90%) are the oxidation of H_2O_2 by ClO^- [32], the thermolysis of some aromatic endoperoxides [7], the decomposition of peracetic acid in basic medium [33], and the thermolysis of phosphite ozonides [34]. However, none of these reactions are suitable as $^1O_2$ source to selectively oxidize organic substrates on a large scale. Indeed, they form $^1O_2$ through stoichiometric reactions and simultaneously release coproducts that complicate the recovery of oxidation products. Also, some of the starting reactants are themselves powerful oxidants (peracetic acid, hypochlorite, ozonide) that may give side reactions with organic substrates. This section focuses on chemical sources of $^1O_2$ based on catalyzed disproportionation of H_2O_2, which releases only water as a coproduct (Eq. (22.3)).

Now, the particular case of the oxidation of H_2O_2 to triplet or singlet oxygen will be considered (Eqs. (22.11) and (22.12)):

$$H_2O_2 \rightleftharpoons {}^3O_2 + 2H^+ + 2e^- \tag{22.11}$$

$$H_2O_2 \rightleftharpoons {}^1O_2 + 2H^+ + 2e^- \tag{22.12}$$

The standard potential associated with the redox couple $^3O_2/H_2O_2$ (Eq. (22.11)) is $E^{\circ}_{11} = +0.69$ V. Thus, the standard potential associated with the redox couple $^1O_2/H_2O_2$ (Eq. (22.12)) is equal to $E^{\circ}_{12} = +1.18$ V. This value is calculated according to Eq. (22.13), where $E\left({}^3\Sigma_g^- \rightarrow {}^1\Delta_g\right)$ is the energy needed for the excitation of ground-state oxygen (94 kJ/mol) and n is the number of electrons involved in the process ($n = 2$) and F is the Faraday (96 500 °C):

$$E^{\circ}_{12} = 0.69 + \frac{E({}^3\Sigma_g^+ \rightarrow {}^1\Delta_g)}{nF} = 1.18\,\text{V} \tag{22.13}$$

If all chemical species are in a standard state, the oxidation of H_2O_2 into $^3O_2$ or $^1O_2$ is thermodynamically allowed with any oxidizer having a standard redox potential higher than 0.69 or 1.18 V, respectively.

However, $^1O_2$ is never in the standard state because it has a short lifetime ($\approx 3\,\mu s$ in H_2O) and its stationary concentration is very low. Thus, $^1O_2$ is best considered as an intermediate lying on the chemical pathway going from H_2O_2 to $^3O_2$. Moreover, pH is seldom in standard state; hence to forecast whether a reaction could provide $^1O_2$, it is more reliable to consider the pseudo-standard potential $E^{\circ}_{12}{}'$ corresponding to half-redox reactions with all the chemical species in standard state except the concentration of H^+, which is maintained constant with a buffer (Eq. (22.14)):

$$E'_{12} = E^{\circ}_{12}{}' - 0.06\,\text{pH} \tag{22.14}$$

The expression (22.14) permits to predict which reactions are energetically allowed when all chemicals are in a standard state except H^+. However, even a thermodynamically allowed reaction may be very slow or provide $^3O_2$ instead of $^1O_2$. For instance, the redox potential of strong oxidants such as MnO_4^-/Mn^{2+}, ClO^-/Cl^-, or H_2O_2/H_2O is higher than $E^{\circ\prime}$ and could potentially oxidize H_2O_2 into $^1O_2$ without significant activation energy. Thus, although H_2O_2 can disproportionate giving $^1O_2$ (Eq. (22.3)), this reaction is very slow at room temperature without catalyst. In contrast, MnO_4^- and ClO^- react rapidly with H_2O_2 leading to $^3O_2$ and $^1O_2$, respectively. Based on this possibility, we screened the entire periodic table to investigate whether some oxides, hydroxides, or oxoanions were able to catalyze the disproportionation of H_2O_2 into $^1O_2$, which was trapped with tetrapotassium rubrenetetracarboxylate [18]. This red compound is very soluble in water and specifically reacts with $^1O_2$ yielding only the characteristic colorless endoperoxide (Eq. (22.15)) [35, 36].

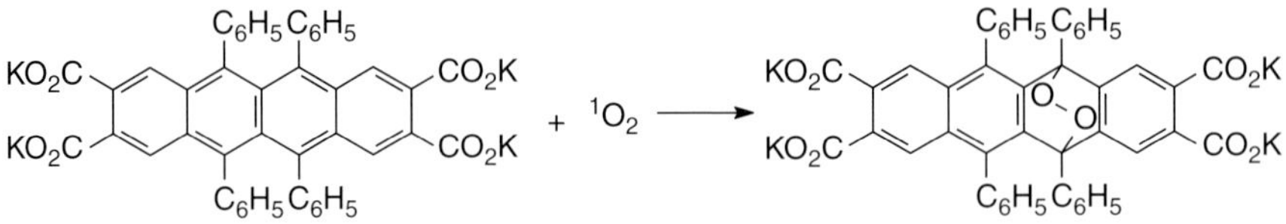

$$\text{[diphenyl-substituted anthracene tetracarboxylate]} + {}^1O_2 \longrightarrow \text{[endoperoxide product]} \tag{22.15}$$

Results obtained were quite unexpected since about one-third of the chemical elements were found to be more or less effective in the generation of $^1O_2$ when mixed with H_2O_2 in basic aqueous solution. Compounds that produce $^1O_2$ with good yields (10–100%) are shown in Figure 22.3 and may be classified into two main categories, that is, strong oxidizers (black boxes) and catalysts (gray boxes).

Compounds belonging to the first family such as ClO^- oxidize H_2O_2 into $^1O_2$ since they have redox potentials higher than $E_{12}^{\circ\prime}$ (Eq. (22.1)). Other mineral compounds (calcium hydroxide, oxides of lanthanides and actinides, oxides, and oxoanions of transition metals in d° configuration) are not strong enough oxidants. Therefore, they act as catalysts that induce a more or less intricate disproportionation of H_2O_2. So far, only reactions involving two homogeneous catalysts (Na_2MoO_4 and Na_2WO_4) [22, 23] and two heterogeneous catalysts ($Ca(OH)_2$ and $La(OH)_3$) [25, 26] were studied in detail.

Each of these chemical sources of $^1O_2$ has its own advantages and limitations. The catalytic system $H_2O_2/Ca(OH)_2$ might appear to be the cheapest and safest one, but it actually requires high concentration of H_2O_2 and it provides $^1O_2$ with a limited yield (25%) [24]. $La(OH)_3$ gives a higher yield of $^1O_2$ (45%) [26]. As it operates in neutral or slightly basic media, this catalyst is appropriate for oxidizing unsaturated amines that react effectively with $^1O_2$ when they are in the basic form but do not when they are in the ammonium form at pH 7. Thus, geranylamine

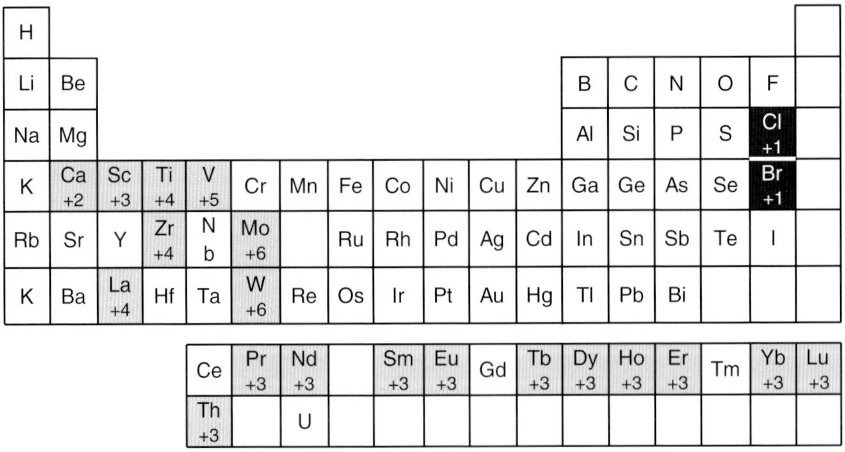

H																	
Li	Be											B	C	N	O	F	
Na	Mg											Al	Si	P	S	Cl +1	
K	Ca +2	Sc +3	Ti +4	V +5	Cr	Mn	Fe	Co	Ni	Cu	Zn	Ga	Ge	As	Se	Br +1	
Rb	Sr	Y	Zr +4	Nb	Mo +6		Ru	Rh	Pd	Ag	Cd	In	Sn	Sb	Te	I	
K	Ba	La +4	Hf	Ta	W +6	Re	Os	Ir	Pt	Au	Hg	Tl	Pb	Bi			
				Ce	Pr +3	Nd +3		Sm +3	Eu +3	Gd	Tb +3	Dy +3	Ho +3	Er +3	Tm	Yb +3	Lu +3
				Th +3		U											

Figure 22.3 Periodic table showing the most effective oxides, hydroxides, and oxoanions able to trigger the formation of $^1O_2$ via the oxidation (black boxes) or the disproportionation (gray boxes) of H_2O_2 in basic aqueous solution.

has been oxidized with both a better selectivity and a higher yield with the catalytic system $H_2O_2/La(OH)_3$ compared with the H_2O_2/Na_2MoO_4 system [26]. Both homogeneous catalysts MoO_4^{2-} and WO_4^{2-} react rapidly with H_2O_2 in basic aqueous media and provide $^1O_2$ with quantitative yields. However, molybdate anion is more effective since it generates $^1O_2$ approximately four times faster than tungstate anion under the same conditions. That is why MoO_4^{2-} is the only catalyst that was applied to industrial manufacture (see Section 22.3.3) [37].

22.2.4
Optimal Generation of $^1O_2$ Through the Catalytic System H_2O_2/MoO_4^{2-}

The most common and thoroughly studied chemical source of $^1O_2$ is the system H_2O_2/MoO_4^{2-} because it is the most efficient in terms of $^1O_2$ yield and reaction rate. Although its ability to release $^1O_2$ was proved only in 1985 [18], the catalytic decomposition of H_2O_2 induced by MoO_4^{2-} has been recognized and investigated since a long time. The first report published by Spitalsky and Funck in 1927 was entitled "about the complicated homogeneous catalytic decomposition of hydrogen peroxide by sodium molybdate" [38]. This telling title is indicative of the difficulties encountered by the authors to explain the mechanism of this catalytic reaction. Nevertheless, they established the main features of this reaction, namely:

- Strong dependence of kinetics upon the pH of the solution
- Complicated effect of H_2O_2 concentration upon the rate of reaction
- First-order dependence with respect to MoO_4^{2-} concentration
- Recovering of unchanged molybdate at the end of the reaction

Baxendale's review proposes diperoxymolybdate $HMoO_6^-$ as the only intermediate [39]. In the 1990s, the mechanism of this reaction in aqueous solution was reexamined with the aim of finding the best experimental conditions leading to $^1O_2$ and identifying the nature of the precursor. The cumulated amount of $^1O_2$ generated through Eq. (22.3) was measured by trapping with the water-soluble rubrene derivative (Eq. (22.15)). It was found that $^1O_2$ yield is 100% irrespective of pH and H_2O_2 concentration. A simple first order for MoO_4^{2-} was confirmed, but much more complex kinetic behavior with respect to pH and H_2O_2 concentration was observed. The rate of decomposition of H_2O_2 sharply increases as the concentration of H_2O_2 increases and then it decreases slowly for very high concentrations of H_2O_2. Such a behavior strongly suggests the formation of an active peroxomolybdate intermediate, which is converted into another inactive peroxo compound for high concentrations of H_2O_2. Reaction rates exhibit strong pH dependence with an asymmetrical bell-shaped curve exhibiting a maximum at pH 10.5. This behavior is readily explained in terms of disappearance of the effective precursor of $^1O_2$ either by protonation in neutral and acid media or by hydrolysis in highly alkaline medium.

^{95}Mo NMR analysis of the reaction medium allowed the identification of the different peroxomolybdates formed depending on the pH and the concentration of H_2O_2 [22]. In basic medium, four distinct peaks are observed by NMR. They appear successively as the concentration of H_2O_2 increases and were assigned to mono-, di-, tri-, and tetraperoxomolybdates with the generic formula $MoO_{4-n}(O_2)_n^{2-}$. Comparing the regions of prevalence of each peroxo compound according to the concentration of H_2O_2 with the evolution of the rate of H_2O_2 disproportionation, it was inferred that the triperoxomolybdate $MoO(O_2)_3^{2-}$ is likely to be the main precursor of $^1O_2$ while tetraperoxomolybdate $Mo(O_2)_4^{2-}$, which forms at high concentration of H_2O_2, is stable and should be regarded as a kinetic dead end. Accordingly, it is essential to maintain the concentration of free H_2O_2 within a narrow range favoring the formation of triperoxomolybdate in order to maximize the rate of $^1O_2$ formation. The same spectroscopic technique allowed rationalizing pH effect on the rate of $^1O_2$ formation and on the selectivity of the reaction. Thus, it was shown that below pH 7 predominantly a dinuclear species $Mo_2O_3(O_2)_4^{2-}$ is formed that does not release $^1O_2$ but is able to effectively epoxidize unsaturated compounds. Therefore, the pH value is another crucial parameter that must be controlled to avoid the occurrence of side reactions. For example, it was shown that the catalytic system H_2O_2/MoO_4^{2-} selectively oxidizes tiglic acid into hydroperoxides via $^1O_2$ when the pH is higher than 9 or provides selectively an epoxide when the pH is lower than 6 (Scheme 22.2) [40].

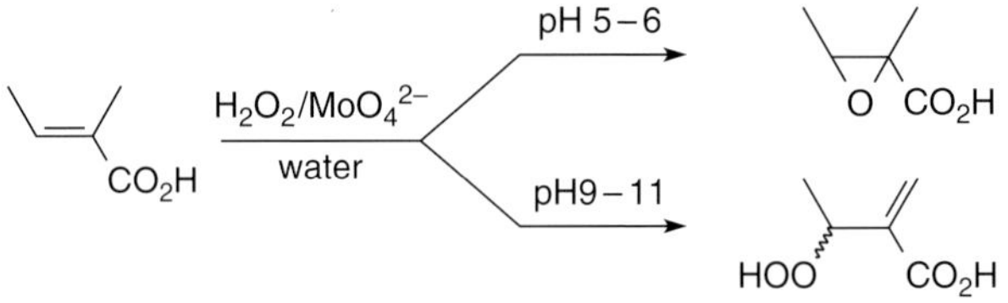

Scheme 22.2 Influence of the pH value on the chemoselectivity of the oxidation of tiglic acid by the system H_2O_2/MoO_4^{2-}.

Taking into account the chemical equilibria between all peroxomolybdates, a predominance diagram showing the prevalent species as a function of pH and free H_2O_2 concentration can be drawn (Figure 22.4). The borderlines separating the different domains correspond to conditions for which the concentrations of the complexes located on each side of the line are equal. The inner zone corresponds to conditions for which the concentration of indicated species is predominant. Finally, to maximize the rate of $^1O_2$ generation, it is necessary to be positioned in the orange area where the triperoxomolybdate $MoO(O_2)_3^{2-}$ predominates. This can be achieved by maintaining the pH of the solution in the range of 9–11 and the concentration of free H_2O_2 between 10^{-2} and 1 M.

22.2.5
Potential Molecular Targets for Singlet Oxygenation

Despite its excited state nature, $^1O_2$ can interact with many organic molecules in solution, thanks to its relatively long lifetime. The use of $^1O_2$ in organic synthesis has enabled the preparation of a large number of natural products and has been the subject of many studies [41]. The first reaction of $^1O_2$ was

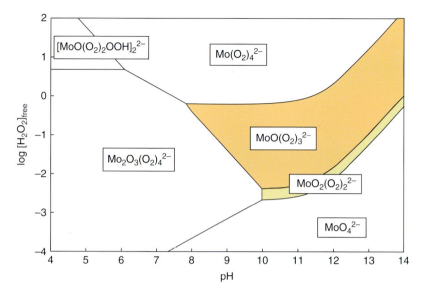

Figure 22.4 Predominance diagram showing prevalent species formed by the interaction between H_2O_2 and MoO_4^{2-} as a function of pH and concentration of free H_2O_2 ($T = 0\,°C$, $[Na_2MoO_4] = 1\,M$). The orange area corresponds to the conditions under which the formation of the main precursor of $^1O_2$ is favored.

reported by Fritzsche in 1867 who described the reaction of naphthacene with oxygen in the presence of light [5]. From the 1950s, studies on the reactivity of $^1O_2$ have increased, though industrial applications of $^1O_2$ are still few [42]. $^1O_2$ reacts with electron-rich organic molecules according to several reaction modes [14]. The most classical and important ones are (i) the [4 + 2] cycloaddition (Eqs. (22.16–22.19)) with conjugated dienes and aromatics and furans leading to endoperoxides or ozonides [43–45]; (ii) the [2 + 2] cycloaddition (Eqs. (22.20–22.22)) with electron-rich alkenes such as enol ethers, enamines, and alkenes devoid of allylic hydrogens, providing unstable 1,2-dioxetanes that generally thermally or photochemically cleave and decompose into carbonyl compounds through a chemiluminescence process [44]; and (iii) the ene or Schenck reaction (Eq. (22.23)) with alkenes bearing allylic hydrogen atoms and giving allylic hydroperoxides, which can be further converted into allylic alcohols after reduction or unsaturated α,β-unsaturated ketones by oxidation [44, 46–48]. Products resulting from the $^1O_2$ addition on unsaturated compounds, that is, dioxetanes, allylic hydroperoxides, and endoperoxides, are particularly valuable intermediates since they are very reactive and can lead to a wide variety of organic oxygenated molecules such as ketones, α-diols, and α-hydroxyketones. G. Ohloff published a review describing the processes of preparation of some flavors and fragrances involving these three types of reaction [6]. $^1O_2$ also oxidizes phenols (Eqs. (22.24) and (22.25)), sulfides (Eq. (22.26)), and many heterocyclic compounds (e.g., furans, pyrroles, indoles, imidazoles, purines, oxazoles, thiazoles, thiophenes). Details on the mechanistic and synthetic aspects can be found in the two excellent reviews of Clennan and Pace [5, 12].

Acyclic 1,3-dienes (22.16)

Cyclic 1,3-dienes (22.17)

Polycyclic aromatics (22.18)

Furans (22.19)

Olefins without allylic H (22.20)

Enamines (22.21)

Enol ethers (22.22)

Olefins with allylic H (22.23)

Nonsubstituted phenols (22.24)

p-Substituted phenols (22.25)

Sulfides (22.26)

22.3
Prospects for Scale-Up

22.3.1
Respective Advantages and Disadvantages of "Dark" and "Luminous" Singlet Oxygenation

The main advantage of photooxidation on lab-scale experiments is its versatility with regard to the choice of solvent and the wide operational temperature range. As photooxidations via $^1O_2$ are usually induced by visible light, the design of the photoreactor and the selection of the light source are easier and less costly than when the photoreaction requires UV light and a quartz photoreactor. On a large scale, immersion photoreactors are preferred since these use all the photons emitted by the light source [49].

More sustainably, modern solar collectors instead of artificial lamps are used [50]. Many solar parabolic trough reactors can be used both on laboratory and industrial scales, for example, at the German Aerospace Centre (DLR) in Cologne. The parabolic trough-facility for organic photochemical syntheses in sunlight (PROPHIS) is suitable for industrial-scale reactions [51]. The concentration factor is about 30 times compared with direct sunlight. Oelgemöller *et al.* [52] have carried out the photooxygenation of citronellol and 1,5-dihydroxynaphthalene at the DLR using the PROPHIS reactor. They [53] have also reported the preparation of juglone by photooxygenation of 1,5-dihydroxynaphtalene using nonconcentrated sunlight in Dublin (Scheme 22.3).

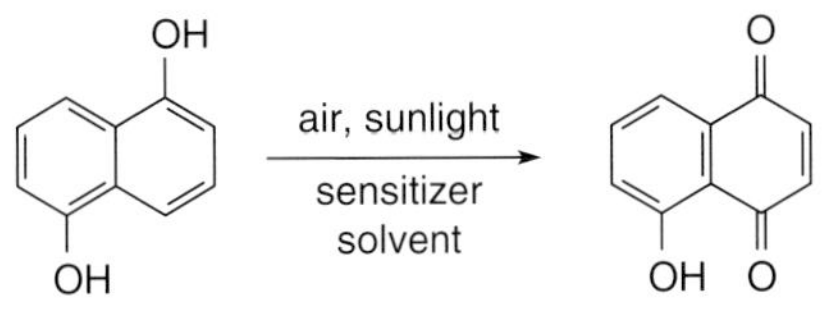

Scheme 22.3 Photooxidation of 1,5-dihydroxynaphthalene into juglone.

Although often claimed to be "green," photochemical processes rarely reach industrial scale. Indeed, the combination of organics, oxygen, light, and electricity entails serious safety problems when the reaction is conducted on large scale. Also, the wear of lamps and the gradual opacification of the photoreactor walls with organic deposits require adjusting the irradiation time so as to secure sufficient conversion rate of the substrate. Moreover, photoreactors suffer from a lack of economy of scale and are rarely available in a multipurpose chemical plant. Photooxidation is sometimes the only singlet oxygen process that can be used. Indeed, when the substrate is likely to interact directly with hydrogen peroxide or with the peroxo intermediates, chemical sources based on H_2O_2 cannot be used.

Recently, scale-up of photochemical reactions has significantly advanced, thanks to the development of microreactor technology. In the field of singlet oxygen, the first example in 2002 was the rose bengal-sensitized photooxidation of α-terpinene in a glass microchip [54]. Eighty percent conversion was obtained within less than 5 s. Recently, the Max Planck Institute applied this type of microreactor to the important antimalarial drug artemisinin [55, 56]. The authors carried out the reaction on a 3-g scale in a specially designed continuous flow reactor and obtained artemisinin in 39% yield. Small-channel microreactors enable optimal light absorption and secure safe processing by a better heat control and low hold-up of highly explosive or toxic chemicals. Moreover, efficient, high-power light-emitting diodes (LEDs) with high photon fluxes, wide range of available wavelengths, and long lifetime also offer new and promising opportunities for photochemistry. Recently, Lapkin et al. reported the application of the microreactor technology to the oxygenation of α-pinene to pinocarvone (Scheme 22.4), a useful building block for antimalarial peroxides, chiral ligands for catalysis, and so on. The authors also generalized the fundamental principles for design and optimization of scalable photochemical reactors [57].

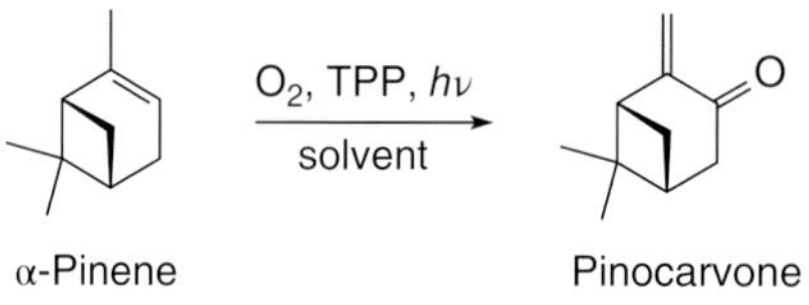

Scheme 22.4 Photooxidation of α-pinene into pinocarvone.

The disadvantages of the photochemical singlet oxygen production can be overcome by "dark" singlet oxygenation (DSO) through catalytic disproportionation of H_2O_2 that can be carried out in commonly available stirred-tank reactors. However, this chemical generation of $^1O_2$ is not efficient at low temperatures or in the absence of polar, preferably protic, solvents. Finally, catalyst separation and recycling need to be addressed.

22.3.2
Choice of the Medium for Dark Singlet Oxygenation

The medium used to perform DSO plays a key role in the efficiency of the singlet oxygenation process. It involves both thermodynamic and kinetic aspects as it must take into account (i) the ability of the peroxo intermediates to form and to decompose readily into $^1O_2$ [58]; (ii) the lifetime of $^1O_2$ in this medium; (iii) the solubility of the organic substrate, which is most often hydrophobic while the catalyst and the oxidant, that is, H_2O_2, are hydrophilic; (iv) the reactivity of the substrate toward $^1O_2$; and finally (v) the accessibility of the substrate by the excited species.

22.3.2.1 Homogeneous Aqueous and Alcoholic Media
DSO based on molybdate-, tungstate-, or lanthanum(III)-catalyzed H_2O_2 disproportionation proceeds most efficiently in water in terms of rate and $^1O_2$ yield. Several water-soluble substrates (e.g., sodium tiglate, sodium naphthalene dipropionate) were thus peroxidized efficiently [40, 59, 60]. As for hydrophobic substrates, several strategies have been developed to chemically oxidize them

on a preparative scale. The simplest one is to use alcohol/water mixtures as reaction media to oxidize low molecular weight compounds such as terpenes, but since the solubility is below the one required for an industrial process (>0.5 M), it is preferable to use a neat organic solvent. An extensive screening revealed that only polar and protic solvents, such as lower alcohols (e.g., methanol or ethanol), allow efficient DSO [61]. Reaction rates are optimized by favoring the formation of the triperoxomolybdate, precursor of $^1O_2$, and avoiding the stable tetraperoxomolybdate (see Section 22.2.4 and Figure 22.4). Although the $^1O_2$ yield from H_2O_2 disproportionation in methanol is reduced to 82% compared with the quantitative yield in water, this negative effect is counterbalanced by an enhanced $^1O_2$ lifetime in MeOH (10 μs) compared with that in water (3 μs). Thus, the overall trapping efficiency of $^1O_2$ by sufficiently reactive substrates is still high enough to allow almost full conversion with an economically acceptable amount of H_2O_2. In addition, the process affords a very convenient recovery of products and catalysts since Na_2MoO_4 is not soluble in alcohols whereas the intermediate peroxomolybdates are. Thus, the catalyst precipitates at the end of the reaction indicating the total consumption of H_2O_2. This method, well adapted to small and polar substrates (e.g., α-terpinene, mesitol), has been industrially implemented using ethylene glycol as a solvent to perform the preparation of rose oxide from β-citronellol as detailed in Section 22.3.3. Unfortunately, it cannot be applied to highly hydrophobic substrates, which are not sufficiently soluble in alcohols.

22.3.2.2 Single-Phase Microemulsions

The compatibility issue between water-soluble reagents and hydrophobic substrates in biphasic systems is often solved by adding phase transfer catalysts. However, this method does not work for DSO since $^1O_2$ catalytically generated in the aqueous droplets from H_2O_2 is physically quenched by water molecules into $^3O_2$ before being able to diffuse from the millimetric aqueous droplets to the organic phase where the substrate resides. One elegant way to overcome the incompatibility between hydrophilic reagents (i.e., H_2O_2, Na_2MoO_4) and hydrophobic substrates consists of using single-phase microemulsions. Such systems are thermodynamically stable submicronic dispersions ($\phi_{droplets} \approx 10-50$ nm) of two immiscible liquids stabilized and separated by a surfactant monolayer [62–64]. They have been used as efficient reaction media for several reactions demonstrating their interest in organic synthesis [62, 65]. Indeed, the nanostructuration of the medium affords a considerable increase of the oil–water interfacial area, and the compartmentalization of reactants provides a better selectivity compared with cosolubilization of all reactants into polar solvents. Single-phase water-in-oil microemulsions are particularly well suited to the DSO of typical organic substrates [66–68]. In these systems, the catalyst MoO_4^{2-} reacts with H_2O_2 to generate $^1O_2$ in the aqueous nanodroplets. This small uncharged and rather hydrophobic excited species then freely diffuses through the interfacial film into the organic phase where it reacts with the hydrophobic substrate. The success of this process is based on the fact that the typical size of the aqueous

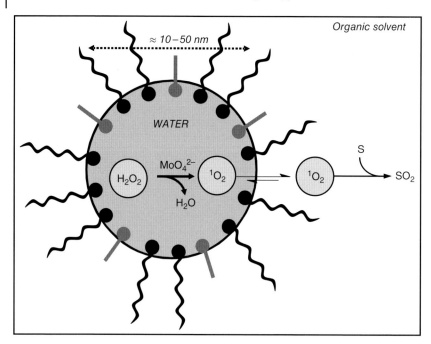

Figure 22.5 Schematic representation of the dark singlet oxygenation of a substrate S in a water-in-oil single-phase μem water/surfactant/cosurfactant/solvent by the chemical source hydrogen peroxide/sodium molybdate.

nanodomains (about 10 nm) is much smaller than the mean travel distance of $^1O_2$ in water (about 200 nm). Typical single-phase microemulsions can be composed of sodium dodecyl sulfate (SDS) as the surfactant, 1-butanol as the cosurfactant, water and methylene chloride, toluene, or cyclohexane as the organic phase (Figure 22.5).

When the substrate is a liquid, it may act as the organic phase providing thus solvent-free microemulsions [69]. Although the single-phase microemulsions have been successfully applied to the DSO of various hydrophobic substrates, these systems suffer from two main drawbacks, which limit industrial implementation (i) lengthy recovery of products because of the high concentration of amphiphiles required to reach the complete water/solvent cosolubilization ($\approx$15–20%) and (ii) demixing after addition of a certain amount of H_2O_2 making them poorly efficient for weakly reactive substrates. Proper formulation of the microemulsion medium can partially relieve these problems, in particular those arising from the large amount of surfactant. Recently, we showed that thermoresponsive single-phase microemulsions with only 6% surfactants can be obtained by combining an anionic catalytic surfactant and a nonionic one resulting in a strong synergistic effect. With such systems, the reaction is performed in the single-phase microemulsion, which switches in a two-phase microemulsion system by lowering the temperature [70].

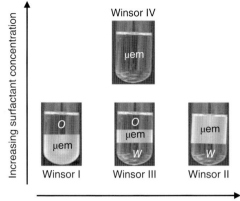

Scheme 22.5 Different types of microemulsion (μem) systems according to the concentration of surfactant and the water/oil interfacial curvature. WI: Winsor I system (O/W μem in equilibrium with excess oil); WII: Winsor II system (W/O μem in equilibrium with excess water); WIII: Winsor III system (μem in equilibrium with excess oil and water); and WIV: Winsor IV (single-phase μem).

22.3.2.3 Multiphase Microemulsions with Balanced Catalytic Surfactants

The problem of large amounts of amphiphiles (i.e., surfactant and cosurfactant) required to form a single-phase microemulsion can be overcome with multiphase microemulsions. Indeed, two- and three-phase microemulsion systems can be obtained with only about 3–5% surfactant (Scheme 22.5).

There are two types of biphasic microemulsion systems: one with an excess oil in equilibrium with an oil-in-water microemulsion (Winsor I system) and one with an excess water phase in equilibrium with a water-in-oil microemulsion (Winsor II system). The WI system is particularly relevant with regard to the ease of workup since the product is extracted into the excess oil phase and the catalyst remains in the microemulsion phase. However, it is still sensitive to dilution by water arising from H_2O_2 disproportionation. This can be avoided by using pervaporation membrane [71]. A further improvement has been brought by three-liquid phase microemulsion systems based on "balanced catalytic surfactants" [72]. This new kind of catalysts is carefully designed in order to provide spontaneously a Winsor III microemulsion system in the presence of water and an appropriate oil, without requiring cosurfactant nor electrolyte. They play a dual role as [1] O_2 generating catalyst and as surfactant stabilizing the bicontinuous microemulsion. From a physicochemical point of view, they must have an "effective packing parameter" close to one [73]. Bis(dimethyldialkylammonium)molybdates are typical examples of such "balanced catalytic surfactants," which lead to a three-phase system as shown in Figure 22.6. In such reaction media, $^1O_2$ is exclusively generated in the aqueous nanodomains of the middle-phase microemulsion where the reaction takes place. Under stirring, the excess oil phase transfers the substrate S to the

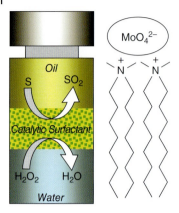

Figure 22.6 Three-liquid phase microemulsion (Winsor III) based on the balanced catalytic surfactant bis(dimethyldioctylammonium)molybdate.

microemulsion phase and extracts the product as it is formed while the excess water phase can be seen as a reservoir for H_2O_2 and H_2O.

Under well-chosen conditions (the so-called optimal formulation), the interfacial tension $\gamma_{water/oil}$ between the aqueous and the oil phases is ultralow and, as a consequence, a Winsor III system appears spontaneously. In that case, it becomes a very effective reaction medium with the following features:

- Maximal water/oil cosolubilization in the microemulsion (μem) phase, thus also allowing efficient DSO of highly lipophilic substrates.
- The oil/water interface increases tremendously since it is ≈10^5 times larger than without surfactant and, therefore, the exchange of substrates and products between the three phases is dramatically accelerated.
- Very fast separation of the three phases (within a few seconds when stirring is stopped), thus allowing easy product recovery and catalyst recycling by simple phase separation.

22.3.3
Examples of Industrialized Singlet Oxygenation

Historically, the most famous example is the synthesis of ascaridole discovered by Schenck and Ziegler [74]. Ascaridole, an anthelmintic drug, which can also be isolated from the leaves of *Chenopodium ambrosioides*, is the first discovered naturally occurring organic peroxide. It was obtained by photooxidation of α-terpinene in the presence of chlorophyll as photosensitizer. After 1945, this reaction was applied in large scale for the production of ascaridole in Germany. Some years later, application of photooxidation was extended to the ene reaction by two perfumery manufacturers: Dragoco in Germany (which became later Symrise) and Firmenich in Switzerland for the production of rose oxide starting from β-citronellol [75–78]. Among the various types of $^1O_2$ reactions, the ene reaction is the most important one for fine chemistry because the allylic hydroperoxides that result from the $^1O_2$ ene reaction are useful synthetic intermediates. Many organic syntheses involving $^1O_2$ have been described in the patent literature but few reached industrial development. We detail later the two recent examples: one uses a photochemical source of $^1O_2$ (artemisinin) and the other a chemical one (rose oxide).

22.3.3.1 Synthesis of Artemisinin

Artemisinin or Qinghaosu (QHS) is a naturally occurring sesquiterpene lactone endoperoxide isolated in 1972 from the leaves of the *Artemisia annua* plant by Chinese scientists. It is the key ingredient of artemisinin-based combination therapies (ACTs), which has been identified by the World Health Organization (WHO) as the most effective malaria treatment. However, its bioavailability is low making prices and quality fluctuating. In 2004, the WHO launched a public–private consortium including Sanofi, University of California Berkeley, and Amyris to develop a new manufacturing process to produce high-quality and nonseasonal artimisinin to supplement the plant-based supply. The project was funded by the Bill & Melinda Gates Foundation. In 2013, the large-scale production of semisynthetic artemisinin was launched by Sanofi with a production capacity of 50–60 t/year. A no-profit, no-loss production model helps to maintain a low price for developing countries. The industrial process for semisynthetic artemisinin consists of the production of artemisinic acid through fermentation, performed by Huve Pharma in Bulgaria (2013 production: 60 t from 590 t of glucose), followed by a synthetic transformation of the artemisinic acid into artemisinin via photochemistry, which is performed at Sanofi's Garessio site. In 2012, Sanofi received the *Pierre Potier* prize for artemisinin. Sanofi's synthesis pathway for artemisinin (Scheme 22.6) is mechanistically very close to that of the *A. annua* plant [79].

The first step is the diastereoselective hydrogenation of artemisinic acid into dihydroartemisinic acid using a chiral homogeneous Ru catalyst developed by Takasago. Subsequently, dihydroartemisinic acid is converted with ethylchloroformate into an O-acylated ethylcarbonate that facilitates the final ring closure. The mixed carbonate then undergoes tetraphenylporphine (TPP)-sensitized photooxidation in the presence of trifluoroacetic acid with methylene chloride as a solvent. Without isolation, the resulting tertiary allylic hydroperoxide is transformed into artemisinin via a complex cascade that includes acid-catalyzed Hock cleavage, enol oxygenation with $^3O_2$, and final ring closure. The overall isolated yield of artemisinin equals 55% over all steps. The safe one-pot approach from the hydroperoxide to artemisinin is the main driver for using the photochemical pathway instead of DSO on large scale. In DSO, the unstable hydroperoxide accumulates, because the acid-triggered cascade cannot be carried out in the alcoholic medium required for (monophasic) DSO and consequently a hazardous solvent switch has to be carried out.

Besides a relatively large number of patents, artemisinin and derivative synthesis has also been the subject of several scientific articles [80, 81], including the continuous flow conversion of dihydroartemisinic acid into artemisinin by a three-step reaction sequence involving photooxidation with singlet oxygen (with TPP in CH_2Cl_2), acid-catalyzed Hock cleavage (with trifluoroacetic acid in CH_2Cl_2),

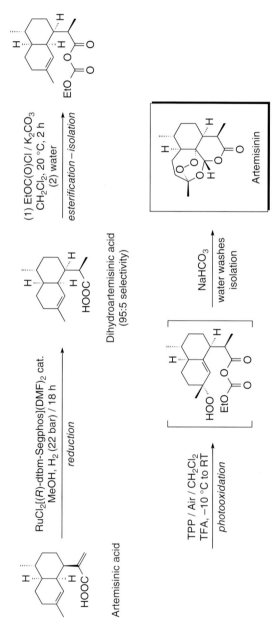

Scheme 22.6 Synthesis of the antimalarial artemisinin from artemisinic acid.

and oxidation with $^3O_2$ (60 °C) [55]. DSO with H_2O_2/Na_2MoO_4 has also been described as an alternative to the photochemical approach toward artemisinin [82–85].

22.3.3.2 Synthesis of Rose Oxide

An example of industrialized DSO is the manufacture of rose oxide from β-citronellol [37]. (−)-*cis*-Rose oxide is a natural fragrance present in flowers and fruits. Since natural (−)-*cis*-rose oxide is expensive (3000 kg of rose blossoms required for 1 kg of rose oil), it finds only limited applications in high-value perfumes. Synthetic rose oxide is produced on large scale as a diastereomeric mixture (rose oxide inactive). It is applied as a "rosy" fragrance and flavor. A large variety of preparative methods have been described in the literature including photochemical processes [86]. A key step in the manufacture of synthetic rose oxide is the ene-type peroxidation of β-citronellol by singlet oxygen (Scheme 22.7). The rose bengal-sensitized photooxidation of β-citronellol in methanol (Symrise procedure) provides a mixture of secondary and tertiary hydroperoxides, which, after reduction with bisulfite, yields the corresponding alcohols. Acid-catalyzed ring closure of the main tertiary product leads to a mixture of the stereoisomeric rose oxides.

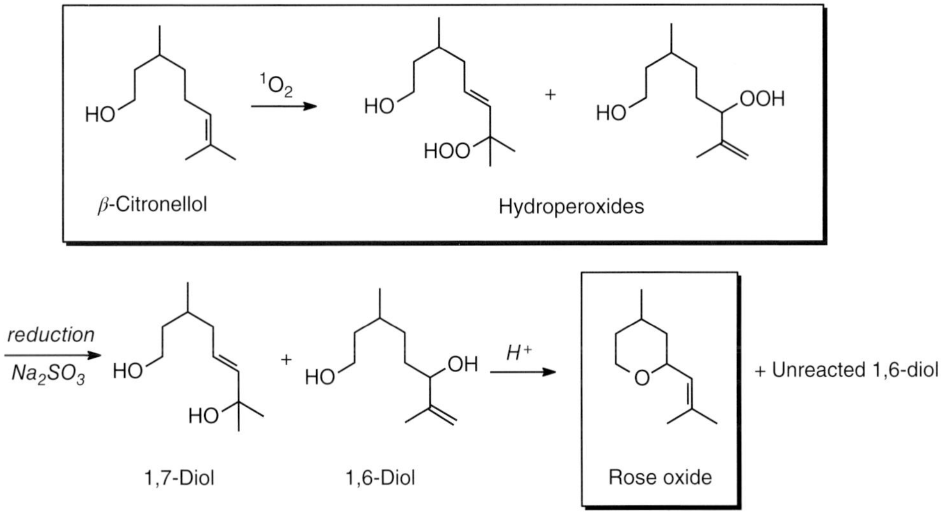

Scheme 22.7 Synthesis of rose oxide from β-citronellol.

Because DSO can be readily carried out in conventional stirred-tank reactors, thus benefiting from a more favorable economy of scale compared with photooxidation, DSM implemented a rose oxide process based on molybdate-catalyzed H_2O_2 disproportionation as $^1O_2$ source. Although on laboratory scale this DSO may be carried out in methanol as a water-miscible solvent, ethylene glycol was used instead on large scale for two reasons. First, ethylene glycol has a much higher oxygen limit concentration (50 vol%) compared with methanol

(8 vol%). By purging nitrogen from the bottom of the reactor, the composition of the headspace was kept well below the OLC, thus ensuring safe processing. Second, ethylene glycol allowed easy catalyst recycling and product isolation by simple extraction with MTBE after reduction of the hydroperoxide products with Na_2SO_3. After distilling-off most of the water, the molybdate-containing aqueous ethylene glycol layer could be reused by simply adding new batches of β-citronellol and H_2O_2. Accumulation of the latter was avoided by working at rather high temperature (55 °C), which, however, is well below the organic hydroperoxide decomposition temperature in this medium (110 °C). Almost full conversion of β-citronellol could be achieved with only a 1.5-fold excess of 50% aqueous H_2O_2 compared with the theoretically required amount. Gratifyingly, the olfactory properties of rose oxide obtained via DSO equaled those obtained via photooxidation of β-citronellol.

22.4 Conclusion

In contrast to ground-state oxygen, $^1O_2$ is a powerful and selective oxidizer that reacts with many electron-rich substrates. Most often, it is prepared through photosensitized oxidation and widely used in biology and organic synthesis on a laboratory scale. The photosensitized generation of $^1O_2$ is very versatile because it can be conducted in a wide range of polar or nonpolar solvents and at low temperature if unstable oxidized products have to be prepared. However, it requires dedicated photochemical reactors, which are not always available in laboratories or industrial plants. This problem of photochemical reactor availability aggravated by the explosion risk due to the simultaneous presence of oxygen, solvent, peroxides, and light has limited its industrial development.

DSO via molybdate-catalyzed disproportionation of H_2O_2 provides a readily scalable alternative to photooxidation. It can be carried out in commonly available stirred-tank reactors. However, the reaction does not work at low temperatures and organic media are limited to alcoholic polar solvents (methanol or the safer ethylene glycol) or to microstructured media such as one-, two-, or three-phase microemulsion systems. The latter based on "balanced catalytic surfactants" advantageously combine low surfactant concentration with easy product isolation and catalyst recycling via simple phase separation. Safe processing may be further enhanced by microreactors, which minimize peroxide hold-up.

Acknowledgments

We thank all the coworkers of (former) DSM Fine Chemicals Austria who were involved in the R&D and the production of rose oxide via DSO, in particular, Walther Jary and Peter Pöchlauer.

References

1. Khan, A.U. and Kasha, M. (1963) *J. Chem. Phys.*, **39**, 2105.
2. Schenck, G.O. (1948) *Naturwissenschaften*, **35**, 28.
3. Moureu, C., Dufraisse, C., and Dean, P.M. (1926) *Compt. Rend.*, **182**, 1440.
4. Prein, M. and Adam, W. (1996) *Angew. Chem. Int. Ed. Engl.*, **35**, 477.
5. Clennan, E.L. (2000) *Tetrahedron*, **56**, 9151.
6. Ohloff, G. (1975) *Pure Appl. Chem.*, **43**, 481.
7. Aubry, J.-M., Pierlot, C., Rigaudy, J., and Schmidt, R. (2003) *Acc. Chem. Res.*, **36**, 668.
8. Dufraisse, C. and Velluz, L. (1942) *Bull. Soc. Chim. Fr.*, **9**, 171.
9. Thomas, M.J. and Foote, C.S. (1978) *Photochem. Photobiol.*, **27**, 683.
10. Jiang, G., Chen, J., Huang, J.-S., and Che, C.-M. (2009) *Org. Lett.*, **11**, 4568.
11. Clennan, E.L. (2001) *Acc. Chem. Res.*, **34**, 875.
12. Clennan, E.L. and Pace, A. (2005) *Tetrahedron*, **61**, 6665.
13. Gollnick, K. and Griesbeck, A. (1985) *Tetrahedron*, **41**, 2057.
14. Montagnon, T., Tofi, M., and Vassilikogiannakis, G. (2008) *Acc. Chem. Res.*, **41**, 1001.
15. Adam, W., Ahrweiler, M., Peters, K., and Schmiedeskamp, B. (1994) *J. Org. Chem.*, **59**, 2733.
16. Richardson, R.J., Wiswall, C.E., Carr, P.A.G., Hovis, F.E., and Lilenfeld, H.V. (1981) *J. Appl. Phys.*, **52**, 4962.
17. Dolmans, D.E.J.G.J., Fukumura, D., and Jain, R.K. (2003) *Nat. Rev. Cancer*, **3**, 380.
18. Aubry, J.M. (1985) *J. Am. Chem. Soc.*, **107**, 5844.
19. Wahlen, J., De Vos, D.E., Jacobs, P.A., and Alsters, P.L. (2004) *Adv. Synth. Catal.*, **346**, 152.
20. Adam, W., Kazakov, D.V., and Kazakov, V.P. (2005) *Chem. Rev.*, **105**, 3371.
21. Dewilde, A., Pellieux, C., Hajjam, S., Wattré, P., Pierlot, C., Hober, D., and Aubry, J.-M. (1996) *J. Photochem. Photobiol., B*, **36**, 23.
22. Nardello, V., Marko, J., Vermeersch, G., and Aubry, J.M. (1995) *Inorg. Chem.*, **34**, 4950.
23. Nardello, V., Marko, J., Vermeersch, G., and Aubry, J.M. (1998) *Inorg. Chem.*, **37**, 5418.
24. Pierlot, C., Nardello, V., Schrive, J., Mabille, C., Barbillat, J., Sombret, B., and Aubry, J.-M. (2002) *J. Org. Chem.*, **67**, 2418.
25. Trokiner, A., Bessière, A., Thouvenot, R., Hau, D., Marko, J., Nardello, V., Pierlot, C., and Aubry, J.-M. (2004) *Solid State Nucl. Magn. Reson.*, **25**, 209.
26. Nardello, V., Barbillat, J., Marko, J., Witte, P.T., Alsters, P.L., and Aubry, J.-M. (2003) *Chem. – Eur. J.*, **9**, 435.
27. Wahlen, J., Vos, D.D., Hertogh, S.D., Nardello, V., Aubry, J.-M., Alsters, P., and Jacobs, P. (2005) *Chem. Commun.*, 927.
28. Wahlen, J., De Vos, D.E., Jacobs, P.A., Nardello, V., Aubry, J.-M., and Alsters, P.L. (2007) *J. Catal.*, **249**, 15.
29. Aubry, J.M. and Cazin, B. (1988) *Inorg. Chem.*, **27**, 2013.
30. Turro, N.J., Chow, M.F., and Rigaudy, J. (1981) *J. Am. Chem. Soc.*, **103**, 7218.
31. Sivéry, A., Barras, A., Boukherroub, R., Pierlot, C., Aubry, J.M., Anquez, F., and Courtade, E. (2014) *J. Phys. Chem. C*, **118**, 2885.
32. Held, A.M., Halko, D.J., and Hurst, J.K. (1978) *J. Am. Chem. Soc.*, **100**, 5732.
33. Evans, D.F. and Upton, M.W. (1985) *J. Chem. Soc., Dalton Trans.*, 1151.
34. CaMinade, A.M., Khatib, F.E., Koenig, M., and Aubry, J.M. (1985) *Can. J. Chem.*, **63**, 3203.
35. Aubry, J.M., Rigaudy, J., and Cuong, N.K. (1981) *Photochem. Photobiol.*, **33**, 149.
36. Aubry, J.M., Rigaudy, J., and Cuong, N.K. (1981) *Photochem. Photobiol.*, **33**, 155.
37. Alsters, P.L., Jary, W., Nardello-Rataj, V., and Aubry, J.-M. (2010) *Org. Process Res. Dev.*, **14**, 259.
38. Spitalskii, E. and Funck, A. (1927) *Z. Phys. Chem.*, **126**, 1.
39. Baxendale, J.H. (1952) in *Advances in Catalysis*, vol. **4** (eds W.G. Frankenburg,

V.I. Komarewsky, and E.K. Rideal), Academic Press, pp. 31–86.
40. Nardello, V., Bouttemy, S., and Aubry, J.-M. (1997) *J. Mol. Catal. Chem.*, **117**, 439.
41. Wasserman, H.H. and Ives, J.L. (1981) *Tetrahedron*, **37**, 1825.
42. DeRosa, M.C. and Crutchley, R. (2002) *J. Coord. Chem. Rev.*, **233–234**, 351.
43. Lesce, M.R. (2005) in *Synthetic Organic Photochemistry* (eds A.G. Griesbeck and J. Mattay), Marcel Dekker, New York, 299–364.
44. Horspool, W.M. and Lenci, F. (eds) (2004) *CRC Handbook of Organic Photochemistry and Photobiology*, 2nd edn, CRC Press, Boca Raton, FL.
45. Adam, W. and Prein, M. (1996) *Acc. Chem. Res.*, **29**, 275.
46. Schenck, G.O., Eggert, H., and Denk, W. (1953) *Justus Liebigs Ann. Chem.*, **584**, 177.
47. Frimer, A.A. (1985) *CRC Singlet Oxygen*, CRC Press.
48. Rappoport, Z. (2007) *The Chemistry of Peroxides*, John Wiley & Sons, Ltd..
49. Braun, A.M., Maurette, M.-T., and Oliveros, E. (1991) *Photochemical Technology*, John Wiley & Sons, Ltd.
50. Oelgemöller, M., Jung, C., Ortner, J., Mattay, J., and Zimmermann, E. (2005) *Green Chem.*, **7**, 35.
51. Jung, C., Funken, K.-H., and Ortner, J. (2005) *Photochem. Photobiol. Sci.*, **4**, 409.
52. Schiel, C., Oelgemöller, M., Ortner, J., and Mattay, J. (2001) *Green Chem.*, **3**, 224.
53. Suchard, O., Kane, R., Roe, B.J., Zimmermann, E., Jung, C., Waske, P.A., Mattay, J., and Oelgemöller, M. (2006) *Tetrahedron*, **62**, 1467.
54. Wootton, R.C.R., Fortt, R., and de Mello, A. (2002) *J. Org. Process Res. Dev.*, **6**, 187.
55. Lévesque, F. and Seeberger, P.H. (2012) *Angew. Chem. Int. Ed.*, **51**, 1706.
56. Kopetzki, D., Lévesque, F., and Seeberger, P.H. (2013) *Chem. – Eur. J.*, **19**, 5450.
57. Loponov, K.N., Lopes, J., Barlog, M., Astrova, E.V., Malkov, A.V., and Lapkin, A.A. (2014) *Org. Process Res. Dev.*, **18**, 1443.
58. Csányi, L.J. (2010) *J. Mol. Catal. Chem.*, **322**, 1.
59. Nardello, V., Aubry, J.M., and Linker, T. (1999) *Photochem. Photobiol.*, **70**, 524.
60. Aubry, J.M., Cazin, B., and Duprat, F. (1989) *J. Org. Chem.*, **54**, 726.
61. Nardello, V., Bogaert, S., Alsters, P.L., and Aubry, J.-M. (2002) *Tetrahedron Lett.*, **43**, 8731.
62. Holmberg, K. (2003) *Curr. Opin. Colloid Interface Sci.*, **8**, 187.
63. Cosgrove, T. (2010) *Colloid Science: Principles, Methods and Applications*, John Wiley & Sons, Inc.
64. Stubenrauch, C. (2009) *Microemulsions: Background, New Concepts, Applications, Perspectives*, John Wiley & Sons, Inc.
65. Häger, M., Currie, F., and Holmberg, K. (2003) *Colloid Chemistry II*, Springer, pp. 53–74.
66. Nardello, V., Caron, L., Aubry, J.-M., Bouttemy, S., Wirth, T., Saha-Möller Chantu, R., and Adam, W. (2004) *J. Am. Chem. Soc.*, **126**, 10692.
67. Aubry, J.-M., Adam, W., Alsters, P.L., Borde, C., Queste, S., Marko, J., and Nardello, V. (2006) *Tetrahedron*, **62**, 10753.
68. Aubry, J.-M. and Bouttemy, S. (1997) *J. Am. Chem. Soc.*, **119**, 5286.
69. Nardello, V., Hervé, M., Alsters, P.L., and Aubry, J.-M. (2002) *Adv. Synth. Catal.*, **344**, 184.
70. Hong, B., Leclercq, L., Collinet-Fressancourt, M., Lai, J., Bauduin, P., Aubry, J.-M., and Nardello-Rataj, V. (2015) *J. Mol. Catal. Chem.*, **397**, 142.
71. Caron, L., Nardello, V., Mugge, J., Hoving, E., Alsters, P.L., and Aubry, J.-M. (2005) *J. Colloid Interface Sci.*, **282**, 478.
72. Nardello-Rataj, V., Caron, L., Borde, C., and Aubry, J.-M. (2008) *J. Am. Chem. Soc.*, **130**, 14914.
73. Bouton, F., Durand, M., Nardello-Rataj, V., Borosy, A.P., Quellet, C., and Aubry, J.-M. (2010) *Langmuir*, **26**, 7962.
74. Schenck, G.O. and Ziegler, K. (1944) *Naturwissenschaften*, **32**, 157.
75. Rojahn, W. and Warnecke, H.U. (1980) *Dragoco-Rep.*, **27**, 159.
76. Ohloff, G., Klein, E., and Schenck, G.O. (1961) *Angew. Chem.*, **73**, 578.

77. Gerhard, S. and Gunther, O. (1966) Cyclic 5-and 6-membered ethers. US Patent 3252998 A, May 24, 1966.
78. Erich, K., Gunther, O., and Otto, S.G. (1968) Mixtures of oxygenated acyclic terpenes. US Patent 3382276 A, May 7, 1968.
79. Turconi, J., Griolet, F., Guevel, R., Oddon, G., Villa, R., Geatti, A., Hvala, M., Rossen, K., Göller, R., and Burgard, A. (2014) *Org. Process Res. Dev.*, **18**, 417.
80. Acton, N. and Roth, R.J. (1992) *J. Org. Chem.*, **57**, 3610.
81. Griesbeck, A.G., Bartoschek, A., El-Idreesy, T.T., Höinck, O., and Miara, C. (2006) *J. Mol. Catal. Chem.*, **251**, 41.
82. Zhu, C. and Cook, S.P. (2012) *J. Am. Chem. Soc.*, **134**, 13577.
83. Paddon, C.J., Westfall, P.J., Pitera, D.J., Benjamin, K., Fisher, K., McPhee, D., Leavell, M.D., Tai, A., Main, A., Eng, D., Polichuk, D.R., Teoh, K.H., Reed, D.W., Treynor, T., Lenihan, J., Jiang, H., Fleck, M., Bajad, S., Dang, G., Dengrove, D., Diola, D., Dorin, G., Ellens, K.W., Fickes, S., Galazzo, J., Gaucher, S.P., Geistlinger, T., Henry, R., Hepp, M., Horning, T., Iqbal, T., Kizer, L., Lieu, B., Melis, D., Moss, N., Regentin, R., Secrest, S., Tsuruta, H., Vazquez, R., Westblade, L.F., Xu, L., Yu, M., Zhang, Y., Zhao, L., Lievense, J., Covello, P.S., Keasling, J.D., Reiling, K.K., Renninger, N.S., and Newman, J.D. (2013) *Nature*, **496**, 528.
84. Chen, H.-J., Han, W.-B., Hao, H.-D., and Wu, Y. (2013) *Tetrahedron*, **69**, 1112.
85. Jin, H.-X., Liu, H.-H., and Wu, Y.-K. (2004) *Chin. J. Chem.*, **22**, 999.
86. Ravelli, D., Protti, S., Neri, P., Fagnoni, M., and Albini, A. (2011) *Green Chem.*, **13**, 1876.

Part VIII
Reactor Concepts for Liquid Phase Aerobic Oxidation

23
Reactor Concepts for Aerobic Liquid phase Oxidation: Microreactors and Tube Reactors

Hannes P. L. Gemoets, Volker Hessel, and Timothy Noël

23.1
Introduction

The growing awareness of sustainable processing has created demand for environmentally friendly, atom efficient, and mild chemical transformations with minimal waste production. In oxidation chemistry, the use of molecular oxygen is of particular interest to industry [1–6]. Being completely ecologically compatible and inexpensive, molecular oxygen is the oxidant of choice when compared with dangerous and noble metal-containing oxidants. In addition, the gaseous nature of oxygen facilitates its separation from the product stream, thereby allowing the use of super-stoichiometric amounts to boost the reaction without having to worry about tedious workup procedures.

Despite the appealing nature of molecular oxygen as a green oxidant, there are still some considerable process limitations. Gas–liquid mass transfer phenomena and general low solubility of oxygen are main considerations when using oxygen. In addition, since mixtures of flammable solvents and oxygen can lead to explosive regimes, considerable safety measures have to be taken into account, especially on larger-scale processes. Past research has shown that accurate control over such parameters is by no means trivial, and therefore, aerobic oxidations are rarely applied for conventional large-scale protocols, except for dedicated bulk chemical applications.

These limitations, which are associated with the use of molecular oxygen, might be overcome by the use of microreactor technology [7–15]. Due to their small inner dimensions, microreactors provide both high safety and enhanced process intensification [16–18]. The high surface-to-volume ratio properties of microchannels (inner diameter 100–1000 μm) are highly beneficial, especially for multiphase reactions. In addition to enhanced gas–liquid interfacial transfer, intense recirculation within the liquid slugs allows for fast renewal of the interfaces

---

Liquid Phase Aerobic Oxidation Catalysis: Industrial Applications and Academic Perspectives,
First Edition. Edited by Shannon S. Stahl and Paul L. Alsters.
© 2016 Wiley-VCH Verlag GmbH & Co. KGaA. Published 2016 by Wiley-VCH Verlag GmbH & Co. KGaA.

and efficient mass transfer within the liquid. As a net result of both effects, transfer limitations are overcome and reaction times can be significantly reduced [19–24]. Furthermore, small hold-up volumes allow safe use of hazardous chemicals.

Today, multiple microreactor designs have been reported for the aerobic oxidation of organic compounds. Most of the reactors are built out of simple polymer or stainless steel tubing, assembled together as prototype reactor concepts. But also multigram to kilogram-scale reactor setups have been reported, with automated control and inline purification methods.

This chapter provides some highlights in the innovative field of aerobic oxidation reactions in continuous flow. Topics include transition metal-catalyzed aerobic oxidations in continuous flow, photosensitized singlet oxygen oxidation in continuous flow, metal-free aerobic oxidations in continuous flow, aerobic coupling chemistry in continuous flow, and general prospects for scale-up. Ozonolysis is not covered in this chapter; hereto, we refer to the literature [25–27].

23.2
Chemistry and Catalysis

23.2.1
Transition Metal-Catalyzed Aerobic Oxidations in Continuous Flow

Transition metal-catalyzed oxidations play a prominent role in both fine chemical synthesis and the manufacture of large-volume petrochemicals in industry. Compared with their noncatalytic counterparts, catalytic oxidations offer the ability for mild and selective transformations, thus leading to a more energy-efficient way of processing. Although much progress has been made, organic chemists face the continuing challenge for creating more environmentally sustainable chemistry. The employment of "green" oxidants is one of these key features. With low variable costs and exclusion of toxic waste, molecular oxygen is the oxidant of choice.

Alcohol oxidation to carbonyl compounds is among the most important chemical transformations in organic chemistry. Kobayashi *et al.* achieved quantitative aerobic oxidation of alcohols to aldehydes or ketones by the use a gold-immobilized microcapillary reactor (50 cm length, 250 µm inner diameter) [28]. By a cross-linking method, gold particles (microencapsulated gold as gold source) were immobilized on a polysiloxane-coated capillary. The substrate (organic) solution was combined with an aqueous K_2CO_3 solution in a T-shaped connector before merging with molecular oxygen in the second T-shaped connector. The combined flow was fed to the gold-immobilized capillary reactor (Figure 23.1). Multiphase aerobic oxidation reactions were carried out for a wide scope of benzylic, aliphatic, allylic, and other alcohols. The complete range of secondary alcohols achieved 99% conversion with excellent yields (89–99%). No leaching of gold was observed for at least 4 days. The oxidation of benzyl alcohol led to a significantly lower yield (53%), but by the use of a bimetallic Au/Pd-immobilized capillary column, the yield was improved to 92%.

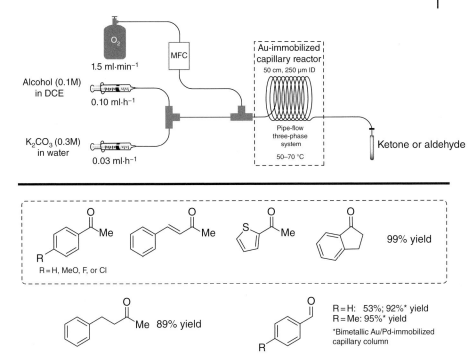

Figure 23.1 Schematic setup of the capillary reactor for gold-catalyzed oxidation reaction of alcohols.

The same group reported a second multiphase flow system for the aerobic oxidation of alcohols, catalyzed by bimetallic nanoclusters (Au–Pt and Au–Pd) in a packed-bed configuration [29]. In addition, the direct oxidative methyl ester formation of various aliphatic and benzylic alcohols was achieved, showing much higher yields and selectivities as compared with its batch counterpart.

Hii et al. developed a highly practical and selective process for the aerobic oxidation of primary and secondary alcohols in a new flow system (differential batch) [30]. A commercially available XCube™ flow reactor was used and readily available Ru/Al$_2$O$_3$ was loaded in the cartridges as a heterogeneous catalyst (Figure 23.2). ICP analysis of products revealed no detectable leaching of catalyst. A solution of alcohol (in toluene) and molecular oxygen was premixed by the use of a gas mixer in order to saturate the solution upon entering the catalyst bed. The single-pass residence time was about 44 s and the product stream was recirculated in a continuous-flow manner until reaction was complete (0.75–7 h). High conversions (94–99%) and excellent selectivities (>95%) were found for a variety of benzylic alcohols (including bulky substituents and/or heteroatoms at the *ortho*-position) and allylic alcohols. The catalyst remained stable in the presence of pyridyl and thienyl functional groups. Primary and secondary aliphatic alcohols were challenging substrates for this system (conversion 64–87%). However, by incorporating a desiccant cartridge (containing

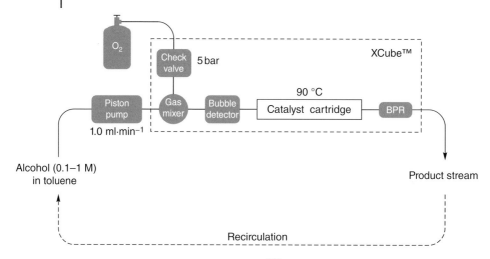

Figure 23.2 General schematic of the XCube™ flow reactor.

MgSO$_4$) into the flow system, negative effects of accumulating water (leading to deactivation of the catalyst) could be diminished. As an illustration, the yield of 2-hexanone (from 2-hexanol) could be improved significantly from 75% to 91%. Moreover, additional rate studies revealed that product inhibition occurs at higher concentrations (0.6–1 M), but the use of higher pressures of molecular oxygen (from 5 to 25 bar) could effectively reduce these effects, restoring the first-order rate. Finally, a telescoping oxidation–Wittig reaction was carried out to demonstrate its potential. In conclusion, good turnover rates, space-time yields, and safe operation ability make this system favorable in comparison to other existing flow systems.

A similar setup was used by Kappe and coworkers (H-Cube™ Pro module) [31]. A heterogeneous iron oxide nanoparticle catalyst was used successfully for the aerobic oxidation of benzyl alcohol. Iron oxide nanocrystals were covalently bound to an aluminosilicate support (Al-SBA15) using a very simple microwave approach, affording Fe/Al-SBA15 (1 wt% Fe) as catalyst. The mesoporous silica/iron oxide catalyst was diluted with silica gel (1 : 3) and loaded into the cartridge. With the aid of TEMPO as cocatalyst, a yield of 42% was achieved for the selective oxidation of benzyl alcohol to benzaldehyde, for a single-pass reaction setup. Continuous-flow recirculation up to 1 h was necessary to achieve full conversion and 95% selectivity. Since the use of noble metals (e.g., Au, Pd) is highly unwanted in pharmaceutical chemistry, this proof-of-principle study shows great potential, since iron oxide is a nontoxic, abundant, and cost-effective heterogeneous catalyst. Note that no leaching of the iron oxide catalyst was detected.

Stahl *et al.* reported a continuous homogeneous copper-catalyzed aerobic oxidation method for primary alcohols [32]. A modified tube reactor was used (Figure 23.3). A dilute oxygen source (9% O$_2$ in N$_2$) was controlled by a mass flow controller and mixed with the substrate and catalyst solution, respectively, by

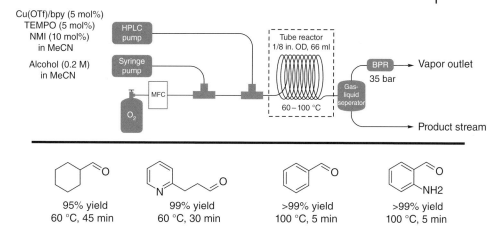

Figure 23.3 Schematic diagram of the tube reactor setup for homogeneous copper-catalyzed oxidation of primary alcohols.

T-mixers, producing a slug flow pattern. The reactor consisted of a stainless steel or PTFE tubing submerged in Paratherm HE heat transfer fluid at a designated temperature (100 °C for benzylic alcohols and 60 °C for aliphatic alcohols). Upon exiting the reactor, remaining gas and liquids were separated using a vapor–liquid separator. Liquids were collected at the bottom and pressure was controlled by a gas pressure relief valve set at 35 bar. This research was an improvement on the former research conducted by the group in continuous flow [33]. The latter research is discussed in Section 23.3. The use of a homogeneous Cu^I/TEMPO catalytic system proved to be superior over the use of homogeneous palladium. The catalyst offered higher catalytic rates, tolerated a wide range of heteroatom substituents without catalyst poisoning, and proved to be highly selective for the conversion of primary alcohols, including aliphatic substrates. Quantitative yields (99%) for activated (benzylic) alcohols and near quantitative yields (95–99%) for primary aliphatic alcohols were achieved within residence times of 5 and 30–45 min, respectively. In addition to its success, a large-scale application of the setup was carried out, with benzyl alcohol as a model substrate. Higher substrate concentration (4 M in MeCN) and lower catalyst loading (2.5 mol% Cu^I(OTf)/2,2'-bipyridine (bpy), 5 mol% N-methylimidazole (NMI), and 0.25 mol% TEMPO) were used. Quantitative formation of benzaldehyde (99%) on a 100 g scale was achieved with 5 min residence. A steady-state profile was sustained for at least 24 h.

Kappe et al. reported a direct aerobic oxidation of 2-benzylpyridines in a gas–liquid continuous-flow regime [34]. A standard two-feed approach was used. Pressurized air was mixed with the substrate liquid phase in a simple T-shaped mixing device (Figure 23.4). The liquid phase was pumped by the use of an HPLC pump. Pressurized air was monitored by a mass flow controller. The two-phase reaction mixture was pumped through a stainless steel coil (0.8 mm inner diameter, 120 m length) and heated with a standard GC oven. Due to

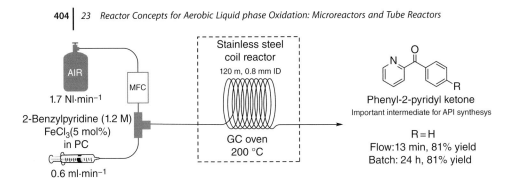

Figure 23.4 Schematic diagram of the gas–liquid continuous-flow reactor for the direct aerobic oxidation of 2-benzylpyridines.

excellent mass transfer characteristics inside the reactor coil, reaction times could be reduced significantly to a minute range protocol (13 min residence time). Moreover, the use of propylene carbonate as green solvent and inexpensive, nontoxic $FeCl_3$ as catalyst, high concentrations of substrate (1.2 M), exclusion of additives, replacement of pure oxygen by pressurized air, and use of high-temperature/high-pressure process windows highlight the sustainable aspect of the conducted research, offering high productivity of the desired ketones in a continuous manner.

23.2.2
Photosensitized Singlet Oxygen Oxidations in Continuous Flow

Photosensitized singlet oxygen oxidation reactions are considered highly promising in terms of green chemistry and sustainable processing [35–37]. More specific, dye-sensitized singlet oxygen generation is recognized as being highly beneficial for pharmaceutical and biomedical chemistry, since heavy metals containing product stream can be avoided [38]. Despite being a clean alternative, singlet oxygen chemistry is rarely used in industrial applications. In typical batch systems, oxygen is bubbled through the reaction mixture while being exposed to light irradiation. This results in low surface-to-volume ratio and poor photon efficiency, since irradiation occurs only at the most outer rim of the batch reactor (Lambert–Beer law). In addition, the short lifetime of singlet oxygen in a solvent environment (e.g., 9.5 µs in methanol) adds up to very low effective concentrations of singlet oxygen, resulting in prolonged reaction times and low selectivity.

One way to overcome such limitation is to implement microflow technology [39–44]. Characterized by their high surface-to-volume ratio, microreactors offer high homogeneous illumination of reaction mixtures, resulting in excellent light penetration throughout the whole medium. In addition, scale-up possibilities (by numbering up) make it possible to increase product throughput to an appealing level for pharmaceutical and biomedical chemistry.

DeMello *et al.* reported the use of a microcapillary film (MCF) reactor for the aerobic oxidation of α-terpinene to ascaridole [45]. The MCF was manufactured

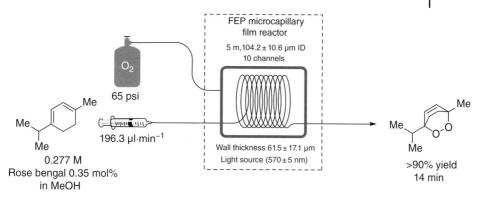

Figure 23.5 Schematic flow setup for the through-wall singlet oxygen oxidation of α-terpinene to ascaridole.

out of fluorinated ethylene propylene (FEP) polymer containing 10 parallel microcapillaries in planar format (Figure 23.5). A reaction mixture of α-terpinene and rose bengal (as catalyst) dissolved in methanol was pumped through the capillaries and oxygen gas was dosed into the solution via a through-wall mass transport. The MCF reactor was irradiated by LED light, which excites the photoredox catalyst and in turn generates singlet oxygen inside the capillaries. In batch, *p*-cymene is observed as the main by-product as a result of a single electron transfer with triplet oxygen, while in the MCF reactor, high selectivity toward ascaridole suggests that the concentration of triplet oxygen is maintained low at all times inside the liquid phase. Within a residence time of 14 min, yields over 90% were achieved for the synthesis of ascaridole.

Another remarkable example is the continuous-flow artemisinin drug synthesis, reported by Seeberger and coworkers [46]. Artemisinin is today's most effective antimalarial drug known and is extracted directly from the plant *Artemisia annua* (sweet wormwood). However, isolation from the plant does not yield sufficient quantities to cope with global demand. Due to its molecular complexity, total synthesis of artemisinin is too laborious (<5% yield over 15 steps) and is, therefore, not a viable alternative. In 2012, Seeberger reported a semisynthetic, continuous-flow conversion of dihydroartemisinic acid (DHAA) to artemisinin in a four-step (singlet oxygen photooxidation, Hock cleavage, triplet oxygen oxidation, and condensation) telescoping process. Especially the first step, photooxidation of DHAA with singlet oxygen, highly benefited from this procedure, since microflow technology allows ideal conditions for efficient light exposure. The starting material DHAA can be found in the plant but in much higher concentrations than artemisinin. In addition, DHAA can be harvested from bioengineered yeast and, therefore, acts as an ideal starting material. The research group was capable of producing 200 g of artemisinin a day with an overall yield of 39%.

The same group greatly improved this process by a novel continuous-flow reactor setup design (Figure 23.6) [47]. For the photooxidation reaction, the

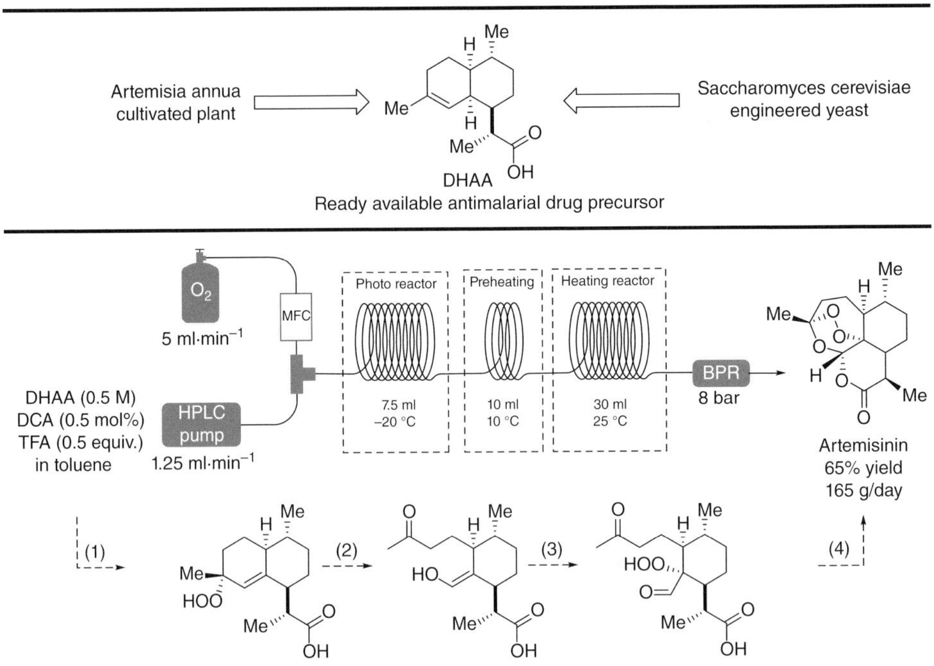

Figure 23.6 Continuous-flow setup for the conversion of dihydroartemisinic acid (DHAA) into artemisinin.

initial 450 W medium-pressure mercury lamp was replaced by 60 high-power LEDs (72 W total), which resulted in increased energy efficiency. Furthermore, the photoredox catalyst was changed to 9,10-dicyanoanthracene (DCA) instead of tetraphenylporphyrin (TPP), as a higher quantum yield and catalyst stability was obtained. Moreover, decreasing temperature (−20 °C) during the photooxidation reaction resulted in higher selectivity. As solvent, DCM was replaced by toluene as being the better choice to cause less environmental and health issues. The acid-catalyzed step (Hock cleavage) and consecutive condensations were conducted at room temperature. All reagents (including the acid for the Hock cleavage) could be added to the initial solution, simplifying the overall process significantly. The process required 11.5 min and allowed the production of 165 g artemisinin a day for a single setup. With the total volume of the setup being 47.5 ml, a space-time yield of 3500 kg·m^{-3}·day^{-1} was acquired.

Disulfides play a prominent role in pharmaceuticals, pesticides, and rubber vulcanization reagents. In biological systems, disulfide bonding is essential for peptides and proteins to fold into their active conformation. Noel et al. reported a mild batch and continuous-flow method to access disulfides by the aerobic oxidation of thiols [48]. Initially, an open flask batch procedure with Eosin Y (1 mol%) as catalyst resulted in near quantitative yields for a variety of thiophenols and heteroaromatic and aliphatic thiols, within 16 h. Moreover, it was found that the addition of the base tetramethylethylenediamine (TMEDA) significantly reduces reaction time to 2–4 h. It was observed that the reaction rate was greatly influenced by the mixing efficiency of gaseous oxygen and the liquid reaction mixture. Therefore, a continuous-flow microreactor setup was built in order to overcome these mass transfer limitations (Figure 23.7). In addition, the confined dimensions of microreactors lead to a highly improved irradiation of the reaction mixture. The setup was constructed out of a high-purity perfluoroalkoxyalkane (PFA) capillary tubing (750 μm inner diameter, volume 950 μl) which was exposed to LED light (3.12 W). Molecular oxygen was fed to a T-mixer by the aid of a mass flow controller and was mixed with the liquid reaction mixture, added by a syringe pump. A segmented flow was established before entering the reactor. Upon exiting the reactor, the mixture was diluted with ethanol to avoid clogging. A variety of thiophenol substrates could by coupled successfully within 20 min with quantitative yields. Notably, difurfuryl disulfide could be exclusively synthesized in the microflow setup with a yield of 87%. To demonstrate the mildness of this setup, the intramolecular S-S coupling of a peptide was carried out to synthesize the hormone oxytocin. Gratifyingly, full conversion could be obtained in water within 200 s residence time. Finally, further studies on kinetic analysis demonstrated a first and second reaction order for thiophenol and oxygen, respectively [49]. In addition, the Hatta number, that is, Ha = 0.06, indicates that gas–liquid mass transfer limitations were eliminated, thereby demonstrating its potential for other gas–liquid photocatalytic reactions.

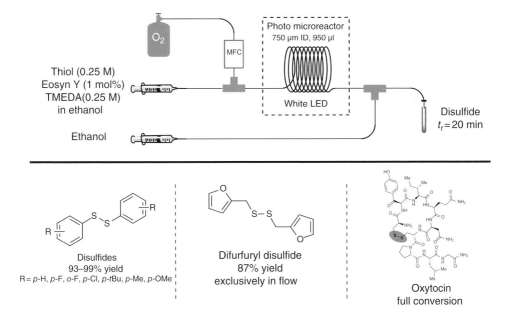

Figure 23.7 Continuous-flow setup for the visible-light photocatalytic aerobic oxidation of thiols to disulfides.

23.2.3
Metal-Free Aerobic Oxidations in Continuous Flow

It is clear that metal-free aerobic oxidations have the advantage over metal-catalyzed aerobic oxidations, in terms of downstream separation and waste treatment.

Hermans and coworkers reported a metal-free aerobic alcohol oxidation in continuous flow [50]. Molecular oxygen was used as terminal oxidant and sub-stoichiometric amounts of HNO_3 as oxygen shuttle. A three-phase flow setup was constructed. A fixed-bed microreactor was loaded with Amberlyst-15, a solid-phase acid catalyst. Molecular oxygen and a liquid solution of the alcohol and HNO_3 (10 mol%) in 1,4-dioxane were merged in a T-mixer before entering the catalyst bed. Upon exiting the reactor, the gaseous and liquid phases were separated and were each monitored by online infrared (IR) spectroscopy for N_2O and product (aldehyde or ketone) concentration consecutively. As anticipated, high gas–liquid mass transfer would allow fast oxidation of NO species, which in turn reduces the formation of N_2O, a strong greenhouse gas. IR spectroscopy of the gaseous phase revealed that significantly less N_2O is formed (~25 times less for 50% conversion) as compared with batch. In addition, the microreactor setup offered high selectivity (>97%) for the oxidation of benzylic, primary aliphatic, and secondary aliphatic alcohols to their respective aldehydes or ketones. Residence times between 8 and 25 s were sufficient to complete this transformation. In the case of benzyl alcohol to benzaldehyde, the space-time yield was increased by 2 orders of magnitude as compared with other reactors (batch, bubble column, and gas loop).

The same group reported a similar transformation of alcohols but with a novel catalytic system [51]. The catalytic system, made out of a commercially available silica-immobilized TEMPO catalyst (instead of the former acidic Amberlyst-15 packing), enabled them to carry out the reaction under much milder conditions. This allowed the possibility for a wider substrate scope to be carried out. The reactor setup was essentially the same as in the previously reported research (Figure 23.8); only now the packed bed was filled with silica-immobilized TEMPO particles. Solvent screening revealed that dichloromethane was a more suitable solvent. The reactor was operated at a mild temperature of 55 °C. These changes allowed the scope to be extended with less stable alcohols. Acid-sensitive alcohols such as prenol and isoprenol could be smoothly oxidized to their corresponding carbonyls within 2 min residence time. Next, 5-hydroxymethylfurfural (HMF), a readily available feedstock from biomass, could selectively be oxidized into its aldehyde, 2,5-diformylfuran (DFF), or carboxylic acid, 2,5-furandicarboxylic acid (FDCA), depending on its residence time (2 or 6 min). Finally, the selective oxidation of lactic acid proved that this method is mild, since 96% yield of pyruvic acid was obtained within 15 s residence time, without decarboxylation of the substrate.

De Bellefon *et al.* reported a safe and straightforward aerobic flow oxidation procedure for aliphatic aldehydes to their corresponding carboxylic acids [52]. The procedure was characterized by its simplicity and mildness. Neither additional

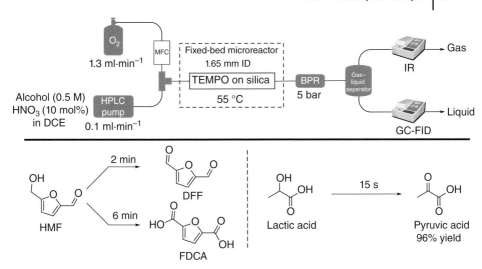

Figure 23.8 Experimental setup for the metal-free aerobic oxidation of alcohols under three-phase flow conditions.

catalysts nor radical initiators were used, and the reactor was set at room temperature. The setup consisted of one T-mixer, merging molecular oxygen and the liquid phase, and a disposable PFA capillary tubing (1000 μm ID), connected to a back pressure regulator (Figure 23.9). Via a sample valve, online GC measurements were taken. Since the reaction rate is limited by oxygen transfer, the gas–liquid transfer efficiency of the microreactor was evaluated. By varying the length of the microreactor, while maintaining a constant residence time (1.8 min), different two-phase superficial velocities were achieved. 2-Ethylhexanal, an industrially important aldehyde, was used as a model substrate. Results indicated that velocities above 0.5 m·min^{-1} ensured a chemically limited regime (optimum of 55% conversion), in the capillary reactor under Taylor flow. Increasing residence time to approximately 15 min, while maintaining 0.5 m·min^{-1} as the minimal superficial flow speed, resulted in greater than 95% conversion. Increasing the concentration of 2-ethylhexanal from 0.8 up to 3.2 M did not affect the conversion or selectivity (75%). Cyclohexanecarboxaldehyde showed a conversion of 90% with a selectivity of over 95% for the same reaction conditions. In the case of benzaldehyde and *n*-octanal, the addition of 100 ppm Mn(II) 2-ethylhexanoate was necessary in order to achieve full conversion, in 17.4 min.

The continuous-flow synthesis of functionalized phenols by aerobic oxidation of aryl Grignard reagents has been demonstrated by Jamison and coworkers [53]. Preliminary research in continuous flow showed significant improvement in terms of high yields within short time frames and enhanced safety procedures. The initial setup was further optimized (PFA tubing 750 μm inner diameter, 2.73 ml, $t_r = 3.4$ min, reactor at −25 °C, air as oxidant and 250 psi back pressure) and a wide scope of substituted phenylmagnesium bromides could be used to prepare functionalized phenols. The scope showed excellent functional-group

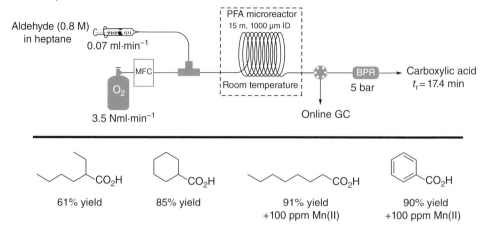

Figure 23.9 Experimental setup for the aldehyde oxidation under Taylor flow conditions.

tolerance and moderate to excellent yields (24–95%). Having demonstrated an efficient flow procedure for the generation of functionalized phenols, the group envisioned that the inline generation of organomagnesium intermediates before the aerobic oxidation would highly benefit the efficiency and practicality of this procedure. The integrated system consisted of a three-step continuous process (Figure 23.10). In the first reactor (R1), a sulfur- or nitrogen-containing nucleophile was deprotonated in the presence of isopropylmagnesium chloride lithium chloride (*i*PrMgCl·LiCl). In the second reactor (R2), the resulting mixture was merged with 1,2-dihalobenzenes at elevated temperature (80–120 °C), resulting in *in situ* benzyne generation and consecutive nucleophilic addition of the organomagnesium nucleophile. In the third reactor (R3), the desired functionalized arylmagnesium intermediate was rapidly cooled to −25 °C and combined with pressurized air to finally yield the *ortho*-functionalized phenols. The overall residence time was only 14 min. Overall yields varied from 33% to 55% for different nucleophiles, including thiophenols, *N*-containing heterocycles, and secondary anilines.

23.2.4
Aerobic Coupling Chemistry in Continuous Flow

Coupling chemistry has played a prominent role in organic chemistry, since it allows efficient formation of carbon–carbon and carbon–heteroatom bonds. Since its introduction in the 1970s, cross-coupling chemistry has witnessed several waves. Most notably was the shift toward the C–H activation pathway. To make cross-coupling chemistry more appealing in terms of sustainability, chemists tried to bypass prefunctionalization of substrates by directly applying carbon–hydrogen bond activation. Cross-dehydrogenative coupling offers the coupling of two simple C–H bonds, eliminating two hydrogen bonds along its way. Despite its atom efficiency, hydrogen gas formation is thermodynamically

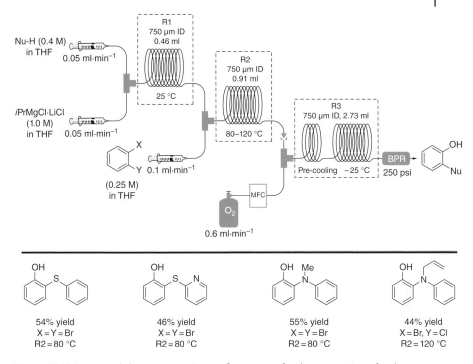

Figure 23.10 Integrated three-step continuous-flow system for the preparation of ortho-functionalized phenols.

unfavored and thus requires an extra driving force, an oxidant. In terms of green chemistry, molecular oxygen is considered most suitable, but its use is often rejected due to process limitations (limited gas–liquid phase transfer and safety consideration on high scale).

Ley et al. demonstrated an innovative continuous-flow procedure for the synthesis 1,3-butadiynes through an aerobic Glaser–Hay acetylene coupling [54]. In order to provide an applicable and consistent system, Ley and coworkers used a semipermeable tube-in-tube reactor to generate homogeneous solutions of gas (Figure 23.11). The inner tube consisted of gas-permeable Teflon AF-2400 membrane tubing (0.8 mm inner diameter and 1.0 mm outer diameter) contained within an outer thick-walled PTFE tubing. Acetonitrile was pumped through the inner tubing of the tube-in-tube reactor. Pressurized oxygen in the outer tubing ensured migration of molecular oxygen through the membrane, generating an oxygenated solvent stream. Next, the saturated solvent stream was mixed with a precombined substrate (alkyne) and catalyst/ligand (CuOTf(MeCN)$_4$/TMEDA) stream. The resulting mixture entered a heating coil (20 ml) at set temperature. After the reaction, inline purification methods (functionalized polymer-supported cartridges) were used to scavenge copper and TMEDA consecutively out of the product stream, allowing product isolation in high purity without the need of extra chromatographic purification.

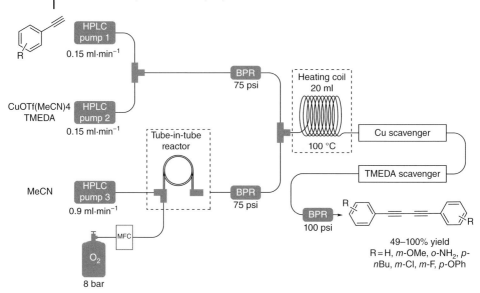

Figure 23.11 Schematic setup for the continuous-flow oxidative acetylene coupling.

For initial investigations, 3-ethynylanisole was coupled to afford 1,4-bis-(3-methoxyphenyl)-buta-1,3-diyne. A quantitative yield (99%) was achieved with the following conditions: 8 bar of O_2 pressure, pump 1 and 2 set at 0.15 ml·min^{-1}, pump 3 set at 0.9 ml·min^{-1}, and heating coil set at 100 °C. An initial scope of electron-rich and electron-deficient aromatic acetylenes resulted in moderate to high yields (49–100%). In the presence of 25 mol% 1,8-diazabicyclo-[5.4.0]-undec-7-ene (DBU), several aliphatic and silyl acetylenes were coupled in good to excellent yields (65–94%). Finally, in order to investigate the scalability and consistency of the reactor, a coupling reaction of 17.4 mmol 3-ethynylanisole (2.30 g) was carried out. Eighty-four percent yield of pure product was obtained without the need of any additional chromatography.

Noel and coworkers developed a fast and straightforward continuous-flow protocol for the aerobic C3-olefination of indoles [55]. With the use of a simple capillary microreactor setup (Figure 23.12), Noel et al. were able to boost intrinsic kinetics of the cross-dehydrogenative Heck coupling to a minute range ($t_r = 10-20$ min) procedure. The microreactor was constructed with FEP polymer capillary tubing (750 μm inner diameter, 4 ml) and set at 110 °C. At first, substrate/catalyst stream and olefin stream were combined in a T-shaped mixer. In a second T-shaped mixer, molecular oxygen was added perpendicular to the liquid stream. A 5 : 1 gas–liquid flow ratio ensured a stable segmented flow regime. The gas–liquid flow regime was found to be crucial in order to successfully reoxidize the reduced catalyst Pd(0) back to its activated Pd(II) format, and the induced vortices inside each segment prevented possible clogging of the microreactor. With optimized conditions in hand, the olefin scope was explored.

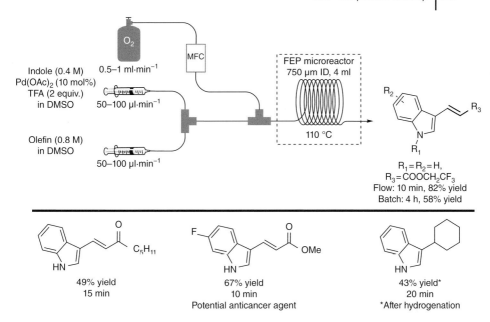

Figure 23.12 Schematic representation of the microflow setup for the aerobic C-3 olefination of indoles.

A wide range of highly activated (acrylates and fluorinated acrylates), activated (N,N-dimethylacrylamide and 1-octen-3-one), and nonactivated (styrene and cyclohexene) olefins were coupled successfully with moderate to excellent yields (27–92%), within a residence time of 10–20 min. In addition, a variety of indole substrates were investigated with ethyl acrylate as a benchmark coupling partner. For substrates with functional groups (both electron withdrawing and electron donating), the reaction proceeded smoothly (66–78%), since exposure to high temperature was kept minimal. Methyl substituents on the C-2 position of the indole were well tolerated (52–62%). Notably, they were able to prepare methyl (E)-3-(6-fluoro-1H-indol-3-yl)acrylate, a potential anticancer agent, with an obtained yield of 67% within 10 min residence time.

23.3
Prospects for Scale-Up

Large-scale aerobic oxidation protocols are often discouraged in pharmaceutical synthesis, because of considerable safety concerns and process constraints. Despite being the cleanest and cheapest oxidant, molecular oxygen is characterized as being unselective, inadequately soluble, and hazardous in the presence of flammable solvents. Using diluted air could successfully eliminate associated safety concerns. However, low concentrations of oxygen produce prolonged reaction times, often resulting in low selectivity and throughput. In addition, since molecular oxygen is a gas, mass transfer limitation in scale-up procedures is of major concern.

On the other hand, chemical processing is under increasing pressure in terms of green chemistry and green engineering practices, and therefore the search for safe and scalable aerobic oxidation methods is of highest interest. One method of choice would be the use of microreactor technology, since microreactors offer scale-up potential, while maintaining the same reaction conditions under steady-state conditions. Maintaining low hold-up volumes on bigger scale offer the ideal environment for using molecular oxygen. Microreactor scale-up possibilities include the use of numbering-up (microreactors in parallel), the use of longer operation times of the device, or the use of higher flow rates in order to increase throughput of each device.

Researchers from Lonza Company have recommended a complex scale-out strategy for liquid phase reactions, which is composed of multiple subsequent steps [56, 57]. These comprise process optimization/intensification strategies as well as reactor scale-out. The first strategy consists of an increase in concentration (commonly giving a factor of 2 increase in productivity) and the use of higher reaction temperature and pressure (2–6× increase in productivity), which are examples of Novel Process Windows. The reactor scale-out consists of a smart increase in dimensions (4–10× increase in productivity), multiscaled processing by adding larger residence time elements (100×), and finally (external) numbering-up (n times increase in productivity with n being the number of reactors). In principle, similar strategies can be employed for gas–liquid aerobic oxidation reactions. However, the numbering-up strategy is more complex due to the fact that a multiphase system needs to be equally distributed over the different channels with major differences in interfacial tension, viscosity, and density.

Several scale-up possibilities for aerobic oxidation in microflow have been reported. Jähnisch *et al.* used a falling-film microreactor for the photooxygenation of olefins with singlet oxygen [58]. The falling-film microreactor was constructed by the Institut für Mikrotechnik Mainz (IMM) and consisted of 32 parallel microchannels (66 mm length, 600 μm width, and 300 μm depth) [59–61]. A liquid solution of cyclopentadiene (4 ml) and Rose Bengal (100 mg) in 250 ml of methanol was pumped into the reactor. Oxygen was fed at 15 l/h cocurrently to the liquid phase. As such, both phases can be considered as continuous, which simplifies substantially the numbering-up procedure. The reactor was kept at 10–15 °C and irradiated by a Xenon lamp to activate the photosensitizer to generate singlet oxygen. The endoperoxide intermediate was reduced immediately resulting in 0.95 g of 2-cyclopenten-1,4-diol (20% yield). This proof-of-principle research clearly demonstrated the feasibility of the falling-film microreactors over conventional falling-film reactors. Liquid films as thin as 20 μm and specific phase interface up to 20 000 m^2/m^3 could be achieved, allowing efficient singlet oxygen formation. In addition, the short hold-up ensured safe handling of explosive endoperoxide intermediate throughout the whole reactor. A further increase for oxidation chemistry was achieved by utilizing a cylindrical falling-film reactor, which contained a 10-fold increase in microchannels compared with the classical falling-film reactor [62]. Interestingly, no performance losses were observed during scale-out.

Stahl and coworkers demonstrated the potential of continuous-flow reactors for the development of safe and scalable palladium-catalyzed aerobic oxidation reactions [33]. Palladium-catalyzed oxidation reactions are known to be rather unforgiving with respect to conventional scaling up, because of unstable catalyst behavior. Temporary periods of low oxygen concentrations can easily lead to irreversible catalyst deactivation (Pd(0) agglomeration) in palladium-catalyzed aerobic oxidations. Stahl et al. focused on the scaling up of the continuous-flow aerobic oxidation of alcohols as a model for similar palladium-catalyzed aerobic oxidations (α,β-dehydrogenation of carbonyl compounds, allylic/vinylic heterofunctionalization, aromatic C–H oxidation, and oxidative cross-coupling), which follow the same catalytic Pd(II)/Pd(0) cycle. A gram-scale continuous-flow setup was built, consisting of a 400 ml coiled tube reactor (0.5 in. OD, stainless steel). The Pd(OAc)$_2$ (5 mol%)/pyridine (20 mol%) catalyst system proved to be most suitable at higher temperature (100 °C). For safety considerations, a diluted oxygen source (8% O$_2$ in N$_2$) was used for scale-up procedures. The oxygen feed was mixed with catalyst and substrate streams consecutively before entering the coiled tube reactor. Upon exiting the reactor (2.5 h residence time), the remaining liquid and gaseous phase were separated in a vapor–liquid separator. Back pressure was applied by inert gas on to the separator. 1-Phenylethanol could be converted into its corresponding ketone in near quantitative yield (98%), on a 25 g scale. Various other alcohols could be converted successfully on gram scale (>20 g), with near quantitative conversions toward the desired product (93–99% GC yield). Finally, a kilogram-scale procedure was carried out (Figure 23.13). The 400 ml coiled reactor was exchanged by a 7 l coil of stainless steel tubing (0.375 in. OD). The transformation of 1-phenylethanol (1 M) was carried out with

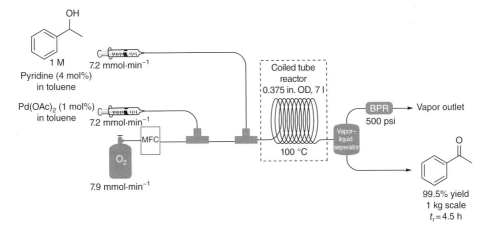

Figure 23.13 Schematic representation of the continuous-flow tube reactor designed for homogeneous Pd-catalyzed aerobic oxidation reactions.

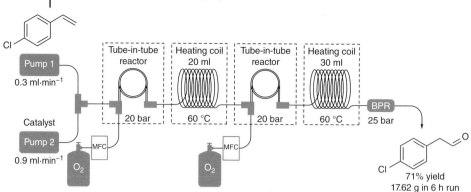

Figure 23.14 Schematic setup for the synthesis of phenylacetaldehydes from styrenes through an aerobic anti-Markovnikov Wacker oxidation.

1 mol% of Pd(OAc)$_2$ and 4 mol% of pyridine at the same temperature (100 °C) and pressure (500 psi) and resulted in near quantitative yield (99.5%) of acetophenone for the oxidation of 1 kg substrate. The residence time of the reactor was 4.5 h. The kilogram-scale application clearly highlights the potential of continuous-flow tube reactors for large-scale aerobic oxidation processes.

Ley et al. constructed a flow setup for the preparation of phenylacetaldehydes from functionalized styrenes, through an aerobic anti-Markovnikov Wacker oxidation [63]. On a multigram scale (21.6 mmol·h^{-1}), a double tube-in-tube gas reactor setup was used, which allowed accurate charging of the reaction mixture with stoichiometric amounts of molecular oxygen (Figure 23.14). This resulted in minimal overoxidation, which would be particularly challenging in batch protocols. In addition, the microreactor has no effective reactor headspace, preventing the formation of possible explosive gas-phase mixtures. The substrate (1.2 M 4-chlorostyrene in toluene: t-BuOH 1:6) and catalyst (5 mol% (MeCN)$_2$PdCl$_2$, 2 mol% CuCl$_2$, and 1.4 equiv. H$_2$O) streams were combined in a T-shaped mixer and pumped into the tube-in-tube gas reactor (AF-2400 inner tubing, 1.5 m, 0.7 ml) with pressurized pure oxygen (20 bar). The oxygenated reaction stream was then passed through a 20 ml stainless steel coiled reactor at 60 °C. Exiting the first reactor coil, the reaction stream was then passed through a second similar tube-in-tube gas reactor and a 30 ml coiled reactor (60 °C) consecutively. After a 40 min overall residence time, the exiting product stream was collected in a stirred beaker containing sodium bisulfite solution. Due to the relative purity of the product stream, simple crystallization (formation of bisulfite adducts) allowed isolation of the desired phenylacetaldehyde without other purification steps. The multigram setup was run over a course of 6 h, obtaining 17.62 g (68 mmol) of pure 4-chlorophenylacetaldehyde sulfite adduct with good yield (71%) and high selectivity (76 : 4 acetaldehyde:ketone).

As mentioned earlier, numbering-up of gas–liquid reactions is not easily achieved. One must ensure that equal reaction conditions are established in different reaction channels. Recently, Schouten et al. have developed a barrier-based micro-/millireactor to allow for numbering-up of gas–liquid reactions [64–68]. Uniform distribution of the gas and liquid flows over the microchannels is achieved when the pressure drop over the barrier channels is 4–25 times the pressure drop over the corresponding T-mixers and microchannel. Gas–liquid channeling is prevented at equal pressures in the gas and liquid manifolds.

23.4 Conclusions

The use of continuous-flow microreactors for aerobic oxidation chemistry has received an increasing amount of attention in recent years. So far, the technology has been mostly applied in research environments where it can be used to safely screen a number of process parameters. The potential to use molecular oxygen in combination with flammable solvents is one of the main drivers behind the success of microreactor technology. We anticipate that in the coming years many improvements will come with regard to the technology itself but also with regard to reaction selection. From a technology perspective, more efficient unit operations will facilitate multistep syntheses in a fully continuous fashion [69]. In addition, improved automation protocols of these processes will allow the reduction of the manual handling required to operate the process [70]. From a chemistry perspective, we expect that this technology can be used to develop new and more efficient aerobic oxidation processes. The ease of implementation of this technology will serve as an incentive for future research. As an example, for metal-catalyzed reactions, the high gas–liquid interfacial areas allow the reduction of the catalyst loadings and to improve the selectivity of the reaction due to shorter contact times. In this chapter, a few examples have been given to exploit the scale-up potential of continuous-flow reactors. However, much more needs to be done and this will require an intense collaboration between chemists and process engineers.

References

1. Campbell, A. and Stahl, S. (2012) *Acc. Chem. Res.*, **45**, 851–863.
2. Wendlandt, A.E., Suess, A.M., and Stahl, S.S. (2011) *Angew. Chem. Int. Ed.*, **50**, 11062–11087.
3. Gulzar, N., Schweitzer-Chaput, B., and Klussmann, M. (2014) *Catal. Sci. Technol.*, **4**, 2778–2796.
4. Cao, Q., Dornan, L.M., Rogan, L., Hughes, N.L., and Muldoon, M.J. (2014) *Chem. Commun.*, **50**, 4524–4543.
5. Allen, S.E., Walvoord, R.R., Padilla-Salinas, R., and Kozlowski, M.C. (2013) *Chem. Rev.*, **113**, 6234–6458.
6. Davis, S.E., Ide, M.S., and Davis, R.J. (2013) *Green Chem.*, **15**, 17–45.
7. Hessel, V., Kralisch, D., Kockmann, N., Noël, T., and Wang, Q. (2013) *ChemSusChem*, **6**, 746–789.
8. McQuade, D.T. and Seeberger, P.H. (2013) *J. Org. Chem.*, **78**, 6384–6389.
9. Pastre, J.C., Browne, D.L., and Ley, S.V. (2013) *Chem. Soc. Rev.*, **42**, 8849–8869.
10. Noël, T. and Buchwald, S.L. (2011) *Chem. Soc. Rev.*, **40**, 5010–5029.
11. Hartman, R.L., McMullen, J.P., and Jensen, K.F. (2011) *Angew. Chem. Int. Ed.*, **50**, 7502–7519.

12. Wegner, J., Ceylan, S., and Kirschning, A. (2011) *Chem. Commun.*, **47**, 4583–4592.
13. Glasnov, T.N. and Kappe, C.O. (2011) *J. Heterocycl. Chem.*, **48**, 11–30.
14. Webb, D. and Jamison, T.F. (2010) *Chem. Sci.*, **1**, 675–680.
15. Jähnisch, K., Hessel, V., Löwe, H., and Baerns, M. (2004) *Angew. Chem. Int. Ed.*, **43**, 406–446.
16. Kumar, V. and Nigam, K.D.P. (2012) *Green Process. Synth.*, **1**, 79–107.
17. Gorak, A. and Stankiewicz, A.I. (2011) *Annu. Rev. Chem. Biomol. Eng.*, **2**, 431–451.
18. Becht, S., Franke, R., Geißelmann, A., and Hahn, H. (2009) *Chem. Eng. Process. Process Intensif.*, **48**, 329–332.
19. Noël, T. and Hessel, V. (2013) *ChemSusChem*, **6**, 405–407.
20. Sobieszuk, P., Aubin, J., and Pohorecki, R. (2012) *Chem. Eng. Technol.*, **35**, 1346–1358.
21. Kashid, M.N. and Kiwi-Minsker, L. (2009) *Ind. Eng. Chem. Res.*, **48**, 6465–6485.
22. Chen, G., Yue, J., and Yuan, Q. (2008) *Chin. J. Chem. Eng.*, **16**, 663–669.
23. Günther, A. and Jensen, K.F. (2006) *Lab Chip*, **6**, 1487–1503.
24. Hessel, V., Angeli, P., Gavriilidis, A., and Löwe, H. (2005) *Ind. Eng. Chem. Res.*, **44**, 9750–9769.
25. Nobis, M. and Roberge, D.M. (2011) *Chim. Oggi*, **29**, 56–58.
26. Irfan, M., Glasnov, T.N., and Kappe, C.O. (2011) *Org. Lett.*, **13**, 984–987.
27. Roydhouse, M.D., Ghaini, A., Constantinou, A., Cantu-Perez, A., Motherwell, W.B., and Gavriilidis, A. (2011) *Org. Process Res. Dev.*, **15**, 989–996.
28. Wang, N., Matsumoto, T., Ueno, M., Miyamura, H., and Kobayashi, S. (2009) *Angew. Chem. Int. Ed.*, **48**, 4744–4746.
29. Kaizuka, K., Lee, K.-Y., Miyamura, H., and Kobayashi, S. (2012) *J. Flow Chem.*, **2**, 1–4.
30. Zotova, N., Hellgardt, K., Kelsall, G.H., Jessiman, A.S., and Hii, K.K. (2010) *Green Chem.*, **12**, 2157–2163.
31. Obermayer, D., Balu, A.M., Romero, A.A., Goessler, W., Luque, R., and Kappe, C.O. (2013) *Green Chem.*, **15**, 1530–1537.
32. Greene, J.F., Hoover, J.M., Mannel, D.S., Root, T.W., and Stahl, S.S. (2013) *Org. Process Res. Dev.*, **17**, 1247–1251.
33. Ye, X., Johnson, M.D., Diao, T., Yates, M.H., and Stahl, S.S. (2010) *Green Chem.*, **12**, 1180–1186.
34. Pieber, B. and Kappe, C.O. (2013) *Green Chem.*, **15**, 320–324.
35. Montagnon, T., Tofi, M., and Vassilikogiannakis, G. (2008) *Acc. Chem. Res.*, **41**, 1001–1011.
36. Margaros, I., Montagnon, T., Tofi, M., Pavlakos, E., and Vassilikogiannakis, G. (2006) *Tetrahedron*, **62**, 5308–5317.
37. Clennan, E. and Pace, A. (2005) *Tetrahedron*, **61**, 6665–6691.
38. Wahlen, J., De Vos, D.E., Jacobs, P.A., and Alsters, P.L. (2004) *Adv. Synth. Catal.*, **346**, 152–164.
39. Su, Y., Straathof, N.J.W., Hessel, V., and Noël, T. (2014) *Chem. Eur. J.*, **20**, 10562–10589.
40. Garlets, Z.J., Nguyen, J.D., and Stephenson, C.R.J. (2014) *Isr. J. Chem.*, **54**, 351–360.
41. Noël, T., Volker, H., and Wang, X. (2013) *Chim. Oggi*, **31**, 10–14.
42. Oelgemoeller, M. (2012) *Chem. Eng. Technol.*, **35**, 1144–1152.
43. Knowles, J.P., Elliott, L.D., and Booker-Milburn, K.I. (2012) *Beilstein J. Org. Chem.*, **8**, 2025–2052.
44. Oelgemöller, M. and Shvydkiv, O. (2011) *Molecules*, **16**, 7522–7550.
45. Elvira, K.S., Wootton, R.C.R., Reis, N.M., Mackley, M.R., and deMello, A.J. (2013) *ACS Sustainable Chem. Eng.*, **1**, 209–213.
46. Lévesque, F. and Seeberger, P.H. (2012) *Angew. Chem. Int. Ed.*, **51**, 1706–1709.
47. Kopetzki, D., Lévesque, F., and Seeberger, P.H. (2013) *Chem. Eur. J.*, **19**, 5450–5456.
48. Talla, A., Wang, X., Driessen, B., Straathof, N.J.W., Milroy, L.-G., Brunsveld, L., Hessel, V., and Noël, T. (2015) *Adv. Synth. Catal.*, **357**, 2180–2186.
49. Su, Y., Volker, H., and Noël, T. (2015) *AIChE J.*, **61**, 2215–2227.
50. Aellig, C., Scholz, D., and Hermans, I. (2012) *ChemSusChem*, **5**, 1732–1736.

51. Aellig, C., Scholz, D., Conrad, S., and Hermans, I. (2013) *Green Chem.*, **15**, 1975–1980.
52. Vanoye, L., Aloui, A., Pablos, M., Philippe, R., Percheron, A., Favre-Réguillon, A., and de Bellefon, C. (2013) *Org. Lett.*, **15**, 5978–5981.
53. He, Z. and Jamison, T.F. (2014) *Angew. Chem. Int. Ed.*, **53**, 3353–3357.
54. Petersen, T.P., Polyzos, A., O'Brien, M., Ulven, T., Baxendale, I.R., and Ley, S.V. (2012) *ChemSusChem*, **5**, 274–277.
55. Gemoets, H.P.L., Hessel, V., and Noël, T. (2014) *Org. Lett.*, **16**, 5800–5803.
56. Kockmann, N., Gottsponer, M., and Roberge, D.M. (2011) *Chem. Eng. J.*, **167**, 718–726.
57. Kockmann, N. and Roberge, D.M. (2009) *Chem. Eng. Technol.*, **32**, 1682–1694.
58. Jähnisch, K. and Dingerdissen, U. (2005) *Chem. Eng. Technol.*, **28**, 426–427.
59. Rebrov, E.V., Duisters, T., Löb, P., Meuldijk, J., and Hessel, V. (2012) *Ind. Eng. Chem. Res.*, **51**, 8719–8725.
60. Ziegenbalg, D., Löb, P., Al-Rawashdeh, M., Kralisch, D., Hessel, V., and Schönfeld, F. (2010) *Chem. Eng. Sci.*, **65**, 3557–3566.
61. Cantu-Perez, A., Al-Rawashdeh, M., Hessel, V., and Gavriilidis, A. (2013) *Chem. Eng. J.*, **227**, 34–41.
62. Vankayala, B.K., Löb, P., Hessel, V., Menges, G., Hofmann, C., Metzke, D., Krtschil, U., and Kost, H.-J. (2007) *Int. J. Chem. Reactor Eng.*, **5**, A91.
63. Bourne, S.L. and Ley, S.V. (2013) *Adv. Synth. Catal.*, **355**, 1905–1910.
64. Al-Rawashdeh, M., Fluitsma, L.J.M., Nijhuis, T.A., Rebrov, E.V., Hessel, V., and Schouten, J.C. (2012) *Chem. Eng. J.*, **181–182**, 549–556.
65. Al-Rawashdeh, M., Yu, F., Nijhuis, T.A., Rebrov, E.V., Hessel, V., and Schouten, J.C. (2012) *Chem. Eng. J.*, **207–208**, 645–655.
66. Al-Rawashdeh, M., Nijhuis, X., Rebrov, E.V., Hessel, V., and Schouten, J.C. (2012) *AIChE J.*, **58**, 3482–3493.
67. Al-Rawashdeh, M., Yue, F., Patil, N.G., Nijhuis, T.A., Hessel, V., Schouten, J.C., and Rebrov, E.V. (2014) *AIChE J.*, **60**, 1941–1952.
68. Al-Rawashdeh, M., Zalucky, J., Müller, C., Nijhuis, T.A., Hessel, V., and Schouten, J.C. (2013) *Ind. Eng. Chem. Res.*, **52**, 11516–11526.
69. Kenig, E.Y., Su, Y., Lautenschleger, A., Chasanis, P., and Grünewald, M. (2013) *Sep. Purif. Technol.*, **120**, 245–264.
70. McMullen, J.P. and Jensen, K.F. (2010) *Annu. Rev. Anal. Chem.*, **3**, 19–42.

Index

a

AA. *see* azelaic acid (AA)
acetaldehyde
– olefin oxidation
– – ethylene oxidation   140, 145–148
– – kinetics and mechanism   140–145
– Wacker process
– – process improvements   151–155
– – single-stage process   148–149
– – two-stage process   149–150
acetophenone (ACP)   18
acetylene-based Reppe process   159, 160
acrolein oxidation   178–180
activated carbons (ACs)   267–275, 283
– dehydrogenation reactions   279
– dehydrogenative coupling reactions   280
– H_2O_2 time   271–272
– and hydrogen peroxide   269
– nitrogen content   272–274
– oxidative cleavage reactions   275–277
– oxygenation reactions   275
– oxygen flow   271
– oxygen pressure   271
– pore size distribution   271, 272
active pharmaceutical ingredient (API)   304, 306
adamantane oxidation   257–258
adamantanols   257
adipic acid   258
aerobic DDQ-catalyzed reactions   229–230
aerobic oxidative esterification reaction
– Au–NiO$_x$ nanoparticle catalyst   216–217
– block flow diagram   214
– Pd–Pb catalyst
– – discovery   210
– – industrial catalyst   213–214
– – intermetallic compounds   210–212
– – Pb role   213

– Pd/SiO$_2$–Al$_2$O$_3$ catalyst   215–216
– reaction mechanism   212
alcohol oxidation
– copper catalysts (*see* copper-catalyzed aerobic alcohol oxidation)
– NO$_x$ cocatalysts (*see* NO$_x$-catalyzed aerobic alcohol oxidation)
alcohol-to-ketone ratio   9
aldehyde oxidation, Taylor flow conditions   410
alkene acetoxylation   117–118
alkoxyl radicals   4
allyl alcohol   118
allylic amines   152
amberlyst-15/NO$_x$-catalyzed aerobic oxidation   241–243, 247–248
amfepramone   81, 82
amino acid-promoted aerobic C–H olefination   130
aminomethylphosphonic acid (AMPA)   269–271
anthraquinone oxidation (AO) process   221–223
– autoxidation step   223–225
– hydrogenation process   225–226
– *vs.* quinone-catalyzed oxidation reactions   229
anti-Markovnikov Wacker oxidation   416
– aldehyde formation   151–152
– continuous-flow microreactors   416
– NO$_x$ cocatalysts   131
API. *see* active pharmaceutical ingredient (API)
arene olefination   126–128
artemisinic acid   389, 390
artemisinin   389–391
Ashland Oil, Inc.   124
Aspergillus niger mold oxidation process   351, 362–364

Liquid Phase Aerobic Oxidation Catalysis: Industrial Applications and Academic Perspectives,
First Edition. Edited by Shannon S. Stahl and Paul L. Alsters.
© 2016 Wiley-VCH Verlag GmbH & Co. KGaA. Published 2016 by Wiley-VCH Verlag GmbH & Co. KGaA.

AstraZeneca process   77, 78
(all-*rac*)-α-tocopherol   106
Au–NiO$_x$ nanoparticle catalyst   216–217
autooxidation reaction   19
– chain-propagation   5
– cyclohexane   7
– induction-initiation   4–5
– initiators   4
– radical pathway   3
– termination reactions   5–6
AZD8926 synthesis   77, 78
azelaic acid (AA)
– applications   331, 343
– biotechnological syntheses   339–341
– oleic acid, synthesis from
– – double bond cleavage   336–337
– – ozonolysis   332–336
– – three-stage conversion   339
– – two-stage conversion   337–339
– oleic acid, synthesis from
– properties   331
– scale-up process   341–342

b

Baeyer–Villiger monooxygenase (BVMO)-mediated enzymatic process
– esomeprazole production   295–298
– mechanism   292–293
balanced catalytic surfactants   387–388
benzene production. *see* cumene route synthesis
benzoquinone   153, 221
benzoylsalicyclic acid   71
benzyl acetates   125
benzyl alcohol oxidation   241–242
benzylic acetoxylation   125–126, 131
2-benzylpyridines oxidations   403–404
bicyclic [3.1.0]proline   299, 300
bimolecular initiation reaction   7, 8
binuclear ethylene–palladium complex   139
bio-based feedstock   313, 320, 325–326
bio-1,4-butanediol production   169
biocatalysis and fermentation
– AA   339–341
– bisulfite oxidation   302–303
– BVMOs
– – esomeprazole production   295–298
– – mechanism   292–293
– CHMO   304, 305
– CYPs   304, 305
– MAO   298–302
– SMOs   304, 305
biorefinery concept   349, 350

bis(dimethyldialkylammonium)molybdates   387, 388
boceprevir   299, 303
BP PTA process. *see* purified terephthalic acid (PTA)
1,4-butanediol   161
– application   159
– 1,3-butadiene-based synthesis
– – oxidative acetoxylation   162–164
– – oxyhalogenation   162
– butane-based process   161
– 3,4-epoxy-1-butene route   169
– Mitsubishi Chemical's production method   160
– – first generation process   165–166
– – second generation process   167–168
– process improvement   168–169
– propylene-based process   161
– Reppe process   159, 160
– Rh–Te–C catalyst   168
t-butyl hydroperoxide   154
BVMO-mediated enzymatic process. *see* Baeyer–Villiger monooxygenase enzymatic process

c

Cabot Norit Activated Carbon   270
CAO enzymes. *see* copper amine oxidase (CAO) enzymes
carbocatalysis   274, 283
carbohydrate oxidation
– disaccharides   358–361
– enzymatic catalysis   363–364
– metal catalysts   364–366
– monosaccharides   354–357
– organic acids, global market of   366
– polysaccharides   361–362
carbon nanotubes (CNTs)   28, 29
– dehydrogenation reactions   279
– oxygenation reactions   278
catalytic aerobic oxidation reactions
– *vs.* AO process   229
– CAO enzymes   231–234
– NO$_x$ cocatalysts   229–230
– Pd-catalyzed acetoxylation   230–231
cellobiose, gold catalyzed oxidation of   359, 360
chain initiation reactions   6–7
chain-propagation reactions   5, 7–10
Chan–Lam coupling reactions   77
chemical oxidation   364–366
chemoenzymatic oxidation   360–361
CHHP. *see* cyclohexyl hydroperoxide (CHHP)
chloranil   221

2-chloroethanol 147–148
CHMO. see cyclohexanone monooxygenase (CHMO)-mediated enzymatic process
(−)-cis-rose oxide 391
β-citronellol 391, 392
Claisen-type rearrangement 100
Co–Br catalyzed oxidation
– p-xylene (pX) oxidation
– – branching sequence 46–47
– – hydrogen abstraction reactions 47, 48
– – reactive intermediates 43, 44
– – thermal homolytic cleavage 44–45
Co(III)-catalyzed acetic acid combustion 56, 57
cofactor recycling systems 305
colloidal metals, deposition/immobilization of 365
Co/Mn/Br-catalyzed oxidation
– furanics oxidation process 315–316
– – gas composition control 322
– – heterogeneous catalysts 316–318
– – homogeneous catalysts 318–320
– – oxygen mass transfer limitations 324
– – reaction pathways 320, 321
– – safety operation 324–325
– – temperature control 323–324
– levulinic acid 325, 326
– lignin 325–326
continuous-flow microreactors
– anti-Markovnikov Wacker oxidation 416
– Glaser–Hay acetylene coupling 411–412
– indoles, C3-olefination of 412–413
  metal free aerobic oxidations
– – *ortho* functionalized phenols 410, 411
– – Taylor flow conditions 409, 410
– – three phase flow conditions 408, 409
– palladium-catalyzed aerobic oxidation 415
– photosensitized singlet oxygen oxidation
– – artemisinin synthesis 405–406
– – ascaridole synthesis 404–405
– – thiol oxidation 406–407
– transition metal-catalyzed aerobic oxidations
– – 2-benzylpyridines 403–404
– – gold-catalyzed oxidation 400–401
– – homogeneous copper-catalyzed oxidation 402–403
copper amine oxidase (CAO) enzymes 231–234
copper-catalyzed aerobic alcohol oxidation
– catalyst types 86
– Cu/TEMPO catalyst system 87, 89
– – BASF adaptation 91, 92
– – DSM 92

– – simplified mechanism 90
– – transition state energies 90
– nitroxyl radicals 87, 88
– scale-up 91–93
– scope 86
– Zeneca application 93
copper-catalyzed aerobic oxidation
– α-amination reactions 80, 81
– DMC formation
– – chloride-free catalysts 73, 74
– – process technology 75, 76
– – reaction mechanism 72–73
– pharmaceutical applications
– – AZD8926 synthesis 77, 78
– – DCPA 77, 78
– – imidazo[1,2-*a*]pyridine 80, 81
– – 1,2,4-triazoles 79
– phenol synthesis
– – chemistry and catalysis 70–72
– – process technology 74–75
copper(I)/DBED-catalyzed aerobic alcohol oxidation 86
copper(II) salts 69
coprecipitation 365
corrosion-resistant reactors 73, 75
coupling chemistry 410
cross-coupling chemistry 410
Cu/ABNO catalyst systems 90
Cu/DBAD-catalyzed aerobic alcohol oxidation
– reaction conditions 89
– Zeneca application 93
cumene hydroperoxide (CHP) 15, 16
– carbon nanotubes (CNT) 28
– thermal decomposition 17–19, 27
cumene oxidation process 15, 262
– axial dispersion model 24
– bubble column reactor 22–23
– compartment reactor model 25–26
– design improvements 29–30
– film reactor model 24–25
– ideally mixed reactor model 23–24
– modification 27–29
– oxidation block diagram 22
– oxidation scheme 19–21
– oxidation side reactions 21
– process overview 16, 21–22
– process safety 26–27
cumene process 15, 16
cumene route phenol synthesis 15, 16
cupric chloride 145–147
Cu/TEMPO-catalyzed alcohol oxidation 87, 89
– BASF adaptation 91, 92
– DSM 92

Cu/TEMPO-catalyzed alcohol oxidation (*contd.*)
– simplified mechanism   90
– transition state energies   90
cyclohexane autoxidation   3, 7, 8, 11–12
cyclohexane oxidation   11–12, 258
– noncatalyzed oxidation process   37–38
– patent publications   38
– process improvements   38–39
– selectivity   34–35
– simplified reaction scheme   34
– traditional catalyzed process   35–37
cyclohexanone   175–177, 189
– general information   33
– production routes   33
– synthesis (*see* cyclohexane oxidation)
cyclohexanone monooxygenase (CHMO)-mediated enzymatic process
– esomeprazole production   296–298
– evolution challenges and strategies   296
– process challenges and strategies   297
– reaction scheme   296
cyclohexene oxidation   176–178
cyclohexylbenzene oxidation   261
cyclohexyl hydroperoxide (CHHP)   34, 35, 37
CYP102A1   305
cytochrome P450s (CYPs)   304, 305

d

dark singlet oxygenation (DSO)
– advantages   384
– artemisinin synthesis   389
– homogeneous aqueous and alcoholic media   384–385
– multiphase microemulsions   387–388
– rose oxide synthesis   391–392
– schematic representation   386
– single-phase microemulsions   385–387
DCPA. *see* dicyclopropylamine (DCPA)
deposition–precipitation (DP) procedure   365
1,4-diacetoxy-2-butene   162–165, 168
dibutyl carbonate   119
dibutyl oxalate   119
2,3-dichloro-5,6-dicyanobenzoquinone (DDQ)   221, 229–230
dicyclopropylamine (DCPA)   77, 78
dihydroartemisinic acid (DHAA)   405, 406
1,5-dihydroxynaphtalene, photooxygenation of   383
3,3-dimethoxy methyl propionate   178, 180–181, 184, 185, 187
dimethylbenzylalcohol (DMBA)   17, 19–21
dimethylcarbonate (DMC)   72–73, 75

– chloride-free catalysts   73, 74
– process technology   75, 76
– reaction mechanism   72–73
2,6-dimethylphenol (DMP)
– oxidative coupling of   97–99
– TMHQ production   108
dimethyl phthalate, Pd-catalyzed oxidative coupling of   121
1,4-dioxaspiro[4,5]decane   176, 177
diphenoquinone (DPQ)   98, 100, 108
diphenyl carbonate (DPC)
– polycarbonate production   189–190
– synthesis (*see* direct DPC process)
direct DPC process
– bromide role   199–201
– catalyst optimization   201–202
– downstream processing and catalyst recovery   203
– GE   190–192
– heterogeneous palladium catalysts   203–204
– patent activity   204
– phenol oxidative carbonylation
– – catalysts   193–195
– – inorganic cocatalysts   196–198
– – mechanism   192–193
– – multicomponent catalytic packages   199
– – organic cocatalysts   196
– water removal   202
disaccharides oxidation   358–361
DMC. *see* dimethylcarbonate (DMC)
DPC. *see* diphenyl carbonate (DPC)

e

Eastman Chemical 1,4-butandiol manufacturing process   169
EMEROX® azelaic acids   333
Enichem/Versalis process   72, 73, 76
enzymatic oxidation   363–364
3,4-epoxy-1-butene   169
esomeprazole   295–298
esterification
– methacrolein   209
– palladium catalyzed aerobic oxidation   123–124
ethylene oxidation
– acetaldehyde synthesis   140, 145–148
– palladium-catalyzed oxidation   173
ExxonMobil   261, 262

f

fermentation   363–364
2,5-furandicarboxylic acid (FDCA)
– Avantium's YXY process   314

– furanics oxidation process   315–316
– – gas composition control   322
– – heterogeneous catalysts   316–318
– – homogeneous catalysts   318–320
– – oxygen mass transfer limitations   324
– – reaction pathways   320, 321
– – safety operation   324–325
– – temperature control   323–324

g

galactose oxidase   89
galactose-6-oxidase enzyme   362
gas–liquid continuous-flow reactor   404
Glaser–Hay acetylene coupling   411–412
glucose oxidation
– chemical intermediates   349
– derived products   349, 350, 354
– enzymatic process *vs.* chemical process   362–366
– gold catalytic reaction   355, 356
– kinetic investigation   353
– molecular mechanism   352
– mono-and bimetallic catalysts   357
– stoichiometry   352
glyphosate production
– DEA–IDA route   267, 268
– glycine route   267, 268
– HCN–IDA route   267, 268
– macro kinetic model   270
– reaction scheme   267, 268
gold catalytic oxidation   364–366
graphene oxide (GeO) catalyst   275, 278–280
graphite oxide (GiO) catalyst   275, 279, 280
graphitic carbon nitride (g-CN) catalyst   274, 275
– dehydrogenation reactions   281–282
– dehydrogenative coupling reactions   282
– oxidative cleavage reactions   281
– oxygenation reactions   280–281
ground-state molecular oxygen ($^3O_2$)   373, 374
guaiacol process   91

h

heterogeneous bromineless oxidation catalysis   62
heterogenized $PdSO_4$-$VOSO_4$-H_2SO_4-on-coal catalyst   153
heteropolyacids   152
high-throughput (HTP) identification strategy   301

Hock-process. *see* cumene oxidation process
homogeneous bromineless catalysis   61–62
homogeneous copper-catalyzed aerobic oxidation   402–403
horseradish peroxidase–Amplex Red spectrophotometric assay   301
HPPO process. *see* hydrogen peroxide propylene oxide (HPPO) process
Hu's $NaNO_2$/Br_2/TEMPO-catalyzed aerobic alcohol oxidation   245, 246
Hydranone® technology   39
hydrocarbon autoxidation
– chain initiation reactions   6–7
– chain-propagation reactions   5, 7–10
– epoxide selectivity   12–13
– mechanistic difficulties   13
– ring-opened by-products   11–12
hydrogen abstraction reactions   5, 10
hydrogen peroxide based propylene oxide (HPPO) process   222–223
hydrogen peroxide synthesis
– AO process (*see* anthraquinone oxidation (AO) process)
– process technology   227–229
– simplified process flow diagram   227
hydrolysis/hydrogenation   165, 314, 359, 365
– acrolein synthesis   178–180
– 1,4-butanediol synthesis   162, 166
– phenol synthesis   75, 123, 124
hydroperoxides
– alkylaromatics, selective oxidation of   260–262
– co-propagation   8–10
hydroquinones   221
5-(hydroxymethyl)furfural (HMF)   314–318, 320, 324, 325

i

ICC. *see* inorganic cocatalysts (ICC)
imidazo[1,2-*a*]pyridine   80, 81
impregnation method   365
industrialized singlet oxygenation
– artemisinin   389–391
– ascaridole   388
– rose oxide synthesis   391–392
industrial oxidative acetoxylation catalyst   163–164
inorganic cocatalysts (ICC)   196–198
Ishii's catalyst   263

j

Jablonski diagram   375
juglone   383

k

Kagan catalytic oxidation method 294–295
KA oil 33, 36, 37
Kharasch's ethylene–Pd complex 143

l

lead-based DPC catalysts 197, 201
levulinic acid (LA) 314, 325, 326
ligand-modulated aerobic oxidation catalysis 128–130
lignin oxidation 325–326
limiting oxygen concentrations (LOCs) 27, 93, 322
linear aliphatic diols 159
liquid phase Cu-catalyzed aerobic oxidation reactions 70
liquid phase Pd-catalyzed process 119
luminous singlet oxygenation 383–384

m

MAO-catalyzed oxidation. *see* monoamine oxidase (MAO)-catalyzed oxidation
MC-catalyzed oxidation
 – bromine species 50–52
 – by-products 56–58
 – Co(III) species cycle 52
 – H_2O concentration effects 52–54
 – hydrocarbon oxidation 54–55
 – manganese precipitation 54
 – Mn(III) species cycle 52
 – oxidation process 58
 – purification process 58–61
mesoporous graphitic carbon nitride (mpg-CN). *see* graphitic carbon nitride
metal catalytic oxidation 364–366
metal-free aerobic oxidations, continuous-flow microreactors
 – *ortho* functionalized phenols 410, 411
 – Taylor flow conditions 409, 410
 – three phase flow conditions 408, 409
methacrolein 209
methane oxidation 132
methanol, Cu-catalyzed oxidative carbonylation of 72–74
5-(methoxymethyl)furfural (MMF) 314–318, 320, 322–324
methyl acrylate oxidation
 – adhesive industry application 180
 – inside reaction temperatures 181
 – large-scale oxidation 184
 – reaction simulation studies 184–186
 – small-scale reaction optimization 181–184
methylene blue 376
methylene diphenyl diisocyanate 120, 121
methyl methacrylate (MMA) 209
α-methylstyrene (AMS) 22
methyl-substituted phenol oxidations 106–109
Minisci's $Mn(NO_3)_2/Co(NO_3)_2$/TEMPO-catalyzed aerobic alcohol oxidation 244–246
Mitsubishi Chemical's 1,4-butanediol manufacturing process 160
 – block flow diagram 166
 – first generation process
 – – hydrogenation step 165–166
 – – hydrolysis step 166
 – – oxidative acetoxylation step 165
 – – process flow diagram 167
 – second generation process 167–168
Mizoroki–Heck reactions 126, 128
MJOD millireactor system. *see* multijet oscillating disk (MJOD) millireactor system
molybdate-catalyzed H_2O_2 disproportionation 379–380, 384, 385, 392
monoamine oxidase (MAO)-catalyzed oxidation 298–302
monosaccharides oxidation
 – chemical intermediates 349
 – derived products 349, 350, 354
 – enzymatic process *vs.* chemical process 362–366
 – gold catalytic reaction 355, 356
 – kinetic investigation 353
 – molecular mechanism 352
 – mono-and bimetallic catalysts 357
 – stoichiometry 352
Monsanto's Roundup process 267, 269, 270
multicomponent catalytic packages 199
multijet oscillating disk (MJOD) millireactor system 260

n

Nagasawa's 1,2,4-triazole synthesis 79
nanoshell carbon (NSC) catalyst 274, 279
NaOCl 85
N-hydroxyphthalimide (NHPI)-catalyzed oxidations
 – adamantane 257–258
 – adsorption/desorption cycles 262, 263
 – alkylaromatics 260–262
 – catalytic cycle 254, 255
 – cyclohexane 258
 – enthalpic effect 256
 – entropic effect 257
 – lipophilic catalyst 262–263
 – olefin epoxidation 259–260
 – PINO self-decomposition 254

– polar effect   256–257
nitrite-modulated Wacker process   131
nitrogen oxide (NO_x) cocatalysts   130–132
nitroxyl/NO_x-catalyzed aerobic alcohol oxidation   248
– ABNO and keto-ABNO   247
– catalytic cycles   244, 245
– 5-F-AZADO   247
– Hu's oxidation scheme   245, 246
– Minisci's oxidation scheme   244–246
– N-oxoammonium salt   243–244
– structures   242, 243
nonbarbotage method   29
noncatalyzed DSM Oxanone® cyclohexane oxidation process   37–38
Norit SXRO activated carbon catalysts   271–273
Noryl® resins   98, 104
NO_x-catalyzed aerobic alcohol oxidation
– amberlyst resin   241–243, 247–248
– applications   249
– Lewis acid-coordinated $LCoNO_2$ complex   241
– nitroxyl radicals   248
– – ABNO and keto-ABNO   247
– – catalytic cycles   244, 245
– – 5-F-AZADO   247
– – Hu's oxidation scheme   245, 246
– – Minisci's oxidation scheme   244–246
– – N-oxoammonium salt   243–244
– – structures   242, 243
– scale-up   247–249
– TEMPO catalysts   244, 245, 248
N-phenyl carbamates   120
N-(phosphonomethyl)iminodiacetic acid (PMIDA)   267–271

o

OA. see oleic acid (OA)
OCC. see organic cocatalysts (OCC)
olefin autoxidation   12–13
olefinic oxidation   140, 141
oleic acid (OA)   332, 336
– double bond cleavage   336–337
– ozonolysis   332–336
– three-stage conversion   339
– two-stage conversion   337–339
omeprazole   293, 294
organic cocatalysts (OCC)   196
oxidative acetoxylation, process flow diagram of   167
oxidative carbonylation
– DPC (see direct DPC process)

– palladium catalyzed aerobic oxidation   118–121
oxidative esterification   209
oxidative Heck reaction   126–128
oxoammonium-catalyzed alcohol oxidation   90
ozonolysis   332–336, 343

p

palladium-catalyzed aerobic oxidation
– alkene acetoxylation   117–118
– arene olefination   126–128
– benzylic acetoxylation   125–126, 131
– continuous-flow microreactors   415
– esterification   123–124
– industrial applications   115, 116
– ligand-modulated catalysis   128–130
– methane oxidation   132
– NO_x cocatalysts   130–132
– oxidative carbonylation   118–121
– oxidative coupling   121–122
– pyridine-based ligands   129
palladium-catalyzed oxidation
– acrolein   178–180
– cyclohexene   174–176
– methyl acrylate oxidation
– – adhesive industry application   180
– – inside reaction temperatures   181
– – large-scale oxidation   184
– – reaction simulation studies   184–186
– – small-scale reaction optimization   181–184
– olefins   176–178
palladium(II)-ethylene oxidation   175
parabolic trough-facility for organic photochemistry (PROPHIS) reactor   383
Pd black formation   174–176
Pd-catalyzed allylic acetoxylation reaction   231
Pd/Cu/Fe-catalyzed oxidation
– cyclohexene   175–176
– methylacrylate   180
– olefins   176–178
Pd–Pb catalyst
– discovery   210
– industrial catalyst   213–214
– intermetallic compounds   210–212
– Pb role   213
Pd_3Pb_1 catalyst   211
Pd–Pb–Ti–NaOH–NaBr–TG catalytic package   202, 203
$Pd/SiO_2–Al_2O_3$ catalyst   215–216
PEF. see polyethylene furandicarboxylate (PEF)
peroxyl radicals   6, 8

phase-transfer catalysis (PTC)   300
phenol   124
– Cu-catalyzed aerobic oxidation synthesis
– – chemistry and catalysis   70–72
– – process technology   74–75
– methyl-substituted phenol oxidations   106–109
– oxidative carbonylation, direct DPC process
– – catalysts   193–195
– – inorganic cocatalysts   196–198
– – mechanism   192–193
– – multicomponent catalytic packages   199
– – organic cocatalysts   196
– polyphenylene oxides
– – chemistry and catalysis   99–102
– – process improvements   104–105
– – process technology   102–104
– – synthesis route   97–98
phenylacetaldehydes   416
photochemical singlet oxygenation. see singlet oxygenation
phthalimide N-oxyl (PINO)   253, 254
α-pinene, photooxidation of   384
pinocarvone   384
Plavix   81, 82
polybutylene terephthalate (PBT)   159
polycarbonate plastic   189–190
poly-2,6-dimethyl-1,4-phenylene ether (PPO™ resin)
– CuCl/pyridine catalytic system   101
– Cu/DTBEDA catalytic system   101–102
– greener solvents   105
– physical properties   99
– process technology   102–104
– production   97–98
– quinone-ketal rearrangement   100–101
– reaction pathways   99–100
– resonance structures   98
– TMEDA/Cu catalytic system   101
polyethylene furandicarboxylate (PEF)   314, 315
polymer-immobilized ligands   73
polyoxometalates (POMs)   342, 343
polypyrrole   155
polysaccharides oxidation   361–362
POMs. see polyoxometalates (POMs)
pravastatin   304, 305
1,3-propanediol   178–180, 187
propene production. see cumene route synthesis
PTA. see purified terephthalic acid (PTA)
p-toluic acid   48
purified terephthalic acid (PTA)   41

– bromineless catalysis   61–62
– Co–Br catalysis
– – branching sequence   46–47
– – hydrogen abstraction reactions   47, 48
– – reactive intermediates   43, 44
– – thermal homolytic cleavage   44–45
– MC catalysis
– – bromine species   50–52
– – by-products   56–58
– – Co(III) species cycle   52
– – H_2O concentration effects   52–54
– – hydrocarbon oxidation   54–55
– – manganese precipitation   54
– – Mn(III) species cycle   52
– – oxidation process   58
– – purification process   58–61
– synergistic effect   48–50
p-xylene (pX) oxidation
– Co–Br catalysis
– – branching sequence   46–47
– – hydrogen abstraction reactions   47, 48
– – reactive intermediates   43, 44
– – thermal homolytic cleavage   44–45
– intermediates   43
– MC catalysis (see MC-catalyzed oxidation)

q

quinone-catalyzed aerobic oxidation reactions   229
quinone-ketal rearrangement   100–101
quinones
– catalytic aerobic oxidation reactions
– – vs. AO process   229
– – CAO enzymes   231–234
– – NO_x cocatalysts   229–230
– – Pd-catalyzed acetoxylation   230–231
– hydrogen peroxide synthesis
– – AO process (see anthraquinone oxidation (AO) process)
– – process technology   227–229
– – simplified process flow diagram   227

r

renewable feedstock oxidation. see carbohydrate oxidation
Reppe process   159
Riedl–Pfleiderer process   222
ring-opened by-products   11–12
rose oxide synthesis   391–392

s

self-accelerating-decomposition temperature (SADT)   23, 27, 30
single-stage Wacker process   148–149

singlet oxygen
- application 371
- chemical sources of, H_2O_2 disproportionation 376–379
- DSO
- – advantages 384
- – artemisinin synthesis 389
- – homogeneous aqueous and alcoholic media 384–385
- – multiphase microemulsions 387–388
- – rose oxide synthesis 391–392
- – schematic representation 386
- – single phase microemulsions 385–387
- vs. ground-state molecular oxygen 373–375
- H_2O_2/MoO_4^{2-} catalytic system 379–380
- industrialized oxygenation
- – artemisinin 389–391
- – ascaridole 388
- – rose oxide synthesis 391–392
- luminous oxygenation 383–384
- molecular targets 380–382
- photosensitized generation 375–376
- ROS 371, 372
SMOs. see styrene monooxygenases (SMOs)
sodium gluconate 363
solvent-cage efficiency 9
styrene monooxygenases (SMOs) 304, 305
succinic acid (SA) 314, 325, 326
sucralose 351
sucrose oxidation 358–359
sugar oxidation 360–361
supercritical carbon dioxide (scCO$_2$) 105

t

TEMPO/NO$_x$-catalyzed aerobic alcohol oxidation 244, 245, 248
Tenax® 105
termination reactions 5–6
tertiary alkylperoxyl radicals 5–6
3,4,3′,4′-tetramethyl biphenyltetracarboxylate 121
tetraphenylporphyrin 376
three-stage oleic acid conversion 339
through-wall singlet oxygen oxidation 405
TMHQ. see 2,3,5-trimethylhydroquinone (TMHQ)
toluene acetoxylation 126
traditional catalyzed cyclohexane oxidation process 35–37
trans–cis isomerization reaction 142

trans-2,3-dideuterio-β-propiolactone 144
transition metal-catalyzed aerobic oxidations, in microreactors
- 2-benzylpyridines 403–404
- gold-catalyzed oxidation 400–401
- homogeneous copper-catalyzed oxidation 402–403
1,2,4-triazoles 79
tricarboxylated sucrose 359
2,3,5-trimethylhydroquinone (TMHQ) 106, 108, 109
two-stage oleic acid conversion 337–339
two-stage Wacker process 149–150
type I photooxidation 375
type II photooxidation 375

u

Ube liquid phase process 119
unimolecular RO–OH dissociation process 8
unsaturated hydrocarbon autoxidation 12–13
Upilex production 121

v

vanillin 91
vegetable oil-based feedstocks. see oleic acid (OA)
vinyl acetate synthesis 117
visible-light photocatalytic aerobic oxidation 407
vitamin A 92
vitamin E 69, 106–109

w

Wacker–Hoechst process 16
Wacker oxidation 175, 176, 182
Wacker process 131
- anti-Markovnikov reaction 152
- process improvements 151–155
- single-stage process 148–149
- two-stage process 149–150
Wigner's rule 374, 375
Winsor III system 388

x

XCube™ flow reactor 402

z

Zeise's salt 139, 143
Zeneca Pharmaceuticals 93
zolimidine 80, 81